Monitoring Forest Damage with Mass Spectrometry-Based Metabolomics Methods

WILEY SERIES ON MASS SPECTROMETRY

Series Editors
Dominic M. Desiderio
Departments of Neurology and Biochemistry
University of Tennessee Health Science Center
Joseph A. Loo
Department of Chemistry and Biochemistry UCLA

Founding Editors
Nico M. M. Nibbering (1938–2014)
Dominic M. Desiderio

John R. de Laeter *Applications of Inorganic Mass Spectrometry*
Michael Kinter and Nicholas E. Sherman *Protein Sequencing and Identification Using Tandem Mass Spectrometry*
Chhabil Dass *Principles and Practice of Biological Mass Spectrometry*
Mike S. Lee *LC/MS Applications in Drug Development*
Jerzy Silberring and Rolf Eckman *Mass Spectrometry and Hyphenated Techniques in Neuropeptide Research*
J. Wayne Rabalais *Principles and Applications of Ion Scattering Spectrometry: Surface Chemical and Structural Analysis*
Mahmoud Hamdan and Pier Giorgio Righetti *Proteomics Today: Protein Assessment and Biomarkers Using Mass Spectrometry, 2D Electrophoresis, and Microarray Technology*
Igor A. Kaltashov and Stephen J. Eyles *Mass Spectrometry in Structural Biology and Biophysics: Architecture, Dynamics, and Interaction of Biomolecules, Second Edition*
Isabella Dalle-Donne, Andrea Scaloni, and D. Allan Butterfield *Redox Proteomics: From Protein Modifications to Cellular Dysfunction and Diseases*
Silas G. Villas-Boas, Ute Roessner, Michael A.E. Hansen, Jorn Smedsgaard, and Jens Nielsen *Metabolome Analysis: An Introduction*
Mahmoud H. Hamdan *Cancer Biomarkers: Analytical Techniques for Discovery*
Chabbil Dass *Fundamentals of Contemporary Mass Spectrometry*
Kevin M. Downard (Editor) *Mass Spectrometry of Protein Interactions*
Nobuhiro Takahashi and Toshiaki Isobe *Proteomic Biology Using LC-MS: Large Scale Analysis of Cellular Dynamics and Function*
Agnieszka Kraj and Jerzy Silberring (Editors) *Proteomics: Introduction to Methods and Applications*
Ganesh Kumar Agrawal and Randeep Rakwal (Editors) *Plant Proteomics: Technologies, Strategies, and Applications*
Rolf Ekman, Jerzy Silberring, Ann M. Westman-Brinkmalm, and Agnieszka Kraj (Editors) *Mass Spectrometry: Instrumentation, Interpretation, and Applications*
Christoph A. Schalley and Andreas Springer *Mass Spectrometry and Gas-Phase Chemistry of Non-Covalent Complexes*
Riccardo Flamini and Pietro Traldi *Mass Spectrometry in Grape and Wine Chemistry*
Mario Thevis *Mass Spectrometry in Sports Drug Testing: Characterization of Prohibited Substances and Doping Control Analytical Assays*
Sara Castiglioni, Ettore Zuccato, and Roberto Fanelli *Illicit Drugs in the Environment: Occurrence, Analysis, and Fate Using Mass Spectrometry*
Angel Garcia and Yotis A. Senis (Editors) *Platelet Proteomics: Principles, Analysis, and Applications*
Luigi Mondello *Comprehensive Chromatography in Combination with Mass Spectrometry*
Jian Wang, James MacNeil, and Jack F. Kay *Chemical Analysis of Antibiotic Residues in Food*
Walter A. Korfmacher (Editor) *Mass Spectrometry for Drug Discovery and Drug Development*
Alejandro Cifuentes (Editor) *Foodomics: Advanced Mass Spectrometry in Modern Food Science and Nutrition*
Christine M. Mahoney (Editor) *Cluster Secondary Ion Mass Spectrometry: Principles and Applications*
Despina Tsipi, Helen Botitsi, and Anastasios Economou *Mass Spectrometry for the Analysis of Pesticide Residues and their Metabolites*
Xianlin Han *Lipidomics: Comprehensive Mass Spectrometry of Lipids*
Marek Smoluch, Giuseppe Grasso, Piotr Suder, and Jerzy Silberring (Editors) *Mass Spectrometry: An Applied Approach, Second Edition*
Carla António (Editor) *Monitoring Forest Damage with Mass Spectrometry-Based Metabolomics Methods*

Monitoring Forest Damage with Mass Spectrometry-Based Metabolomics Methods

Edited by
Carla António
Plant Metabolomics Lab Portugal
Forest Research Centre (CEF)
School of Agriculture (ISA)
University of Lisbon (ULisboa)
Lisbon
Portugal

WILEY

Published by John Wiley & Sons, Inc., Hoboken, New Jersey.
Published simultaneously in Canada.

For general information on our other products and services or for technical support, please contact our Customer Care Department within the United States at (800) 762-2974, outside the United States at (317) 572-3993 or fax (317) 572-4002.

Wiley also publishes its books in a variety of electronic formats. Some content that appears in print may not be available in electronic formats. For more information about Wiley products, visit our web site at www.wiley.com.

Library of Congress Cataloging-in-Publication Data

Name: António, Carla, editor.
Title: Monitoring forest damage with mass spectrometry-based metabolomics methods / edited by Carla António.
Description: Hoboken, New Jersey : Wiley, 2024. | Series: Wiley series on mass spectrometry | Includes index.
Identifiers: LCCN 2023032892 (print) | LCCN 2023032893 (ebook) | ISBN 9781119868729 (hardback) | ISBN 9781119868736 (adobe pdf) | ISBN 9781119868743 (epub)
Subjects: LCSH: Forests and forestry–Research. | Metabolites. | Mass spectrometry.
Classification: LCC SD356.4 .M66 2024 (print) | LCC SD356.4 (ebook) | DDC 634.9072–dc23/eng/20231006
LC record available at https://lccn.loc.gov/2023032892
LC ebook record available at https://lccn.loc.gov/2023032893

Cover Design: Wiley
Cover Image: Courtesy of Ana Margarida Rodrigues

Set in 10/12pt WarnockPro by Straive, Chennai, India

Contents

List of Contributors

Ilka Nacif Abreu
Department of Plant Biochemistry
Albrecht-von-Haller-Institute for Plant Sciences
University of Goettingen
Goettingen
Germany

Juan Manuel Acién
Departament de Biologia
Bioquímica i Ciències Naturals
Universitat Jaume I
Castelló de la Plana
Spain

Sara Adrián Lopez de Andrade
Department of Plant Biology
Institute of Biology
University of Campinas
Campinas
Brazil

Carla António
Plant Metabolomics Lab Portugal
Centro de Estudos Florestais (CEF)
Instituto Superior de Agronomia (ISA)
Universidade de Lisboa (ULisboa)
Lisboa
Portugal

Ismael Aranda
Instituto de Ciencias Forestales (ICIFOR-INIA)
CSIC
Madrid
Spain

Vicent Arbona
Departament de Biologia
Bioquímica i Ciències Naturals
Universitat Jaume I
Castelló de la Plana
Spain

Pierluigi Bonello
Department of Plant Pathology
The Ohio State University
Columbus
USA

Federico Brilli
Institute for Sustainable Plant Protection (IPSP)
National Research Council of Italy (CNR)
Sesto Fiorentino (Florence)
Italy

Ilara Gabriela Frasson Budzinski
Department of Genetics
Max Feffer Laboratory of Plant Genetics
'Luiz de Queiroz' College of Agriculture
University of São Paulo
Piracicaba
SP
Brazil

Eva Cañizares
Departament de Biologia
Bioquímica i Ciències Naturals
Universitat Jaume I
Castelló de la Plana
Spain

Francisco Javier Cano
Instituto de Ciencias Forestales (ICIFOR-INIA)
CSIC
Madrid
Spain

and

ARC Centre of Excellence for Translational Photosynthesis
Hawkesbury Institute for the Environment
Western Sydney University
Penrith
NSW
Australia

Isabel Carrasquinho
Instituto Nacional de Investigação Agrária e Veterinária I. P. Avenida da República
Oeiras
Portugal

and

LEAF—Linking Landscape Environment, Agriculture and Food Research Centre
Associated Laboratory TERRA
Instituto Superior de Agronomia
Universidade de Lisboa
Lisboa
Portugal

Fernanda Rezende Castro-Moretti
Department of Plant Pathology
School of Agriculture Luiz de Queiroz
University of São Paulo
Piracicaba
SP
Brazil

Thaís Regiani Cataldi
Department of Genetics
Max Feffer Laboratory of Plant Genetics
'Luiz de Queiroz' College of Agriculture
University of São Paulo
Piracicaba
SP
Brazil

Anna O. Conrad
Northern Research Station
USDA Forest Service
West Lafayette
USA

Jorge Domínguez
Department of Natural Systems and Resources
Research Group
Functioning of Forest Systems in a Changing Environment
Universidad Politécnica de Madrid
Madrid
Spain

Pia Guadalupe Dominguez
Instituto de Agrobiotecnología y Biología Molecular (IABIMO)
Instituto Nacional de Tecnología Agropecuaria (INTA)
Consejo Nacional de Investigaciones Científicas y Técnicas (CONICET)
Hurlingham
Buenos Aires
Argentina

Daniela Feltrim
Department of Plant Biology
Institute of Biology
University of Campinas
Campinas
SP
Brazil

Andrea Ghirardo
Research Unit Environmental Simulation (EUS)
Helmholtz Zentrum München
Neuherberg
Germany

Luis Gil
Department of Natural Systems and Resources
Research Group
Functioning of Forest Systems in a Changing Environment
Universidad Politécnica de Madrid
Madrid
Spain

Elsa Gonçalves
LEAF—Linking Landscape Environment, Agriculture and Food Research Centre
Associated Laboratory TERRA
Instituto Superior de Agronomia
Universidade de Lisboa
Lisboa
Portugal

Miguel González-Guzmán
Departament de Biologia
Bioquímica i Ciències Naturals
Universitat Jaume I
Castelló de la Plana
Spain

Maria Kenosis Emmanuelle Galingay Lachica
Division of Biological Sciences
Nara Institute of Science and Technology (NAIST) – Graduate School of Science and Technology
Ikoma
Nara
Japan

Rosana López
Department of Natural Systems and Resources
Research Group
Functioning of Forest Systems in a Changing Environment
Universidad Politécnica de Madrid
Madrid
Spain

Juan Antonio Martín
Department of Natural Systems and Resources
Research Group
Functioning of Forest Systems in a Changing Environment
Universidad Politécnica de Madrid
Madrid
Spain

Paulo Mazzafera
Department of Plant Biology
Institute of Biology
University of Campinas
Campinas
Brazil

Thomas Moritz
Department of Forest Genetics and Plant Physiology
Umeå Plant Science Centre
Swedish University of Agricultural Sciences
Umeå
Sweden

Vinícius Henrique de Oliveira
Department of Plant Biology
Institute of Biology
University of Campinas
Campinas
Brazil

Ana Margarida Rodrigues
Plant Metabolomics Lab Portugal
Centro de Estudos Florestais (CEF)
Instituto Superior de Agronomia (ISA)
Universidade de Lisboa (ULisboa)
Lisboa
Portugal

Jesús Rodríguez-Calcerrada
Research Group Functioning of Forest Systems in a Changing Environment
Department of Natural Systems and Resources
Universidad Politécnica de Madrid
Madrid
Spain

Marta Nunes da Silva
CBQF – Centro de Biotecnologia e Química Fina – Laboratório Associado
Escola Superior de Biotecnologia
Universidade Católica Portuguesa
Porto
Portugal

María Brígida Fernández de Simón
Instituto de Ciencias Forestales (ICIFOR-INIA)
CSIC
Madrid
Spain

Juan Sobrino-Plata
Research Group Functioning of Forest Systems in a Changing Environment
Department of Natural Systems and Resources
Universidad Politécnica de Madrid
Madrid
Spain

and

Department of Genetics
Physiology and Microbiology
Universidad Complutense de Madrid
Madrid
Spain

Takayuki Tohge
Division of Biological Sciences
Nara Institute of Science and Technology (NAIST) – Graduate School of Science and Technology
Ikoma
Nara
Japan

Marta Vasconcelos
CBQF – Centro de Biotecnologia e Química Fina – Laboratório Associado
Escola Superior de Biotecnologia
Universidade Católica Portuguesa
Porto
Portugal

Caterina Villari
Warnell School of Forestry
University of Georgia
Athens
USA

Mutsumi Watanabe
Division of Biological Sciences
Nara Institute of Science and Technology (NAIST) – Graduate School of Science and Technology
Ikoma
Nara
Japan

Preface

Many of the world's forest ecosystems are severely damaged by climate change. Not only are warmer and drier climates expected to increase the risks of droughts (abiotic forest damage) and forest insect pest outbreaks (biotic forest damage), but warmer and rainier climates are also expected to increase unhealthy pathogen interactions. This book covers the latest challenges and resources provided by highly sensitive mass spectrometry (MS)-based metabolomics methods to support forest tree research and explore the mechanisms and metabolic landscapes of forest tree species (i.e. primary and secondary metabolites, including phytohormones) in response to damage associated with the adverse effects of global climate change. Although much work is still needed to better understand the molecular basis underlying growth, development, and tree tolerance to environmental fluctuations, unraveling the pathways that fuel tree-specific resistance mechanisms is essential for the protection of our forests and, ultimately, ourselves.

Chapters 1–5 introduce the existing challenges facing forest tree metabolomics research, including challenges in the experimental design and sample preparation workflows prior to the extraction of metabolites from forest tree tissues and MS-based metabolomics analysis. The planning of the experimental design is the first crucial step of the entire metabolomics workflow and should include (in great detail) all the technical parameters that ensure data consistency and reproducibility of the generated data and metadata among metabolomics datasets for reliable biological interpretation. Moreover, metabolites are organized in a wide metabolic network, itself composed of multi-biochemical pathways that are dependent on many genetic and signaling networks for their regulation. A glimpse on systems biology frameworks that integrate MS-based metabolomics data with other Omics technologies (e.g. genomics, transcriptomics, proteomics) is explored to help uncover interdisciplinary links within the complex forest tree systems.

Chapters 6–8 cover recent advances in leading MS-based methodologies in forest tree metabolomics, including the analysis of volatile organic compounds

(VOCs), specific secondary (specialized) metabolites, and phytohormones. Chapters provide an overview of the methodology, general aspects of instrumentation, analytical approaches, sample preparation, and the latest trends relevant to forest tree research.

Chapters 9–13 examine how comprehensive MS-based metabolomics has helped advance our understanding of forest tree adaptations in response to a changing climate through studies related to (i) forest tree metabolite responses to abiotic and/or combined abiotic stresses and (ii) forest tree diseases caused by biotic disturbances (insect pest and pathogen outbreaks). Particular focus is given to the mechanisms and metabolite adjustments involved in central metabolism and infection-related metabolic pathways in metabolomics studies of economically and ecologically important forest tree species, such as *Populus, Pinus, Picea, Eucalyptus, Quercus,* and *Ulmus* spp.

I hope that both new and well-established academics and students in the forest tree research field find in this book helpful guidelines and valuable insights for running forest tree metabolomics experiments.

I would like to express my heartfelt thanks to all contributors who made this book possible and Professors Dominic M. Desiderio and Joseph A. Loo (Wiley Series on Mass Spectrometry Editors) for the opportunity to take on the role of editor for what I hope will be an important source of informed MS-based metabolomics methods to monitor forest damage.

Last but not least, next time you walk in the forest, I hope you feel the fertile pulse beneath your feet. And as you wander in contemplation of this vital source, may each breath you take be a reminder of the beauty that surrounds you.

"Every walk in the forest is like taking a shower in oxygen."
—Peter Wohlleben, The Hidden Life of Trees

Lisbon, Portugal
21 June 2023

Editor
Carla António

1

Forest Tree Metabolomics Under a Changing Climate

Ana Margarida Rodrigues and Carla António

Plant Metabolomics Lab Portugal, Forest Research Centre, School of Agriculture, University of Lisbon, Lisboa, Portugal

1.1 Introduction

Forests are complex ecosystems and the vital source of the air we breathe. Forests cover about 31% of the total land area (1) and provide habitat for more than 75% of the terrestrial biodiversity, shelter, outdoor recreation spaces, food, medicines, and essential raw materials of great economic importance to humankind. Forests also provide important environmental services, such as soil erosion and flood control, air quality improvement, and play a direct role in climate change mitigation through carbon sequestration and storage (2). Maintaining and enhancing our forests' health, integrity, and resilience to the threats of climate change is therefore crucial for them to continue providing these ecological, environmental, and socio-cultural benefits.

A forest is usually defined by the presence of trees; however, its designation is slightly more complex and also depends on biophysical and land-use criteria (3). The FAO Global Forest Resources Assessment 2020 (FAO-FRA 2020) defines forest as the "land spanning more than 0.5 hectares with trees higher than 5 meters and a canopy cover of more than 10 percent, or trees able to reach these thresholds in situ" (4). According to this definition, a forest area includes regeneration sites after harvesting, understocked sites due to clear-cut and (natural) disasters, nurseries, forest roads and firebreaks, rubber-wood, cork oak and Christmas tree plantations, areas with bamboo and palms (if height and canopy cover criteria are met), but excludes any agricultural or other specific non-forest land use (i.e. fruit tree and oil palm plantations, olive orchards, or agroforestry systems). This definition has been openly criticized for failing to distinguish between land covered by natural forests and forest plantations; forest plantations cause a huge impact on the protection of natural forests because when these areas are cleared and replaced with plantations, no net loss of forest cover is reported (3, 5).

Monitoring Forest Damage with Mass Spectrometry-Based Metabolomics Methods, First Edition.
Edited by Carla António.

Research in forest trees has been primarily focused on ecologically and economically important tree species, such as *Populus, Pinus, Picea, Eucalyptus, Quercus,* and *Ulmus* spp. However, progress in this area has been slow mainly due to the long life cycles of trees, the extremely large genome sizes, and lack of genomic tools (6–8). Efforts by the forest scientific community motivated the development of a model tree system for functional genomics. The rapid growth and relatively small genome size of *Populus* trichocarpa (black cottonwood) compared to other trees (9–11) led to the sequencing of the first tree genome (12). The sequencing of the *Populus* genome allowed the first insight into the functional biology of a tree and enabled the development of valuable genetic and genomic resources that greatly facilitated research in forest trees (13). Since then, the application of omics technologies to forest tree species (i.e. genomics, transcriptomics, proteomics, and metabolomics) in studies of responses to abiotic and biotic stresses, tree growth and development, and elucidation of wood formation has been key to advance our understanding of complex traits of economic and ecological relevance that allow forest trees survive during long periods of time under threat associated with global climate change (6–8, 14, 15). Moreover, the growing availability of multi-omics data and emergent integrative systems biology approaches provide new opportunities for studying forest tree resilience across multiple levels of omics information (i.e. from genome, transcriptome, and proteome to metabolome). By comprehensively identifying novel genes and pathways associated with target traits underlying enhanced resistance or susceptibility to abiotic and biotic stresses, important biological insights into the complex defense mechanisms involved in forest tree resilience can be achieved. This systems-level integration provides powerful information to developing strategies (e.g. through conventional breeding, genetic engineering, and genome editing) that can help forest ecosystems mitigate the damaging effects of climate change and, ultimately, restore forest health (16–20).

1.2 Forest Damage

During the lifetime of forests, trees are exposed to many different environmental factors that can threaten their survival. This is particularly true for tree species that, given their long life cycle, must withstand regular unfavorable weather conditions over many changing seasons. To respond to recurring fluctuations in environmental conditions, trees rely on evolved adaptive mechanisms that allow them to survive long periods of time and maintain their cellular homeostasis. However, this homeostasis can be greatly disrupted by the occurrence of a wide range of abiotic (e.g. drought, heat, salinity, and cold) and biotic (e.g. herbivory, pathogen infection) stresses, which often occur in combination under natural field conditions (Figure 1.1) (21–24).

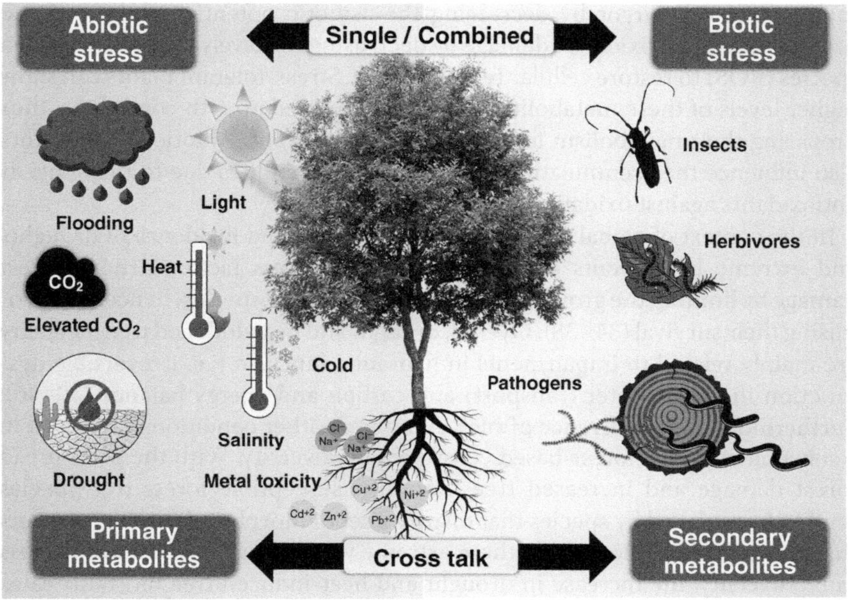

Figure 1.1 Forest tree metabolite responses to abiotic (e.g. light, heat, drought, cold, flooding, salinity, metals, and elevated CO_2) and biotic stresses (e.g. pests, such as insects and herbivores, and pathogens) are often related to adjustments in the primary and secondary metabolism, respectively. However, these events can also be observed in combination, which further compromises tree survival. Metabolite regulation during stress response (single or combined) and the activation of resistance mechanisms are strongly dependent on the cross talk between primary and secondary metabolism (e.g. primary metabolite adjustments to provide energy for the activation of defense responses). In the context of global climate change, drought and heat stress, often in combination with elevated atmospheric CO_2 concentration, are the most frequently abiotic-stress factors being studied in forest metabolomics. Source: blueringmedia / Adobe Stock; KenStock / 94 images / Pixabay; David Pires / T. Michael Keesey / Public Domain.

1.2.1 Abiotic Forest Damage

Plant responses to abiotic stress are finely regulated by molecular networks, and significant advances using omics technologies have accelerated the discovery of stress-responsive genes, proteins, and metabolites involved in these cascades (7, 25, 26). The damage caused by abiotic-stress factors can deeply affect many aspects of plant physiology and metabolism, and further negatively impair plant growth and development.

In the presence of an abiotic-stress event, the accumulation of phytohormones (e.g. abscisic acid [ABA]) and osmolytes (e.g. branched-chained amino acids, soluble sugars, raffinose family of oligosaccharides, and polyamines) is well documented. Osmolyte accumulation has an important role in maintaining cell turgor by decreasing the osmotic potential of the cell and protection against oxidative damage by decreasing the levels of reactive oxygen species (ROS) to restore cellular redox balance. Stress-tolerant plants can show higher levels of these metabolites even under normal growth conditions, thus preparing their metabolism for an adverse stress event. Abiotic-stress factors also influence the accumulation of secondary metabolites due to their role as antioxidants against oxidative damage (25, 27–33).

In the context of global climate change, the increased incidence of droughts and extreme heat events are the main abiotic-stress factors driving forest damage by limiting the growth and productivity of tree species; hence, compromising their survival (34–38). Overall, drought and heat-induced tree mortality are mainly related to impairments in hydraulic function (i.e. irreversible dysfunction in xylem water transport) and carbon and energy balance (39–46). Furthermore, the occurrence of such extreme weather conditions can result in a substantial loss of forest-based ecosystem biodiversity. With the widespread forest damage and increased tree mortality, susceptible forest tree species tend to be replaced by species that display a set of morpho-physiological traits that confer higher tolerance to these extreme weather events. Besides this loss in biodiversity, the increase in drought and heat-induced tree mortality rates can also cause profound changes in global carbon and water cycles, thereby leading to a shift in global forest ecosystems from carbon sinks to carbon sources (47–49).

As a result of human activity and the rapid and intensifying global climate change, the negative consequences of abiotic stress events are undeniably expected to worsen (50, 51). The prevalence of more intense drought periods, in combination with the increased frequency of extreme heat events, has already been responsible for the amplified vulnerability of forest ecosystems all over the world, affecting several economically important fast-growing forest tree species, mainly *Pinus*, *Populus*, and *Eucalyptus* (52–54).

1.2.2 Biotic Forest Damage

The combined effect of biotic and abiotic stresses has been associated with major forest damage and, consequently, extensive tree mortality (20, 52). Forest tree species are especially prone to insect pests and pathogen attacks because their long life cycle impairs trees to match the fast-evolutionary rates of these organisms (55, 56). In addition, in the context of global climate change and the increased frequency and intensity of extreme weather events, tree vulnerability facilitates the spread, reproduction, and development of these biotic agents (57–59).

Forest pests and pathogens can also be subjected to the negative effects of climate change; however, further studies are urgently needed to better understand the dynamics between trees, pests and pathogens, and extreme weather conditions and their role in infection and disease outbreaks (20). For example, the effect of drought in combination with heat stress can have different effects depending on the type of disease, the affected tissue, and the severity of the abiotic stress (60, 61). Drought has been shown to significantly increase the damage caused by leaf pathogens and reduce that of root and stem pathogens (61). In addition, the interaction between drought, heat, and biotic agents often triggers physiological mechanisms of tree death, such as carbon starvation and hydraulic failure (62). However, studies in the literature are still scarce and often do not reflect accurate field conditions.

To deal with the constant threat of biotic agents, trees have evolved effective defense resistance mechanisms that allow them to survive during their long life cycle (63). These defense mechanisms associated with tree responses to biotic-stress events include: (i) inducible chemical defenses (e.g. secondary metabolites such as terpenoids, phenolic compounds, and alkaloids); (ii) inducible protein-based defenses (e.g. proteinase inhibitors); (iii) anatomical defenses acting as structural barriers (e.g. enhanced lignification); and (iv) indirect defenses (e.g. attraction of natural enemies) (64, 65). Phytohormones also have a crucial role in regulating plant defense responses against pests and pathogens, in particular jasmonic acid (JA), salicylic acid (SA), and ethylene (ET) (66). SA is generally involved in the activation of defense responses against biotrophic and hemi-biotrophic pathogens, whereas JA and ET are responsible for defense responses against necrotrophic pathogens and herbivorous insects. Plant defense signaling pathways also involve the cross talk between primary and secondary metabolism. Primary metabolic pathways act as a source of energy to support defense responses and also as a source of signaling molecules to directly or indirectly trigger defense responses (67, 68).

1.3 Forest Tree Metabolomics

Metabolomics is the omics technology that best reflects the interaction between the genetic traits and environmental factors, thus being considered the molecular phenotype of a living organism (69, 70). Metabolomics studies greatly benefit from the development of next-generation sequencing technologies and the availability of genome sequence data for the establishment of gene-metabolite correlations that allow for a broader metabolome characterization and additional insights on the various functions of metabolites in biological processes, such as plant growth and development, and abiotic and biotic stress responses (71–77). After major genomics breakthroughs in forest tree research, that is the availability of the early reference genomes of *Populus* (the first tree genome) (12) and *Eucalyptus* (78) followed by the sequence of other important woody plant genomes (8, 79, 80), metabolomics studies in forest tree species have generated increased interest.

When compared to other plants, metabolomics studies using forest tree species are characterized by additional challenges. These include the need for an experimental design that takes into account the long life cycle and the genetic variability of forest tree species as well as the presence of interferents that can require additional processing steps during sample preparation. One strategy to overcome the effect of genetic variability of forest trees when evaluating quantitative traits (e.g. tree productivity and metabolite production) is the use of clonal stands. The introduction of clonal stands and clonal orchards can rapidly speed the breeding progress and plant selection of desired quantitative traits. However, in field conditions, the introduction of new clonal material among established natural forest ecosystems should not compromise the existent genetic or genotypic diversity, which could in turn lead to negative effects, such as increased competition and increased susceptibility to pests and pathogens (81). Other challenges concern the reporting standards because details of the experimental metadata (e.g. parental original, geographical location, field growth conditions, biological growth stages, and phenological parameters) that allow re-use of the data are often not described (8, 82); see Chapters 2–5 for further discussions. Despite the struggles, continuous efforts from the metabolomics scientific community have been made to ensure forest metabolomics data and metadata reproducibility between laboratories and to promote the availability of curated databases and repositories containing high-quality data (including dedicated woody species platforms).

Current plant metabolomics studies rely on the application of mass spectrometry (MS)- and nuclear magnetic resonance (NMR)-based methods, and several detailed protocols are available (83). MS-based methods, namely liquid chromatography mass spectrometry (LC–MS) and gas chromatography mass spectrometry (GC–MS), can deliver highly sensitive qualitative and quantitative data that provide a detailed description of the biochemical pathways that are influenced by environmental changes. In this context, although much work is still needed, the use of MS-based metabolomics methods in forest tree research has been crucial to advance our understanding of the metabolic landscapes, that is primary and secondary (specialized) metabolites (including phytohormones), of forest trees in response to environmental cues, such as changing weather patterns (abiotic forest damage) and pathogen interactions (biotic forest damage) (84–92; to name a few). For further examples and methods, see Chapters 6–13.

By contrast, despite requiring minimal sample preparation and being highly reproducible, NMR-based methods are only able to detect highly abundant metabolites (93–95). Consequently, due to the lower high-throughput relative to MS, NMR has been largely outperformed by MS-based methods for a broader metabolite coverage (31, 75–77, 96). Nevertheless, NMR has demonstrated to be a powerful complementary technique in plant metabolomics, being typically employed in natural products research (e.g.

structure elucidation of unknown compounds) (94, 95). In forest tree research, NMR-based methods have been used to explore the industrial processing of wood (e.g. pulp and paper industry or biofuel production), particularly in studies of the structure elucidation and composition of lignin in forest tree species, including *Populus* (97), *Eucalyptus* (98), and *Quercus* (99).

1.4 Conclusion and Future Perspectives

Many of the world's forest ecosystems are severely threatened by global climate change. Not only warmer and drier climates are expected to increase the risks of droughts and forest insect pest outbreaks, but also warmer and wetter climates are expected to increase unhealthy pathogen interactions. In addition, different forest tree species display a wide array of specific secondary (specialized) metabolites that might be produced for defense against abiotic and biotic-stress factors only in specific environmental circumstances. More integrated research studies that combine multi-level information from omics technologies and systems biology approaches are also needed to provide new insights into the complex molecular mechanisms involved in responses of forest trees to abiotic and biotic stresses. A better understanding of the specific mechanisms of resistance that fuel forest tree-defense responses to abiotic and biotic threats is urgent to develop strategies for more resilient natural forest ecosystems around the world and to, ultimately, help them regenerate and fight climate change.

References

1 FAO. 2020. Global Forest Resources Assessment 2020 — Key findings. Rome.
2 IUCN. 2021. Forests and climate change. IUCN Issues Brief [Internet]. Available from (http://www.iucn.org/resources/issues-brief/forests-and-climate-change), accessed 20 January 2023.
3 Chazdon RL, Brancalion PHS, Laestadius L, Bennett-Curry A, Buckingham K, Kumar C, et al. When is a forest a forest? Forest concepts and definitions in the era of forest and landscape restoration. *Ambio*. 2016; 45: 538–550.
4 FAO. 2020. Global Forest Resources Assessment 2020 — Terms and Definitions. Rome.
5 Mackey B, Skinner E, Norman P. A Review of Definitions, Data, and Methods for Country-level Assessment and Reporting of Primary Forest. Griffith Climate Action Beacon Discussion Paper. Brisbane (AU): Griffith University; 2021. pp. 1–35.

6 Neale DB, Kremer A. Forest tree genomics: growing resources and applications. *Nat Rev Genet.* 2011; 12(2): 111–122.
7 zu Castell W, de Ernst D. Experimental 'omics' data in tree research: facing complexity. *Trees.* 2012; 26: 1723–1735.
8 Rodrigues AM, Miguel C, Chaves I, António C. Mass spectrometry-based forest tree metabolomics. *Mass Spectrom Rev.* 2021; 40(2): 126–157.
9 Bradshaw H, Ceulemans R, Davis J, Stettler R. Emerging model systems in plant biology: poplar (*Populus*) as a model forest tree. *J Plant Growth Regul.* 2000; 19: 306–313.
10 Brunner AM, Busov VB, Strauss SH. Poplar genome sequence: functional genomics in an ecologically dominant plant species. *Trends Plant Sci.* 2004; 9: 49–56.
11 Tuskan GA, DiFazio SP, Teichmann T. Poplar genomics is getting popular: the impact of the poplar genome project on tree research. *Plant Biol.* 2004; 6: 2–4.
12 Tuskan GA, Difazio S, Jansson S, Bohlmann J, Grigoriev I, Hellsten U, et al. The genome of black cottonwood, *Populus trichocarpa* (Torr. & Gray). *Science.* 2006; 313: 1596–1604.
13 Wullschleger SD, Weston DJ, DiFazio SP, Tuskan GA. Revisiting the sequencing of the first tree genome: *Populus trichocarpa. Tree Physiol.* 2013; 33(4): 357–364.
14 Plomion C, Bastien C, Bogeat-Triboulot MB, Bouffier L, Déjardin A, Fady B, et al. Forest tree genomics: 10 achievements from the past 10 years and future prospects. *Ann For Sci.* 2016; 73: 77–103.
15 Neale DB, Martínez-García PJ, De La Torre AR, Montanari S, Wei XX. Novel insights into tree biology and genome evolution as revealed through genomics. *Annu Rev Plant Biol.* 2017; 68: 457–483.
16 Sniezko RA, Koch J. Breeding trees resistant to insects and diseases: putting theory into application. *Biol Invasions.* 2017; 19: 3377–3400.
17 Bewg WP, Ci D, Tsai C-J. Genome editing in trees: from multiple repair pathways to long-term stability. *Front Plant Sci.* 2018; 9: 1732.
18 Naidoo S, Slippers B, Plett JM, Coles D, Oates CN. The road to resistance in forest trees. *Front Plant Sci.* 2019; 10: 273.
19 Polle A, Chen SL, Eckert C, Harfouche A. Engineering drought resistance in forest trees. *Front Plant Sci.* 2019; 9: 1875.
20 Cortés AJ, Restrepo-Montoya A, Bedoya-Canas LE. Modern strategies to assess and breed forest tree adaptation to changing climate. *Front Plant Sci.* 2020; 11: 583323.
21 Jones JDG, Dangl JL. The plant immune system. *Nature.* 2006; 444: 323–329.
22 Atkinson NJ, Urwin PE. The interaction of plant biotic and abiotic stresses: from genes to the field. *J Exp Bot.* 2012; 63(10): 3523–3543.

23 Suzuki N, Rivero RM, Shulaev V, Blumwald E, Mittler R. Abiotic and biotic stress combinations. *New Phytol.* 2014; 203: 32–43.

24 Teshome DT, Zharare GE, Naidoo S. The threat of the combined effect of biotic and abiotic stress factors in forestry under a changing climate. *Front Plant Sci.* 2020; 11: 601009.

25 Ahuja I, de Vos RC, Bones AM, Hall RD. Plant molecular stress responses face climate change. *Trends Plant Sci.* 2010; 15(12): 664–674.

26 Harfouche A, Meilan R, Altman A. Molecular and physiological responses to abiotic stress in forest trees and their relevance to tree improvement. *Tree Physiol.* 2014; 34: 1181–1198.

27 Krasensky J, Jonak C. Drought, salt and temperature stress-induced metabolic rearrangements and regulatory networks. *J Exp Bot.* 2012; 63: 1593–1608.

28 Obata T, Fernie AR. The use of metabolomics to dissect plant responses to abiotic stresses. *Cell Mol Life Sci.* 2012; 69(19): 3225–3243.

29 Arbona V, Manzi M, Ollas C, Gómez-Cadenas A. Metabolomics as a tool to investigate abiotic stress tolerance in plants. *Int J Mol Sci.* 2013; 14: 4885–4911.

30 Ensminger I, Chang CY, Bräutigam K. Tree responses to environmental cues. In: Adam-Blondon A, & Plomion C, editors. *Advances in Botanical Research.* Volume 74. London (UK): Elsevier; 2015. p. 229–264.

31 Jorge TF, Rodrigues JA, Caldana C, Schmidt R, van Dongen J, Thomas-Oates J, António C. Mass spectrometry-based plant metabolomics: metabolite responses to abiotic stress. *Mass Spectrom Rev.* 2016; 35: 620–649.

32 Jorge TF, Mata AT, António C. Mass spectrometry as a quantitative tool in plant metabolomics. *Phil Trans R Soc A.* 2016; 374: 20150370.

33 Mata AT, Jorge TF, Pires MV, António C. Drought stress tolerance in plants: insights from metabolomics. In: Hossain MA, Wani SH, Bhattacharjee S, Burritt DJ, & Tran LSP, editors. *Drought Stress Tolerance in Plants Volume 2. Molecular and Genetic Perspectives.* Cham (CH): Springer; 2016. p. 187–216.

34 Allen CD, Macalady AK, Chenchouni H, Bachelet D, McDowell N, Vennetier M, et al. A global overview of drought and heat-induced tree mortality reveals emerging climate change risks for forests. *For Ecol Manag.* 2010; 5(4): 259, 660–684.

35 Choat B, Jansen S, Brodribb T, Cochard H, Delzon S, Bhaskar R, et al. Global convergence in the vulnerability of forests to drought. *Nature.* 2012; 491: 752–755.

36 Anderegg WRL, Kane JM, Anderegg LDL. Consequences of widespread tree mortality triggered by drought and temperature stress. *Nat Clim Chang.* 2013; 3: 30–36.

37 Clifford MJ, Royer PD, Cobb NS, Breshears DD, Ford PL. Precipitation thresholds and drought-induced tree die-off: insights from patterns of *Pinus edulis* mortality along an environmental stress gradient. *New Phytol.* 2013; 200(2): 413–421.

38 McDowell NG, Allen CD, Anderson-Teixeira K, Aukema BH, Bond-Lamberty B, Chini L, et al. Pervasive shifts in forest dynamics in a changing world. *Science.* 2020; 368: 964.

39 McDowell N, Pockman WT, Allen CD, Breshears DD, Cobb N, Kolb T, et al. Mechanisms of plant survival and mortality during drought: why do some plants survive while others succumb to drought? *New Phytol.* 2008; 178: 719–739.

40 McDowell NG. Mechanisms linking drought, hydraulics, carbon metabolism, and vegetation mortality. *Plant Physiol.* 2011; 155: 1051–1059.

41 Anderegg WRL, Berry JA, Smith DD, Sperry JS, Anderegg LDL, Field CB. The roles of hydraulic and carbon stress in a widespread climate-induced forest die-off. *Proc Natl Acad Sci USA.* 2011; 109(1): 233–237.

42 Camarero JJ, Gazol A, Sangüesa-Barreda G, Oliva J, Vicente-Serrano SM. To die or not to die: early warnings of tree dieback in response to a severe drought. *J Ecol.* 2015; 103(1): 44–57.

43 Adams HD, Zeppel MJB, Anderegg WRL, Hartmann H, Landhausser SM, Tissue DT, et al. A multi-species synthesis of physiological mechanisms in drought-induced tree mortality. *Nat Ecol Evol.* 2017; 1: 1285–1291.

44 Xiong D, Nadal M. Linking water relations and hydraulics with photosynthesis. *Plant J.* 2020; 101: 800–815.

45 Mantova M, Herbette S, Cochard H, Torres-Ruiz JM. Hydraulic failure and tree mortality: from correlation to causation. *Trends Plant Sci.* 2022; 27(4): 335–345.

46 Li X, Xi B, Wu X, Choat B, Feng J, Jiang M, et al. Unlocking drought-induced tree mortality: physiological mechanisms to modeling. *Front Plant Sci.* 2022; 13: 835921.

47 Brienen RJW, Phillips OL, Feldpausch TR, Gloor E, Baker TR, Lloyd J, et al. Long-term decline of the Amazon carbon sink. *Nature.* 2015; 519: 344–348.

48 Cavaleri MA, Coble AP, Ryan MG, Bauerle WL, Loescher HW, Oberbauer SF. Tropical rainforest carbon sink declines during El Niño as a result of reduced photosynthesis and increased respiration rates. *New Phytol.* 2017; 216(1): 136–149.

49 Hisano M, Searle EB, Chen HYH. Biodiversity as a solution to mitigate climate change impacts on the functioning of forest ecosystems. *Biol Rev.* 2018; 93: 439–456.

50 IPCC. Climate Change 2022: Impacts, Adaptation and Vulnerability. In: Pörtner HO, Roberts DC, Tignor M, Poloczanska ES, Mintenbeck K, Alegría A, et al., editors. Contribution of Working Group II to the Sixth

Assessment Report of the Intergovernmental Panel on Climate Change. Cambridge (UK) and New York (NY, USA): Cambridge University Press; 2022.

51 IPCC. Climate Change 2022: Mitigation of Climate Change. In: Shukla PR, Skea J, Slade R, Khourdajie AA, van Diemen R, McCollum D, et al., editors. Contribution of Working Group III to the Sixth Assessment Report of the Intergovernmental Panel on Climate Change. Cambridge (UK) and New York (NY, USA): Cambridge University Press; 2022.

52 McDowell NG, Beerling DJ, Breshears DD, Fisher RA, Raffa KF, Stitt M. The interdependence of mechanisms underlying climate-driven vegetation mortality. *Trends Ecol Evol*. 2011; 26: 523–532.

53 Menezes-Silva PE, Loram-Lourenço L, Alves RDFB, Sousa LF, Almeida SEDS, Farnese FS. Different ways to die in a changing world: consequences of climate change for tree species performance and survival through an ecophysiological perspective. *Ecol Evol*. 2019; 9(20): 11979–11999.

54 Schuldt B, Buras A, Arend M, Vitasse Y, Beierkuhnlein C, Damm A, et al. A first assessment of the impact of the extreme 2018 summer drought on Central European forests. *Basic Appl Ecol*. 2020; 45: 86–103.

55 Kohler A, Rinaldi C, Duplessis S, Baucher M, Geelen D, Duchaussoy F, et al. Genome-wide identification of NBS resistance genes in *Populus trichocarpa*. *Plant Mol Biol*. 2008; 66(6): 619–636.

56 Naidoo S, Külheim C, Zwart L, Mangwanda R, Oates CN, Visser EA, et al. Uncovering the defence responses of *Eucalyptus* to pests and pathogens in the genomics age. *Tree Physiol*. 2014; 34: 931–943.

57 Sturrock RN, Frankel SJ, Brown AV, Hennon PE, Kliejunas JT, Lewis KJ, et al. Climate change and forest diseases. *Plant Pathol*. 2011; 60: 133–149.

58 Pureswaran DS, Roques A, Battisti A. Forest insects and climate change. *Curr For Rep*. 2018; 4: 35–50.

59 Matsuhashi S, Hirata A, Akiba M, Nakamura K, Oguro M, Takano KT, et al. Developing a point process model for ecological risk assessment of pine wilt disease at multiple scales. *For Ecol Manag*. 2020; 463: 118010.

60 Desprez-Loustau M-L, Marçais B, Nageleisen L-M, Piou D, Vannini A. Interactive effects of drought and pathogens in forest trees. *Ann For Sci*. 2006; 63: 597–612.

61 Jactel H, Petit J, Desprez-Loustau ML, Delzon S, Piou D, Battisti A, et al. Drought effects on damage by forest insects and pathogens: a meta-analysis. *Glob Chang Biol*. 2012; 18: 267–276.

62 Anderegg WR, Hicke JA, Fisher RA, Allen CD, Aukema J, Bentz B, et al. Tree mortality from drought, insects, and their interactions in a changing climate. *New Phytol*. 2015; 208(3): 674–683.

63 Veluthakkal R, Dasgupta MG. Pathogenesis-related genes and proteins in forest tree species. *Trees*. 2010; 24: 993–1006.

64 Witzell J, Martín JA. Phenolic metabolites in the resistance of northern forest trees to pathogens —past experiences and future prospects. *Can J For Res.* 2008; 38: 2711–2727.
65 Eyles A, Bonello P, Ganley R, Mohammed C. Induced resistance to pests and pathogens in trees. *New Phytol.* 2010; 185(4): 893–908.
66 Bari R, Jones JDG. Role of plant hormones in plant defence responses. *Plant Mol Biol.* 2009; 69: 473–488.
67 Bolton MD. Primary metabolism and plant defense — fuel for the fire. *Mol Plant-Microbe Interact.* 2009; 22(5): 487–497.
68 Rojas CM, Senthil-Kumar M, Tzin V, Mysore KS. Regulation of primary plant metabolism during plant-pathogen interactions and its contribution to plant defense. *Front Plant Sci.* 2014; 5: 17.
69 Fiehn O. Metabolomics — the link between genotypes and phenotypes. *Plant Mol Biol.* 2002; 48: 155–171.
70 Fernie AR, Trethewey RN, Krotzky AJ, Willmitzer L. Metabolite profiling: from diagnostics to systems biology. *Nat Rev Mol Cell Biol.* 2004; 5: 763–769.
71 Sumner LW, Mendes P, Dixon RA. Plant metabolomics: large-scale phytochemistry in the functional genomics era. *Phytochemistry.* 2003; 62: 817–836.
72 Bino RJ, Hall RD, Fiehn O, Kopka J, Saito K, Draper J, et al. Potential of metabolomics as a functional genomics tool. *Trends Plant Sci.* 2004; 9(9): 418–425.
73 Weckwerth W. Green systems biology — from single genomes, proteomes and metabolomes to ecosystems research and biotechnology. *J Proteome.* 2011; 75(1): 284–305.
74 Tohge T, de Souza LP, Fernie AR. Genome-enabled plant metabolomics. *J Chromatogr B.* 2014; 966: 7–20.
75 Fernie AR, Tohge T. The genetics of plant metabolism. *Annu Rev Genet.* 2017; 51: 287–310.
76 Alseekh S, Fernie AR. Metabolomics 20 years on: what have we learned and what hurdles remain? *Plant J.* 2018; 94(6): 933–942.
77 Aharoni A, Goodacre R, Fernie AR. Plant and microbial sciences as key drivers in the development of metabolomics research. *Proc Natl Acad Sci USA.* 2023; 120(12): e2217383120.
78 Myburg AA, Grattapaglia D, Tuskan GA, Hellsten U, Hayes RD, Grimwood J, et al. The genome of *Eucalyptus grandis*. *Nature.* 2014; 510: 356–362.
79 Neale D, Langley C, Salzberg S, Wegrzyn J. Open access to tree genomes: the path to a better forest. *Genome Biol.* 2013; 14: 120.
80 Tuskan GA, Groover AT, Schmutz J, DiFazio SP, Myburg A, Grattapaglia D, et al. Hardwood tree genomics: unlocking woody plant biology. *Front Plant Sci.* 2018; 9: 1799.

81 Ingvarsson PK, Dahlberg H. The effects of clonal forestry on genetic diversity in wild and domesticated stands of forest trees. *Scand J For Res.* 2018; 34: 370–379.
82 Rodrigues AM, Ribeiro-Barros AI, António C. Experimental design and sample preparation in forest tree metabolomics. *Metabolites.* 2019; 9(12): 285.
83 António C. *Plant Metabolomics: Methods and Protocols.* Methods in Molecular Biology. Volume 1778. New York (NY): Humana Press; 2018.
84 Warren CR, Aranda I, Cano FJ. Metabolomics demonstrates divergent responses of two *Eucalyptus* species to water stress. *Metabolomics.* 2012; 8: 186–200.
85 Budzinski IGF, Moon DH, Lindén P, Moritz T, Labate CA. Seasonal variation of carbon metabolism in the cambial zone of *Eucalyptus grandis. Front Plant Sci.* 2016; 7: 932.
86 Mokochinski JB, Mazzafera P, Sawaya ACHF, Mumm R, de Vos RCH, Hall RD. Metabolic responses of *Eucalyptus* species to different temperature regimes. *J Integr Plant Biol.* 2018; 60(5): 397–411.
87 Rodríguez-Calcerrada J, Rodrigues AM, Perdiguero P, António C, Atkin OK, Li M, et al. A molecular approach to drought-induced reduction in leaf CO_2 exchange in drought-resistant *Quercus ilex. Physiol Plant.* 2018; 162: 394–408.
88 Amaral J, Correia B, António C, Rodrigues AM, Gómez-Cadenas A, Valledor L, et al. *Pinus* susceptibility to pitch canker triggers specific physiological responses in symptomatic plants: an integrated approach. *Front Plant Sci.* 2019; 10: 509.
89 Rodrigues AM, Langer S, Carrasquinho I, Bergström E, Larson T, Thomas-Oates J, et al. *Pinus pinaster* early hormonal defence responses to pinewood nematode (*Bursaphelenchus xylophilus*) infection. *Metabolites.* 2021; 11: 227.
90 Rodrigues AM, Carrasquinho I, António C. Primary metabolite adjustments associated with pinewood nematode resistance in *Pinus pinaster. Front Plant Sci.* 2021; 12: 777681.
91 Rodríguez-Calcerrada J, Rodrigues AM, António C, Perdiguero P, Pita P, Collada C, et al. Stem metabolism under drought stress — a paradox of increasing respiratory substrates and decreasing respiratory rates. *Physiol Plant.* 2021; 172: 391–404.
92 Bertić M, Schroeder H, Kersten B, Fladung M, Orgel F, Buegger F, et al. European oak chemical diversity — from ecotypes to herbivore resistance. *New Phytol.* 2021; 232(2): 818–834.
93 Kim HK, Verpoorte R. Sample preparation for plant metabolomics. *Phytochem Anal.* 2010; 21: 4–13.
94 Kim HK, Choi YH, Verpoorte R. NMR-based metabolomic analysis of plants. *Nat Protoc.* 2010; 5: 536–549.

95 Kim HK, Choi YH, Verpoorte R. NMR-based plant metabolomics: where do we stand, where do we go? *Trends Biotechnol.* 2011; 29(6): 267–275.

96 Emwas A-HM. The strengths and weaknesses of NMR spectroscopy and mass spectrometry with particular focus on metabolomics research. *Methods Mol Biol.* 2015; 1277: 161–193.

97 Hedenström M, Wiklund-Lindström S, Oman T, Lu F, Gerber L, Schatz P, et al. Identification of lignin and polysaccharide modifications in *Populus* wood by chemometric analysis of 2D NMR spectra from dissolved cell walls. *Mol Plant.* 2009; 2(5): 933–942.

98 Rencoret J, Gutiérrez A, Nieto L, Jiménez-Barbero J, Faulds CB, Kim H, et al. Lignin composition and structure in young versus adult *Eucalyptus globulus* plants. *Plant Physiol.* 2011; 155(2): 667–682.

99 Lourenço A, Rencoret J, Chemetova C, Gominho J, Gutiérrez A, Del Río JC, et al. Lignin composition and structure differs between xylem, phloem and phellem in *Quercus suber* L. *Front Plant Sci.* 2016; 7: 1612.

2

Experimental Methodology for Clonal Forest Research

Isabel Carrasquinho[1,2] *and Elsa Gonçalves*[2]

[1] *Instituto Nacional de Investigação Agrária e Veterinária I. P. Avenida da República, Oeiras, Portugal*
[2] *LEAF—Linking Landscape, Environment, Agriculture and Food Research Center, Associated Laboratory TERRA, Instituto Superior de Agronomia, Universidade de Lisboa, Lisboa, Portugal*

2.1 Introduction

Forestry research usually involves the implementation of laboratory, nursery, and field trials. Selecting species, provenances, or individuals, understanding genetic differences and physiological processes that control tree growth and performance, analyzing interactions among species in agroforestry environments, such as competition, synergies, or others, are some examples frequently evaluated in trials (1). All genetic traits are related to physiological processes that interact with the environment and result in phenotypic expression. The genetic correlation between a phenotypic trait and some biochemical properties can be strong enough to be considered in the breeding strategies (2). For instance, differences in metabolism among genotypes under identical conditions can be useful in tree breeding to help to identify genetically superior trees or families, best-adapted provenances, clones, and even to perform indirect selection.

Experiments begin long before data analysis, with the previous design of the research strategy. The experiments are important tools as they allow cause-and-effect hypotheses to be tested. Decisions need to be based on solid evidence, and those experiments are vital for generating this evidence. Therefore, it is important to be familiar with the principles of experimental design, which involves representativeness, randomization, replication, and error control. According to Giesbrecht and Gumpertz (3), these four principles of the experimental design allow inference without bias and with enough precision to be meaningful. Representativeness refers to the degree to which the experimental material is representative of the units or material from which inferences are to be made. The purpose of randomization in controlled

Monitoring Forest Damage with Mass Spectrometry-Based Metabolomics Methods, First Edition.
Edited by Carla António.

scientific experiments is to separate causation from correlation. Part of the precision of estimates and the power of hypothesis tests depend on the number of replications as well as on the magnitude of unexplained variation, such as differences in soil conditions (fertility, moisture, or topography), microclimate, genetic variation among plants or errors in measurements, or even from experimental techniques application.

This chapter is organized according to the adequate steps of a research strategy focused on the quantification of the genetic variability within a forest species using clonal field trials to study non-wood forest products. In Section 2.2, the approach to defining the objectives of an experiment is discussed. Section 2.3 describes the appropriate sampling strategies to prospect the genetic material to be tested. In Section 2.4, the topics of planning and establishing the experimental design are addressed. Examples of the implementation of field trials using clones to quantify genetic variability within *Pinus pinaster* and *Pinus pinea* are given in Section 2.5. In Section 2.6, remarks on statistical analysis and evaluated quantitative genetic parameters are made. And finally, in Section 2.7, a conclusion about the strategy to quantify genetic variability is provided.

2.2 Defining the Objectives of an Experiment

The sustainability of forests depends on the genetic variability existing in all their components. In general, forest tree species with a long-life cycle show significant adaptive capacity, achieved through evolutionary processes (e.g. genetic flow and natural selection). The existence of genetic variability for a characteristic of interest is a necessary condition for the success of the genetic improvement program. In fact, the greater this variability, the greater the genetic gains associated with the improvement programs. Therefore, the quantification of this variability constitutes an essential task on which the efforts of the breeder must be concentrated. Genetic tests using clones or seeds from different provenances or families (half-sibs or full-sibs), established with appropriate experimental designs, constitute the main approach to quantify this genetic variability (e.g. adaptive and neutral traits).

The purpose of this work is to encourage the wider use of clonal field trials to study the genetic variability within a species concerning non-wood forest products, such as resin production in maritime pine and cone production in Mediterranean stone pine. Using this strategy allows an effective reduction in the time required for genetic selection because the experimental process for the evaluation of the breeding value is longer than the one required for the evaluation of the genetic value. Moreover, by using this strategy, the time needed to obtain data to perform selection is also reduced. This approach is particularly effective for stone pine, where grafting material is increasingly used in the establishment of the orchards for kernel production. For instance, the usage of the grafting technique permits a reduction of around 12 years to obtain the first cone production.

2.3 Sampling Strategies to Represent the Species

In general, the quality of the forest reproduction material used for afforestation depends on a vast set of conditions, the first of which is related to the provenance of the material to be tested. In the case of clonal collections with a representative sample of a forest species, the directions of prospection that will be proposed are based on the methods first described in works with autochthonous grapevine varieties (4, 5). In Portugal, the same methodology started to be adapted to forest species in the scope of national forest projects. Using the provenance regions and the updated forest inventory as the bases for the identification of representative stands, the key rules to obtain a representative sample of the genetic variability within a forest species, can be summarized as follows: (i) to prospect trees in natural stands in the provenance regions from across the natural range of the species; (ii) the total number of trees should come from the largest possible number of provenance regions; (iii) in each provenance region, trees should come from the largest possible number of unrelated natural stands; (iv) within each stand, only a few unrelated trees should be sampled (separated from one another by at least 100 m), and should be marked randomly (except in cases of serious diseases of a systemic type); and (v) the minimum sample size needed to represent the species in a provenance region is approximately 70 trees, and thus, because there are several provenance regions, samples with 400 or more trees are frequently considered.

2.4 Planning and Establishing the Experimental Design

The overall planning of the field trial is a key point to guarantee the success of the experimental process. First, it is essential to be demanding when choosing the location of the field trial. The choice of a suitable site is needed to ensure that the conditions are as homogeneous as possible. A site containing considerable local variation, such as mounds, holes, and rocks, must be avoided. However, any systematic trends may be controlled by the choice of an appropriate experimental design and the corresponding statistical model for the data analysis.

A field trial to study the genetic variability within a species contains a large number of different genotypes (hundreds). Therefore, a large square or rectangular area of land (4–13 ha) is needed, comprising at least one buffer row around the outside of the trial. The criteria to establish the spacing between trees at planting are defined to avoid competition between adjacent trees and depend on the species and the objective of the field trial.

The choice of an adequate experimental design is the next crucial step. Well-designed field trials are required to provide a reliable evaluation of the genetic variability within a species for important quantitative traits. Many of the important principles of experimental design were developed in the 1920s and 1930s, in particular by Fisher (6). In fact, Fisher's experimental design principles of randomization, replication, and blocking to control extraneous variation are continuously recognized in many texts related to experimental designs in agriculture, forests, and other sciences (3,7–10).

Replication is essential to allow the estimation of the error variance and the precision of an experiment increases with the number of replications. In fact, without repetitions, there is no basis to quantify genetic variability within a species nor to perform the selection of quantitative traits because there is no way to assess the error and genetic variance estimates nor the predicted genetic values. There is the need to have enough replicates to provide adequate degrees of freedom for the error term used for testing the parameters of interest in the study. Although, the number of replicates also depends on practical feasibility. According to Williams et al. (7), for forest breeding trials, namely for species or provenance trials, the usually recommended number of replicates is four to six. Concerning the number of trees per plot (pseudo-repetitions), it should not be too high so as not to increase variability within a resolvable replicate, but adequate to ensure that there are enough trees in each plot to protect against missing plots and to reduce, to some extent, competition from adjacent plots. In practice, plots with two to three trees are currently used.

Randomization ensures that all experimental units are equally likely to receive any treatment, thereby minimizing systematic errors or bias induced by the researcher. Consequently, it is essential to provide a valid estimation of the error variance. As previously mentioned, notwithstanding randomization being a basic principle in research, its importance is not always fully respected. About this topic, the work of Piepho et al. (11) provides a general reminder that field experiments should be randomized.

The efficiency of the experimental designs depends on the randomization process used to control several types of non-genetic variation. It is very rare for a forest field trial that covers a large area to be completely uniform. Therefore, a completely randomized design is not appropriate to handle this type of experiment. Blocking plays a key role in controlling different sources of natural variation among experimental units, such as slope, soil changes, prevailing wind, and site preparation techniques. In this type of field trial, even the randomized complete block design is not recommended because its ability to account for the site variability is limited. As a consequence, designs with multiple blocking factors have been developed. Among these are: (i) the experimental designs with incomplete but smaller blocks, such as alpha designs (12), which constitute a particular class of generalized lattice designs; (ii) the row–column designs with blocks in both rows and columns to control for environmental gradients in both directions (13), which correspond to groups of more complex Latin square designs; (iii) the Latinized row–column designs for use when replicates are contiguous and form long blocks of plots (14); and (iv) the resolvable spatial row–column designs (15). In all the multiple blocking designs mentioned, the incomplete blocks are grouped into complete blocks, and therefore, they are called resolvable designs.

Generating efficient experimental designs is a current research topic (16, 17). The overall objective is to optimize neighbor balance, ensuring that the number of pairwise adjacencies is as equal as possible across pairs of treatments over the field layout. The generation of experimental designs with higher efficiency in well neighbor-balanced properties has been continuously studied and discussed (18, 19), and more recent approaches have been implemented in experimental design packages (20).

The advantages of the experimental designs described above over randomized complete block designs have been demonstrated recurrently (21). In forestry, they have been progressively applied in genetic trials (7, 22–25). For instance, based on simulation studies, Fu et al. (26) and Fu et al. (27) list a number of advantages to implementing incomplete block designs in genetic trials, such as high statistical efficiency of evaluating family performance, the possibility of testing more families in single trials, and the flexibility in the choice of planting site on irregular surface areas.

2.5 Examples of the Implementation of Field Trials to Quantify Genetic Variability within a Species

2.5.1 *Pinus pinea* L. in Portugal

Mediterranean stone pine (*P. pinea* L.) is one of the ten *Pinus* species of the Mediterranean basin (28). Since ancient times, this pine has been planted along the Mediterranean basin, from Portugal to Turkey, for its edible seeds and useful timber. In the second half of the nineteenth century, reforestation programs along the Mediterranean area used this species to control erosion. The Portuguese National Forest Inventory (IFN6) indicates that Mediterranean stone pine currently occupies about 193,600 ha representing over 20% of the global distribution area (29). The increase in the area reveals the interest of the forest owners because of the policies implemented by the UE frameworks. These policies contributed to change the composition and structure of the stone pine stands and made the pinecone and kernel its main product of economic exploitation. Besides, this species is very tolerant to summer droughts and has the ability to cope with occasional low winter temperatures (30). It can be found in zones where annual precipitation is less than 300–350 mm. According to Ribeiro (31), seven provenance regions and two restricted areas were defined.

The genetic diversity of *P. pinea* L. has been measured using molecular markers and quantitative genetic methodologies. Several authors report that this species has low genetic diversity when molecular markers were applied (32–34). Concerning quantitative genetic methodologies, several studies have been conducted to evaluate the adaptive traits of *P. pinea* L. In the south of France, Court-Picon et al. (30) found that germination capacity, seed size and weight, and height growth varied among provenances. Mutke et al. (35) demonstrated significant genetic variability in the cone yield of the clonal bank. Carneiro et al. (36) found significant differences between provenances for plant height at ages 2 and 13, but they did not detect significant differences for diameter at breast height (DBH) at age 13. Based on data collected over a period of several years and from three field sites from a *P. pinea* L. provenance trial established in Portugal in 1993, Carrasquinho et al. (37) showed significant provenance genetic variability in terms of height at different planting ages, DBH at age 13 years, and survival at different ages. Mutke et al. (38) analyzed survival, growth height, and phenology in a provenance trial with 34 different populations. No significant differences in survival were found between accessions, while growth height was slightly significantly different. Sánchez-Gómez et al. (39) found variability in morphological and physiological traits using 20 clones from five different Spanish populations. The reported level of *P. pinea* L. genetic variability seems to be dependent on individual study methodologies. In fact, it is important to emphasize that well-established field tests and adequate models for data analysis are critical to obtain accurate genetic estimates of variability and to increase the reliability of the resulting performance.

Regarding genetic variability for cone production, a clonal field trial was proposed to be established in an area at *National Forest of Alva de Pataias*, Leiria, Portugal. In this field trial, 297 genotypes randomly identified along the seven provenance regions (40–45 trees per provenance region) are represented. The experimental design was generated in CycDesigN 7.0 software (https://www.vsni.co.uk/software/cycdesign), according to a set of conditions. The planting spacing is 6 m × 6 m and each experimental unit (plot) consists of 2 × 1 trees. The genotypes to be tested are 297 ($v = 297$), 6 resolvable replicates will be established ($r = 6$), each one with 33 columns ($s = 33$), and 9 of rows ($k = 9$). Excluding the border columns and rows, the experimental site consists of non-contiguous $k \times s$ array of plots (9 × 33) (Figure 2.1). The layout of a suitable row–column design using the option "spatial" in CycDesigN is shown in Figure 2.2. The design has a row–column average efficiency factor of 0.845098 (upper bound of 0.851766).

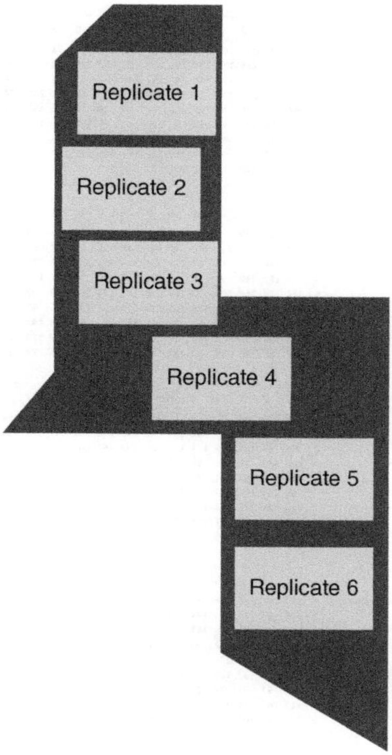

Figure 2.1 Layout of the replicates for the Mediterranean stone pine clonal field trial established at *National Forest of Alva de Pataias*, Leiria, Portugal. Each replicate has an array of 9 rows × 33 columns.

2.5.2 *Pinus pinaster* Aiton in Portugal

Maritime pine (*P. pinaster* Aiton) is a widespread conifer original from the western Mediterranean basin, occurring in the Iberian Peninsula, South France, West Italy, the western Mediterranean Isles, North Morocco, Algeria, and Tunisia (40). In Portugal, maritime pine is the third-most important native forest species, occupying approximately 22.6% of the Portuguese forest area. This species can be found throughout almost all of the mainland, favoring the coastal area. Based on ecological conditions, seven regions of provenance and four restricted areas were delimited in the country (31).

Several studies have been conducted to assess the genetic variability of maritime pine. Provenance tests detected genetic variability between populations for some adaptive characteristics, namely, height growth, diameter, trunk shape, survival, cyclicity of the height growth characteristic (polycyclism), water use efficiency (41,42), and for neutral characteristics (43,44). In progeny trials, genetic variability has been identified among half-sibling families for adaptive traits, such as total height and diameter at breast height, DBH (45), for wood quality (46,47) and also, more recently, for resistance to pine wilt disease, using young plants in a controlled environment (48). Concerning the resin

		1	2	3	4	5	6	7	8	9	10	11	12	13	14	15	16	17	18	19	20	21	22	23	24	25	26	27	28	29	30	31	32	33
Replicate 1	1	15	90	217	65	32	151	276	92	234	131	71	43	140	296	174	258	10	27	178	222	91	11	47	224	283	240	16	286	162	50	188	86	160
	2	95	245	265	125	181	262	203	9	2	295	223	98	60	169	51	248	275	242	180	26	13	204	73	112	39	165	145	269	237	19	200	232	70
	3	219	67	100	235	126	249	157	148	267	261	113	264	284	5	123	213	282	41	62	53	137	136	192	210	97	195	144	124	135	257	206	244	1
	4	208	79	290	205	17	159	115	141	189	57	4	231	74	251	20	285	117	255	216	58	191	158	182	132	82	274	167	110	227	129	138	211	185
	5	154	230	78	48	147	250	68	94	28	14	201	152	104	202	218	221	187	209	176	77	183	31	184	246	44	293	260	228	172	72	66	35	173
	6	83	56	107	45	287	103	207	134	76	108	24	179	121	254	197	88	111	215	22	25	105	233	168	99	247	114	214	128	46	64	102	243	96
	7	280	119	55	164	238	266	256	133	18	198	3	161	163	229	143	109	153	7	239	166	297	54	80	294	212	81	106	130	29	75	270	253	155
	8	156	59	85	93	52	190	268	291	279	278	171	63	196	84	87	277	6	33	220	61	259	49	37	12	194	89	40	139	199	69	34	241	120
	9	21	149	175	116	42	101	177	36	118	38	127	193	263	225	273	30	8	271	186	252	281	122	236	170	272	226	292	150	289	288	142	146	23

		1	2	3	4	5	6	7	8	9	10	11	12	13	14	15	16	17	18	19	20	21	22	23	24	25	26	27	28	29	30	31	32	33
Replicate 2	1	170	11	215	23	263	86	105	80	245	274	97	123	287	68	145	246	252	294	122	277	229	154	62	190	48	148	239	177	2	183	6	140	283
	2	93	248	143	237	222	44	185	16	139	27	28	61	243	111	203	267	270	202	147	257	275	197	249	47	281	228	65	151	261	120	163	3	226
	3	210	278	127	43	87	255	131	54	231	266	125	126	242	29	4	150	35	116	21	232	284	69	76	58	256	199	218	91	174	40	292	59	84
	4	37	137	250	109	72	56	136	205	130	196	225	164	82	88	289	291	128	171	34	161	217	254	119	282	1	115	181	112	45	221	83	212	180
	5	273	238	244	209	94	169	176	223	214	204	89	92	15	96	234	200	162	240	182	9	276	39	53	153	8	31	206	118	247	179	224	132	20
	6	121	168	172	260	138	191	51	173	259	208	75	12	30	124	33	264	100	108	98	286	192	55	167	79	71	186	149	166	201	81	66	207	135
	7	189	117	158	22	63	26	85	129	187	67	106	142	52	262	280	178	295	195	24	104	46	235	36	265	110	42	279	146	241	113	184	271	152
	8	258	102	290	288	220	18	211	25	32	227	157	114	13	38	253	64	194	70	14	101	230	141	60	213	41	95	293	251	297	216	57	5	107
	9	77	296	10	268	99	188	165	219	156	78	175	73	159	49	134	90	74	272	160	144	155	236	19	198	103	7	285	50	233	17	193	269	133

		1	2	3	4	5	6	7	8	9	10	11	12	13	14	15	16	17	18	19	20	21	22	23	24	25	26	27	28	29	30	31	32	33
Replicate 3	1	46	208	201	83	213	5	112	27	81	61	11	135	191	154	231	195	197	42	260	15	211	266	296	233	175	38	200	140	96	291	110	189	19
	2	127	50	219	218	82	246	71	247	216	251	267	109	155	102	180	23	139	55	210	69	33	53	142	183	122	164	57	17	37	207	54	10	148
	3	259	43	105	169	163	167	20	238	162	268	269	221	171	271	104	77	225	56	156	41	32	178	4	95	261	143	80	243	235	282	28	256	275
	4	66	185	58	245	31	107	253	258	60	215	174	295	270	98	16	229	137	286	153	99	29	146	52	67	70	113	248	39	151	128	133	199	209
	5	236	241	263	141	78	88	126	114	117	64	22	49	232	294	244	168	48	89	65	177	44	2	193	130	119	21	252	262	84	47	14	108	138
	6	86	204	106	281	51	198	7	150	242	134	34	222	132	283	274	9	230	237	30	111	103	173	45	118	40	121	157	217	293	280	190	176	116
	7	76	26	276	161	279	8	93	62	250	124	94	129	145	272	194	170	239	123	85	188	186	165	257	3	152	224	91	264	74	181	172	277	25
	8	147	284	120	12	214	240	97	159	166	184	13	18	6	223	149	265	63	158	249	205	87	115	212	35	227	144	1	226	203	273	90	288	278
	9	220	101	254	179	59	125	187	228	68	297	289	292	131	136	160	196	182	285	73	79	206	234	192	92	287	75	36	290	72	202	255	100	24

		1	2	3	4	5	6	7	8	9	10	11	12	13	14	15	16	17	18	19	20	21	22	23	24	25	26	27	28	29	30	31	32	33
Replicate 4	1	233	147	285	18	31	289	128	161	284	170	131	177	73	216	262	3	200	56	33	194	195	217	75	84	104	227	235	47	95	201	105	53	214
	2	278	40	224	215	124	37	129	270	169	7	249	89	8	158	62	13	63	159	114	205	32	44	4	98	192	248	140	1	50	199	280	66	34
	3	246	239	245	242	141	222	112	265	228	27	269	185	231	6	225	103	152	145	294	21	109	183	188	46	273	38	286	76	173	41	19	45	78
	4	162	207	203	232	93	154	149	263	16	42	208	213	221	102	184	125	297	111	43	260	99	193	148	176	287	229	171	209	110	26	14	54	164
	5	120	51	117	276	28	238	24	136	288	60	218	135	87	65	244	123	115	121	85	134	257	266	146	80	52	39	255	142	220	202	240	175	91
	6	20	64	57	107	258	127	292	247	48	254	86	230	210	15	271	166	49	36	70	290	267	282	293	144	190	259	165	281	153	119	186	163	100
	7	151	97	59	189	22	167	198	172	81	160	251	223	55	196	61	234	126	137	118	68	283	295	10	181	25	72	122	211	243	156	179	277	150
	8	116	241	58	212	155	279	250	268	272	138	9	187	130	2	29	180	30	11	132	261	252	274	197	92	69	256	168	264	12	96	67	90	5
	9	236	83	101	291	226	108	191	71	94	133	88	174	219	17	79	23	77	35	296	178	113	182	204	139	237	206	157	106	74	82	143	253	275

		1	2	3	4	5	6	7	8	9	10	11	12	13	14	15	16	17	18	19	20	21	22	23	24	25	26	27	28	29	30	31	32	33
Replicate 5	1	154	269	67	190	74	121	288	50	193	236	239	24	178	31	83	100	68	8	46	253	16	75	120	242	119	283	162	174	200	207	9	238	211
	2	127	235	260	115	44	202	196	277	137	15	142	243	25	252	117	112	213	105	184	169	206	98	29	89	232	155	208	41	139	160	133	233	40
	3	217	43	183	3	223	148	296	203	227	28	261	273	145	259	194	159	116	81	7	69	246	97	153	244	66	285	11	123	59	53	42	180	257
	4	224	79	4	175	36	165	262	290	109	179	219	176	281	263	126	215	247	286	191	214	47	279	251	34	197	60	199	149	272	6	72	230	77
	5	287	212	111	101	51	231	226	113	275	61	163	33	96	258	209	106	144	147	220	136	141	110	129	95	132	289	198	168	186	48	91	188	55
	6	65	94	181	37	152	88	76	130	201	185	13	5	78	10	90	195	12	222	39	62	150	170	56	292	23	143	64	270	205	107	266	157	278
	7	87	158	103	131	164	172	80	27	291	30	173	245	20	284	254	2	26	192	267	293	171	146	274	19	124	63	102	17	265	35	271	21	177
	8	268	38	18	99	128	52	187	225	216	140	255	49	161	280	167	218	182	82	54	276	166	210	125	229	228	241	1	85	151	237	70	189	92
	9	114	156	57	264	256	118	14	138	86	282	84	58	248	294	204	122	22	134	240	295	135	297	45	71	73	108	104	234	221	32	93	249	250

		1	2	3	4	5	6	7	8	9	10	11	12	13	14	15	16	17	18	19	20	21	22	23	24	25	26	27	28	29	30	31	32	33
Replicate 6	1	232	293	139	113	193	140	84	273	248	10	209	220	26	105	54	214	283	97	114	242	124	289	9	270	272	251	129	230	55	65	79	103	115
	2	234	155	154	235	265	48	42	249	177	87	2	57	3	122	121	296	160	109	282	255	176	25	82	204	36	12	259	240	253	254	162	64	34
	3	182	286	46	69	206	146	1	95	119	246	45	17	151	207	133	189	40	127	261	263	169	161	111	187	27	216	88	13	49	145	290	194	32
	4	120	83	59	107	93	238	143	268	94	18	131	223	231	196	74	123	30	51	33	178	252	183	150	144	192	39	61	47	110	199	19	66	241
	5	7	62	203	297	11	292	132	125	274	168	43	237	72	188	211	173	156	67	175	164	118	262	24	291	28	108	215	106	284	149	218	213	8
	6	99	44	165	52	130	227	75	142	278	96	280	90	100	116	147	101	276	141	279	198	71	228	6	138	174	221	60	271	180	77	137	158	260
	7	172	288	166	86	185	219	245	78	243	126	159	68	134	267	295	171	98	104	229	197	285	153	233	186	53	91	41	195	225	212	226	37	128
	8	5	20	258	117	224	22	222	191	136	16	170	208	89	73	15	29	250	200	35	70	21	257	256	80	181	190	217	148	76	50	201	179	163
	9	239	58	281	31	167	112	247	81	184	269	135	210	102	244	23	63	56	85	205	294	266	152	264	287	38	4	92	157	202	14	277	275	236

Figure 2.2 Layout of a row–column design, generated in CycDesigN 7.0 software, for the clonal field trial with 297 Mediterranean stone pine genotypes at *National Forest of Alva de Pataias*, Leiria, Portugal, with six non-contiguous replicates.

production, and other important maritime pine non-wood natural products, the first phenotypic selections in Portugal were carried out in the fifties of the twentieth century. At that time, resin tapping in Portuguese pine forests was being performed on a large scale. However, this genetic improvement activity was not continued and only in 2018, it was again considered a priority (49).

Aiming to reinforce the productive capacity of Portuguese pinewoods through the production of resin, a breeding program has been initiated, using an innovative methodology based on predicted genotypic values to obtain superior quality material with an effective reduction of the time interval required for genetic selection.

To evaluate the genetic variability for resin production, a clonal field trial was established at *National Forest of Alva de Pataias*, Leiria, Portugal. In this field trial, 420 genotypes, randomly identified along the seven provenance regions (around 60 trees per provenance region), are represented. The experimental design was generated in CycDesigN 7.0 software (https://www.vsni.co.uk/software/cycdesign). The planting spacing is 3 m × 3 m and each experimental unit (plot) consists of 3 × 1 trees. The genotypes to be tested are 420 ($v = 420$), 8 resolvable replicates were established ($r = 8$), each one with 21 columns ($s = 21$), and 20 of rows ($k = 20$). Excluding the border columns and rows, the experimental site consists of $kr \times s$ array of plots (160 × 21), that is, the eight replicates are contiguous and form long blocks (or columns) of plots (Figure 2.3). The layout of a Latinized row–column design, Latinized by columns, using the option "spatial" in CycDesigN, is shown in Figure 2.4. The design has a row–column average efficiency factor of 0.898759 (upper bound of 0.903620).

Figure 2.3 Field layout for the eight contiguous replicates for testing 420 maritime pine genotypes at *National Forest of Alva de Pataias*, Leiria, Portugal.

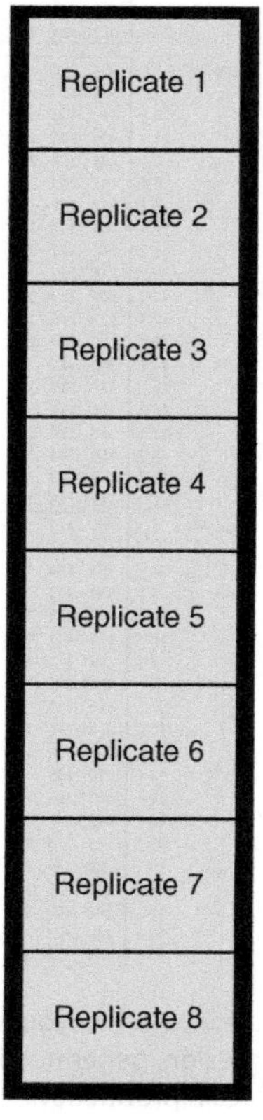

		1	2	3	4	5	6	7	8	9	10	11	12	13	14	15	16	17	18	19	20	21
Replicate 1	1	41	91	307	288	131	397	385	242	153	260	96	239	37	315	68	194	301	157	254	129	188
	2	407	228	42	24	318	189	249	217	408	392	294	115	138	235	420	416	191	317	23	107	5
	3	106	253	352	154	90	121	331	134	252	244	313	358	2	347	419	86	280	282	414	74	100
	4	356	236	147	84	345	324	190	180	125	215	186	12	243	367	139	162	354	34	259	312	110
	5	128	89	182	406	35	258	319	247	310	250	277	297	418	339	83	306	9	155	73	26	353
	6	57	400	1	342	103	178	19	327	109	25	104	389	375	276	274	225	117	281	174	187	369
	7	411	361	119	140	326	409	144	161	78	143	413	79	402	393	122	206	7	80	163	61	322
	8	386	379	32	370	164	123	378	398	171	220	124	38	145	262	231	60	272	299	152	176	341
	9	387	218	77	39	302	268	285	373	112	396	382	137	159	148	49	28	227	127	205	95	150
	10	53	351	338	196	27	286	13	168	135	241	219	177	88	359	211	368	99	293	405	395	391
	11	269	298	69	222	3	214	267	20	314	169	394	412	55	295	346	350	94	167	18	97	75
	12	165	67	170	333	40	325	238	93	47	230	364	51	22	251	321	283	195	185	46	66	291
	13	366	232	202	184	54	384	256	56	16	374	200	403	105	381	273	113	417	226	72	332	10
	14	265	216	81	237	108	362	149	234	133	335	209	31	255	197	114	376	289	33	229	183	82
	15	404	141	278	132	17	30	355	383	334	279	344	101	271	221	192	240	120	337	142	85	304
	16	92	330	181	203	309	58	130	179	287	320	151	365	172	48	257	63	245	388	156	166	160
	17	118	116	198	126	311	98	62	266	6	329	399	43	415	8	264	102	377	263	52	390	308
	18	201	363	212	199	146	71	15	76	328	208	87	340	248	14	360	296	316	193	372	292	343
	19	290	371	223	210	305	44	59	246	65	323	233	300	224	111	45	29	136	410	261	173	284
	20	36	401	64	207	158	270	349	70	275	357	380	336	4	204	303	21	11	348	213	50	175
Replicate 2	1	240	61	122	363	89	245	40	316	42	13	335	108	190	365	290	263	184	174	96	230	77
	2	218	180	150	129	46	138	1	199	381	399	264	404	175	277	215	280	166	322	315	214	103
	3	22	24	281	285	198	200	278	411	357	295	231	168	29	76	239	112	98	354	26	156	321
	4	167	260	124	246	100	208	330	67	258	110	393	102	317	255	104	192	261	183	36	319	88
	5	127	38	165	182	268	364	210	406	368	151	62	48	405	350	176	191	373	366	244	235	243
	6	81	226	73	308	20	351	92	375	313	101	193	154	59	94	237	397	367	274	408	326	50
	7	206	213	275	131	196	86	291	267	115	133	58	401	109	45	302	134	145	8	2	83	249
	8	152	282	216	303	172	119	155	409	369	407	207	194	269	238	179	105	292	135	412	220	241
	9	256	416	99	298	265	187	146	52	270	296	306	186	202	229	78	323	34	242	116	121	137
	10	395	419	390	54	233	346	333	49	55	383	153	149	14	337	162	178	222	228	353	41	4
	11	114	130	37	72	272	126	389	160	139	400	128	5	66	380	53	212	170	111	132	147	266
	12	82	120	97	352	11	185	398	300	203	402	227	360	362	334	9	414	394	370	420	28	56
	13	287	262	371	294	248	234	382	27	232	283	374	205	329	125	204	31	106	21	293	318	30
	14	71	224	252	177	69	23	377	74	195	310	396	342	314	320	311	378	361	344	312	273	347
	15	385	64	43	25	403	304	163	171	418	201	217	325	161	299	259	219	336	251	3	289	173
	16	35	84	19	7	85	157	254	279	93	276	181	359	158	189	253	387	16	379	44	415	169
	17	107	118	247	271	113	148	33	87	307	301	355	305	79	117	372	140	341	32	10	65	388
	18	141	225	345	410	221	343	123	136	386	18	12	250	39	332	413	188	349	327	338	358	91
	19	309	15	197	57	392	297	143	68	142	331	75	95	60	236	384	348	284	286	6	51	376
	20	223	324	328	47	80	144	159	339	90	17	164	417	257	288	391	356	211	340	63	70	209
		...										...										...
Replicate 7	1	63	256	5	367	369	414	279	222	8	134	114	214	130	136	382	161	181	4	123	407	51
	2	362	389	295	317	154	337	96	166	308	35	32	377	125	144	243	353	137	196	335	93	171
	3	413	408	152	241	153	34	246	158	253	139	316	188	339	291	19	237	334	95	366	89	126
	4	400	396	104	145	106	329	78	194	286	3	85	45	33	364	226	386	190	300	172	24	99
	5	283	255	250	254	175	387	121	111	261	187	249	47	179	312	228	43	80	15	118	345	313
	6	285	211	55	232	365	163	197	9	358	294	322	87	378	128	406	40	71	56	102	112	401
	7	217	162	113	230	275	340	361	66	76	176	397	276	91	287	280	231	138	380	410	258	52
	8	405	347	116	305	324	363	221	277	84	238	41	373	108	156	143	223	53	346	281	259	21
	9	146	327	288	50	420	124	356	289	267	415	225	207	46	383	341	236	133	29	218	359	182
	10	16	140	74	384	323	350	115	328	248	174	185	180	98	77	338	183	149	330	17	303	355
	11	348	233	368	315	292	321	391	31	301	39	25	26	213	298	119	58	260	101	81	127	273
	12	173	296	191	68	252	418	274	393	210	120	216	293	390	371	11	220	18	271	302	333	404
	13	195	107	234	304	37	12	269	198	257	155	311	375	100	184	70	13	352	44	186	160	398
	14	343	417	290	151	379	103	94	142	36	409	159	202	395	282	148	394	251	318	297	374	178
	15	117	263	326	212	235	177	205	419	82	351	412	97	132	403	240	268	357	385	109	242	215
	16	224	411	227	306	332	110	209	92	325	42	392	203	64	372	30	266	122	381	219	86	167
	17	83	129	192	20	193	48	299	169	1	284	310	23	270	349	200	6	27	272	79	57	105
	18	150	244	14	354	376	65	360	402	344	265	67	54	28	206	314	157	247	170	262	38	309
	19	147	75	336	49	399	204	22	62	388	72	90	88	208	370	131	278	342	73	61	59	189
	20	239	245	135	229	319	320	264	2	165	10	168	141	7	307	201	199	60	164	331	69	416
Replicate 8	1	359	124	40	343	138	84	323	209	3	264	57	82	377	50	337	250	161	247	238	153	368
	2	171	18	420	226	266	310	380	117	363	345	107	376	268	286	340	342	291	233	55	280	393
	3	361	128	190	96	370	170	42	241	127	330	51	129	65	112	388	234	324	212	201	339	302
	4	364	243	401	301	95	289	326	41	338	66	282	13	146	203	180	299	143	63	278	379	265
	5	384	283	367	109	261	332	105	71	85	182	33	144	298	62	195	160	358	231	391	205	346
	6	111	409	157	163	219	75	184	120	103	137	11	16	25	165	81	415	296	319	154	45	179
	7	97	177	341	114	230	294	30	59	162	136	402	260	135	80	390	77	378	115	387	198	417
	8	305	208	279	334	204	239	100	312	335	263	308	99	56	218	206	292	119	23	225	139	333
	9	369	1	86	220	408	52	321	276	37	113	189	382	149	246	164	354	344	118	9	349	412
	10	374	94	173	14	134	172	252	200	262	70	373	15	277	389	229	413	307	325	27	132	397
	11	202	6	155	176	419	79	213	404	141	185	242	46	36	316	5	256	255	295	90	320	365
	12	76	259	47	98	269	191	306	396	73	322	187	2	181	273	89	123	48	383	348	392	199
	13	399	300	31	236	281	183	102	159	121	130	131	411	152	351	221	196	314	215	416	54	352
	14	142	257	350	88	32	315	193	336	407	237	272	371	331	108	178	357	35	39	148	104	224
	15	29	174	270	318	288	93	253	410	403	58	122	49	313	38	385	347	400	216	145	110	87
	16	28	188	327	418	355	167	166	78	284	44	405	311	24	60	92	297	275	125	303	22	228
	17	293	235	156	69	101	4	150	83	398	106	356	258	61	394	116	67	68	158	12	140	19
	18	192	168	366	21	214	8	133	43	227	395	223	34	304	271	244	328	72	406	91	267	317
	19	249	309	248	186	7	240	251	353	274	254	232	20	211	74	210	285	207	53	381	362	329
	20	245	26	151	169	372	360	386	414	126	64	222	147	197	217	175	10	290	375	287	194	17

Figure 2.4 Layout of four from the eight contiguous replicates of the Latinized row–column design, generated in CycDesigN 7.0 software, to evaluate 420 maritime pine genotypes for resin production.

2.6 Statistical Analysis and Quantification of Genetic Variability within a Species

2.6.1 Statistical Analysis

The theory of mixed models is the basis for the analysis of genetic variability within a species. For data analysis, a model including genotypic effects and all design effects, in accordance with the randomization scheme of the experimental design, should be fitted. In the context of this work, since a random sample of genotypes of the species is prospected and studied, the genotypic effects are assumed to be random and a key objective in the study is to assess the estimate of the genotypic variance component. For example, a very common experimental design in the context described is a resolvable row–column design (see Section 2.5.1). For this type of design, the model can be described as follows:

$$y_{ijlm} = \mu + u_{\mathrm{g}_i} + u_{\mathrm{r}_j} + u_{\mathrm{col(r)}_{jl}} + u_{\mathrm{row(r)}_{jm}} + e_{ijlm}$$

for $i = 1, \cdots, v$; $j = 1, \cdots, r$; $l = 1, \cdots, s$; and $m = 1, \cdots, k$ (v, number of genotypes; r, number of resolvable replicates; s, number of columns; k, number of rows). In this model, y_{ijlm} is the response (the mean of the trait of each plot) of the ith genotype in the jth replicate, lth column and mth row; μ is the general intercept; u_{g_i} is the ith genotype effect; u_{r_j} is the jth replicate effect; $u_{\mathrm{col(r)}_{jl}}$ is the lth column effect within the jth replicate; $u_{\mathrm{row(r)}_{jm}}$ is the mth row effect within the jth replicate; and e_{ijlm} is the random error corresponding to the observation y_{ijlm}.

Model effects (with the exception of μ) are assumed to be independent and identically distributed normal variables with zero mean and variances σ^2_{g}, σ^2_{r}, $\sigma^2_{\mathrm{col(r)}}$, $\sigma^2_{\mathrm{row(r)}}$, and σ^2_e, respectively. All random effects are assumed to be mutually independent. If a resolvable row–column design, latinized by column is used (see Section 2.5.2), the latinized column effects should be added to the model, also considering to be independent and identically distributed normal variables with zero mean and respective variance. If instead of the mean of the trait of each plot, individual tree data of the plot is used, then in the model, the effects of the plots should be added. In the same vein, they should be assumed to be independent and identically distributed normal variables with zero mean and respective variance.

The residual maximum likelihood (REML) estimation method (50) is used for the estimation of the covariance parameters. In practice, linear mixed models can be fitted using, among others, the ASReml-R package (51) within R (52), or Proc Mixed of SAS (53).

The fitting of the model is followed by the diagnosis of the residuals, namely, the spatial correlation not accounted for the design effects should be checked. Spatial models may be used, when necessary, as a complement to the experimental design. Several theoretical approaches to this latter issue have been discussed (54–57). In practice, spatial methods have been used to study patterns of site variation to overcome problems of spatial heterogeneity in forest genetic trials, such as for *Pinus* spp., *Picea* spp., *Peudotsuga menziesii*, and *Castanea sativa* (37, 38, 58–65).

Specifically, using data from maritime pine progeny trial with a randomized complete design, Costa e Silva et al. (60) showed that there was significant spatial variation. Adding a correlation error term to the model increased the efficiency of data analysis. For Mediterranean stone pine provenance trials, Mutke et al. (38) used the nearest neighbor adjustment by the iterated Papadakis method to find significant differences in adaptive traits between provenances. For the same species, Carrasquinho et al. (37) demonstrated the advantages of using a mixed spatial model for the evaluation of provenance genetic variability for growth traits.

2.6.2 Genetic Parameters

2.6.2.1 Quantification of Genetic Variability within a Species

The estimate of the genetic variance component ($\widehat{\sigma}_{\mathrm{g}}^2$) obtained with REML estimation is the basis for the quantification of the genetic variability within a species.

The genetic variance component is tested ($H_0 : \sigma_{\mathrm{g}}^2 = 0$ versus $H_1 : \sigma_{\mathrm{g}}^2 > 0$) using a residual maximum likelihood ratio test. Because the null hypothesis is on the boundary of the parameter space, the p value of the test is usually assumed to be half of the reported p value from the chi-squared distribution with one degree of freedom (66, 67). For the evaluation of the precision of the quantification of the genetic variability within a species, the ratio $\widehat{\sigma}_{\mathrm{g}}^2/\mathrm{SE}$ can be considered (where $\widehat{\sigma}_{\mathrm{g}}^2$ is the estimate of the genetic variance of the evaluated trait and SE is the standard error associated with this estimate). The higher the value for this ratio, the higher the precision associated with the quantification of the genetic variability within a species.

For a clearer interpretation of the results of the genetic variability within the species, a relative measure of the genetic variability can also be used – the genotypic coefficient of variation, which is the estimate of the genotypic standard deviation of the evaluated trait divided by the mean of the trait in the field trial.

2.6.2.2 Broad-Sense Heritability

When quantitative traits from clonal field trials are analyzed, broad-sense heritability is obtained. In classical models of quantitative genetics (i.e. balanced data with no random effects other than those associated with genotypes and error and diagonal variance–covariance matrices), broad-sense heritability is the ratio between genotypic variance and phenotypic variance (68). Several studies have considered the problem of defining the heritability for more complex models (69–73). In experimental designs with multiple blocking factors, the prediction error variances differ for different genotypes. As a consequence, a generalized measure of broad-sense heritability should be used to evaluate how much phenotypic variability is due to genetic causes and to judge the efficiency of selection. For the case of clonal field trials, a generalized measure of broad-sense heritability can be calculated according to Gonçalves et al. (73) and Gonçalves et al. (74).

2.6.2.3 Selection and Genetic Gain

The Empirical Best Linear Unbiased Predictors (EBLUPs) of the genotypic effects of the evaluated quantitative traits are obtained through the mixed model equations (75, 76). The part of the difference between the phenotypic mean of the genotype and the overall mean of the field trial, which is explained by genetic causes, is the EBLUP of the genotypic effect.

The EBLUPs of the genotypic effects are the basis for the selection and prediction of genetic gains. In a selection program in the scope of clonal field trials, the EBLUPs of the genotypic effects for a given trait (for instance, resin production in maritime pine, or cone production in Mediterranean stone pine) are ranked, and the top-ranked genotypes are selected (usually, 20 top genotypes).

In classical models of quantitative genetics, the prediction of genetic gain is obtained through the product of the differential of selection with the heritability (68). However, under the experimental designs and models described, the conditions to apply the classical expression are no longer met. Therefore, the predicted genetic gain of selection should be computed as the mean of the EBLUPs of the genotypic effects of the trait of the top-ranked selected genotypes.

2.7 Conclusions

The existence of genetic variability for a certain characteristic of interest is an important condition for the success of a breeding program. The establishment of clonal genetic field trials constitutes an important approach to quantify this genetic variability within a species in the context of non-wood forest products due to an effective reduction in the time required for genetic selection.

This chapter provided a methodology to implement a breeding strategy, covering the adequate steps to achieve the proposed objectives, with a particular focus on the importance of performing adequate sampling of the available genetic variability as well as using efficient experimental designs to control non-genetic variability and respective statistical analysis. In Portugal, this methodological approach is being implemented to study the genetic variability for resin production in maritime pine and cone production in Mediterranean stone pine. The objective of this work is to encourage its wider use in studying the genetic variability within this and other species concerning non-wood forest products.

Acknowledgments

This work was supported by the funding of the projects "Pinheiro-manso: Conservação e melhoramento dos recursos genéticos" (PDR2020-785-063761) and TRANSFORM-P1.1 "Melhoramento genético e materiais florestais de reprodução" (PRR-C644865735 – 00000007).

References

1 Gassner A, Coe R. Principles of field experiment design. In: Dawson I, Harwood C, Jamnadass R, & Beniest J, editors. *Agroforestry Tree Domestication: A Primer. The World Agroforestry Centre.* Nairobi (KE): World Agroforestry Centre; 2012. p. 55–62.
2 von Weissenberg K. Indirect selection for improvement of desired traits. In: Miksche JP, editor. *Modern Methods in Forest Genetics.* Spring-Verlag; 1976. p. 217–228.
3 Giesbrecht FG, Gumpertz ML. *Planning, Construction, and Statistical Analysis of Comparative Experiments.* 1st Edition. New York (NY, USA): Wiley; 2004.
4 Martins A, Gonçalves E. Grapevine breeding programmes in Portugal. In: Reynolds AG, editor. *Grapevine Breeding Programs for the Wine Industry: Traditional and Molecular Techniques.* Woodhead Publishing Elsevier; 2015. p. 159–182.
5 Gonçalves E, Martins A. Optimizing conservation and evaluation of intra-varietal grapevine diversity. In: Costa JM, Catarino S, Escalona JM, & Comuzzo P, editors. *Sustainable Viticulture and Winemaking Practices.* Amsterdam (NL): Academic Press; 2022. p. 45–64.
6 Fisher RA. *The Design of Experiments.* Edinburgh (UK): Oliver & Boyd; 1935.
7 Williams ER, Matheson AC, Harwood CE. *Experimental Design and Analysis for Tree Improvement.* 2nd Edition. Collingwood: CSIRO Publishing; 2002.
8 Mead R, Gilmour SG, Mead A. *Statistical Principles for the Design of Experiments: Applications to Real Experiments.* Cambridge (UK): Cambridge University Press; 2012.
9 Welham SJ, Gezan SA, Clark SJ, Mead A. *Statistical Methods in Biology.* Boca Raton (FL): CRC Press; 2015.
10 Verdooren LR. History of the statistical design of agricultural experiments. *J Agric Biol Environ Stat.* 2020; 25(4): 457–486.
11 Piepho HP, Möhring J, Williams ER. Why randomize agricultural experiments? *J Agron Crop Sci.* 2013; 199: 374–383.

12 Patterson HD, Williams ER. A new class of resolvable incomplete block designs. *Biometrika*. 1976; 63: 83–92.
13 Williams ER, John JA. Construction of row and column designs with contiguous replicates. *Appl Stat*. 1989; 38: 149–154.
14 Williams ER, John JA, Whitaker D. Applications: example of block designs for plant and tree breeding trials. *Aust N Z J Stat*. 1999; 41(3): 277–284.
15 Williams ER, John J, Whitaker D. Construction of resolvable spatial row–column designs. *Biometrics*. 2006; 62(1): 103–108.
16 Piepho HP. Generating efficient designs for comparative experiments using the SAS procedure OPTEX. *Commun Biometry Crop Sci*. 2015; 10: 96–114.
17 Mramba LK, Peter GF, Whitaker VM, Gezan SA. Generating improved experimental designs with spatially and genetically correlated observations using mixed models. *Agronomy*. 2018; 8: 40.
18 Piepho HP, Michel V, Williams ER. Neighbor balance and evenness of distribution of treatment replications in row–column designs. *Biom J*. 2018; 60: 1172–1189.
19 Williams ER, Piepho HP. Error variance bias in neighbour balance and evenness of distribution designs. *Aust N Z J Stat*. 2019; 61(4): 466–473.
20 Piepho HP, Williams ER, Michel V. Generating row–column field experimental designs with good neighbour balance and even distribution of treatment replications. *J Agron Crop Sci*. 2021; 207(4): 745–753.
21 Williams ER, Forde CG, Imaki J, Oelkers K. Experimental design in practice: the importance of blocking and treatment structures. *Aust N Z J Stat*. 2021; 63(3): 455–467.
22 Fu YB, Clarke PY, Namkoong G, Yanchuck AD. Incomplete block designs for genetic testing: statistical efficiencies of estimating family means. *Can J For Res*. 1998; 28: 977–986.
23 Fu YB, Yanchuck AD, Namkoong G. Spatial patterns of tree height variations in a series of Douglas-fir progeny trials: implications for genetic testing. *Can J For Res*. 1999; 29: 714–723.
24 Gezan SA, White TL, Huber DA. Comparison of experimental designs for clonal forestry using simulated data. *For Sci*. 2006; 52(1): 108–116.
25 Shaliz M, Salvador G, McKeand S, Sherrill J, Cumbie W, Whetten R, et al. Correspondence between breeding values of the same *Pinus taeda* L. genotypes from clonal trials and half-sib seedling progeny trials. *For Sci*. 2020; 66: 600–611.
26 Fu YB, Yanchuk AD, Namkoong G. Incomplete block designs for genetic testing: some practical considerations. *Can J For Res*. 1999; 29: 1871–1878.
27 Fu YB, Yanchuck AD, Namkoong G. Incomplete block designs for genetic testing: accuracy of ranking families and individuals. *For Genet*. 2000; 7(4): 287–293.

28 Barbéro M, Loisel R, Quézel P, Richardson DM, Romane F. Pines of Mediterranean basin. In: Richardson DM, editor. *Ecology and Biogeography of Pinus*. 1st Edition. Cambridge (UK): Cambridge University Press; 1998. p. 153–170.

29 Uva JS (2015). IFN6: 6° Inventário Florestal. Relatório Final. Instituto da Conservação da Natureza e das Florestas [Internet]. (6th Forest Inventory. Final Report). Available from (https://www.icnf.pt/api/file/doc/c8cc40b3b7ec8541), accessed 11 March 2023.

30 Court-Picon M, Gadbin-Henry C, Guibal F, Roux M. Dendrometry and morphometry of *Pinus pinea* L. in Lower Provence (France): adaptability and variability of provenances. *For Ecol Manag*. 2004; 194: 319–333.

31 Ribeiro D. (2012). Regiões de proveniência. Portugal. Autoridade Florestal Nacional [Internet]. (Portuguese provenance regions). Available from (https://www.icnf.pt/api/file/doc/7eba73fd7d9976f1), accessed 11 March 2023.

32 Fallour D, Fady B, Lefevre F. Study on isozyme variation in *Pinus pinea* L.: evidence for low polymorphism. *Silvae Genet*. 1997; 46(4): 201–207.

33 Evaristo I, Santos S, Tenreiro R, Costa R. Comparison of genetic structure assessed by amplified fragment length polymorphism and retrotransposon-based sequence-specific amplification polymorphism for Portuguese populations of *Pinus pinea* L. *Silvae Genet*. 2008; 57(3): 93–100.

34 Vendramin GG, Fady B, González-Martínez SC, Hu FS, Scott I, Sebastiani F, et al. Genetically depauperate but widespread: the case of an emblematic Mediterranean pine. *Evolution*. 2008; 62(3): 680–688.

35 Mutke S, Gordo J, Gil L. Cone yield characterization of a stone pine (*Pinus pinea* L.) clone bank. *Silvae Genet*. 2005; 54(4–5): 189–197.

36 Carneiro A, Navalho E, Fernandes L, Carrasquinho I. Provenance variation for growth traits of *Pinus pinea* L. in Portugal. In: Proceedings of the IUFRO Division 2 Joint Conference. Low input breeding and genetic conservation of forest tree species. Antalya (Turkey); 2006. p. 29–32.

37 Carrasquinho I, Gonçalves E. Genetic variability among *Pinus pinea* L. provenances for survival and growth traits in Portugal. *Tree Genet Genomes*. 2013; 9: 855–866.

38 Mutke S, Gordo J, Chambel MR, Prada MA, Álvarez D, Iglesias S, Gil L. Phenotypic plasticity is stronger than adaptative differentiation among Mediterranean stone pine provenances. *For Syst*. 2010; 19(3): 354–366.

39 Sánchez-Gómez D, Velasco-Conde T, Cano-Martín FJ, Guevara MA, Cervera MT, Aranda I. Inter-clonal variation in functional traits in response to drought for a genetically homogeneous Mediterranean conifer. *Environ Exp Bot*. 2011; 70: 104–109.

40 Abad Viñas R, Caudullo G, Oliveira S, de Rigo D. *Pinus pinaster* in Europe: distribution, habitat, usage and threats. In: San-Miguel-Ayanz J, de Rigo D,

Caudullo G, Houston Durrant T, & Mauri A, editors. *European Atlas of Forest Tree Species*. Luxembourg: Publ. Off. EU; 2016. p. e012d59+.

41 Correia I, Almeida MH, Aguiar A, Alía R, David TS, Pereira JS. Variation in growth, survival and carbon isotope composition (δ^{13}C) among *Pinus pinaster* populations of different geographic origins. *Tree Physiol*. 2008; 28: 1545–1552.

42 Correia I, Alía R, Yan W, David T, Aguiar A, Almeida MH. Genotype × environment interactions in *Pinus pinaster* at age 10 in a multi-environment trial in Portugal: a maximum likelihood approach. *Ann For Sci*. 2010; 67: 612.

43 Ribeiro MM, Plomion C, Petit R, Vendramin GG, Szmidt AE. Variation in chloroplast single-sequence repeats in Portuguese maritime pine (*Pinus pinaster* Ait.). *Theor Appl Genet*. 2001; 102: 97–103.

44 Ribeiro MM, Mariette S, Vendramin GG, Szmidt AE, Plomion C, Kremer A. Comparison of genetic diversity estimates within and among populations of maritime pine using chloroplast simple-sequence repeat and amplified fragment length polymorphism data. *Mol Ecol*. 2002; 11: 869–877.

45 Aguiar A, Almeida MH, Borralho N. Genetic control of growth, wood density and stem characteristics of *Pinus pinaster* in Portugal. *Silva Lusitana*. 2003; 11(2): 131–139.

46 Gaspar MJ, Louzada JL, Silva ME, Aguiar A, Almeida MH. Age trends in genetic parameters of wood density components in 46 half-sibling families of *Pinus pinaster*. *Can J For Res*. 2008; 38: 1470–1477.

47 Gaspar MJ, Alves A, Louzada JL, Morais J, Santos A, Fernandes C, et al. Genetic variation of chemical and mechanical traits of maritime pine (*Pinus pinaster* Aiton). Correlations with wood density components. *Ann For Sci*. 2011; 68: 255–265.

48 Carrasquinho I, Lisboa A, Inácio ML, Gonçalves E. Genetic variation in susceptibility to pine wilt disease of maritime pine (*Pinus pinaster* Aiton) half-sib families. *Ann For Sci*. 2018; 75: 85.

49 PROGEN. Programa operacional da administração pública para a conservação e melhoramento dos recursos genéticos florestais. INIAV e ICNF. Ministério da Agricultura, Florestas e Desenvolvimento Rural. Documento anexo à Portaria n°232/2019. Portugal; (Portuguese Operational Program for the Conservation and Improvement of Forest Genetic Resources (PROGEN)) 2018 [Internet]. Available from (https://www.iniav.pt/images/Investigacao_Inovacao/sistemas-florestais/progen_pdr_7_8_iniav_icnf_2018.pdf), accessed 11 March 2023.

50 Patterson HD, Thompson R. Recovery of inter-block information when block sizes are unequal. *Biometrika*. 1971; 58: 545–554.

51 Butler DG, Cullis BR, Gilmour AR, Gogel, BJ, Thompson R. ASReml-R reference manual, version 4. Wollongong (NSW, AU): University of Wollongong; 2018.
52 R Core Team. R: A language and environment for statistical computing. R Foundation for Statistical Computing, Vienna, Austria; 2022.
53 SAS Institute Inc. SAS/STAT® 14.1 User's Guide. Cary, NC: SAS Institute Inc; 2015.
54 Cullis BR, Gleeson AC. Spatial analysis of field experiments-an extension to two dimensions. *Biometrics*. 1991; 47(4): 1449–1460.
55 Stroup WW. Power analysis based on spatial effects mixed models, a tool for comparing design and analysis strategies in the presence of spatial variability. *JABES*. 2002; 7: 491–511.
56 Piepho HP, Richter C, Williams ER. Nearest neighbour adjustment and linear variance models in plant breeding trials. *Biom J*. 2008; 50: 164–189.
57 Piepho HP, Boer MP, Williams ER. Two-dimensional P-spline smoothing for spatial analysis of plant breeding trials. *Biom J*. 2022; 64(5): 835–857.
58 Anekonda TS, Libby WJ. Effectiveness of nearest-neighbor data adjustment in a clonal test of redwood. *Silvae Genet*. 1996; 45(1): 46–51.
59 Kusnandar D, Galwey N. A proposed method for estimation of genetic parameters on forest trees without raising progeny: critical evaluation and refinement. *Silvae Genet*. 2000; 49: 15–21.
60 Costa e Silva JC, Dutkowski GW, Gilmour AR. Analysis of early tree height in forest genetic trials is enhanced by including a spatially correlated residual. *Can J For Res*. 2001; 31: 1887–1893.
61 Joyce D, Ford R, Fu YB. Spatial patterns of tree height variations in black spruce farm-field progeny test and neighbors-adjusted estimations of genetic parameters. *Silvae Genet*. 2002; 51(1): 13–18.
62 Dutkowski GW, Costa e Silva JC, Gilmour AR, Lopez GA. Spatial analysis methods for forest genetic trials. *Can J For Res*. 2002; 32: 2201–2214.
63 Dutkowski GW, Costa e Silva J, Gilmour AR, Wellendorf H, Aguiar A. Spatial analysis enhances modelling of a wide variety of traits in forest genetic trials. *Can J For Res*. 2006; 36: 1851–1870.
64 Ye TZ, Jayawickrama KJS. Efficiency of using spatial analysis in first-generation coastal Douglas-fir progeny tests in the US Pacific Northwest. *Tree Genet Genomes*. 2008; 4: 677–692.
65 Gapare WJ, Ivkovic M, Dutkowski GW, Spencer DJ, Buxton P, Wu HX. Genetic parameters and provenance variation of *Pinus radiata* D. Don. 'Eldridge collection' in Australia 1: growth and form traits. *Tree Genet Genomes*. 2012; 8(2): 391–407.
66 Self SG, Liang KY. Asymptotic properties of maximum likelihood estimators and likelihood ratio tests under nonstandard conditions. *JSTOR*. 1987; 82(398): 605–610.

67 Stram DO, Lee JW. Variance components testing in the longitudinal mixed effects model. *Biometrics.* 1994; 50: 1171–1177.
68 Falconer DS, Mackay TFC. *An Introduction to Quantitative Genetics.* 4th Edition. London (UK): Prentice Hall; 1996.
69 Cullis BR, Smith AB, Coombes NE. On the design of early generation variety trials with correlated data. *JABES.* 2006; 11: 381–393.
70 Oakey H, Verbyla A, Pitchford W, Cullis B, Kuchel H. Joint modeling of additive and non-additive genetic line effects in single field trials. *Theor Appl Genet.* 2006; 113: 809–819.
71 Piepho HP, Möhring J. Computing heritability and selection response from unbalanced plant breeding trials. *Genetics.* 2007; 177: 1881–1888.
72 Welham SJ, Gogel BJ, Smith AB, Thompson R, Cullis BR. A comparison of analysis methods for late-stage variety evaluation trials. *Aust N Z J Stat.* 2010; 52(2): 125–149.
73 Gonçalves E, Carrasquinho I, St. Aubyn A, Martins A. Broad-sense heritability in the context of mixed models for grapevine initial selection trials. *Euphytica.* 2013; 189: 379–391.
74 Gonçalves E, Carrasquinho I, Martins A. Fully and partially replicated experimental designs for evaluating intravarietal variability in grapevine. *Aust J Grape Wine Res.* 2022; 2022: 5293298.
75 Henderson C. Best linear unbiased estimation and prediction under a selection model. *Biometrics.* 1975; 31: 423–447.
76 Searle S, Casella G, McCulloch C. *Variance Components.* 1st Edition. Hoboken (NJ, USA): Wiley; 1992.

3

Sample Preparation for Forest Tree Metabolomics

Pia Guadalupe Dominguez[1], *Thaís Regiani Cataldi*[2], *Ilka Nacif Abreu*[3], *Thomas Moritz*[4], *and Ilara Gabriela Frasson Budzinski*[2]

[1] *Instituto de Agrobiotecnología y Biología Molecular (IABIMO), Instituto Nacional de Tecnología Agropecuaria (INTA), Consejo Nacional de Investigaciones Científicas y Técnicas (CONICET), Hurlingham, Buenos Aires, Argentina*
[2] *Department of Genetics, Max Feffer Laboratory of Plant Genetics, 'Luiz de Queiroz' College of Agriculture, University of São Paulo, Piracicaba, SP, Brazil*
[3] *Department of Plant Biochemistry, Albrecht-von-Haller-Institute for Plant Sciences, University of Goettingen, Goettingen, Germany*
[4] *Department of Forest Genetics and Plant Physiology, Umeå Plant Science Centre, Swedish University of Agricultural Sciences, Umeå, Sweden*

3.1 Experimental Design for Metabolomics

Plant metabolism involves the coordination of multiple biochemical processes to ensure plant growth, development, and survival. Because metabolites are spatiotemporal and environment dependent (1), the use of metabolomics has been a powerful approach to characterize and optimize the phenotypic plasticity, stress tolerance, defense metabolism, microbial interaction, and nutritional availability of important forest trees (2–4). Forest tree metabolomics studies can be even more challenging when compared to other plants due to the tree's long life cycle, genetic variability, large genome size, and presence of interferents (e.g. oleoresins) (3). In this context, in a metabolomics study, the first step is to establish the experimental design. This is a critical point that aims to make the analysis and its biological interpretation as clear as possible to, thereby, ensure that the data generated from the sample reflects the biological hypothesis under investigation (3). Before setting up the right experimental design, the researcher must have a clear research hypothesis and/or biological question (3, 5). Preliminary decisions are made within the scope of the experimental design and must consider:

i. **Plant material**: Genotype (wild type, lines), developmental stage (seeds, plantlets, mature plants), age, and tissue.

Monitoring Forest Damage with Mass Spectrometry-Based Metabolomics Methods, First Edition.
Edited by Carla António.

ii. **Growth conditions**: Whether the experiment will be carried out in greenhouses (pots, type of subtract, hydroponic), growth chambers (pots, plates, lights), or open fields (various interferences).

iii. **Replicates, randomization, and blocking**: The number of biological replicates determinates the power of statistical analysis. The idea that "the more biological replicates, the better" is not always applicable due to the availability of material, greenhouse space, but also for analytical reasons (i.e. sample stability over time, the performance of mass spectrometers can be compromised after analyzing too many samples). Generally, six biological replicates are the minimum necessary to provide sufficient power for statistical analysis (3). Technical replicates can also be used in liquid chromatography mass spectrometry (LC–MS)-based measurements (where the same sample is injected several times) as a way of anticipating possible injection failures, their average values representing a biological replicate. Once the number of plants and the treatment factors have been defined, consideration should be given to excluding external factors that could potentially interfere with the results (e.g. differences in natural light in the greenhouse, corners not covered by the automatic irrigation system). In this sense, randomization is a practice that can be used to reduce experimental error. Different strategies can be used, such as plant rotation in growth chambers. However, in experiments performed in a greenhouse or an open field, plants can be arranged in blocks. The experimental block design is a strategy to subdivide plants into homogeneous groups to control the local variations, for example the presence of slopes or shaded corners. In such a design, treatments are randomized within each block. Any variations under such an experimental design can be calculated later using appropriate statistical analysis (a detailed discussion can be found in Chapter 2) (3, 6).

iv. **Treatments**: Treatments support the experimental purpose or hypothesis, and can have three types of structure, namely, simple, factorial, or nested (6). Single factor is used to compare genetic material (e.g. transgenics, mutant lines), abiotic stress (e.g. wound, nutrients, heat, salt, heavy metals), nitrogen source, or biotic stress/interactions (e.g. pathogen, insect, mycorrhiza). The factorial factor combines more than one treatment and is used when it is possible to verify the interaction between them (e.g. comparison of performance of different genetic materials under different levels of NO_3 and NH_4). When the treatment structure is hierarchical (often due to biology), the nested factor is frequently used for the selection of lines in breeding experiments (e.g. selection of lines based on phenotype for resistance, early/late development). Regardless of the treatment structure used, the control treatment is mandatory in any experimental design.

v. **Phenotyping**: Before sampling, different physiological measurements can be performed to ensure that plants achieve a desired biological process; for example the determination of the water potential should be considered in drought-stress experiments as well as the photosynthetic performance (7).

3.2 Sampling and Quenching of Tree Tissue Material

In this section, we first discuss general aspects of sampling and quenching techniques, including their advantages and disadvantages. We also present techniques to achieve cell-level separation and novel techniques that don't require cell or tissue separation. To achieve sample state representativeness in metabolomics experiments, metabolism should be efficiently interrupted during sampling to provide a snapshot of the metabolic status (8). Thus, sampling should be carried out as fast as possible to avoid metabolite degradation and to prevent changes in metabolite levels caused by fast turnover metabolites (e.g. pyruvate and other glycolytic intermediates) (9, 10). The complete stop of cell metabolism is known as quenching, which is achieved by rapidly freezing plant tissues, after sampling, with liquid nitrogen (shock freezing), and subsequent storage at −80 °C prior to metabolite extraction (11). Considering the importance of sampling in metabolomics experiments, detailed sampling methods for different tree tissues are described below:

3.2.1 Leaf Sampling

Leaf metabolism is highly influenced by several factors that need to be standardized when sampling, including age, phenology, diurnal cycles, and season. Leaf greatest variation by age is due to the change from being a sugar importer to being photosynthetically active, with other metabolic changes accompanying this transformation (12–15). Besides, leaf metabolism has a circadian cycle (16) that affects (among other compounds) the pool of starch and soluble sugars (17–19), and as a result, metabolite levels can change by sampling at different times. Several anatomic factors affect metabolism, including leaf type, position of fruiting growth terminals with respect to leaves, leaf height, angle, and position within the canopy (20–22). Finally, climate has a direct impact on the phenology of the leaves, mainly in trees from temperate and cold climates that need to adapt to low winter temperatures (16, 23). While deciduous trees lose their leaves in autumn to avoid freezing, perennial trees have developed strategies for cold acclimatization, including changes in chlorophyll (24), starch (25), phenols (26), non-structural carbohydrates (24, 26), and various metallic elements (Cd, Cu, Zn, Pb, Ni, and Fe) (27). There are three basic leaf sampling techniques: with scissors from the ground, climbing trees, and shooting; the first technique being the most widely used (22).

3.2.2 Wood Tissue Sampling

In trees, wood is one of the tissues of greatest interest because it is their main commodity. Wood consists of an axial and a radial cell system (28). The axial system is parallel to the trunk and is made up of vessels, parenchyma cells, and fibers, and it provides trees with support and water and nutrient transport. The radial system is perpendicular to the trunk and contains the ray cells,

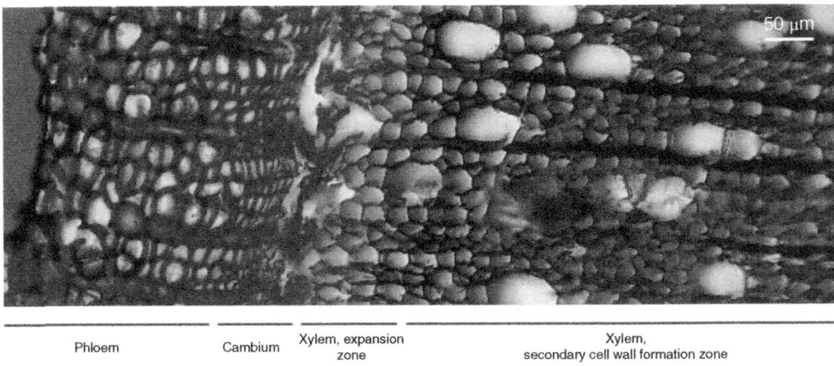

Figure 3.1 Cross section of a hybrid aspen stem, stained with Safranin O and Alcian blue.

which are involved in nutrient transport and carbon storage. These two systems are interconnected mainly through the symplast (29–32). The axial and radial cellular systems comprise the trunk tissues, which are oriented radially and concentrically to the trunk or stem axis, namely, phloem, vascular cambium, and xylem. The different tissues that compose wood can be distinguished by optical microscopy using a combination of the dyes Safranin O, which stains lignin of secondary cell walls in red-fuchsia color, and Alcian Blue, which stains polysaccharides mainly found in primary cell walls in blue-green color. Therefore, when the degree of xylem lignification increases, the red color of the stain will also increase, while non-lignified tissues (e.g. the cambium) will stain blue (Figure 3.1). Although the combination of Safranin O and Alcian Blue is widely used because it simultaneously distinguishes lignified and non-lignified tissues, there are multiple dyes that detect different compounds and/or tissues of wood. A detailed review can be found in Hubbe et al. (33).

The spatial distribution of the trunk tissues is radial to the trunk axis and some of these tissues (e.g. the cambium) are only a few micrometers wide. A usual way to obtain tissue resolution is through cryo-sectioning performed with a cryotome (or cryostat microtome) in the transverse plane of the trunk. This allows the stem to be sectioned into samples a few micrometers wide, thus separating the different tissues. Unlike the microtome technique, the use of the cryotome allows tissue slices to be cut at low temperature (e.g. −35 °C), which in turn allows cellular metabolism to be in a quiescent state while preserving the structure of cell walls and subcellular structures (34). Because it is not necessary to previously cut the tissue with paraffin, as in traditional histological techniques, the original characteristic of the tissue is preserved, particularly the distribution of metabolites and water. Tissues must be previously frozen, which is achieved by immersing them in liquid nitrogen at the time of sampling. One of the disadvantages of the cryotome is the formation of ice crystals, which might distort the cytoarchitecture.

The sampling strategy employing cryotome wood transversal sectioning combined with radial sectioning for microscopical tissue identification with Safranin O/Alcian Blue has been successfully applied to obtain a dynamical insight into the transcriptome (35, 36), the proteome (37), and biochemical analyses (36, 38, 39) of wood tissues. For wood sampling, if the tree is young (e.g. greenhouse-grown trees), the whole stem or a segment of it is sampled. But if the tree is larger, a block (typically 2×1×10 cm) can be removed from the tree using different cutting-tools (e.g. scalpel, hammer, and chisel) (40).

3.2.3 Phloem Sap Sampling

The phloem is the system that transports nutrients and signal molecules from the leaves to the rest of the plant. The phloem exudate contains several molecules, including metabolites (mainly carbohydrates, sugar alcohols, and amino acids), proteins, RNA, hormones, and ions that account for the transport of substances, the nutritional and health status of trees, and communication between different tissues (41). Studying them is key to integrating the physiology of the leaves with that of the sink tissues as well as understanding carbon partitioning.

Exudate sampling methods often have contaminants/artifacts, and two or more techniques are often combined to obtain more reliable results (41). Existing methods to collect phloem sap are described in several manuscripts (41–43), and a summary containing relevant methods is presented in Table 3.1.

3.2.4 Xylem Sap Sampling

The xylem exudate mainly transports water, minerals, some metabolites, and proteins from the roots to the leaves. Its study allows us to deepen into the solute transport from the roots toward the rest of the plant and into possible signals between sink and source tissues. The composition of the xylem exudate undergoes seasonal variations in species that have growth-dormancy transitions (56–59). Several methods to collect xylem sap have been described by Alexeu and Peuke (60) and are presented in Table 3.2.

3.2.5 Root Sampling

Roots are a complex system that evolves over time and spatially (i.e. in different areas of the root system) (68). Different root orders are functionally differentiated; it is generally considered that roots of orders 1–3 in trees have the function of absorbing nutrients and water, while those of a higher order would be associated with structure and transport (68, 69). Although it is often assumed that roots with 2 mm diameter (or less) are responsible for absorption, this is not always the case and should be verified (69, 70).

Table 3.1 Methods used to sample phloem exudate.

Method	Method description	Application
Exudate with water	The stem is cut, immediately placed into a recipient with water, which collects the content of the phloem sap. The exudate comes from the entire sieve element-companion cell complex, and exudation problems may occur if callose/P-protein (molecules used by plants to heal wounds) deposits form on the sieve plates close to the wounds. Metabolites are diluted in the collection water to an unknown extent.	– European beech (*Fagus sylvatica*) (44). – *Pinus pinaster* (45). – *P. trichocarpa* × *P. deltoides* (46).
Exudate with EDTA	The petiole of a leaf (or a cut stem) is placed in a solution of K2-EDTA (or another chelating agent), which prevents the accumulation of callose and P-protein in the phloem sieve, so large volumes of exudate can be obtained. As in the previous method, the sap comes from the entire sieve element–companion cell complex, and the metabolites are diluted by the EDTA solution to unknown levels.	– European beech (44). – *Fagus sylvatica*, *Magnolia kobus*, and *Gnetum gnemon* (47).
Incision/ bleeding	In some species, the phloem content will sprout spontaneously when making a cut in the phloem tissue. Its use is limited to some species because, in most trees, either the cut induces the accumulation of callose and P-protein that limits bleeding or they do not have enough pressure on the sieve elements.	– *Eucalyptus* spp. (48,49). – Coconut palm tree (*Cocos nucifera*) (50). – Paper mulberry (*Broussonetia papyrifera*) (51). – European beech (44).
Centrifugation	The bark is removed from the tree, and a piece of it is placed vertically in a tube sieved at the bottom, which in turn is placed inside another tube. It is centrifuged so that the phloem exudate falls into the outer tube due to the action of centrifugation. It is a simple method with less possibility of contamination by cellular components.	– *Pinus pinaster* (45). – Orange (*Citrus sinensis)*, orange jasmine (*Murray a paniculate*), and curry leaf tree (*Bergera koenegii*) (43,52).

Table 3.1 (Continued)

Method	Method description	Application
Stylectomy	Aphids (*Aphidoidea*) feed on the phloem exudate. To access the phloem, aphids have a stylet in their mouthparts, capable of traversing the epidermis to reach the phloem exudate. The technique consists of allowing aphids to feed on the plants. The stylet is cut while the insect is feeding on the plant. Next, the content of the stylet is purified using a microcapillary. This method is limited to species that are affected by aphids and is technically challenging. The exudate may be contaminated by components of the insect saliva or plant xylem because many insects can also occasionally consume xylem exudate, although contamination with other plant tissues is less frequent than with other techniques.	– Peach (*Prunus persica*) (53,54). – *Quercus robur* (55). – *Fraxinus excelsior* (55). – *Fagus sylvatica* and *Magnolia kobus* (47).

Some authors suggest classifying roots according to their function (or their order) instead of based on their diameter (68, 69, 71).

There are multiple ways to sample roots, although none works universally. According to Freschet and Roumet (68), to decide which sampling technique to apply, it is necessary to evaluate the root function to be studied and which are the root categories (or orders) involved in such functions. Freschet et al. (71) elaborated a detailed report on different root sampling techniques that include: sampling of the whole root by excavation, monolith excavation, soil core sampling, and growth cores. In all cases, roots must be finely cleaned and classified once sampled.

3.2.6 Techniques for Separating Cell Types in Complex Tissues

The advent of new techniques makes it possible to study tissues at single-cell resolution. These techniques allow the identification of new cell types and the detailed characterization of individual cell functions.

3.2.6.1 Laser-assisted Microdissection

Laser-assisted microdissection (LAM) is possibly the most widely used technique to isolate tree cells today. Cell isolation is achieved with a laser microdissection microscope, which is equipped with a laser to cut the cells of interest

Table 3.2 Methods for collecting xylem sap.

Method	Method description	Application
Root pressure	This method is based on the exudation of the sap caused by the pressure exerted inside the roots by the absorption of water and minerals. The stem is cut near the base of the roots. In this way, the exudate will sprout spontaneously and can be collected with a pipette or a wad of cotton (or of other alternative material) from which the metabolites can be later extracted. The procedure can last for hours or days, depending on the species.	– Proteome of hybrid poplar (*Populus trichocarpa* × *Populus deltoides*) xylem sap (46).
Scholander-Hammel	A cut section of the stem is introduced into a Scholander-Hammel chamber through a hole, leaving one of its extremes outside the chamber. Subsequently, gas is injected into the chamber. The increased pressure compensates for the negative pressure in the vessels and pushes the exudate to the end of the stem section that was left outside the chamber. In this way, the exudate can be collected with a pipette while pouring outward.	– Nitrogen composition of xylem sap in wild cherry (*Prunus avium*) and poplar (*Populus trichocharpa* × *Populus balsamifera*) (61).
Passioura pressure vessel	The entire plant is placed in a chamber, and pneumatic pressure is applied to the roots to overcome the tension of the xylem and push the xylem exudate outward. Then, samples are collected from different points of the plant. Plants must be grown in special pots placed inside the chamber.	– Nutrient remobilization in hybrid poplar (*Populus tremula* × *Populus alba*) (62).
Vacuum pump	A branch (or stem section) is placed in a sample collection vial, and a vacuum is created with a pistol, causing the sap to be exudated into the vial. This method is only applicable in trees with abundant xylem exudate.	– Metabolite seasonal variation in *Populus nigra* (63).
Pressure/ decompression	This technique was originally developed for dry wood by generating a pressure difference between the inner side and the surface of a piece of wood (64), which drives the xylem sap out of the wood piece.	– Metabolite content of xylem sap of maple trees (*Acer platanoides*) (65).

(Continued)

Table 3.2 (Continued)

Method	Method description	Application
Xylem sap feeding insect	This method can potentially be used to extract xylem sap. However, it is not used much because few insects feed on xylem sap due to its low nutritional content.	– Xylem sap recovery from *Prunus salicina* (66), olive trees (*Olea europea*), and *Acacia saligna* (67) employing common meadow spittlebugs (e.g. *P. spumarius*).
Centrifugation method	This method is similar to the one described for obtaining phloem sap by centrifugation (Table 3.1). The xylem of the stem is introduced into a perforated tube, which in turn is introduced into a larger tube and centrifuged. The obtained sap is mainly xylem sap.	– Metabolites in *Citrus* spp. (43).

and a mechanism to capture them. There are essentially two main types of instruments: laser capture microdissection (LCM) and laser cutting microdissection (LMD) (72, 73). In LCM, the selected cells are cut by a laser and catapulted to the lid of a tube containing a special substance to retain them. In LMD, cells are also cut by laser, but they fall by gravity into a collection tube. Sample preparation for LAM may include tissue staining prior to using laser microdissection to facilitate identification of the cells as well as techniques to fix samples and embed them in paraffin. However, given that the use of solvents can alter the metabolic composition of cells (72,73), some authors have developed techniques that do not involve preparing samples with paraffin but rather preparing them through cryo-sectioning with minimal use of solvents (74) or drying them by freeze-drying, even in woody tissues (75).

The advantages of LAM are that it can be applied to any species and tissue, as long as the cells of interest can be microscopically observed, and that it does not require cell markers or transgenics (72). However, LAMs tend to have low throughput, lack a generalized protocol for all tissues and species, and often require prior tissue fixation, which may limit their uses (72). Nevertheless, this technique has been applied to tree tissues to achieve the separation of a small number of cell groups, namely, cambial cells (76, 77), cortical resin ducts (76), phellogen cells (78), phloem (79, 80), ray cells (75, 81, 82), tracheary elements (75, 83), fibers (75, 82–84), bark (85), and developing xylem (77, 86). Some of these studies included gymnosperms, such as white spruce (*Picea glauca*), Norway spruce (*Picea abies*), and maritime pine (*Pinus pinaster*) (75, 76, 79, 83, 85, 87), which is interesting given the experimental difficulties usually found when working with these types of trees. Tissue-specific cells have been used mainly for gene expression studies (76–78, 80–87), followed by metabolic studies (39, 75, 79).

3.2.6.2 Fluorescence-activated Cell Sorting

In the fluorescence-activated cell sorting (FACS) technique, cell types of interest are marked with a fluorescent reporter (e.g. GFP), whose expression is regulated by the promoter of a gene that is specifically expressed in that cell type; thus, fluorescence indicates a particular cell type (88, 89). Consequently, transgenic plants carrying a GFP (or other fluorescent reporter gene) construct whose expression is specific to the cell type of interest is a requirement, which can sometimes prove to be a limitation, particularly in trees. The GFP-expressing tissues are used to generate protoplasts, which are subsequently separated according to their GFP signal in the FACS apparatus: single cell-containing droplets are generated in the cytometer and pass through a fluorescence detector; if they are fluorescent, they are separated, and a culture of identical cells is generated from them. The selected cells correspond to a single type and transcriptomics, proteomics, and metabolomics analyses can be performed on them. In plants, the greatest application of FACS is the separation of Arabidopsis root cells, for which there are well-established protocols to generate protoplasts (89). One of the biggest challenges of protoplast generation is the need for their optimization, since they are not directly applicable to other tissues and species (90). Therefore, FACS is still a technique that requires further development for its application in trees and whose main limitations are difficulties in generating protoplasts, particularly from woody tissues, lack of knowledge of tree tissue-specific promoters, and the amount of time required to generate transgenic plants.

3.2.7 Measuring Metabolites Without Cell or Tissue Separation

The measurement of metabolites without cell or tissue separation encompasses a group of emerging techniques that provide spatial information to help interpret the physical context of the data obtained from a cell while having the advantage of reducing sample preparation work. These techniques can be referred to as spatial metabolomics or mass spectrometry imaging (MSI), and one of the main objectives of using these techniques is to obtain data while preserving as much as possible the structure of tissues. MSI includes several different types of techniques (91, 92) bearing different degrees of resolution and ability to identify compounds (Table 3.3).

Table 3.3 Most commonly used techniques in spatial metabolomics or mass spectrometry imaging (MSI) studies.

Spatial metabolomics technique	Technique description	Application
Matrix Assisted Laser Desorption Ionization (MALDI)	MALDI, a soft ionization technique, is the most popular MSI method that is commonly used for direct measurement of molecular ions. A laser energy-absorbing matrix is placed onto the sample to induce the ionization of the metabolites. In general, it only allows the identification of molecules with *m/z* greater than 500 because at low *m/z* values, there is an abundance of matrix ions; this problem is often solved by coupling MALDI to ultra-high-resolution mass spectrometers such as an orbitrap or Fourier-transform ion cyclotron resonance mass spectrometry (FTICR–MS), or MALDI trapped ion mobility spectrometry coupled to quadrupole time-of-flight mass spectrometry (timsTOF–MS). Samples must be dried before analysis.	– Heartwood and sapwood of *Cedrus deodara* (93). – Cellulose and hemicellulose in *Populus* (94). – Monolignol glucosides in the conifers *Chamaecyparis obtusa* and *Cryptomeria japonica* (95). – Aromatic compounds in wood decaying produced by the fungi *Rhodonia placenta* (96).
Secondary Ion Mass Spectrometry (SIMS)	The ionization of the sample occurs through ion beams, and it has a high resolution (less than 2 μm). The distance below the surface on which the ionization works is only a few nanometers, which limits the analysis to metabolites on the surface of the sample. Because ionization tends to fragment analytes, SIMS tends to produce better results when analyzing molecules with low molecular weight or with known fragmentation patterns. The ionization is performed at a high vacuum, which implies that the tissues must be fixed cryogenically or by chemical methods.	– Method development for wood (97,98). – Analyses of the spatial distribution of γ-lactone compounds in ray cells and oil cells in the tree *Sextonia rubra* (99). – Wood extractives in European larch (*Larix decidua*) (100). – Tryptamine in the wood of *Dicorynia guianensis* (101). – Diterpene phenol ferruginol in the heartwood of *Cryptomeria japonica* (102). – Lignin in aspen (*Populus trichocarpa*), maple (*Acer micranthum*), and *Pinus* wood (103–105).

(Continued)

Table 3.3 (Continued)

Spatial metabolomics technique	Technique description	Application
Desorption Electrospray Ionization (DESI)	Samples are ionized by electrospray, without much sample preparation. The method uses solvents to solubilize the analytes. This limits the molecules to be detected and the spatial resolution, which is usually low. Alternatively, NanoSIMS was developed to achieve better resolution (50–200 μm).	– Leaves of *Citrus sinensis* grafted on *Citrus limonia* to uncover metabolic alterations produced by *Xylella fatidiosa* (106).
Laser Ablation Electrospray Ionization (LAESI)	It combines laser ablation of the sample followed by electrospray ionization. Ionization takes place at room temperature, which avoids the desiccation or cryopreservation required by other techniques. It has a resolution of 200–300 μm.	– Achlorophyllous cells in oil glands of sour orange (*Citrus aurantium*) leaves (107).

3.3 Labeling of Tree Tissues

The traditional way of studying metabolism is through measurements of total metabolic pools (i.e. the content of a given metabolite at a given time). However, this type of analysis can be insufficient to fully explain metabolism. Occasionally, metabolic pools remain stable while active conversions between different metabolites (the so-called metabolic turnover) occur due to the modification of enzyme activities (108). A foundational example in plants that exemplifies this is the work of Stitt (109), where intense changes in photosynthesis did not produce effects on metabolic levels. Several methodologies are used to study metabolic turnover, including: (i) the use of stable isotope labeling experiments; (ii) the measurement of enzymatic activities; and (iii) the use of fluorescence resonance energy transfer (FRET) to follow, in real time, the fate of a metabolite (110). The use of stable isotope labeling of metabolites is particularly widespread due to the flexibility of its applications, among which is the study of the partition of an element or metabolites into different tissues over time (111), metabolic pathway elucidation (112), protein synthesis and degradation rate (113), metabolic pathway modeling (114), polluting compounds or waste forestry metabolic studies (115, 116), and metabolite annotation (117).

Isotopic labeling of metabolites is a way of tracking the fate of a metabolite through a metabolic pathway by labeling an element of its structure with a stable isotope of higher atomic weight than the most abundant isotope in nature. The most abundant elements that are generally labeled are ^{1}H, ^{12}C, ^{14}N, ^{16}O, and ^{32}S, whose respective less-abundant stable isotopes with higher atomic weight are ^{2}H, ^{13}C, ^{15}N, ^{18}O, and ^{34}S (118). The most widely used methods to identify isotopes are mass spectrometry (MS), which detects differences at the molecular level, and nuclear magnetic resonance (NMR) as an atomic-level approach, while isotope-ratio mass spectrometry (IRMS) is used to detect differences at element level (i.e. to track C or N without identifying the molecules that contain them).

The first step in labeling experiments is to introduce the stable isotope into cells, tissues, or organisms. This isotope will form part of a molecule (or tracer) that can be incorporated by the tree or tissue through a pre-existing physiological mechanism (e.g. $^{13}CO_2$ through photosynthesis). Once inside the cells, tracers are incorporated into the metabolic pool and begin their journey through the cell metabolic framework. Some tracer introduction strategies use the most common physiological metabolite incorporation pathways to avoid disturbing the system and to obtain physiologically and biochemically reliable results. Other strategies do not employ the most common physiological pathways because they only focus on metabolic pathways. Next, the most commonly used methods to incorporate labeled C (^{13}C) and/or labeled N (^{15}N) into trees are described.

3.3.1 C-labeling

Carbon (C)-labeling can be performed using the normal route of C entry in plants (i.e. through photosynthesis) using labeled CO_2 ($^{13}CO_2$). The origin of the $^{13}CO_2$ can be commercial $^{13}CO_2$ injected with an injection device, such as a syringe or a gas flask, or it can be generated *in situ* from a labeled carbonate (e.g. sodium ^{13}C-bicarbonate or barium ^{13}C-carbonate that release $^{13}CO_2$ after reacting with an acid). The main difficulty in using $^{13}CO_2$ is to create a closed environment for the $^{13}CO_2$ to concentrate and avoid mixing it with $^{12}CO_2$. Instead of $^{13}CO_2$, other organic molecules with labeled C atoms that are absorbed by the roots, such as ^{13}C-sucrose, can be used. This does not reflect the physiological C incorporation by plants, but it is a simpler way to proceed with the labeling. There are various strategies for incorporating ^{13}C into trees, and new experimental possibilities are continually being published.

3.3.1.1 Labeling of Whole Trees

i. **Using individual chambers for each plant**: Any device able to isolate individual plants can be used, such as chambers specifically designed to contain whole trees (119, 120), or bags to wrap seedlings (121). Individual chambers are also used for labeling large trees in forests to obtain *in situ* labeling data. For example, Thomas et al. (122) pitched a tent around ashes

(*Fraxinus excelsior* L.) and beeches (*Fagus sylvatica* L.), Gao et al. (123) wrapped the crowns of *Pinus sylvestris* with transparent plastic sheets, while Epron et al. (124) set up a chamber around the canopy of *Eucalyptus grandis* L. to label them. It is a procedure that can be adapted to different types of trees, although it is slow, and perhaps unsuitable if labeling many plants is required.

ii. **Using chambers where a group of plants is labeled**: There are commercially available chambers designed to work with CO_2, such as some Conviron® chambers (125) or designed *ad hoc* (126–129). These chambers have the advantage that several trees can be labeled simultaneously while the growing conditions can be controlled.
iii. **Introducing the label into a whole tree**: This strategy can be performed either through a petiole attached to the plant (130) or through a wound (131).
iv. **Placing the plant into a liquid or solid culture medium**: This strategy contains an organic molecule with labeled C atoms that can be absorbed by the plants; although the culture medium does not mimic the physiological absorption of C, it still generates biochemically relevant data. Ghirardo et al. (119) carried out labeling assays on grey poplar (*Populus × canescens*) employing a ^{13}C-glucose enriched liquid medium, while Kreuzwieser et al. (132) also used ^{13}C-glucose to label *Quercus robur* trees. Jardine et al. (133) employed aqueous solutions of ^{13}C-labeled C1 and C2 intermediates (including ^{13}C-methanol, ^{13}C-formaldehyde, and ^{13}C-formate) to study the C1 pathway and its integration with the central metabolism in guava (*Inga edulis*).

3.3.1.2 Labeling of Tree Tissues

The use of an isolated tissue does not guarantee that it closely reflects the physiological conditions of the label incorporation, but it can be used for studies focused on the biochemistry of the reactions:

i. **Placing a leaf or a piece of leaf in a system in which $^{13}CO_2$ is released**: The origin of $^{13}CO_2$ can be generated *in situ* from ^{13}C-carbonate or a pulse of injected $^{13}CO_2$.
ii. **Incubation of a tissue in a medium with an organic molecule with labeled C atoms**: A strategy similar to that used for whole plants (e.g. ^{13}C-glucose and ^{13}C-sucrose) (134). Cell cultures can also be used, which results in a fast and homogeneous labeling (135).

3.3.2 N-labeling

The most frequent route of nitrogen (N) absorption is the root system. This allows ^{15}N delivery to be generally as $^{15}NH_4^+$ or $^{15}NO_3^-$ through the root substrate. However, a small percentage of N can be absorbed by the leaves (136).

3.3.2.1 Labeling Through the Canopy of Whole Trees

i. **Use of ^{15}N gas in chambers**: Nussbaum et al. (137) used $^{15}NO_2$ to label Norway spruce (*Picea abies*), while Lhuissier et al. (116) used ^{15}NO to label pollen isolated from birch (*Betula verrucosa* Ehrh.) and cocksfoot (*Dactylis glomerata* L.).
ii. **^{15}N canopy spray**: ^{15}N-urea or ^{15}N-enriched KNO_3 are generally used to spray the canopy. The spraying can be done manually, with the advantage of being easily applicable (138,139). Zeller et al. (115) used ^{15}N-labeled urea on the foliage of a beech stand to generate labeled wood, which they later used for decomposition studies of ^{15}N-labeled beech litter. $(^{15}NH_4)_2SO_4$ has also been used to spray a large field area with a helicopter (140).

3.3.2.2 Labeling Through Roots of Whole Trees

i. **Irrigation of the soil surrounding trees or injection of a ^{15}N solution into the soil**: These are possibly the most common N-labeling procedures due to their simplicity and the fact that they are easily applied evenly on the surface (141–155). ^{15}N is applied as $^{15}NO_3^-$, $^{15}NH_4^+$, or ^{15}N-urea.
ii. **Through the rhizosphere**: In a particular case of a labeling experiment, Mayerhofer et al. (156) labeled mycorrhizal hyphae with ^{15}N to study the exchange of metabolites with the roots of beech trees (*Fagus sylvatica*).

3.3.2.3 Stem Injection

Proe et al. (157) injected ^{15}N-labeled ammonium sulfate directly into stems to study nitrogen dynamics in the canopy of *Pinus radiata*. Nair et al. (158) injected a solution of $^{15}NH_4{}^{15}NO_3$ into the trunk of Sitka spruce (*Picea sitchensis*) to obtain strongly labeled tissues. N-labeling can also be carried out in tissues, tissue pieces, or cell cultures (159), with the limitations already mentioned for the labeling of isolated tissues with ^{13}C.

3.3.3 Dual Labeling of C and N

Given the integration of C and N metabolism (160), many studies incorporate both types of tracers simultaneously. For example, Komatsu et al. (161) used plates containing agar enriched in ^{13}C-glucose and ^{15}N-nitrate to label *Jatropha curcas*. Norris et al. (121) labeled metabolites of trembling poplar (*Populus tremuloides* Michx.) using a combination of gaseous $^{13}CO_2$ and liquid $K^{15}NO_3$. Rehschuh et al. (120) labeled *Pinus sylvestris* (Scots pine) seedlings using $^{13}CO_2$ and soil-applied doubly labeled $^{15}NH_4{}^{15}NO_3$. Silva et al. (128) applied labeled gases ($^{13}CO_2$ and $^{15}NH_3$) to label *Coffea arabica* trees. Bennett and Prescott (162) used ^{15}N- and ^{13}C-labeled glutamic acid to study N and C metabolism, respectively, in Western red cedar (*Thuja plicata* Donn.), western hemlock (*Tsuga heterophylla* Raf. Sarge) and salal (*Gaultheria shallon* Pursh) while Pfautsch et al. (163) used it to label *Eucalyptus regnans* and *Acacia* spp.

Multiple variables must be considered in labeling experiments, and the decision on how to perform a labeling experiment must be in agreement with the aim of the experiment. Labeling can be dynamic (i.e. in which the metabolite labeling has not reached a plateau at the time of the sampling) or steady state (i.e. the labeling and the pools have reached a plateau and no longer change over time, so the labeling is independent of the metabolite concentration) (164). On the other hand, labeling experiments can be done through pulses (i.e. the metabolism quenching occurs immediately after the labeling) or through pulse-chase experiments (i.e. the labeling is followed by a time in which the label is allowed to be replaced by unlabeled metabolites to have information about how long that replacement takes). The labeling time depends on the tracer, the introduction method, the tissue under study, and must be determined experimentally; if the introduction route of ^{13}C is through $^{13}CO_2$, then the leaves will be labeled more quickly than sink tissues, such as stem and roots (125, 165). The metabolite turnover rates must also be considered; if they are very fast, then the quenching must be carried out immediately after labeling to avoid the tracer replacement with unlabeled metabolites. A typical example of this occurs with the Calvin cycle metabolites (114).

Although labeling tree seedlings have the same challenges as found in other plants, in the case of larger trees, extra efforts to design large-scale labeling experiments must be considered as well as the long labeling times that are required to study sink tissues. The choice of the labeling method must be made according to the objectives of the work while considering the age of the trees, the tissue under study, the metabolites to be analyzed, the type of analysis to be carried out (descriptive or modeling), and whether it is necessary to emulate the physiological conditions, among other reasons.

3.4 Metabolite Extraction and Mass Spectrometry-Based Metabolite Analysis

A single plant species is estimated to produce tens of thousands of metabolites, far more than those produced by other organisms (1, 166). Those metabolites are present at a wide range of concentrations and are produced at different developmental stages and in a diversity of tissues or as a response to biotic and abiotic environmental stress. No single extraction method or analytical technique can capture the complete metabolome of a plant because of the physicochemical diversity of their metabolites (e.g. aqueous solubility, degree of ionization, degree of electronegativity, volatility, stability, and molecular size/complexity). Thus, multiple extraction systems and analytical platforms are required for a broader coverage of the metabolome. Attempting 100% coverage would be technically impractical, regardless of the array of methods employed.

Metabolite extraction is a crucial step in metabolomics analysis because it can directly interfere with the quality of the final data. An efficient extraction procedure must guarantee: (i) the maximum recovery of metabolites; (ii) no alteration in the metabolite chemical stability; (iii) the removal of interferents (e.g. salts and proteins) that could affect the analysis; (iv) reproducibility; and (v) compatibility with the chosen analytical technique (3, 5, 9, 167, 168). As mentioned before, metabolites are highly diverse in terms of chemical composition, polarity, different concentration levels, and molecular weight. Thus, it is not possible to extract the whole range of metabolites using a single solvent (169, 170). The choice of solvent is a critical step in a metabolomics extraction procedure. As a strategy to identify the most efficient solvent or solvent mixture for the sample extraction, preliminary studies can be performed. Studies on the optimization of metabolite extraction from plant samples have already been performed (171–176). Some considerations should be done while selecting a solvent, such as toxicity, solubilization power (defined as the power or ability of the solvent to dissolve a solute), selectivity (defined as the ability of the solvent to dissolve specific compounds, when their polarities are not very different), dissolution rate, chemical reactivity, and pH (177). The choice of adequate solvent (for sample extraction or resuspension) must also meet the specific requirements of the analytical platform used, such as LC-MS and/or gas chromatography mass spectrometry (GC–MS) (178). The exception is the use of headspace extraction (e.g. solid-phase microextraction [SPME]) that does not require the use of solvents (179).

Based on lipophilicity, a physicochemical property referred to as Log P, metabolites can be classified as follows: (i) polar (Log $P < 0$); (ii) medium polar ($0 < \text{Log } P < 5$); and (iii) non-polar (Log $P > 5$) (180). Non-polar compounds (i.e. main lipid classes) are typically extracted with two-phase extraction methods, namely, methanol:water:chloroform or methanol:water:methyl-tert-butyl ether (MTBE), where lipids are present in the non-polar phase (chloroform or MTBE) and polar metabolites, such as amino acids and carbohydrates are present in the aqueous polar phase (methanol:water). Non-polar compounds are better separated by reverse-phase liquid chromatography (RPLC) (180–182), while polar compound separation can be improved by hydrophilic interaction liquid chromatography (HILIC) or ion-pairing (IP) chromatography. Medium polar metabolites (e.g. phenolics) are often extracted as polar metabolites and can be separated by either HILIC or RPLC (180). Organic solvents are included in extraction mixtures to reduce their polarity and extract polar metabolites. One of the main advantages of using organic solvents rather than water is that more diverse types of metabolites can be extracted; that is using methanol:water:chloroform as a one-phase solvent system, polar and non-polar compounds can be extracted simultaneously (Figure 3.2).

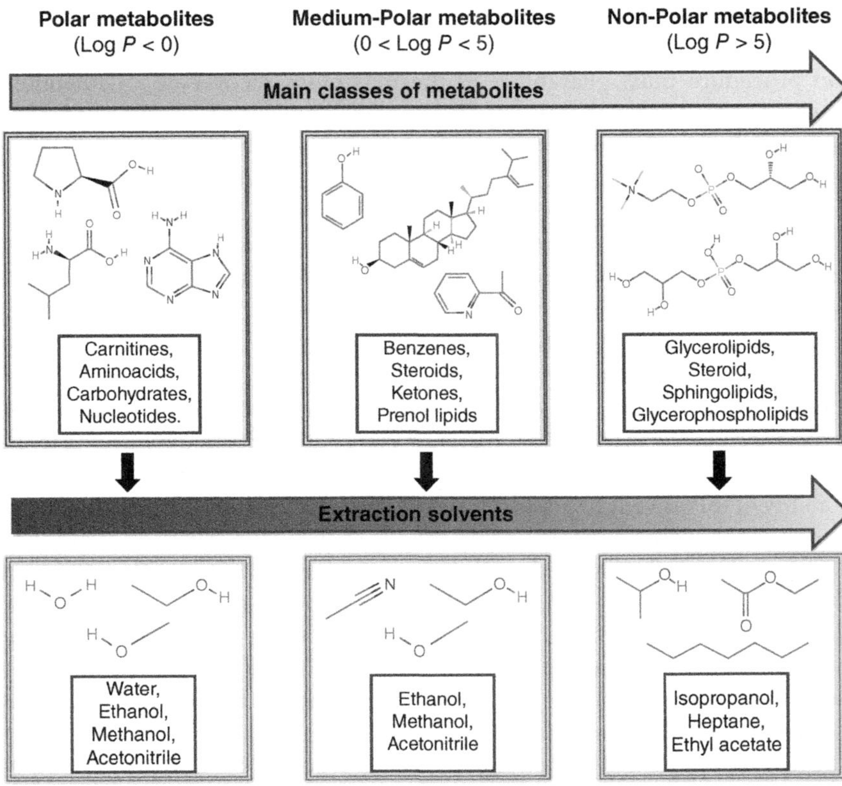

Figure 3.2 General scheme for metabolite extraction highlighting the main chemical classes of metabolites and the typical solvents used in metabolomics for each polarity range.

Other advantages include the compatibility of most organic solvents with different analytical platforms, the ease of solvent evaporation, the absence of salt precipitates, and the increased stability of the extracted metabolites. Among the solvent options, alcohols, mainly methanol, ethanol, and to a lesser extent isopropanol, are often used alone or more frequently in mixtures with varying proportions of water. Being hydrophilic due to their hydrogen-bonding capabilities by their hydroxyl groups, alcohols are also lipophilic, a property that increases with their chain length; the combination of these features confers a high solubilization power that allows them to extract a wide range of metabolites. Commonly used extraction methods include protein precipitation (PPT), liquid–liquid extraction (LLE), and solid phase extraction (SPE). In addition, SPME, monolithic spin column extraction, microextraction by packed sorbent (MEPS), supported liquid extraction (SLE), and micro-solid phase extraction methods have also been used in sample preparation (8).

3.4.1 Untargeted and Targeted Metabolomics

Metabolomics methods typically fall into two distinct groups: (i) untargeted metabolomics, a global approach aiming at the comprehensive analysis of all the measurable analytes in a sample, including chemical unknowns and (ii) targeted metabolomics, the measurement of defined groups of chemically characterized metabolites (183).

Untargeted metabolomics does not need any prior information. This approach is closer to the goal of metabolomics: to provide qualitative and quantitative information for as many metabolites as possible at the same time. However, scoring this goal is very challenging. In metabolomics untargeted studies, mixtures of methanol and water are the most popular combinations due to their capability to extract a wide range of metabolites, namely, sugars, amino acids, organic acids, and phenolic compounds (178). Their "non-selectivity" makes them particularly suited to plant metabolomics applications. Acidified water (0.1% formic acid) in pre-cooled 75% methanol was reported as a very efficient method to extract a wide range of specialized metabolites (flavonoids, phenolic acids, and alkaloids) (172). Untargeted metabolomics data are highly dependent on the acquisition platform, whether it is LC–MS, GC–MS, or capillary electrophoresis mass spectrometry (CE–MS) as well as on the mass analyzer used. Although there is an overlap in the detection of metabolites between these MS-based acquisition platforms, each platform is more suitable for specific metabolite classes. GC–MS is well suited to identify and quantify small polar metabolites (<650 Da), such as a range of primary metabolites, while LC–MS can also cover large hydrophobic metabolites, such as alkaloids, terpenoids, and phenols (184, 185).

In targeted metabolomics studies, a relatively small number of metabolites are selected in advance, and the precursor ions and their fragmentation patterns are already well understood. Multiple reaction monitoring (MRM) is subsequently employed in tandem MS experiments (MS/MS) to specifically analyze the selected ions, which significantly increases the selectivity and sensitivity to detect the small set of targeted metabolites. This approach, combined with the use of isotopically labeled internal standards, is commonly used for the accurate absolute quantification of metabolites. A targeted metabolomics approach would be more applicable when a class of metabolites (or a specific pathway of interest) is investigated. For example, Abreu et al. (39) performed targeted quantification of low-abundant phytohormones (cytokinin and indole 3-acetic acid) across the developing phloem and wood-forming tissues of *Populus tremula* trees.

3.4.2 Chemical Derivatization Methods

Derivatization involves a chemical reaction between an analyte and a reagent to change the chemical and physical properties of the analyte, thereby generating a new compound with properties more amenable to analytical methods. The main uses of derivatization in LC–MS and GC–MS are to: (i) improve detectability; (ii) change the molecular structure or polarity of the analyte for better chromatography; (iii) increase volatility; (iv) change the matrix for better separation; and (v) stabilize an analyte (186). Ideally, a derivatization reaction should be rapid, quantitative, and produce minimal by-products. The excess reagent should not interfere with the analysis or should be easily removed from the reaction matrix (187).

Chemical derivatization is often the last step when developing a method since it increases the analysis time and it can be a source or error (i.e. analyte and/or matrix reaction by-product generation). Despite these issues, chemical derivatization is still a powerful technique for the separation and detection of trace amounts of substances in a complex matrix. Ion suppression negatively affects the analysis of complex mixtures using MS, particularly in electrospray ionization (ESI), which can be overcome employing LC. However, LC columns have some limitations; namely, RPLC columns suffer from poor retention of very polar or ionic compounds and HILIC columns have relatively low separation performance (188). Derivatization strategies can be performed to overcome these issues and improve the separation of complex samples. Additionally, derivatization reactions introduce chemical moieties into the original molecules, resulting in larger metabolites with enhanced ionization efficiency, thereby providing an improvement in the metabolite detection (e.g. low-molecular weight metabolites) (188). In GC–MS, chemical derivatization is required to increase metabolite volatility because most plant primary metabolites (e.g. carbohydrates, organic, and amino acids) are nonvolatile. Besides, the derivatization increases thermal stability, improves peak symmetry, and resolution to, thereby, provide highly efficient metabolite separations (189).

In plants, the standard derivatization method for GC–MS is well established and includes two chemical reactions: methoxyamination followed by silylation (186, 189–191). This reaction converts carbonyl groups (aldehyde or keto groups) into metoxime derivatives ($CH_3ON=$) with the addition of methoxyamine and pyridine, which acts as solvent and catalyst (192). The methoxyamination introduces a carbon-nitrogen double bond (C=N) that generally forms two stereoisomers, that are *syn* or *E* and *anti* or *Z* depending on the C=N orientation, which are chromatographically resolvable with GC–MS (193). In the silylation chemical reaction, trimethylsilane [$(CH_3)_3SiH$] derivatives react with polar compounds containing active hydrogens, such as –OH, –COOH, –SH, and –NH (186, 191). The active hydrogen of hydroxyl, thiol, amine, and carboxylic acid groups can be replaced by an alkyl-silyl group forming trimethylsilyl (TMS) derivatives, such as TMS-ether, TMS-ester, TMS-amine, and TMS-sulfide (186, 191). The reactivity of functional groups toward silylation can be described as follows: alcohols > phenols > carboxylic acids > amines > amides (186); the most common silylation reagents used in plant metabolomics being MSTFA (*N*-methyl-*N*-trimethylsilyl-trifluoroacetamide) and BSTFA (*N*,*O*-bis-trimethylsilyl-trifluoroacetamide) (11, 192).

Silylation must be performed under anhydrous conditions to achieve complete active hydrogen replacement and to avoid degradation products (11, 186). The derivatization reaction is generally performed in glass vials, although possible polytetrafluoroethylene (PTFE-line) leakage from screw caps may occur at elevated temperatures. Derivatization reactions can be performed in two configurations: offline (manual) or online (automated). General challenges with offline methods include: (i) experimental errors (e.g. loss of analyte through evaporation and resuspension steps, high variability among the samples if working with a large batch of samples); (ii) sample contamination; (iii) moisture interference during work-up; and (iv) being time-consuming (192, 194, 195). Online derivatization methods allow to: (i) minimize sample processing times; (ii) analyze large sample batches; (iii) decrease the required reagent amounts; and (iv) increase the efficiency of the analysis (192, 194). In any case, the results should be highly reproducible once they are operator independent (192).

3.5 Conclusions

In this chapter, we thoroughly discussed suitable experimental designs to minimize biological variations in the metabolomics results and to provide accurate answers to biological questions/hypotheses. Undoubtedly, metabolomics is a technology that can assist our understanding of forest tree metabolic adjustments due to internal or external changes. Nonetheless, forest tree samples represent a complicated biological matrix and care must be taken during the critical initial steps of the metabolomics workflow (i.e. sample preparation) to ensure a high-quality metabolomics analysis.

References

1 Fang C, Fernie AR, Luo J. Exploring the diversity of plant metabolism. *Trends Plant Sci.* 2019; 24: 83–98.

2 Feussner I, Polle A. What the transcriptome does not tell – proteomics and metabolomics are closer to the plants' patho-phenotype. *Curr Opin Plant Biol.* 2015; 26: 26–31.

3 Rodrigues AM, Ribeiro-Barros AI, António C. Experimental design and sample preparation in forest tree metabolomics. *Metabolites.* 2019; 9(12): 285.

4 Kasper K, Abreu IN, Feussner K, Zienkiewicz K, Herrfurth C, Ischebeck T, et al. Multi-omics analysis of xylem sap uncovers dynamic modulation of poplar defenses by ammonium and nitrate. *Plant J*. 2022; 111(1): 282–303.
5 Martins MCM, Caldana C, Wolf LD, de Abreu LGF. The importance of experimental design, quality assurance, and control in plant metabolomics experiments. In: António C, editor. *Plant Metabolomics: Methods and Protocols. Methods in Molecular Biology*. Volume 1778. New York (NY): Humana Press; 2018. p. 3–17.
6 Mathews KL, Crossa J. Experimental design for plant improvement. In: Reynolds MP, & Braun HJ, editors. *Wheat Improvement*. Cham (CH): Springer; 2022. p. 215–235.
7 Baccari S, Elloumi O, Chaari-Rkhis A, Fenollosa E, Morales M, Drira N, et al. Linking leaf water potential, photosynthesis and chlorophyll loss with mechanisms of photo- and antioxidant protection in Juvenile olive trees subjected to severe drought. *Front Plant Sci*. 2020; 11: 614144.
8 Álvarez-Sánchez B, Priego-Capote F, de MDL C. Metabolomics analysis II. Preparation of biological samples prior to detection. *Trends Anal Chem*. 2010; 29(2): 120–127.
9 Jorge TF, Mata AT, António C. Mass spectrometry as a quantitative tool in plant metabolomics. *Philos Trans A Math Phys Eng Sci*. 2016; 374(2079): 20150370.
10 Pinu FR, Villas-Boas SG, Aggio R. Analysis of intracellular metabolites from microorganisms: quenching and extraction protocols. *Metabolites*. 2017; 23(4): 53.
11 Jorge TF, Rodrigues JA, Caldana C, Schmidt R, van Dongen JT, Thomas-Oates J, et al. Mass spectrometry-based plant metabolomics: Metabolite responses to abiotic stress. *Mass Spectrom Rev*. 2016; 35(5): 620–649.
12 Li L, Zhao J, Zhao Y, Lu X, Zhou Z, Zhao C, et al. Comprehensive investigation of tobacco leaves during natural early senescence via multi-platform metabolomics analyses. *Sci Rep*. 2016; 6: 37976.
13 Kanojia A, Gupta S, Benina M, Fernie AR, Mueller-Roeber B, Gechev T, et al. Developmentally controlled changes during Arabidopsis leaf development indicate causes for loss of stress tolerance with age. *J Exp Bot*. 2020; 71(20): 6340–6354.
14 Kanojia A, Shrestha DK, Dijkwel PP. Primary metabolic processes as drivers of leaf ageing. *Cell Mol Life Sci*. 2021; 78(19–20): 6351–6364.
15 Guo Y, Ren G, Zhang K, Li Z, Miao Y, Guo H. Leaf senescence: progression, regulation, and application. *Mol Hortic*. 2021; 1(1): 1–25.
16 Hartmann H, Trumbore S. Understanding the roles of nonstructural carbohydrates in forest trees – from what we can measure to what we want to know. *New Phytol*. 2016; 211(2): 386–403.

17 Zhou R, Sicher R, Quebedeaux B. Diurnal changes in carbohydrate metabolism in mature apple leaves. *Aust J Plant Physiol.* 2001; 28(11): 1143–1150.
18 Duan W, Fan PG, Wang LJ, Li WD, Yan ST, Li SH. Photosynthetic response to low sink demand after fruit removal in relation to photoinhibition and photoprotection in peach trees. *Tree Physiol.* 2008; 28(1): 123–132.
19 Haldimann P, Gallé A, Feller U. Impact of an exceptionally hot dry summer on photosynthetic traits in oak (*Quercus pubescens*) leaves. *Tree Physiol.* 2008; 28(5): 785–795.
20 Gall LF. Leaflet position influences caterpillar feeding and development. *Oikos.* 1987; 49(2): 172–176.
21 Bar M, Ori N. Leaf development and morphogenesis. *Development.* 2014; 141(22): 4219–4230.
22 Bussotti F, Pollastrini M. Evaluation of leaf features in forest trees: methods, techniques, obtainable information and limits. *Ecol Indic.* 2015; 52: 219–230.
23 Delpierre N, Vitasse Y, Chuine I, Guillemot J, Bazot S, Rutishauser T, et al. Temperate and boreal forest tree phenology: from organ-scale processes to terrestrial ecosystem models. *Ann For Sci.* 2016; 73(1): 5–25.
24 Ramírez-Briones E, Rodríguez-Macías R, Salcedo-Pérez E, Martínez-Gallardo N, Tiessen A, Molina-Torres J, et al. Seasonal variation in non-structural carbohydrates, sucrolytic activity and secondary metabolites in deciduous and perennial Diospyros species sampled in Western Mexico. *PLoS One.* 2017; 12(10): 1–24.
25 Sperling O, Kamai T, Tixier A, Davidson A, Jarvis-Shean K, Raveh E, et al. Predicting bloom dates by temperature mediated kinetics of carbohydrate metabolism in deciduous trees. *Agric For Meteorol.* 2019; 276–277: 107643.
26 Riipi M, Ossipov V, Lempa K, Haukioja E, Koricheva J, Ossipova S, et al. Seasonal changes in birch leaf chemistry: are there trade-offs between leaf growth and accumulation of phenolics? *Oecologia.* 2002; 130(3): 380–390.
27 Luyssaert S, Raitio H, Vervaeke P, Mertens J, Lust N. Sampling procedure for the foliar analysis of deciduous trees. *J Environ Monit.* 2002; 4(6): 858–864.
28 Ek M, Gellerstedt G, Henriksson G. Pulping chemistry and technology. In: Ek M, Gellerstedt G, & Henriksson G, editors. *Pulp and Paper Chemistry and Technology.* Volume 2. Berlin (DE): De Gruyter; 2009.
29 Chaffey N, Barlow P. The cytoskeleton facilitates a three-dimensional symplasmic continuum in the long-lived ray and axial parenchyma cells of angiosperm trees. *Planta.* 2001; 213(5): 811–823.
30 Ehlers K, van Bel AJE. Dynamics of plasmodesmal connectivity in successive interfaces of the cambial zone. *Planta.* 2010; 231(2): 371–385.

31 Sokolowska K, Zagórska-Marek B. Symplasmic, long-distance transport in xylem and cambial regions in branches of *Acer pseudoplatanus* (Aceraceae) and *Populus tremula* × *P. tremuloides* (*Salicaceae*). *Am J Bot*. 2012; 99(11): 1745–1755.
32 Pfautsch S, Hölttä T, Mencuccini M. Hydraulic functioning of tree stems – Fusing ray anatomy, radial transfer and capacitance. *Tree Physiol*. 2015; 35(7): 706–722.
33 Hubbe MA, Chandra RP, Dogu D, Van Velzen STJ. Analytical staining review. *BioResources*. 2019; 14(3): 7387–7464.
34 Kitin P, Nakaba S, Hunt CG, Lim S, Funada R. Direct fluorescence imaging of lignocellulosic and suberized cell walls in roots and stems. *AoB Plants*. 2020; 12(4): 1–19.
35 Sundell D, Street NR, Kumar M, Mellerowicz EJ, Kucukoglu M, Johnsson C, et al. Aspwood: High-spatial-resolution transcriptome profiles reveal uncharacterized modularity of wood formation in *Populus tremula*. *Plant Cel*. 2017; 29(7): 1585–1604.
36 Alonso-Serra J, Safronov O, Lim KJ, Fraser-Miller SJ, Blokhina OB, Campilho A, et al. Tissue-specific study across the stem reveals the chemistry and transcriptome dynamics of birch bark. *New Phytol*. 2019; 222(4): 1816–1831.
37 Bygdell J, Srivastava V, Obudulu O, Srivastava MK, Nilsson R, Sundberg B, et al. Protein expression in tension wood formation monitored at high tissue resolution in *Populus*. *J Exp Bot*. 2017; 68(13): 3405–3417.
38 Roach M, Arrivault S, Mahboubi A, Krohn N, Sulpice R, Stitt M, et al. Spatially resolved metabolic analysis reveals a central role for transcriptional control in carbon allocation to wood. *J Exp Bot*. 2017; 68(13): 3529–3539.
39 Abreu IN, Johansson AI, Sokołowska K, Niittylä T, Sundberg B, Hvidsten TR, et al. A metabolite roadmap of the wood-forming tissue in *Populus tremula*. *New Phytol*. 2020; 228(5): 1559–1572.
40 Uggla C, Sundberg B. Sampling of cambial region tissues for high resolution analysis. In: Chaffey NJ, editor. *Wood Formation in Trees: Cell and Molecular Techniques*. London: CRC Press; 2002. p. 215–228.
41 Smith MR, Merchant A. Limitations to using phloem sap to assess tree water and nutrient status. *Tree Physiol*. 2018; 39(2): 332–339.
42 Dinant S, Kehr J. Sampling and analysis of phloem sap. In: Maathuis FJM, editor. *Plant Mineral Nutrients: Methods and Protocols. Methods in Molecular Biology*. Volume 953. Humana Totowa (NJ); 2013. p. 185–194.
43 Killiny N. Collection of the phloem sap, pros and cons. *Plant Signaling Behav*. 2019; 14(8): 13.
44 Gessler A, Rennenberg H, Keitel C. Stable isotope composition of organic compounds transported in the phloem of European beech – Evaluation of different methods of phloem sap collection and assessment of gradients

in carbon isotope composition during leaf-to-stem transport. *Plant Biol.* 2004; 6(6): 721–729.

45 Devaux M, Ghashghaie J, Bert D, Lambrot C, Gessler A, Bathellier C, et al. Carbon stable isotope ratio of phloem sugars in mature pine trees throughout the growing season: comparison of two extraction methods. *Rapid Commun Mass Spectrom.* 2009; 23(16): 2511–2518.

46 Dafoe NJ, Zamani A, Ekramoddoullah AKM, Lippert D, Bohlmann J, Constabel CP. Analysis of the poplar phloem proteome and its response to leaf wounding. *J Proteome Res.* 2009; 8(5): 2341–2350.

47 Fink D, Dobbelstein E, Barbian A, Lohaus G. Ratio of sugar concentrations in the phloem sap and the cytosol of mesophyll cells in different tree species as an indicator of the phloem loading mechanism. *Planta.* 2018; 248(3): 661–673.

48 Pate J, Shedley E, Arthur D, Adams M. Spatial and temporal variations in phloem sap composition of plantation-grown *Eucalyptus globulus. Oecologia.* 1998; 117(3): 312–322.

49 Merchant A, Buckley TN, Pfautsch S, Turnbull TL, Samsa GA, Adams MA. Site-specific responses to short-term environmental variation are reflected in leaf and phloem-sap carbon isotopic abundance of field grown *Eucalyptus globulus. Physiol Plant.* 2012; 146(4): 448–459.

50 Nakamura S-i, Watanabe A, Hattori H, Chino M, Chongpraditnun P, Suzui N, et al. Analysis of phloem exudate collected from fruit-bearing stems of coconut palm: palm trees as a source of molecules circulating in sieve tubes. *Soil Sci Plant Nutr.* 2004; 50(5): 739–745.

51 Shi JT, Liu HC, Luo JY, Cai L. Seasonal variations of metabolite content in phloem sap from *Broussonetia papyrifera. Pak J Bot.* 2020; 52(5): 1763–1768.

52 Killiny N. Metabolomic comparative analysis of the phloem sap of curry leaf tree (*Bergera koenegii*), orange jasmine (*Murraya paniculata*), and Valencia sweet orange (*Citrus sinensis*) supports their differential responses to Huanglongbing. *Plant Signaling Behav.* 2016; 11(11): e1249080.

53 Moing A, Carbonne F, Zipperlin B, Svanella L, Gaudillère JP. Phloem loading in peach: symplastic or apoplastic? *Physiol Plant.* 1997; 101(3): 489–496.

54 Nadwodnik J, Lohaus G. Subcellular concentrations of sugar alcohols and sugars in relation to phloem translocation in *Plantago major, Plantago maritima, Prunus persica,* and *Apium graveolens. Planta.* 2008; 227(5): 1079–1089.

55 Öner-Sieben S, Lohaus G. Apoplastic and symplastic phloem loading in *Quercus robur* and *Fraxinus excelsior. J Exp Bot.* 2014; 65(7): 1905–1916.

56 Sauter JJ. Seasonal variation of sucrose content in the xylem sap of salix. *Z Pflanzenphysiol.* 1980; 98(5): 377–391.

57 Alves G, Decourteix M, Fleurat-Lessard P, Sakr S, Bonhomme M, Améglio T, et al. Spatial activity and expression of plasma membrane H+-ATPase in stem xylem of walnut during dormancy and growth resumption. *Tree Physiol.* 2007; 27(10): 1471–1480.

58 Ito A, Sakamoto D, Moriguchi T. Carbohydrate metabolism and its possible roles in endodormancy transition in Japanese pear. *Sci Hortic.* 2012; 144: 187–194.

59 Schenk HJ, Michaud JM, Mocko K, Espino S, Melendres T, Roth MR, et al. Lipids in xylem sap of woody plants across the angiosperm phylogeny. *Plant J.* 2021; 105(6): 1477–1494.

60 Alexou M, Peuke AD. Methods for xylem sap collection. *Plant Miner Nutr.* 2013; 953(July): 195–207.

61 Millard P, Wendler R, Grassi G, Grelet GA, Tagliavini M. Translocation of nitrogen in the xylem of field-grown cherry and poplar trees during remobilization. *Tree Physiol.* 2006; 26(4): 527–536.

62 Siebrecht S, Herdel K, Schurr U, Tischner R. Nutrient translocation in the xylem of poplar – Diurnal variations and spatial distribution along the shoot axis. *Planta.* 2003; 217(5): 783–793.

63 Furukawa J, Abe Y, Mizuno H, Matsuki K, Sagawa K, Kojima M, et al. Seasonal fluctuation of organic and inorganic components in xylem sap of *Populus nigra*. *Plant Root.* 2011; 5: 56–62.

64 Schill V, Korger V, Stahl W. Mechanical drying of wood—an alternative to seasoning drying? In: *Proceedings of the 3rd IUFRO Conference on Wood Drying. Understanding the wood drying process: a synthesis of theory and practice.* Vienna (Austria); 1992. p. 18–21.

65 Schill V, Hartung W, Orthen B, Weisenseel MH. The xylem sap of maple (*Acer platanoides*) trees – Sap obtained by a novel method shows changes with season and height. *J Exp Bot.* 1996; 47(294): 123–133.

66 Bergman EA, Green EL, Matthews PGD. The cibarial pump of the xylem-feeding froghopper *Philaenus spumarius* produces negative pressures exceeding 1 MPa. *Proc Biol Sci.* 1954; 2021(288): 20210731.

67 Cornara D, Saponari M, Zeilinger AR, de Stradis A, Boscia D, Loconsole G, et al. Spittlebugs as vectors of *Xylella fastidiosa* in olive orchards in Italy. *J Pest Sci.* 2017; 90(2): 521–530.

68 Freschet GT, Roumet C. Sampling roots to capture plant and soil functions. *Funct Ecol.* 2017; 31(8): 1506–1518.

69 McCormack ML, Dickie IA, Eissenstat DM, Fahey TJ, Fernandez CW, Guo D, et al. Redefining fine roots improves understanding of below-ground contributions to terrestrial biosphere processes. *New Phytol.* 2015; 207(3): 505–518.

70 Zadworny M, McCormack ML, Żytkowiak R, Karolewski P, Mucha J, Oleksyn J. Patterns of structural and defense investments in fine roots of

Scots pine (*Pinus sylvestris* L.) across a strong temperature and latitudinal gradient in Europe. *Global Change Biol.* 2017; 23(3): 1218–1231.

71 Freschet GT, Pagès L, Iversen CM, Comas LH, Rewald B, Roumet C, et al. A starting guide to root ecology: strengthening ecological concepts and standardising root classification, sampling, processing and trait measurements. *New Phytol.* 2021; 232(3): 973–1122.

72 Nelson T, Tausta SL, Gandotra N, Liu T. Laser microdissection of plant tissue: What you see is what you get. *Annu Rev Plant Biol.* 2006; 57: 181–201.

73 Gautam V, Sarkar AK. Laser assisted microdissection, an efficient technique to understand tissue specific gene expression patterns and functional genomics in plants. *Mol Biotechnol.* 2015; 57(4): 299–308.

74 Barcala M, Fenoll C, Escobar C. Laser microdissection of cells and isolation of high-quality RNA after cryosectioning. In: Hailing J, & Gassmann W, editors. *RNA Abundance Analysis: Methods and Protocols. Methods in Molecular Biology*. Volume 883. New York (NY): Humana Press; 2012. p. 87–95.

75 Blokhina O, Valerio C, Sokolowska K, Zhao L, Kärkönen A, Niittylä T, et al. Laser capture microdissection protocol for xylem tissues of woody plants. *Front Plant Sci.* 2017: 7.

76 Abbott E, Hall D, Hamberger B, Bohlmann J. Laser microdissection of conifer stem tissues: Isolation and analysis of high quality RNA, terpene synthase enzyme activity and terpenoid metabolites from resin ducts and cambial zone tissue of white spruce (*Picea glauca*). *BMC Plant Biol.* 2010; 10: 106.

77 Song D, Shen J, Li L. Characterization of cellulose synthase complexes in *Populus* xylem differentiation. *New Phytol.* 2010; 187(3): 777–790.

78 Teixeira RT, Fortes AM, Bai H, Pinheiro C, Pereira H. Transcriptional profiling of cork oak phellogenic cells isolated by laser microdissection. *Planta.* 2018; 247(2): 317–338.

79 Li SH, Nagy NE, Hammerbacher A, Krokene P, Niu XM, Gershenzon J, et al. Localization of phenolics in phloem parenchyma cells of Norway Spruce (*Picea abies*). *ChemBioChem.* 2012; 13(18): 2707–2713.

80 Velada I, Menéndez E, Teixeira RT, Cardoso H, Peixe A. Laser microdissection of specific stem-base tissue types from olive microcuttings for isolation of high-quality RNA. *Biology (Basel).* 2021; 10(3): 209.

81 Larisch C, Dittrich M, Wildhagen H, Lautner S, Fromm J, Polle A, et al. Poplar wood rays are involved in seasonal remodeling of tree physiology. *Plant Physiol.* 2012; 160(3): 1515–1529.

82 Wang JP, Chuang L, Loziuk PL, Chen H, Lin YC, Shi R, et al. Phosphorylation is an on/off switch for 5-hydroxyconiferaldehyde O-methyltransferase activity in poplar monolignol biosynthesis. *Proc Natl Acad Sci USA.* 2015; 112(27): 8481–8486.

83 Blokhina O, Laitinen T, Hatakeyama Y, Delhomme N, Paasela T, Zhao L, et al. Ray parenchymal cells contribute to lignification of tracheids in developing xylem of Norway Spruce. *Plant Physiol.* 2019; 181(4): 1552–1572.
84 Chen HC, Song J, Wang JP, Lin YC, Ducoste J, Shuford CM, et al. Systems biology of lignin biosynthesis in *Populus trichocarpa*: heteromeric 4-coumaric acid: coenzyme A ligase protein complex formation, regulation, and numerical modeling. *Plant Cel.* 2014; 26(3): 876–893.
85 Luchi N, Capretti P, Fossdal CG, Pazzagli M, Pinzani P. Laser microdissection on Norway spruce bark tissue: a suitable protocol for subsequent real-time reverse transcription-polymerase chain reaction (RT–PCR) analysis. *Plant Biosyst.* 2012; 146(1): 92–98.
86 Yu J, Zhou C, Li D, Li S, Jimmy Lin YC, Wang JP, et al. A PtrLBD39-mediated transcriptional network regulates tension wood formation in *Populus trichocarpa. Plant Commun.* 2022; 3(1): 100250.
87 Cañas RA, Canales J, Gómez-Maldonado J, Ávila C, Cánovas FM. Transcriptome analysis in maritime pine using laser capture microdissection and 454 pyrosequencing. *Tree Physiol.* 2014; 34(11): 1278–1288.
88 Carter AD, Bonyadi R, Gifford ML. The use of fluorescence-activated cell sorting in studying plant development and environmental responses. *Int J Dev Biol.* 2013; 57(6–8): 545–552.
89 Antoniadi I, Skalický V, Sun G, Ma W, Galbraith DW, Novák O, et al. Fluorescence activated cell sorting—A selective tool for plant cell isolation and analysis. *Cytom Part A.* 2022; 101(9): 725–736.
90 Conde D, Triozzi PM, Balmant KM, Doty AL, Miranda M, Boullosa A, et al. A robust method of nuclei isolation for single-cell RNA sequencing of solid tissues from the plant genus *Populus. PLoS One.* 2021; 16(5): e0251149.
91 Spraker JE, Luu GT, Sanchez LM, Park M. Imaging mass spectrometry for natural products discovery: Review of ionization methods. *Nat Prod Rep.* 2021; 37(2): 150–162.
92 de Souza LP, Borghi M, Fernie A. Plant single-cell metabolomics-challenges and perspectives. *Int J Mol Sci.* 2020; 21(23): 8987.
93 Hu W, Nie H, Wang Y, Li N, Di S, Pan Q, et al. Tracing the migration and transformation of metabolites in xylem during wood growth by mass spectrometry imaging. *Analyst.* 2022; 147(8): 1551–1558.
94 Lunsford K, Peter G, Yost R. Direct matrix-assisted laser desorption/ionization mass spectrometric imaging of cellulose and hemicellulose in Populus tissue. *Anal Chem.* 2011; 83(17): 6722–6730.
95 Yoshinaga A, Kamitakahara H, Takabe K. Distribution of coniferin in differentiating normal and compression woods using MALDI mass spectrometric imaging coupled with osmium tetroxide vapor treatment. *Tree Physiol.* 2016; 36(5): 643–652.

96 Veličković D, Zhou M, Schilling JS, Zhang J. Using MALDI–FTICR–MS imaging to track low-molecular-weight aromatic derivatives of fungal decayed wood. *J Fungi*. 2021; 7(8).

97 Aoki D, Fukushima K. Topochemical analysis of cell wall components by TOF–SIMS. In: De Lucas M, & Etchells J, editors. *Xylem: Methods and Protocols. Methods in Molecular Biology*. Volume 1544. New York (NY): Humana Press; 2017. p. 249–256.

98 Aoki D, Matsushita Y, Fukushima K. Cryo secondary ion mass spectrometry for wood component visualization: a mini review. *Holzforschung*. 2022; 76(2): 145–154.

99 Fu T, Touboul D, Della-Negra S, Houël E, Amusant N, Duplais C, et al. Tandem mass spectrometry imaging and in situ characterization of bioactive wood metabolites in Amazonian tree species *Sextonia rubra*. *Anal Chem*. 2018; 90(12): 7535–7543.

100 Fu T, Elie N, Brunelle A. Radial distribution of wood extractives in European larch *Larix decidua* by TOF–SIMS imaging. *Phytochemistry*. 2018; 150: 31–39.

101 Vanbellingen QP, Fu T, Bich C, Amusant N, Stien D, Della-Negra S, et al. Mapping *Dicorynia guianensis* Amsh. wood constituents by submicron resolution cluster–TOF–SIMS imaging. *J Mass Spectrom*. 2016; 51(6): 412–423.

102 Imai T, Tanabe K, Kato T, Fukushima K. Localization of ferruginol, a diterpene phenol, in *Cryptomeria japonica* heartwood by time-of-flight secondary ion mass spectrometry. *Planta*. 2005; 221(4): 549–556.

103 Goacher R, Jeremic D, Master E. Expanding the library of secondary ions that distinguish lignin and polysaccharides in time-of-flight secondary ion mass spectrometry analysis of wood. *Anal Chem*. 2011; 18(3): 804–812.

104 Zhou C, Li Q, Chiang V, Lucia L, Griffis D. Chemical and spatial differentiation of syringyl and guaiacyl lignins in poplar wood via time-of-flight secondary ion mass spectrometry. *Anal Chem*. 2011; 83(18): 7020–7026.

105 Saito K, Watanabe Y, Shirakawa M, Matsushita Y, Imai T, Koike T, et al. Direct mapping of morphological distribution of syringyl and guaiacyl lignin in the xylem of maple by time-of-flight secondary ion mass spectrometry. *Plant J*. 2012; 69(3): 542–552.

106 de Moraes Pontes JG, Vendramini PH, Fernandes LS, de Souza FH, Pilau EJ, Eberlin MN, et al. Mass spectrometry imaging as a potential technique for diagnostic of Huanglongbing disease using fast and simple sample preparation. *Sci Rep*. 2020; 10: 13457.

107 Shrestha B, Patt JM, Vertes A. In situ cell-by-cell imaging and analysis of small cell populations by mass spectrometry. *Anal Chem*. 2011; 83(8): 2947–2955.

108 Fernie AR, Geigenberger P, Stitt M. Flux an important, but neglected, component of functional genomics. *Curr Opin Plant Biol.* 2005; 8(2): 174–182.
109 Stitt M. Metabolic regulation of photosynthesis. In: Baker N, editor. *Advances in Photosynthesis.* Environmental Stress and Photosynthesis. Volume 3. Academic Press; 1996.
110 Kruger NJ, Ratcliffe RG. Fluxes through plant metabolic networks: measurements, predictions, insights and challenges. *Biochem J.* 2015; 465: 27–38.
111 Hartmann H, McDowell NG, Trumbore S. Allocation to carbon storage pools in Norway spruce saplings under drought and low CO_2. *Tree Physiol.* 2015; 35(3): 243–252.
112 Arroo RRJ, Bhambra AS, Hano C, Renda G, Ruparelia KC, Wang MF. Analysis of plant secondary metabolism using stable isotope-labeled precursors. *Phytochem Anal.* 2021; 32(1): 62–68.
113 Ishihara H, Obata T, Sulpice R, Fernie AR, Stitt M. Quantifying protein synthesis and degradation in arabidopsis by dynamic $^{13}CO_2$ labeling and analysis of enrichment in individual amino acids in their free pools and in protein. *Plant Physiol.* 2015; 168(1): 74–93.
114 Szecowka M, Heise R, Tohge T, Nunes-Nesi A, Vosloh D, Huege J, et al. Metabolic fluxes in an illuminated Arabidopsis rosette. *Plant Cel.* 2013; 25(2): 694–714.
115 Zeller B, Colin-Belgrand M, Dambrine E, Martin F, Bottner P. Decomposition of 15N-labelled beech litter and fate of nitrogen derived from litter in a beech forest. *Oecologia.* 2000; 123: 550–559.
116 Lhuissier F, Lefebvre F, Gibouin D, Demarty M, Thellier M, Ripoll C. Secondary ion mass spectrometry imaging of the fixation of ^{15}N- labeled NO in pollen grains. *J Microsc.* 2000; 198(2): 108–115.
117 Doppler M, Bueschl C, Kluger B, Koutnik A, Lemmens M, Buerstmayr H, et al. Stable isotope–assisted plant metabolomics: combination of global and tracer-based labeling for enhanced untargeted profiling and compound annotation. *Front Plant Sci.* 2019; 10: 1366.
118 Freund D, Hegeman A. Recent advances in stable isotope-enabled mass spectrometry-based plant metabolomics. *Curr Opin Biotechnol.* 2017; 43: 41–48.
119 Ghirardo A, Gutknecht J, Zimmer I, Brüggemann N, Schnitzler JP. Biogenic volatile organic compound and respiratory CO_2 emissions after ^{13}C-labeling: online tracing of C translocation dynamics in poplar plants. *PLoS One.* 2011; 6(2): 2–5.
120 Rehschuh R, Rehschuh S, Gast A, Jakab AL, Lehmann MM, Saurer M, et al. Tree allocation dynamics beyond heat and hot drought stress reveal changes in carbon storage, belowground translocation and growth. *New Phytol.* 2022; 233(2): 687–704.

121 Norris CE, Quideau SA, Landhäusser SM, Bernard GM, Wasylishen RE. Tracking stable isotope enrichment in tree seedlings with solid-state NMR spectroscopy. *Sci Rep.* 2012; 2: 719.
122 Thoms R, Köhler M, Gessler A, Gleixner G. Above and below ground carbohydrate allocation differs between ash (*Fraxinus excelsior* L.) and beech (*Fagus sylvatica* L.). *PLoS One.* 2017; 12(9): e0184247.
123 Gao D, Joseph J, Werener R, Brunner I, Zurcher A, Hug C, et al. Drought alters the carbon footprint of trees in soils—tracking the spatio-temporal fate of ^{13}C-labeled assimilates in the soil of an old-growth pine forest. *Global Change Biol.* 2021; 27: 2491–2506.
124 Epron D, Cabral OMR, Laclau JP, Dannoura M, Packer AP, Plain C, et al. In situ $^{13}CO_2$ pulse labeling of field-grown eucalypt trees revealed the effects of potassium nutrition and throughfall exclusion on phloem transport of photosynthetic carbon. *Tree Physiol.* 2015; 36(1): 6–21.
125 Dominguez PG, Donev E, Derba-Maceluch M, Bünder A, Hedenström M, Tomášková I, et al. Sucrose synthase determines carbon allocation in developing wood and alters carbon flow at the whole tree level in aspen. *New Phytol.* 2021; 229(1): 186–198.
126 Mahboubi A, Ratke C, Gorzsás A, Kumar M, Mellerowicz EJ, Niittylä T. Aspen SUCROSE TRANSPORTER3 allocates carbon into wood fibers. *Plant Physiol.* 2013; 163(4): 1729–1740.
127 Mahboubi A, Linden P, Hedenström M, Moritz T, Niittylä T. ^{13}C tracking after $^{13}CO_2$supply revealed diurnal patterns of wood formation in aspen. *Plant Physiol.* 2015; 168(2): 478–489.
128 Silva LCR, Salamanca-Jimenez A, Doane TA, Horwath WR. Carbon dioxide level and form of soil nitrogen regulate assimilation of atmospheric ammonia in young trees. *Sci Rep.* 2015; 5: 13141.
129 Sperling O, Silva LCR, Tixier A, Théroux-Rancourt G, Zwieniecki MA. Temperature gradients assist carbohydrate allocation within trees. *Sci Rep.* 2017; 7: 3265.
130 Dethloff F, Orf I, Kopka J. Rapid in situ 13C tracing of sucrose utilization in Arabidopsis sink and source leaves. *Plant Methods.* 2017; 13(1): 1–19.
131 Amiard V, Morvan-Bertrand A, Billard JP, Huault C, Prud'homme MP. Fate of fructose supplied to leaf sheaths after defoliation of Lolium perenne L.: Assessment by ^{13}C-fructose labeling. *J Exp Bot.* 2003; 54(385): 1231–1243.
132 Kreuzwieser J, Graus M, Wisthaler A, Hansel A, Rennenberg H, Schnitzler JP. Xylem-transported glucose as an additional carbon source for leaf isoprene formation in *Quercus robur*. *New Phytol.* 2002; 156(2): 171–178.
133 Jardine KJ, de Souza VF, Oikawa P, Higuchi N, Bill M, Porras R, et al. Integration of C1 and C2 metabolism in trees. *Int J Mol Sci.* 2017; 18(10): 2045.

134 Beshir WF, Mbong VBM, Hertog MLATM, Geeraerd AH, Van Den Ende W, Nicolaï BM. Dynamic labeling reveals temporal changes in carbon re-allocation within the central metabolism of developing apple fruit. Front. *Plant Sci.* 2017; 8: 1785.

135 Alonso AP, Piasecki RJ, Wang Y, LaClair RW, Shachar-Hill Y. Quantifying the labeling and the levels of plant cell wall precursors using ion chromatography tandem mass spectrometry. *Plant Physiol.* 2010; 153(3): 915–924.

136 Hu Y, Fernández V, Ma L. Nitrate transporters in leaves and their potential roles in foliar uptake of nitrogen dioxide. *Front Plant Sci.* 2014; 5: 360.

137 Nussbaum S, von Ballmoos P, Gfeller H, Schlunegger U, Fuhrer J, Rhodes D, et al. Incorporation of atmospheric $^{15}NO_2$-nitrogen into free amino acids by Norway spruce *Picea abies* (L.). *Karst Oecologia.* 1993; 94: 408–414.

138 Tagliavini M, Millard P, Quartieri M. Storage of foliar-absorbed nitrogen and remobilization for spring growth in young nectarine (*Prunus persica* var. nectarina) trees. *Tree Physiol.* 1998; 18(3): 203–207.

139 Jalonen R, Nygren P, Sierra J. Transfer of nitrogen from a tropical legume tree to an associated fodder grass via root exudation and common mycelial networks. *Plant, Cell Environ.* 2009; 32(10): 1366–1376.

140 Dail BD, Hollinger DY, Davidson EA, Fernandez I, Sievering HC, Scott NA, et al. Distribution of nitrogen-15 tracers applied to the canopy of a mature spruce-hemlock stand, Howland, Maine. *USA Oecologia.* 2009; 160(3): 589–599.

141 Mead DJ, Preston CM. Distribution and retranslocation of ^{15}N in lodge-pole pine over eight growing seasons. *Tree Physiol.* 1994; 14(4): 389–402.

142 Buchmann N, Schulze D, Gebauer G. ^{15}N-ammonium and ^{15}N-nitrate uptake of a 15-year-old. *Oecologia.* 1995; 102(102): 361–370.

143 Dyckmans J, Flessa H. Influence of tree internal N status on uptake and translocation of C and N in beech: A dual ^{13}C and ^{15}N labeling approach. *Tree Physiol.* 2001; 21(6): 395–401.

144 Frak E, Millard P, Le Roux X, Guillaumie S, Wendler R. Coupling sap flow velocity and amino acid concentrations as an alternative method to ^{15}N labeling for quantifying nitrogen remobilization by walnut trees. *Plant Physiol.* 2002; 130(2): 1043–1053.

145 Quiñones A, Bañuls J, Millo EP, Legaz F. Effects of ^{15}N application frequency on nitrogen uptake efficiency in *Citrus* trees. *J Plant Physiol.* 2003; 160(12): 1429–1434.

146 Warren CR, Livingston NJ, Turpin DH. Response of Douglas-fir seedlings to a brief pulse of ^{15}N- labeled nutrients. *Tree Physiol.* 2003; 23(17): 1193–1200.

147 Feng Z, Dyckmans J, Flessa H. Effects of elevated carbon dioxide concentration on growth and N 2 fixation of young *Robinia pseudoacacia. Tree Physiol.* 2004; 24(3): 323–330.

148 Frak E, Roux XLE, Millard P, Guillaumie S, Wendler R. Nitrogen availability, local light regime and leaf rank effects on the amount and sources of N allocated within the foliage of young walnut (*Juglans nigra* x *regia*) trees. *Tree Physiol.* 2006; 26(1): 43–49.

149 El Zein R, Bréda N, Gérant D, Zeller B, Maillard P. Nitrogen sources for current-year shoot growth in 50-year-old sessile oak trees: an in situ ^{15}N labeling approach. *Tree Physiol.* 2011; 31(12): 1390–1400.

150 van Huysen TL, Harmon ME, Perakis SS, Chen H. Decomposition and nitrogen dynamics of ^{15}N-labeled leaf, root, and twig litter in temperate coniferous forests. *Oecologia.* 2013; 173(4): 1563–1573.

151 Veerman L, Kalbitz K, Schoorl JC, Tietema A. Split-root labeling to investigate ^{15}N rhizodeposition by *Pinus sylvestris* and *Picea abies. Isot Environ Health Stud.* 2018; 54(1): 16–27.

152 Henriksson N, Lim H, Marshall J, Franklin O, McMurtrie RE, Lutter R, et al. Tree water uptake enhances nitrogen acquisition in a fertilized boreal forest – but not under nitrogen-poor conditions. *New Phytol.* 2021; 232(1): 113–122.

153 Wessel WW, Boxman AW, Cerli C, van Loon EE, Tietema A. Long-term stabilization of ^{15}N-labeled experimental NH_4^+ deposition in a temperate forest under high N deposition. *Sci Total Environ.* 2021; 768: 144356.

154 Zhu F, Dai L, Hobbie EA, Qu Y, Huang D, Gurmesa GA, et al. Quantifying nitrogen uptake and translocation for mature trees: an in situ whole-tree paired ^{15}N labeling method. *Tree Physiol.* 2021; 41(11): 2109–2125.

155 Gurmesa GA, Wang A, Li S, Peng S, de Vries W, Gundersen P, et al. Retention of deposited ammonium and nitrate and its impact on the global forest carbon sink. *Nat Commun.* 2022; 13(1).

156 Mayerhofer W, Schintlmeister A, Dietrich M, Gorka S, Wiesenbauer J, Martin V, et al. Recently photoassimilated carbon and fungus-delivered nitrogen are spatially correlated in the ectomycorrhizal tissue of *Fagus sylvatica. New Phytol.* 2021; 232(6): 2457–2474.

157 Proe MF, Mead DJ, Byrne D. Effect of pruning on nitrogen dynamics within crowns of *Pinus radiata. Tree Physiol.* 2000; 20(10): 653–661.

158 Nair R, Weatherall A, Perks M, Mencuccini M. Stem injection of 15N-NH_4NO_3 into mature Sitka spruce (*Picea sitchensis*). *Tree Physiol.* 2014; 34(10): 1130–1140.

159 Vega-Mas I, Cukier C, Coleto I, González-Murua C, Limami AM, González-Moro MB, et al. Isotopic labeling reveals the efficient adaptation of wheat root TCA cycle flux modes to match carbon demand under ammonium nutrition. *Sci Rep.* 2019; 9: 8925.

160 Baslam M, Mitsui T, Sueyoshi K, Ohyama T. Recent advances in carbon and nitrogen metabolism in C_3 plants. *Int J Mol Sci*. 2021; 22(1): 318.

161 Komatsu T, Ohishi R, Shino A, Akashi K, Kikuchi J. Multi-spectroscopic analysis of seed quality and 13C-stable-isotopologue monitoring in initial growth metabolism of *Jatropha curcas* L. *Metabolites*. 2014; 4(4): 1018–1033.

162 Bennett JN, Prescott CE. Organic and inorganic nitrogen nutrition of western red cedar, western hemlock and salal in mineral N-limited cedar-hemlock forests. *Oecologia*. 2004; 141(3): 468–476.

163 Pfautsch S, Rennenberg H, Bell TL, Adams MA. Nitrogen uptake by *Eucalyptus regnans* and *Acacia* spp. – Preferences, resource overlap and energetic costs. *Tree Physiol*. 2009; 29(3): 389–399.

164 Buescher JM, Antoniewicz MR, Boros LG, Burgess SC, Brunengraber H, Clish CB, et al. A roadmap for interpreting ^{13}C metabolite labeling patterns from cells. *Curr Opin Biotechnol*. 2015; 34: 189–201.

165 Furze ME, Drake JE, Wiesenbauer J, Richter A, Pendall E. Carbon isotopic tracing of sugars throughout whole-trees exposed to climate warming. *Plant Cell Environ*. 2019; 42(12): 3253–3263.

166 Kessler A, Kalske A. Plant secondary metabolite diversity and species interactions. *Annu Rev Ecol Evol Syst*. 2018; 49(1): 115–138.

167 t'Kindt R, Morreel K, Deforce D, Boerjan W, Van Bocxlaer J. Joint GC–MS and LC–MS platforms for comprehensive plant metabolomics: repeatability and sample pre-treatment. *J Chromatogr B Analyt Technol Biomed Life Sci*. 2009; 877(29): 3572–3580.

168 Mushtaq MY, Choi YH, Verpoorte R, Wilson EG. Extraction for metabolomics: access to the metabolome: extraction for metabolomics: access to the metabolome. *Phytochem Anal*. 2014; 25(4): 291–306.

169 Kim HK, Choi YH, Verpoorte R. NMR-based metabolomic analysis of plants. *Nat Protoc*. 2010; 5(3): 536–549.

170 Kim HK, Verpoorte R. Sample preparation for plant metabolomics: sample preparation for plant metabolomics. *Phytochem Anal*. 2010; 21(1): 4–13.

171 Lisec J, Schauer N, Kopka J, Willmitzer L, Fernie AR. Gas chromatography mass spectrometry-based metabolite profiling in plants. *Nat Protoc*. 2006; 1(1): 387–396.

172 De Vos RC, Moco S, Lommen A, Keurentjes JJ, Bino RJ, Hall RD. Untargeted large-scale plant metabolomics using liquid chromatography coupled to mass spectrometry. *Nat Protoc*. 2007; 2(4): 778–791.

173 Roessner U, Dias DA. Plant tissue extraction for metabolomics. *Methods Mol Biol*. 2013; 1055: 21–28.

174 Jones WP, Kinghorn AD. Extraction of plant secondary metabolites. *Methods Mol Biol*. 2012; 864: 341–366.

175 Salem MA, Yoshida T, Perez de Souza L, Alseekh S, Bajdzienko K, Fernie AR, et al. An improved extraction method enables the comprehensive analysis of lipids, proteins, metabolites and phytohormones from a single sample of leaf tissue under water-deficit stress. *Plant J.* 2020; 103(4): 1614–1632.

176 Feussner K, Abreu IN, Klein M, Feussner I. Metabolite fingerprinting: a powerful metabolomics approach for marker identification and functional gene annotation. *Methods Enzymol.* 2023; 680: 325–350.

177 Schripsema J. Application of NMR in plant metabolomics: techniques, problems and prospects: application of NMR in plant metabolomics. *Phytochem Anal.* 2010; 21(1): 14–21.

178 Mushtaq MY, Choi YH, Verpoorte R, Wilson EG. Extraction for metabolomics: access to the metabolome: extraction for metabolomics: access to the metabolome. *Phytochem Anal.* 2014; 25(4): 291–306.

179 Tikunov YM, Verstappen FWA, Hall RD. Metabolomic profiling of natural volatiles: headspace trapping: GC–MS. *Methods Mol Biol.* 2007; 358: 39–53.

180 Roca M, Alcoriza MI, Garcia-Cañaveras JC, Lahoz A. Reviewing the metabolome coverage provided by LC–MS: Focus on sample preparation and chromatography-A tutorial. *Anal Chim Acta.* 2021; 1147: 38–55.

181 Bligh EG, Dyer WJ. A rapid method of total lipid extraction and purification. *Can J Biochem Physiol.* 1959; 37(8): 911–917.

182 Matyash V, Liebisch G, Kurzchalia TV, Shevchenko A, Schwudke D. Lipid extraction by methyl-tert-butyl ether for high-throughput lipidomics. *J Lipid Res.* 2008; 49(5): 1137–1146.

183 Roberts LD, Souza AL, Gerszten RE, Clish CB. Targeted metabolomics. *Curr Protoc Mol Biol.* 2012; 98: 30.2.1–30.2.24.

184 Fiehn O. Metabolomics by Gas Chromatography-Mass Spectrometry: Combined Targeted and Untargeted Profiling. *Curr Protoc Mol Biol.* 2016; 114: 30.4.1–30.4.32.

185 Salem MA, Yoshida T, Perez de Souza L, Alseekh S, Bajdzienko K, Fernie AR, et al. An improved extraction method enables the comprehensive analysis of lipids, proteins, metabolites and phytohormones from a single sample of leaf tissue under water-deficit stress. *Plant J.* 2020; 103(4): 1614–1632.

186 Villas-Bôas SG, Mas S, Akesson M, Smedsgaard J, Nielsen J. Mass spectrometry in metabolome analysis. *Mass Spectrom Rev.* 2005; 24(5): 613–646.

187 David V, Moldoveanu SC, Galaon T. Derivatization procedures and their analytical performances for HPLC determination in bioanalysis. *Biomed Chromatogr.* 2021; 35: e5008.

188 Zhao S, Li L. Chemical derivatization in LC–MS-based metabolomics study. *Trends Anal Chem.* 2020; 131: 115988.

189 Halket JM, Waterman D, Przyborowska AM, Patel RKP, Fraser PD, Bramley PM. Chemical derivatization and mass spectral libraries in metabolic profiling by GC/MS and LC/MS/MS. *J Exp Bot.* 2005; 56(410): 219–243.

190 Fiehn O, Kopka J, Dörmann P, Altmann T, Trethewey RN, Willmitzer L. Metabolite profiling for plant functional genomics. *Nat Biotechnol.* 2000; 18(11): 1157–1161.

191 Gullberg J, Jonsson P, Nordström A, Sjöström M, Moritz T. Design of experiments: an efficient strategy to identify factors influencing extraction and derivatization of *Arabidopsis thaliana* samples in metabolomic studies with gas chromatography/mass spectrometry. *Anal Biochem.* 2004; 331(2): 283–295.

192 Fritsche-Guenther R, Gloaguen Y, Bauer A, Opialla T, Kempa S, Fleming CA, et al. Optimized workflow for on-line derivatization for targeted metabolomics approach by gas chromatography-mass spectrometry. *Metabolites.* 2021; 11(12): 888.

193 Kopka J. Current challenges and developments in GC–MS based metabolite profiling technology. *J Biotechnol.* 2006; 124(1): 312–322.

194 Docherty KS, Ziemann PJ. On-line, inlet-based trimethylsilyl derivatization for gas chromatography of mono- and dicarboxylic acids. *J Chromatogr A.* 2001; 921(2): 265–275.

195 Marsol-Vall A, Balcells M, Eras J, Canela-Garayoa R. Dispersive liquid-liquid microextraction and injection-port derivatization for the determination of free lipophilic compounds in fruit juices by gas chromatography-mass spectrometry. *J Chromatogr A.* 2017; 1495: 12–21.

4

Systems Biology as a Tool to Uncover Interdisciplinary Links within the Complex Forest Tree System

Pia Guadalupe Dominguez[1], Ilka Nacif Abreu[2], Thomas Moritz[3], and Ilara Gabriela Frasson Budzinski[4]

[1] *Instituto de Agrobiotecnología y Biología Molecular (IABIMO), Instituto Nacional de Tecnología Agropecuaria (INTA), Consejo Nacional de Investigaciones Científicas y Técnicas (CONICET), Hurlingham, Buenos Aires, Argentina*
[2] *Department of Plant Biochemistry, Albrecht-von-Haller-Institute for Plant Sciences, University of Goettingen, Goettingen, Germany*
[3] *Department of Forest Genetics and Plant Physiology, Umeå Plant Science Centre, Swedish University of Agricultural Sciences, Umeå, Sweden*
[4] *Department of Genetics, Max Feffer Laboratory of Plant Genetics, "Luiz de Queiroz" College of Agriculture, University of São Paulo, Piracicaba, SP, Brazil*

4.1 Systems Biology

Systems biology is a holistic approach to understand complex biological systems through an interdisciplinary task force by integrating quantitative molecular measurements obtained by multiple omics techniques, such as genomics, transcriptomics, proteomics, and metabolomics. The aim of the systems biology approach is to combine omics layers to build models that explain system behavior in terms of spatiotemporal interactions and that have predictive power to propose the outcome of mutation or modification of specific biological steps through the application of computational and mathematical modeling (1). The combination of methods to integrate data from different omics layers aiming at gaining insight into how the different biomolecules are interconnected and how the flow of biological information occurs is named multi-omics analysis (2).

The massive increase in high-throughput technologies and reductions in the analysis costs have enabled multi-omics studies to be routinely performed. Although each omics platform allows a comprehensive survey of a particular molecular layer, the crosstalk among them cannot be assessed from a reductionist perspective (3). Instead, the integration of these omics layers by system biology approaches reduces the dimensionality of global data, thereby allowing the improvement of our understanding of interrelation and combined

Monitoring Forest Damage with Mass Spectrometry-Based Metabolomics Methods, First Edition. Edited by Carla António.

influences among molecules (3, 4). Despite major advances in biomolecule state-of-the-art detection and quantification technologies (e.g. DNA sequencing and resequencing, transcriptomics, proteomics, and metabolomics) through mass spectrometry (MS)-based approaches, the integration of multi-omics measurements is a major challenge due to their inherent data differences (4).

4.2 Strategies for Data Integration and Network Analysis

Advances in omics technologies prompted the integration of the multiple layers of cell information, including genes, mRNA, proteins, and metabolites, to identify relationships among them (5) and enhance the readout of cellular processes and molecular programs (6).

While there are multiple ways to perform data integration, including qualitative and quantitative approaches (5, 7, 8), a well-defined methodological scheme for multi-omics integration (MOI) in plant systems biology has been proposed by Jamil et al. (8). Qualitative analysis, also named as conceptual integration (7), refers to multiple omics datasets being analyzed separately and matched without further statistical analysis (8). The objective is to identify the convergence of the biological interpretations from individual analyses (9). Although this approach can produce valuable insights, Jamil et al. (8) argue that this is an arbitrary connection and should not be considered a part of the MOI approach. The authors have therefore defined the MOI workflow into three main integration levels, with increasing complexity: (i) element-based integration; (ii) pathway-based integration; and (iii) mathematical integration. Below we summarize the levels of metabolomics, transcriptomics, and proteomics data integration as defined by Jamil et al. (8). The integration of genomics and metabolomics data will be presented in Section 4.3.

4.2.1 Element-Based Integration

This unbiased integration is the first level of MOI approach. It seeks statistical associations between elements from the omics datasets of interest and includes mainly three strategies: correlation, clustering, and multivariate analyses based on projection methods. Importantly, finding statistical associations between variables does not necessarily mean actual biological associations, so further experiments should be performed to prove them.

Correlation seeks to find correlative links among molecules (transcripts, proteins, and metabolites) (7). Its main advantages are simplicity, intuitiveness, and that they do not require prior knowledge of the pathways (7, 8). The most widely used parametric correlation method is Pearson's correlation, while the most widely used non-parametric method is Spearman's correlation (7).

Clustering is the classification of samples or features (e.g. multi-omics data) into groups (clusters) based on how closely associated they are (10, 11). In practice, clustering analysis aims to determine coherent groups of unclassified samples in the data to uncover patterns, so that similar samples are within a group, and dissimilar samples are in different groups (10). Clustering analysis using multi-omics data for the same set of samples can reveal fine-tuned structures that are not revealed when using single omics data (12). It can also indicate the unknown functionalities of variables when compared with those in the same clusters with known functions (13). Clustering has two main approaches, hierarchical and non-hierarchical methods. Hierarchical clusters are formed in a predetermined order rendering nested clusters, while in non-hierarchical clustering, clusters are formed by partitioning samples to a predefined number of clusters according to a criterion (e.g. maximizing inter-cluster distances while minimizing within-cluster distances) (14). The latter approach is usually applied in the integration of multi-omics data especially using machine learning algorithms, such as k-means and random forest (8).

Multivariate analyses based on projection methods allow making general inferences to predict different aspects or trends of datasets, including the discovery of variance or covariance associations (15). Principal component analysis (PCA) and partial least squares (PLS) are routinely applied. These two methods are very popular for studying omics data because they work particularly well when the data is collinear (e.g. transcriptomics and metabolomics data) (7). Another widely-used analysis is the orthogonal partial least squares discriminant analysis (OPLS–DA) (16), which has the advantage of improving the predictive power of other techniques. The "mixOmics" R package is specially designed to perform multivariate analyses with different methods, such as sparse principal component analysis (sPCA), and sparse partial least squares discriminant analysis (sPLS–DA) (17).

4.2.2 Pathway-Based Integration

The integration of omics datasets to exploit pathway analysis can be done by mapping-based and co-expression. In this type of integration, statistical associations are obtained between elements from different data sets (8).

Pathway mapping analysis relies on mapping omics data to existing metabolic pathway databases. The Kyoto Encyclopedia of Genes and Genomes (KEGG) (18) is one of the main databases used for plant metabolic mapping despite not being plant-specific. Other plant-based databases are available, such as the Plant Metabolic Network (PMN) (19), which contains a multi-species reference database including forest trees, such as *Eucalyptus grandis* (EgrandisCyc),

Populus trichocarpa (PoplarCyc), *Quercus lobata* (VALLEYOAKCYC), and *Quercus suber* (CORKOAKCYC). The aim of most pathway mapping tools is to obtain the metabolic pathway enrichment of a dataset, which allows to infer which metabolic pathways are over- or under-represented in a given experimental treatment using, for example Pageman (20), MetaboAnalyst (21), Gene Ontology (GO) enrichment analysis (22), among other mapping tools (see (7, 8)).

Co-expression is a powerful strategy to investigate candidate molecules involved in specific biological processes and to assign the biological function to unknown molecules (23). These data can be transformed into the so-called weighted networks by employing tools, such as the weighted gene correlation network analysis (WGCNA); an R package that allows the identification of clusters (modules) (24). This strategy can be used to identify modules for molecules related to specific pathways. Despite correlation being highly used as the base of co-expression analyses, there are other methodologies that go beyond simple correlation analyses and that allow studying cases of multicollinear systems, for example, partial correlations and Bayesian networks (5, 25).

4.2.3 Mathematical Integration

As suggested by its name, mathematical integration aims at generating system-level metabolic models by means of mathematical analyses, which can be used to test hypotheses (8). They bear high-standard requirements, including deep data coverage, pathway topology, and regulation knowledge, and mathematical equations describing the system. This level includes the subclasses: differential and genome-scale analyses.

Differential analyses employ differential equations to describe the mass-balance kinetics of a given metabolic pathway (8). They are used to perform flux analyses and to predict metabolite levels. Although they can be targeted (developed for specific pathways) or non-targeted (developed for biomolecule global scales), they have proven to be highly adequate for single and well-annotated targeted pathways, which allows them to be applied to both model and non-model organisms. Several algorithms are available to perform differential analyses, including E-flux and metabolic adjustment by differential expression (MADE) (reviewed in (26)).

Genomic-scale models aim to build systemic metabolic models employing annotated genomes to reconstruct genome-wide metabolism, thus considering all the reactions simultaneously occurring in the system of interest. Its best-known application is flux balance analysis (FBA). For plant metabolic reconstruction, PlantSEED (27) is available; a framework covering primary metabolism for 39 species is used for automated annotation and metabolic reconstruction from private transcriptome data and to simulate metabolic activity under various conditions using FBA.

4.2.4 Integration of Transcriptomics and Metabolomics Data

Over the past few decades, with the developments in sequencing technology, transcriptomics tools have evolved from the traditional chip hybridization platform (microarrays) to RNA-sequencing technology (RNA-Seq) using high-throughput techniques (28). RNA-seq has emerged as an efficient tool to generate functional genomic-level information even for organisms without genomic reference. In the absence of a genome of reference, the transcriptome analysis can be done by *de novo* assembling. This strategy can be complicated for forest trees with high levels of heterozygosity and/or polyploidy (29). In addition, RNA-seq enables to assess and quantify the expression levels of differentially expressed genes (DEGs) and/or to detect previously unidentified genes and splice variants (30). Generally speaking, RNA-seq analysis requires appropriate sample collection, high-quality RNA extraction, amplification, sequencing, and bioinformatics analyses (31, 32).

In forest tree research, RNA-seq technology has been applied to identify DEGs associated with abiotic and biotic stresses (33–38), wood formation and lignification (39–42), seasonal variation (43, 44), among others. Furthermore, RNA-seq has also been used as a strategy to improve the sequencing and annotation of tree species, such as *Pinus* spp. (45, 46), *Ulmus minor* (47), *Populus pruinosa* (48). In addition to the studies on mRNA, the deep coverage achieved with RNA-seq has facilitated the study of other RNA molecules in forest trees, such as microRNAs (miRNAs) and regulatory long intervening noncoding RNAs (lincRNAs) (29).

Transcriptomics and metabolomics data cannot be directly associated. However, their integration represents a powerful combination that allows us to anchor the information obtained by transcriptome measurements with the phenotypic measurements provided by metabolomics (7). To our knowledge, the first successful integrative analysis between transcripts and metabolites in plants was performed almost two decades ago by Urbanczyk-Wochniak et al. (49). In this study, the authors combined metabolic and RNA profiling to distinguish between different potato tuber systems by performing non-parametric Spearman's rank-order correlation analysis. Since then, transcript-metabolite integration has been used to gain insights into many different plant traits, including in forest trees (Table 4.1), such as *Eucalyptus* spp. (50, 56, 58), *Populus* spp. (16, 50, 54, 57, 59), *Pinus* spp. (53, 55), and *Prunus mume* (52). However, most of the omics analyses in forest tree species still focus on single omics platforms, including studies dedicated to biotic stress (60–63), abiotic stresses (64–72), secondary metabolism (73–76), among others.

In the forest tree field, most of the publications reporting transcriptomics-metabolomics integration have performed Pearson's correlation (50–53, 55, 56) (Table 4.1). An interesting approach integrated DEGs and differentially

Table 4.1 Summary of transcript-metabolite integration studies in tree species.

Species	Integration type	Method	Study purpose	Analytical platforms	References
Eucalyptus urophylla	Correlation	Pearson correlation	Defense responses of *Eucalyptus* to *Ralstonia solanacearum*	Illumina Novaseq LC–MS/MS	(50)
Populus alba × *Populus glandulosa*	Correlation	Pearson correlation	Anthocyanin Biosynthesis Pathway	Illumina HiSeq2000 LC–MS/MS	(51)
Prunus mume	Correlation	Pearson correlation	Regulation mechanism of color formation in stem	Illumina HiSeq2500 LC–MS/MS	(52)
Pinus taeda	Correlation Pathway mapping	Pearson correlation and Kyoto Encyclopedia of Genes and Genomes (KEGG)	Terpenoid and flavonoid biosynthesis	Illumina Novaseq6000 LC–MS/MS	(53)
Populus tremula	Qualitative	Kyoto Encyclopedia of Genes and Genomes (KEGG)	Wood forming tissues	AspWood database (Illumina HiSeq2000) LC–MS/MS GC–MS	(54)
Pinus massoniana	Correlation	Pearson correlation	Resin biosynthesis	GC–MS Illumina Hiseq2500	(55)

Eucalyptus grandis	Correlation Multivariate	Pearson correlation Multi-block sPLS–DA	Potassium fertilization and water availability	Illumina Hiseq3000 LC–MS/MS GC–MS	(56)
Populus balsamifera	Pathway mapping	Kyoto Encyclopedia of Genes and Genomes (KEGG)	Response to drought and time of day	Microarray GC–MS	(57)
Eucalyptus grandis	Qualitative	Gene Ontology enrichment and Mapman	Plant defense system against insect pests	Illumina GC–MS	(58)
Populus euphratica	Qualitative	Kyoto Encyclopedia of Genes and Genomes (KEGG)	Salt stress tolerance mechanisms	Microarray FT–ICR–MS	(59)
Populus tremula · *Populus tremuloides*	Multivariate	O2PLS	Short-day induced effects	Microarray GC–MS	(60)

Abbreviations: FT–ICR–MS, Fourier transform ion cyclotron resonance mass spectrometry; sPLS–DA, sparse partial least squares discriminant analysis; O2PLS, two-way orthogonal partial least squares.

accumulated metabolites (DAMs) in different tissues of *Pinus taeda* using Pearson's correlation and pathway mapping (53). By comparing total DEGs and DAMs, authors observed that most of them were enriched in the same KEGG pathways (18), including phenylpropanoid, flavonoid, flavone, flavonol, aminoacyl-tRNA, and amino acids biosynthesis pathways, among others (53). Correlation analysis showed that 249 DEGs were significantly correlated with 45 DAMs. From these, 59 DEGS and 28 DAMs were enriched in flavonoid biosynthesis. Moreover, the authors described for the first time two CYP family genes significantly positively correlated with four flavonoids. Altogether, these results suggest a strong flavonoid metabolic activity in *P. taeda* and that CYP gene family might regulate flavonoid accumulation in this specie (53). In another study, the response of *Eucalyptus urophylla* to the different stages of infestation by *Ralstonia solanacearum* was investigated using transcriptomics (RNA-seq) and untargeted liquid chromatography tandem mass spectrometry (LC–MS/MS) metabolite analysis (50). By performing Pearson's correlation, 30 significant correlated DEGs–DAMs pairs were identified being involved in the phenylpropanoid biosynthesis. Among them, a positive correlation between 12 highly abundant DEGs–DAM pairs was observed, suggesting they may play an important role in disease resistance at the first stage of infection (50).

The integration of metabolomics with transcriptomics data is a powerful approach that can provide an overall picture of metabolism regulation, contributing to the identification of functional genes, key metabolites, metabolic pathways of interest, and in a broad perspective, the global transcriptome information can be anchored to the phenotypic information provided by metabolomics.

4.2.5 Integration of Proteomics and Metabolomics Data

Proteomics is defined as the scientific discipline or methodology focused on the study of the proteome, which represents the total set of proteins, including the proteoforms (the different products of a specific gene due to posttranscriptional and posttranslational modifications), present in a biological unit (subcellular fraction, cell, tissue, organ, individual, ecosystem, or purified extracts), in a certain developmental stage, and under specific environmental conditions (reviewed in Ref.(76)). Despite the relevance of RNA-seq analyses to uncover molecular changes in the plant field, mRNA levels do not provide a complete picture of cellular function, including multiple interactions of proteins and their metabolites in stress responses (51), protein expression levels dependence on translational efficiency and degradation rates (77), and post-translational modifications (PTMs) that cannot be predicted from transcript expression levels or from the genomic sequence (51). Thus, the study of the proteome can reveal PTM, protein–protein interactions, protein accumulation, and stability (78).

The six major steps in proteomics studies include: (i) sample collection; (ii) protein extraction; (iii) enzymatic digestion of proteins into peptides; (iv) separation/fractionation using LC–MS/MS approaches; (v) peptide and protein identification and quantification; and (vi) bioinformatics analyses (31, 79). Protein extraction and sample preparation are critical steps. Due to the complexity and diversity of proteins (molecular weight, charge state, hydrophobicity, dynamic range, modifications, and cellular localization), any single extraction protocol will incorporate biases with respect to specific classes of proteins (51). Although proteomics studies include other techniques (reviewed in 76), this section focuses on mass spectrometry (MS)-based proteomics analyses because it has become the preferred method for proteomics (80).

In the last decades, the proteomics field has moved from two-dimensional polyacrylamide gel electrophoresis (2D-PAGE) analysis to system-wide screening approaches with quantitative steps that take advantage of: (i) label-based approaches (e.g. stable isotope labeling with amino acids in cell culture [SILAC], ^{18}O-stable isotope labeling, isobaric tagging for relative and absolute quantitation [iTRAQ], tandem mass tags [TMT]); (ii) label-free approaches; and (iii) gel-free approaches (reviewed in Ref. (76, 77)). In addition to gel or gel-free and label-based or label-free approaches, MS-based proteomics can also be classified as top-down or bottom-up analysis (i.e. in the former, intact proteins are analyzed, and in the latter, derived peptides are analyzed after digestion), holistic (i.e. the whole protein set of a cell extract) or hypothesis-driven, also named targeted (i.e. focused on specific proteins) (reviewed in Ref. (76)). In terms of protein quantification, relative quantification based on individual peak abundance is used in the comparison of different experimental conditions, while absolute quantification can be done by stable isotope labeling and by using a separate standard curve of the molecule of interest. Label-free absolute quantification of proteins can be performed based on spectral counts, protein abundance index, and absolute protein expression (81).

Multiple analysis tools are required to efficiently classify proteins and metabolites based on their specific functions, chemical structures, and related pathways. Commonly used analysis tools are: KEGG pathway database (18), PMN (19), MetaboAnalyst (21), Lipid Maps (82), Classyfire (83), Metabolomics Workbench (84), agriGO v2.0 (85), ShinyGO (86), UniProt database (87), among others.

The first publication performing proteomics in forest trees comes from maritime pine (88). In this work, two-dimensional polyacrylamide gel electrophoresis (2D–PAGE) was used to compare the protein patterns of 56 haploid megagametophytes from a single maritime pine tree. Since then, the number of proteomics approaches performed for forest tree species has been increasing (reviewed in Refs. (78, 89, 90)), and most of the publications are related to

Table 4.2 Summary of transcript-metabolite integration studies in tree species.

Species	Integration type	Method	Study purpose	References
Castanea sativa	Pathway mapping	Weighted z test	The molecular mechanisms of *Phytophthora cinnamomi* attack on *C. sativa*	(91)
Pinus pinaster	Multivariate	sPLS	Apical bud maturation	(97)
Populus × *canescens*	Qualitative	Kyoto Encyclopedia of Genes and Genomes (KEGG)	*P. canescens responses to colonization by a Paxillus involutus*	(92)
Eucalyptus grandis	Correlation network	Weighted Gene Co-Expression Network Analysis	Early responses to rust infection	(93)
Pinus radiata	Multivariate	sPLS	Ultraviolet stress responses and adaptation mechanisms	(94)
Eucalyptus grandis	Qualitative	Kyoto Encyclopedia of Genes and Genomes (KEGG)	Seasonal variation impacting cambial zone	(95)
Eucalyptus globulus	Multivariate	sPLS	Mechanisms that account for drought recovery	(96)

Abbreviations: sPLS, sparse partial least squares.

tree responses to environmental stresses. However, there are few publications dedicated to proteome-metabolome integrative analyses, focusing on tree responses to biotic stresses (91–93), abiotic stresses (94–96), and bud maturation (97) (Table 4.2). Most of the studies have employed multivariate analyses for data integration. For example, proteomics (LC-Orbitrap–MS) and metabolomics (LC–MS) data were integrated through a sPLS-based network to characterize the apical bud maturation in *Pinus pinaster* (97). The authors found that specific dynamics of proteins and metabolites related to epigenetic regulation, proteome remodeling, hormonal signaling, and abiotic stress response had an essential role during bud maturation (97). Another study using sPLS-based network was performed to investigate stress responses and adaptation mechanisms of *Pinus radiata* to ultraviolet (UV) irradiation by integrating physiological measurements, proteomics (LC–Orbitrap–MS), and gas chromatography mass spectrometry (GC–MS) metabolomics data: sPLS network indicated the presence of two main UV-responsive clusters, the first one links metabolites EPOX (arbamazepine-10,11-epoxide) and 1NAA (1-naphthalenecarboxylic acid) with proteins acting on primary metabolism. The second cluster links glycolate with photorespiratory pathway and light-dependent signaling kinase proteins (94). Another example focused on the investigation of the molecular mechanism associated with the interaction of the fungi-like organism *Phytophthora cinnamomic* with the stem of *Castanea sativa* by integrating proteomics (LC–Orbitrap–MS), metabolomics (GC–MS), and targeted hormone (GC–MS) analyses (91). To pinpoint the pathways underlying *P. cinnamomi* pathogenicity on *C. sativa*, proteins and metabolites significantly altered upon infection were mapped together for functional enrichment and pathway topology analysis using the "Joint Pathway Analysis" tool implemented in MetaboAnalyst 4.0 (98). Phenylpropanoid biosynthesis (phenylalanine, tyrosine, and tryptophan biosynthesis), galactose metabolism, alanine, aspartate, and glutamate metabolism, and TCA cycle represented the pathways with the highest impact in the stem section proximal to the infection site (91).

Together, the proteome and metabolome represent the end products of cellular metabolism, providing a set of quantitative traits that respond to both genetic and environmental factors. The integration of these datasets can reveal critical networks and signaling pathways due to their complementary molecular dependencies that go beyond genetic or transcriptional regulation.

4.2.6 Integration of Multi-omics Data

The coupling of metabolomics with genomics, transcriptomics, and proteomics data provides an integrated portrait of functions, spanning the genome to phenotypic interactions with the environment (99). In addition to studies reporting the integration of two omics datasets (e.g. transcriptomics-metabolomics), the multi-omics integration approach aims at integrating more than three omics datasets. Despite being a new approach, it has already been reported in forest trees, such as *Populus* spp. (100–103) and *Quercus ilex* (104) (Table 4.3).

An interesting example is the work of Wang et al. (101), in which authors performed a multi-omics quantitative integrative analysis to study the lignin biosynthesis regulation during wood formation in *P. trichocarpa*. Using transgenic *P. trichocarpa* with alterations in the 21 genes involved in the lignin synthesis pathway, authors integrated five levels of regulation (genomics, transcriptomics, proteomics, fluxomics, and phenomics) to quantify the effects of gene expression of the monolignol biosynthetic pathway in 25 wood traits related to wood properties and use. Three main components were used to perform the integrative analysis: (i) simple linear regressions

Table 4.3 Summary of multi-omics integration studies in tree species.

Species	Integration type	Method	Study purpose	References
Populus canescens	Clustering	Marvis	Nitrogen responses	(103)
Populus spp.	Multivariate	OnPLS	Wood formation	(102)
Quercus ilex	Pathway mapping	MapMan	Metabolic pathway reconstruction with multi-tissues	(104)
Populus trichocarpa	Differential analysis	Ordinary differential equation/Multiple linear regression	Lignin biosynthesis	(101)
Populus plants	Multivariate	OnPLS	Oxidative stress responses	(100)

Abbreviation: OnPLS, multiple-block orthogonal projections to latent structures.

to represent transcript/protein relations; (ii) mass-balance kinetic equations to calculate metabolite levels and fluxes; and (iii) multiple linear regression equations to predict the phenotypes. With this approach, they were able to predict improvements for the 25 traits individually or in combination through metabolic engineering of the genes studied.

4.3 Integration of Genomics and Metabolomics Data

The integration of genomics with phenotypic data is a very potent way of revealing the genetic bases of phenotypic variations. The various technical possibilities for integrating genomics with metabolic data have been considered by researchers in the plant area for some time (105). Unlike in other plant species, the genetic bases of population metabolic variation have not been extensively studied in trees despite their rich chemical composition (106). In this section, we will focus on the results obtained through: (i) linkage analysis, (ii) genome-wide association studies, and (iii) genomic selection.

4.3.1 Linkage Analysis

Linkage analysis or quantitative trait locus (QTL) mapping is a forward genomics technique where the phenotype reveals the genomic location that affects a trait (107, 108). This methodology requires genetic maps with well-distributed markers throughout the genome and a high number of segregating individuals for both molecular markers and traits of interest (107). It is based on the cross between two parents with contrasting traits, and on the genetic and phenotypic analysis of the segregating progeny (107), although it has also been applied in tree pedigrees of whom neither parent showed contrasting growth characters (108). There are different types of segregation (backcross/testcross, intercross, outcross) that can be used, each of which will vary in the information they provide (109). The phenotypic values of the segregating population are statistically associated with the genotypes to identify the markers that segregate together with the QTL. A statistical association would indicate that there is a linkage between the marker and a gene or group of genes that control the trait of interest (107). Several different statistical analyses can be applied to perform linkage analysis, such as the Log of the ratio of the likelihoods (LOD), multiple regression, marker regression, or maximum-likelihood methods (107, 110).

Although widely used, this technique has limited applications. In general, the mapping resolution is low since there are few opportunities for recombination between families and pedigrees with known ancestry since analyses are performed with F1 or F2 populations (111). Therefore, the genomic region that is associated with a trait is usually large enough to contain many genes, resulting in only a small number of QTLs being identified at the gene level (111, 112). Although the fragments can be sequenced, the fact that they contain many genes implies that further studies are required to identify the gene that is specifically responsible for the trait of interest. On the other hand, the use of biparental populations means that the information on the regulation of a given trait will always be partial since only the alleles that are different between parents will segregate (112). Due to statistical biases, the detected QTLs are usually those with the greatest effect and therefore genes with minor effects are underestimated (107, 110). Finally, many QTLs are environment dependent (107).

Linkage mapping application in tree metabolism is relatively recent. Most of the publications focus on wood chemical composition, targeted metabolite classes and metabolomics strategies (Table 4.4). The work of Casey et al. (113) studies the genetic architecture of the phenylpropanoid metabolism in *Populus* spp. They used *Populus alba*, *Populus tremula* and their hybrids, and were able to detect 41 QTLs for chlorogenic acids, salicinoids, and flavonoids. The flavonoid metabolism presented a great geographical structure and species' specificity. Laoué's (115) work focused on the detection of QTLs associated with phenolic compounds in white spruce (*Picea glauca*), which are involved in the defense of biotic and abiotic stresses as a strategy to find candidate genes that may enable trees to cope with climate change. They found 17 QTLs associated with 8 of the measured metabolites. In particular, one QTL explained up to 91.3% of the variance of neolignan-2. They also combined the measurement of phenolic compounds with RNAseq data, which enriched the obtained results since 50 DEGs were associated with the phenylpropanoid biosynthesis. A third work by Teh et al. (114) followed an untargeted metabolomics strategy in grapevine stems in segregating F2 populations. Stilbenes are phenols derived

Table 4.4 Linkage studies for metabolic traits in trees.

Phenotype	Population	Main findings	References
Metabolomics studies			
Phenylpropanoids measured by UHPLC–MS	*Populus alba*, *Populus tremula*, and their hybrids	41 QTL for chlorogenic acids, salicinoids, and flavonoids were detected	(113)
Untargeted metabolomics by LC–HRMS	Grapevine segregating F2 populations	5 mQTLs related to different patterns of stilbenoid oligomerization were found	(114)
Phenolic compounds by LC–MS/MS	Full sib-siblings of *Picea glauca*	A total of 17 QTLs were detected for eight metabolites, including one major QTL explaining up to 91.3% of the neolignan-2 variance	(115)
Wood chemical studies			
Cellulose, galactan, mannan, xylan, lignin	Three populations from two generations outbred pedigrees of *Pinus taeda*	18 candidate genes related to lignin biosynthesis and cell wall structure were mapped genetically	(116)
Five-carbon hemicellulose sugar (C5), Six-carbon cellulose sugar (C6), Ratio C6/C5, Syringyl lignin monomer, Guaiacyl lignin monomer, Ratio S-lignin/G-lignin, Total lignin, Ratio C6/lignin	Interspecific pseudo back cross pedigree of *Populus*	63 QTLs were identified for 20 traits	(117)
Cellulose	*Eucalyptus nitens* full-sib family	COBL4A is the most strongly associated marker in a QTL region for cellulose content	(118)

S/G ratio, lignin, C5 sugars, C6 sugars	Interspecific *Populus trichocarpa* × *Populus deltoides* pseudo-backcross mapping pedigree and two partially overlapping populations of unrelated *Populus trichocarpa* genotypes	6 QTLs were identified as associated with the measured phenotypes	(119)
Cellulose, total lignin	Biparental population of *Eucalyptus camaldulensis* × *Eucalyptus tereticornis*	All the traits showed low to moderate heritability (0.294-0.439), with cellulose having the highest H2 (0.439)	(120)
Other metabolic studies			
Flavonoids	Two full-sib poplar families: *Populus deltoides* × *Populus nigra*, and *Populus deltoides* × *Populus trichocarpa*	4 mQTL associated with rate-limiting steps in flavonoid biosynthesis were found	(121)
Oil content traits expressed as oil to bunch (O/B) and oil to dry mesocarp (O/DM)	*Elaeis guineensis* cross between Deli *Dura* and Ghana *Pisifera*	One significant and three potential QTL for O/B and O/DM were found	(122)
Natural rubber	Full-sib rubber tree population (created from the cross of two high-production rubber cultivars, PR 255 and PB 217).	11 QTLs associated with latex production were found	(123)

Abbreviations: UHPLC–MS, ultra-high performance liquid chromatography mass spectrometry; LC–HRMS, liquid chromatography high-resolution mass spectrometry; LC–MS/MS, liquid chromatography tandem mass spectrometry; QTL, quantitative trait locus; COBL4A, COBRA-like gene.

from the phenylpropanoid pathway present in a limited number of families, and their synthesis is associated with defense against biotic and abiotic stresses. Following this strategy, the authors were able to detect five metabolic QTLs related to different patterns of stilbenoid oligomerization.

4.3.2 Genome-wide Association Studies

Genome-wide association studies (GWAS) allow us to associate the natural variation of markers with specific phenotypes. Generally single nucleotide polymorphisms (SNPs) spread throughout the genome with specific phenotypes. The technique is called mGWAS when it involves studying the metabolic phenotype. There are multiple applicable models (described in 124, 125), although the most widely used is mixed linear model (MLM) because it is efficient in controlling the effects of population stratification (125).

The main advantage of GWAS is that they can be carried out in natural populations with unknown ancestry since they rely on historical recombination events (111). GWAS are also applied to artificial populations, such as nested association mapping (NAM) and multiparent advanced generation inter-cross (MAGIC) populations (124, 125), whose use solve the problems associated with population structure and the presence of rare variants in natural populations that reduce the statistical power of the analyses (126). Furthermore, trees have undergone little domestication and, consequently, usually have a low-extent linkage disequilibrium, which indicates that markers are close to the variants that cause associations, thereby facilitating the discovery of the exact gene underlying the trait of interest (124).

The main disadvantages of GWAS are that they require a high number of individuals, an ability to accurately genotype genomes (which is challenging in trees with large and repetitive genomes) and that they have difficulties in detecting the contribution of rare variants because of their limiting statistical power (although this depends on the model and/or species used) (125, 126).

A summary of the publications where the genetic bases of tree metabolic traits have been studied by GWAS is found in Table 4.5. Most of the analyses focus on wood chemical studies using specific techniques to measure cellulose, hemicelluloses, and lignin (Table 4.5). Some other works have been carried out performing limited metabolite measurements of natural rubber (136), hormones (139), and chlorogenic acid (140), among other metabolic variables (Table 4.5). There are only a few works where GWAS were applied to GC–MS metabolomics data of forest trees (Table 4.5). Eckert et al. (127) evaluated loblolly pine trees (*Pinus taeda*) from different areas of the United States cultivated in a common garden with the aim of knowing the genetic bases of the metabolome of this species. The authors concluded that the concentration of many primary metabolites is heritable and that multi-SNP models (where each metabolite is evaluated against a set of SNPs) explained more than 50% of that heritability. As a follow-up of that work, Lu et al. (128) re-evaluated

Table 4.5 Summary of genome-wide association studies (GWAS) of metabolic traits in tree species.

Phenotype	Species	Analysis method	References
Metabolomics studies			
GC–MS metabolite profiling	*Pinus taeda*	MLM, GLM	(127,128)
GC–MS metabolite profiling	*Populus trichocarpa*	EMMA	(129)
GC–MS metabolite profiling	*Populus trichocarpa*	MLM	(130)
Wood chemical studies			
Lignin, cellulose	*Populus trichocarpa*	GLM	(131)
Total lignin, Klason lignin, extractives	*Eucaliptus globulus*	Unified mixed-model	(132)
5C sugars, 6C sugars, lignin	*Populus deltoides*	Single-variant and multiple-variant associations on GLM	(133)
Holocellulose, α-cellulose, hemicelluloses, lignin	*Populus tomentosa*	MLM, Kempthorne model	(134)
Holocellulose, α-cellulose, hemicelluloses, lignin	*Populus tomentosa*	MLM	(135)
Other metabolic studies			
Natural rubber	*Hevea brasiliensis*	GLM, MLM	(136)
Lipids measured by GC with standards	*Populus tremula*	GEMMA	(137)
Fatty acid content in oil palm	*Elaeis guineensis*	MLM	(138)
Phytohormone-related metabolites and their secondary metabolites	*Populus tomentosa*	MLM	(139)
Chlorogenic acid	*Populus tomentosa*	MLM	(140)
Oleoresin	*Pinus elliottii*	MLM	(141)

Abbreviations: GLM, simple general linear model; MLM, mixed linear model; EMMA, efficient mixed-model association; GEMMA, genome-wide efficient mixed-model association.

some of the genotypes through GWAS, studying both metabolites and gene expression levels. Guerra et al. (130) conducted a metabolomics study of poplar wood (*P. trichocarpa*) and selected five metabolites with the highest estimates of heritability for GWAS; despite this, no significant associations between metabolites and markers were found. Zhang et al. (129) evaluated the metabolome of a *P. trichocarpa* population through GWAS and found out that the chlorogenic acid content (part of the phenylpropanoid pathway, where the precursors of lignin synthesis are formed) was associated with the enzyme hydroxycinnamoyl-CoA:shikimate hydroxycinnamoyl transferase 2 (PtHCT2).

4.3.3 Genomic Selection

Genomic selection (GS) is a relatively new technique that consists of evaluating the set of all the effects of QTLs using molecular markers by estimating the genomic estimated breeding value (GEBV) of each individual separately (142–145). Its objective is to predict a general breeding value instead of identifying genes associated with a trait of interest as in the techniques described above. It is based on the use of high-coverage genome-wide SNPs such that each marker is in linkage disequilibrium with a locus of a trait. One of the main problems of forest tree improvement is that trees have a very long juvenile phase and, therefore, the evaluation of the phenotypic traits of interest (e.g. wood quality) is performed on young trees rather than on adults, while sometimes there is no correlation between the same trait at different stages (144). GS, for its part, has the advantage of substantially reducing breeding times compared to traditional techniques; according to Resende et al. (146), the breeding time for *Eucalyptus* spp. through the use of GS can be reduced from 18 to 9 years.

In GS studies, the study population is divided into a "training population" and a "test population," which are genotyped and phenotyped. The data obtained from the training population are used to generate a prediction model, which allows the phenotype to be predicted from the genotype. The predictive model is validated in the test population, which is related to the training population but is not used in the estimation of the effects of the markers. A model with good accuracy has a high correlation between the observed and predicted breeding values in the test population. Next, models are used to predict the breeding values of individuals, for which phenotypic data are not available, based on their genotype. Finally, trees with high GEBV values will be selected to be used in breeding. For a detailed description of the methodology, see (142–145).

There are multiple prediction models that can be applied in genomic selection (described in 142, 144). These can be parametric (e.g. ridge regression-best linear unbiased predictor [RR-BLUP], Bayes A, Bayes B), or non-parametric. The choice of the different models will depend on the genetic architecture of the individuals. Parametric models evolved into other types of models, such as genomic BLUP (GBLUP), single-step GBLUP (ssGBLUP), Bayes C, and Bayes LASSO. Non-parametric methods (e.g. random forest and neural networks) have the advantage of being able to evaluate non-additive genetic effects (e.g. epistasis). Nevertheless, they have been scarcely applied to trees so far.

Most tree genomic selection studies have been applied to a very limited range of species and traits (143–145, 147), being *Eucalyptus* spp. the most studied species. In general, GS studies on trees focus on studying the precision of these analyses and on the comparison between different models and/or with other methods, while their application in practice is still scarce. A summary of the published works focusing on tree metabolism can be found in Table 4.6.

Table 4.6 Genomic selection studies performed for metabolic traits of trees.

Phenotype	Species	Models	Main findings	References
Metabolism				
Acetophenone aglycones (AAs) (e.g. picein, piceol, and pungenol) in needles	*Picea glauca*	GBLUP, ABLUP, Bayes $C\pi$	The GBLUP accuracy to predict the AA content was high and comparable with the results obtained for ABLUP and Bayes $C\pi$ models.	(148)
Wood Chemical traits				
Cellulose, lignin	*Pinus taeda*	BLUP	The accuracy of estimated genomic breeding values was 0.61–0.83 for wood lignin and cellulose, being the accuracies based on markers similar to those based on pedigree studies.	(149)
S/G lignin ratio, lignin, pentosan	Families derived from crosses of *Eucalyptus grandis,* or *E. grandis* × *E. camaldulensis*	RR-BLUP, Bayes B	The predictive abilities for this population were similar for both of the used models, with the highest predictive ability for pentosan content.	(150)
Holocellulose, total lignin	*Eucaliptus robusta*	RR-BLUP, BayesB	The low correlation between variables prevented an improvement in the accuracy of a multi-trait GS compared to the single-trait GS.	(151)

(Continued)

Table 4.6 (Continued)

Phenotype	Species	Models	Main findings	References
Cellulose, S/G lignin ratio, total lignin	Hybrids derived from full-sib families comprised of *Eucalyptus grandis*, *Eucalyptus Urophylla*, and *Eucalyptus camaldulensis*	ssGBLUP, GBLUP, ABLUP	ssGBLUP had higher predictive ability (37–75%) and lower predictive bias (21–73%) than the other models.	(152)
Holocellulose, lignin	*Eucaliptus urophylla* × *Eucaliptus grandis*	BLUP	The additive correlations between the stem volume and holocellulose and lignin were weak but positive.	(153)
Cellulose, S/G lignin ratio	*Eucaliptus grandis*	ABLUP, GBLUP	In general, the genetic gain of GS and traditional breeding was higher for GS for all the traits evaluated in a period of 17 years.	(154)

Abbreviations: S/G lignin ratio, syringyl/guaiacyl ratio; BLUP, best linear unbiased prediction; GBLUP, genomic BLUP; ABLUP, additive BLUP; RR-BLUP, ridge regression BLUP; ssGBLUP, single-step GBLUP; GS, genomic selection.

In general, most of the works related to metabolism evaluate the wood chemical characteristics like cellulose, lignin, and hemicelluloses. The only work that studies other types of metabolites is that of Beaulieu et al. (148) (Table 4.6). The authors characterized acetophenone aglycones (AAs) (e.g. picein, piceol, and pungenol) in needles, associated with budworm resistance in white spruce (*Picea glauca*). The AAs showed moderate to high genetic control and little correlation with other measured traits, while the used model (GBLUP) was considered suitable to perform early selection and to predict breeding values in the evaluated population. Although there are multiple challenges for GS application in forest species (143), it is expected that in the near future, GS applications in forestry will increase, given its capacity to accelerate breeding time.

The integration of genomics with metabolomics data aiming at understanding and improving the metabolism of forest trees has not been extensively applied yet. However, a few applications have proven to be successful (Tables 4.4–4.6), which suggests that these strategies could be further applied in the future, especially considering the role of metabolism in stress pathways.

4.4 Systems Biology to Provide Clues for Metabolite Annotation in Different Tree Species in Recent Years

Due to metabolism's closeness to cellular and tissue phenotypes, they are often used for integration with multi-omics data. However, major concerns rely on the lack of plant metabolite databases (155) which in many cases contain hundreds of unknown metabolites, the so-called "dark metabolome" (156). One of the reasons is the secondary (specialized) biosynthetic pathways, which vary among genera, and show plasticity under different environments (157); for example, salicinoid glucosides in *Populus* (54), pentacyclic triterpenoids in *Quercus* (158), formylated phloroglucinols in *Eucalyptus* (159), and stilbenoids in *Pinus* (160).

Specialized metabolites are often analyzed through metabolomics approaches, typically with liquid chromatography coupled with high-resolution mass spectrometry (LC–HRMS)- or nuclear magnetic resonance (NMR)-based approaches. Due to its sensitivity and high coverage of metabolites (different polarities in a single analysis), MS is currently the most used analytical method for a general overview of metabolism (161), while NMR has been primarily used to characterize the structure of phenolic oligomers present in the wood of several forest tree species, such as *Pinus* (162), *Populus* (163), *Quercus* (164), and *Eucalyptus* (165).

The high-resolution TOF, QTOF, and Orbitrap mass spectrometers provide high-accuracy mass-to-charge ratios (*m/z*) to four decimal places, being very close and comparable to the theoretical monoisotopic masses (<1 ppm) of known metabolites. The ion mass fragmentation (MS/MS) via QTOF or

Orbitrap MS analysis results in highly accurate mass fragments and provides reliable annotation of unknown metabolites. Therefore, MS and MS/MS platforms have been developed to facilitate metabolite annotation based on different strategies (149), including the comparison of experimental masses with the monoisotopic mass provided by the KEGG pathway database (18), the PMN (19, 166), and the Plant Reactome database (167), the comparison of experimental MS/MS spectra with those available at MassBank (168) and Metlin (169), employment of MS/MS spectral fragment networks for grouping metabolites according to fragmentation patterns and possible similar biosynthetic origin (Global Natural Products Social Molecular Networking, GNPS) (170), *in silico* MS/MS spectral database (The Network Annotation Propagation, NAP) (171), MetWork (172), and SIRIUS (173).

The integration of metabolomics data in multi-omics studies is essential for functional genomics, which should allow high-throughput functional annotation of the many unknown genes (54) that have been obtained via genome sequencing or also to gain insight into the annotation of unknown metabolites (157). Forest omics platforms are currently available for the search of gene expression in different tissues as well as gene interaction with the environment for different commercial species, such as *Populus, Quercus, Pinus, Eucalyptus,* and Norway spruce. For example, the Plant Genome Integrative Explorer platform (PlantGenIE.org) (174, 175) is a multi-species database that includes genome and transcriptome online tools of model and woody plant species, including Norway spruce (ConGenIE) (174,176), *Populus* and *Salix* (PopGenIE) (174, 177), and *Eucalyptus* (EucGenIE) (174, 178). NorWood (39) is a tool for generating evo-devo insights into the divergence of wood formation between angiosperms and gymnosperms, and for advancing the understanding of the regulation of wood development in *Picea abies.* AspWood (40) provides a high spatial resolution transcriptome profile based on RNA-seq data from the secondary phloem, vascular cambium, and wood-forming tissues of *P. tremula*; it is also an interactive tool for exploring the co-expression network. TreeGenes (179) is a forest tree genome database that offers a curated and integrated web-based relational database, mainly related to the *Pinaceae* family. CorkOakDB (180) is an integrative platform containing cork oak (*Quercus suber*) gene annotation and mRNA expression data from five different tissues (leaf, xylem, inner bark, phellem, and pollen). OakContigDF159.1 (181) is a reference library of differential gene expression of *Quercus robur* under root-ectomycorrhizal symbiosis. The Plant Comparative Genomics portal (Phytozome) (182) allows users to access, visualize, and analyze sequenced plant genomes. All genes present in the Phytozome platform have been annotated with eukaryotic orthologous group analysis (KOG) (183), the KEGG pathway database (184), Protein Families database (Pfam) (185), and Panther classification (184). Moreover, it constitutes an excellent platform for *Eucalyptus, Quercus, Castanea,* and *Populus* species.

4.5 Challenges in Integrating Metabolomics and Other Omics

Multiple challenges may occur when dealing with omics integration ranging from the quality and validity of the experimental design to the nature and the quality of the data acquired from the analytical platforms (4).

Pinu et al. (4) highlighted six requirements for high-quality multi-omics studies: (i) proper experimental design; (ii) appropriate selection, preparation, and storage of biological samples; (iii) careful collection of quantitative multi-omics data and associated meta-data; (iv) tools for integration and interpretation of the data; (v) agreed minimum standards for multi-omics method and meta-data management; and (vi) generation of new resources for the deposition of intact multi-omics datasets.

Jendoubi (6) listed four major points that require attention in omics integration, as follows: (i) omics molecules are not fully characterized; metabolite profiling analysis relies on a complex chemical heterogeneity where metabolites are typically not easily identified, mainly in the case of specialized metabolites. In addition, unknown metabolic features may also represent chemicals that have not been reported in the literature, in both cases, data integration would be rather limited to almost purely statistical analysis. Similarly, there are many genomic, transcriptomic, and proteomic variables not fully characterized (6); (ii) each omics dataset has its own limitations and requires complex analysis pipelines prior to performing data integration (e.g. experimental and inherent biological noise, reproducibility, differences among analytical platforms and detection bias) (186); (iii) intrinsically, differences might be present in individual omics datasets due to platform-systematic bias. Thus, some observed patterns, for example, in metabolomic profiles, might not be the result of perturbations in the biosystem or phenotype of interest (6); instead, they can be attributed to platform-specific systematic variation. To alleviate these individual variations, multivariate methods (e.g. two-way orthogonal partial least squares [O2PLS]) that allow the separation of systematic variation overlapping across analytical platforms from specific systematic variation can be used to integrate two omics datasets; and (iv) metabolomics datasets are characterized by high correlation structures, meaning that, for example, two highly correlated metabolites might not be directly related. Instead, the high correlation can be the result of complex interactions with other metabolites in common (6).

Besides the four key points listed by Jendoubi (6), two additional facts should be considered. First, that individual qualitative and quantitative variations in omics datasets can be compensated by increasing the number of biological replicates, which is mandatory for statistical analyses and correct biological interpretation of the results (although taking into account the run stability limits) (78). And second, that multi-omics as well as individual omics analyses are still expensive to obtain (4) in many countries.

To help with multi-omics integration, there are many available open-access tools and databases that provide assistance with data quality checks, normalization, transformation, scaling, interpretation, statistics, and visualization, including methods to generate and view interactive correlation or association maps and metabolic and signaling pathways (reviewed in ref. (1, 4, 8, 187, 188)). Nevertheless, the use of some open-access tools and databases might not be an easy task for researchers with no bioinformatics/programming skills.

4.6 Conclusion and Future Perspectives

Systems biology and multi-omics analyses have contributed to a better understanding of forest tree species biology than individual approaches. The combination of multi-omics ultimately allows to elucidate the genotype–phenotype linkage and strongly influences systems biology approaches. Regardless of the advances in high-throughput sequencing, the integration of multiple omics platforms remains an ongoing challenge. In this context, a combination of multidisciplinary efforts is necessary to improve the analysis and interpretation of omics-integration data: from experimental design and sequencing platforms to integrative mathematical tool. Important take-home messages from this chapter are to carefully: (i) evaluate the quality and validity of your multi-omics data; (ii) take into account possible bias; (iii) consider the differences inherent to each omics level; and (iv) assess which integrative approach would be best.

References

1 Pazhamala LT, Kudapa H, Weckwerth W, Millar AH, Varshney RK. Systems biology for crop improvement. *Plant Genome.* 2021; 14(2): e20098.
2 Hasin Y, Seldin M, Lusis A. Multi-omics approaches to disease. *Genome Biol.* 2017; 18: 1.
3 Chen R, Mias GI, Li-Pook-Than J, Jiang L, Lam HYK, Chen R, et al. Personal omics profiling reveals dynamic molecular and medical phenotypes. *Cell.* 2012; 148(6): 1293–1307.
4 Pinu FR, Beale DJ, Paten AM, Kouremenos K, Swarup S, Schirra HJ, et al. Systems biology and multi-omics integration: viewpoints from the metabolomics research community. *Metabolites.* 2019; 9(4): 76.
5 Wanichthanarak K, Fahrmann JF, Grapov D. Genomic, proteomic, and metabolomic data integration strategies. *Biomark Insights.* 2015; 10(Suppl 4): 1–6.
6 Jendoubi T. Approaches to integrating metabolomics and multi-omics data: a primer. *Metabolites.* 2021; 11(3): 184.

7 Cavill R, Jennen D, Kleinjans J, Briedé JJ. Transcriptomic and metabolomic data integration. *Briefings Bioinf*. 2016; 17(5): 891–901.

8 Jamil IN, Remali J, Azizan KA, Nor Muhammad NA, Arita M, Goh HH, et al. Systematic multi-omics integration (MOI) approach in plant systems biology. *Front Plant Sci*. 2020; 11: 944.

9 Rai A, Saito K, Yamazaki M. Integrated omics analysis of specialized metabolism in medicinal plants. *Plant J*. 2017; 90(4): 764–787.

10 Jain A, Murty M, Flynn P. Data clustering: a review. *ACM Comput Surv*. 1999; 31(3): 264–323.

11 Zhang X, Zhou Z, Xu H, Liu C-T. Integrative clustering methods for multi-omics data. *Wiley Interdiscip Rev Comput Stat*. 2022; 14(3): e1553.

12 Rappoport N, Shamir R. Multi-omic and multi-view clustering algorithms: review and cancer benchmark. *Nucleic Acids Res*. 2018; 46(20): 10546–10562.

13 Teran Hidalgo SJ, Ma S. Clustering multilayer omics data using MuNCut. *BMC Genomics*. 2018; 19: 1–13.

14 Chalise P, Koestler DC, Bimali M, Yu Q, Fridley BL. Integrative clustering methods for high-dimensional molecular data. *Transl Cancer Res*. 2014; 3(3): 202–216.

15 Meng C, Zeleznik OA, Thallinger GG, Kuster B, Gholami AM, Culhane AC. Dimension reduction techniques for the integrative analysis of multi-omics data. *Briefings Bioinf*. 2016; 17(4): 628–641.

16 Bylesjö M, Rantalainen M, Cloarec O, Nicholson JK, Holmes E, Trygg J. OPLS discriminant analysis: combining the strengths of PLS-DA and SIMCA classification. *J Chemom*. 2007; 20(8–10): 341–351.

17 Rohart F, Gautier B, Singh A, Lê Cao KA. mixOmics: an R package for 'omics feature selection and multiple data integration. *PLoS Comput Biol*. 2017; 13(11): 1–19.

18 Kanehisa M, Goto S. KEGG: kyoto encyclopedia of genes and genomes. *Nucleic Acids Res*. 2000; 28(1): 27–30.

19 Hawkins C, Ginzburg D, Zhao K, Dwyer W, Xue B, Xu A, et al. Plant Metabolic Network 15: a resource of genome-wide metabolism databases for 126 plants and algae. *J Integr Plant Biol*. 2021; 63(11): 1888–1905.

20 Usadel B, Nagel A, Steinhauser D, Gibon Y, Bläsing OE, Redestig H, et al. PageMan: An interactive ontology tool to generate, display, and annotate overview graphs for profiling experiments. *BMC Bioinf*. 2006; 7: 1–8.

21 Xia J, Psychogios N, Young N, Wishart DS. MetaboAnalyst: a web server for metabolomic data analysis and interpretation. *Nucleic Acids Res*. 2009; 37(2): 652–660.

22 Ashburner M, Ball CA, Blake JA, Botstein D, Butler H, Cherry JM, et al. Gene ontology: tool for the unification of biology. The Gene Ontology Consortium. *Nat Genet*. 2000; 25(1): 25–29.

23 Rao X, Dixon RA. Co-expression networks for plant biology: why and how. *Acta Biochim Biophys Sin (Shanghai)*. 2019; 51(10): 981–988.

24 Zhang B, Horvath S. A general framework for weighted gene co-expression network analysis. *Stat Appl Genet Mol Biol.* 2005; 4(1).

25 Hawe JS, Theis FJ, Heinig M. Inferring interaction networks from multi-omics data. *Front Genet.* 2019; 10: 535.

26 Fukushima A, Kanaya S, Nishida K. Integrated network analysis and effective tools in plant systems biology. *Front Plant Sci.* 2014; 5: 598.

27 Seaver SMD, Lerma-Ortiz C, Conrad N, Mikaili A, Sreedasyam A, Hanson AD, et al. PlantSEED enables automated annotation and reconstruction of plant primary metabolism with improved compartmentalization and comparative consistency. *Plant J.* 2018; 95(6): 1102–1113.

28 Mironova VV, Weinholdt C, Grosse I. RNA-seq data analysis for studying abiotic stress in horticultural plants. In: Kanayama Y, & Kochetov A, editors. *Abiotic Stress Biology in Horticultural Plants.* Tokyo (JP): Springer; 2015. p. 197–220.

29 López de Heredia U, Vázquez-Poletti JL. RNA-seq analysis in forest tree species: bioinformatic problems and solutions. *Tree Genet Genomes.* 2016; 12(2).

30 Wang Z, Gerstein M, Snyder M. RNA-seq: a revolutionary tool for transcriptomics. *Nat Rev Genet.* 2009; 10(1): 57–63.

31 Misra BB, Langefeld CD, Olivier M, Cox LA. Integrated omics: tools, advances, and future approaches. *J Mol Endocrinol.* 2018: R21–R45.

32 Hess JF, Kohl TA, Kotrová M, Rönsch K, Paprotka T, Mohr V, et al. Library preparation for next generation sequencing: a review of automation strategies. *Biotechnol Adv.* 2020; 41: 107537.

33 Chen M, Yin Y, Zhang L, Yang X, Fu T, Huo X, et al. Metabolomics and transcriptomics integration of early response of *Populus tomentosa* to reduced nitrogen availability. *Front Plant Sci.* 2021; 12: 769748.

34 Tao R, Liu Y, Jing W. Response and regulatory network analysis of roots and stems to abiotic stress in *Populus trichocarpa*. *Forests.* 2022; 13(8): 1300.

35 Ulloa JL, Aguayo P, Conejera D, Rubilar R, Balocchi C, Valenzuela S. Transcriptomic response in foliar and root tissues of a drought-tolerant *Eucalyptus globulus* genotype under drought stress. *Trees (Berl West).* 2022; 36(2): 697–709.

36 Gao C, Ren L, Wang M, Wang Z, Fu N, Wang H, et al. Full-length transcriptome sequencing-based analysis of *Pinus sylvestris* var. mongolica in Response to *Sirex noctilio Venom. Insects.* 2022; 13(4): 338.

37 Zhan H, Liu H, Wang T, Liu L, Ai W, Lu X. Selection and validation of reference genes for quantitative real-time PCR of *Quercus mongolica* Fisch. ex Ledeb under abiotic stresses. *PLoS One.* 2022; 17(4): e0267126.

38 Zhu C, Wang W, Chen Y, Zhao Y, Zhang S, Shi F, et al. Transcriptomics and antioxidant analysis of two Chinese chestnut (*Castanea mollissima* BL.) varieties provides new insights into the mechanisms of resistance to gall wasp *Dryocosmus kuriphilus* infestation. *Front Plant Sci.* 2022; 13: 874434.

39 Jokipii-Lukkari S, Sundell D, Nilsson O, Hvidsten TR, Street NR, Tuominen H. NorWood: a gene expression resource for evo-devo studies of conifer wood development. *New Phytol.* 2017; 216(2): 482–494.

40 Sundell D, Street NR, Kumar M, Mellerowicz EJ, Kucukoglu M, Johnsson C, et al. AspWood: high-spatial-resolution transcriptome profiles reveal uncharacterized modularity of wood formation in *Populus tremula*. *Plant Cell.* 2017; 29(7): 1585–1604.

41 Zhang L, Xie S, Yang C, Cao D, Fan S, Zhang X. Comparative transcriptome analysis reveals candidate genes and pathways for potential branch growth in elm (*Ulmus pumila*) cultivars. *Biology (Basel).* 2022; 11(5): 711.

42 Yu J, Zhou C, Li D, Li S, Jimmy Lin YC, Wang JP, et al. A PtrLBD39-mediated transcriptional network regulates tension wood formation in *Populus trichocarpa*. *Plant Commun.* 2021; 3(1): 100250.

43 Bag P, Lihavainen J, Delhomme N, Riquelme T, Robinson KM, Jansson S. An atlas of the Norway spruce needle seasonal transcriptome. *Plant J.* 2021; 108(6): 1815–1829.

44 Fernández-Piñán S, Boher P, Soler M, Figueras M, Serra O. Transcriptomic analysis of cork during seasonal growth highlights regulatory and developmental processes from phellogen to phellem formation. *Sci Rep.* 2021; 11(1): 12053.

45 Wegrzyn JL, Liechty JD, Stevens KA, Wu LS, Loopstra CA, Vasquez-Gross HA, et al. Unique features of the loblolly pine (*Pinus taeda* L.) megagenome revealed through sequence annotation. *Genetics.* 2014; 196(3): 891–909.

46 Visser EA, Wegrzyn JL, Steenkmap ET, Myburg AA, Naidoo S. Combined de novo and genome guided assembly and annotation of the *Pinus patula* juvenile shoot transcriptome. *BMC Genomics.* 2015; 16(1): 1057.

47 Perdiguero P, Venturas M, Cervera MT, Gil L, Collada C. Massive sequencing of *Ulmus minor*'s transcriptome provides new molecular tools for a genus under the constant threat of Dutch elm disease. *Front Plant Sci.* 2015; 6: 541.

48 Yang W, Wang K, Zhang J, Ma J, Liu J, Ma T. The draft genome sequence of a desert tree *Populus pruinosa*. *Gigascience.* 2017; 6(9): 1–7.

49 Urbanczyk-Wochniak E, Luedemann A, Kopka J, Selbig J, Roessner-Tunali U, Willmitzer L, et al. Parallel analysis of transcript and metabolic profiles: a new approach in systems biology. *EMBO Rep.* 2003; 4(10): 989–993.

50 Xiaohui Y, Jie H, Huixiao Y, Huanqin L, Fang X, Baozhu Z, et al. Transcriptome and metabolome profiling in different stages of infestation of *Eucalyptus urophylla* clones by *Ralstonia solanacearum*. *Mol Genet Genomics.* 2022; 297(4): 1081–1100.

51 Yan S, Bhawal R, Yin Z, Thannhauser TW, Zhang S. Recent advances in proteomics and metabolomics in plants. *Mol Horticulture.* 2022; 2(1).

52 Qiu L, Zheng T, Liu W, Zhuo X, Li P, Wang J, et al. Integration of transcriptome and metabolome reveals the formation mechanism of red stem in *Prunus mume. Front Plant Sci.* 2022; 13: 884883.

53 Mao J, Huang L, Chen M, Zeng W, Feng Z, Huang S, et al. Integrated analysis of the transcriptome and metabolome reveals genes involved in terpenoid and flavonoid biosynthesis in the loblolly pine (*Pinus taeda* L.). Front. *Plant Sci.* 2021; 12: 729161.

54 Abreu IN, Johansson AI, Sokołowska K, Niittylä T, Sundberg B, Hvidsten TR, et al. A metabolite roadmap of the wood-forming tissue in *Populus tremula. New Phytol.* 2020; 228(5): 1559–1572.

55 Bai Q, He B, Cai Y, Lian H, Zhang Q. Transcriptomic and metabolomic analyses reveal several critical metabolic pathways and candidate genes involved in resin biosynthesis in *Pinus massoniana. Mol Genet Genomics.* 2020; 295(2): 327–341.

56 Ployet R, Veneziano Labate MT, Regiani Cataldi T, Christina M, Morel M, San Clemente H, et al. A systems biology view of wood formation in *Eucalyptus grandis* trees submitted to different potassium and water regimes. *New Phytol.* 2019; 223(2): 766–782.

57 Hamanishi ET, Barchet GLH, Dauwe R, Mansfield SD, Campbell MM. Poplar trees reconfigure the transcriptome and metabolome in response to drought in a genotype- and time-of-day-dependent manner. *BMC Genomics.* 2015; 16(1): 329.

58 Oates CN, Külheim C, Myburg AA, Slippers B, Naidoo S. The transcriptome and terpene profile of *Eucalyptus grandis* reveals mechanisms of defense against the insect pest, *Leptocybe invasa. Plant Cell Physiol.* 2015; 56(7): 1418–1428.

59 Janz D, Behnke K, Schnitzler J-P, Kanawati B, Schmitt-Kopplin P, Polle A. Pathway analysis of the transcriptome and metabolome of salt sensitive and tolerant poplar species reveals evolutionary adaption of stress tolerance mechanisms. *BMC Plant Biol.* 2010; 10(1): 150.

60 Lovat CA, Donnelly DJ. Mechanisms and metabolomics of the host-pathogen interactions between chestnut (*Castanea* species) and chestnut blight (*Cryphonectria parasitica*). *Forest Pathology.* 2019; 49: e12562.

61 Mangeot-Peter L, Tschaplinski TJ, Engle NL, Veneault-Fourrey C, Martin F, Deveau A. Impacts of soil microbiome variations on root colonization by fungi and bacteria and on the metabolome of *Populus tremula* × *alba. Phytobiomes J.* 2020; 4(2): 142–155.

62 Dove NC, Carrell AA, Engle NL, Klingeman DM, Rodriguez M, Wahl T, et al. Relationships between *Sphaerulina musiva* infection and the *Populus* microbiome and metabolome. *mSystems.* 2022; 7(4): e0012022.

63 Escandón M, Meijón M, Valledor L, Pascual J, Pinto G, Cañal MJ. Metabolome integrated analysis of high-temperature response in *Pinus radiata*. *Front Plant Sci*. 2018; 9: 485.

64 Correia B, Hancock RD, Amaral J, Gomez-Cadenas A, Valledor L, Pinto G. Combined drought and heat activates protective responses in *Eucalyptus globulus* that are not activated when subjected to drought or heat stress alone. *Front Plant Sci*. 2018; 9: 819.

65 Mokochinski JB, Mazzafera P, Sawaya ACHF, Mumm R, de Vos RCH, Hall RD. Metabolic responses of *Eucalyptus* species to different temperature regimes. *J Integr Plant Biol*. 2018; 60(5): 397–411.

66 Tschaplinski TJ, Plett JM, Engle NL, Deveau A, Cushman KC, Martin MZ, et al. *Populus trichocarpa* and *Populus deltoides* exhibit different metabolomic responses to colonization by the symbiotic fungus *Laccaria bicolor*. *Mol Plant-Microbe Interact*. 2014; 27(6): 546–556.

67 Almeida T, Pinto G, Correia B, Gonçalves S, Meijón M, Escandón M. In-depth analysis of the *Quercus suber* metabolome under drought stress and recovery reveals potential key metabolic players. *Plant Sci*. 2020; 299(110606): 110606.

68 de Simón BF, Sanz M, Cervera MT, Pinto E, Aranda I, Cadahía E. Leaf metabolic response to water deficit in *Pinus pinaster* Ait. relies upon ontogeny and genotype. *Environ Exp Bot*. 2017; 140: 41–55.

69 Xiao J, Gu C, He S, Zhu D, Huang Y, Zhou Q. Widely targeted metabolomics analysis reveals new biomarkers and mechanistic insights on chestnut (*Castanea mollissima* Bl.) calcification process. *Food Res Int*. 2021; 141(110128): 110128.

70 Dewhirst RA, Handakumbura P, Clendinen CS, Arm E, Tate K, Wang W, et al. High temperature acclimation of leaf gas exchange, photochemistry, and metabolomic profiles in *Populus trichocarpa*. *ACS Earth Space Chem*. 2021; 5(8): 1813–1828.

71 Saunier A, Greff S, Blande JD, Lecareux C, Baldy V, Fernandez C, Ormeño E. Amplified drought and seasonal cycle modulate *Quercus pubescens* leaf metabolome. *Metabolites*. 2022; 12(4): 307.

72 Topal M. Secondary metabolites of ethanol extracts of *Pinus sylvestris* cones from eastern Anatolia and their antioxidant, cholinesterase and α-glucosidase activities. *Rec Nat Prod*. 2019; 14(2): 129–138.

73 Movahedi A, Almasi Zadeh Yaghuti A, Wei H, Rutland P, Sun W, Mousavi M, et al. Plant secondary metabolites with an overview of *Populus*. *Int J Mol Sci*. 2021; 22(13): 6890.

74 Li Z, Rubert-Nason KF, Jamieson MA, et al. Root secondary metabolites in *Populus tremuloides*: effects of simulated climate warming, defoliation, and genotype. *J Chem Ecol*. 2021; 47: 313–321.

75 Tienda-Parrilla M, López-Hidalgo C, Guerrero-Sanchez VM, Infantes-González Á, Valderrama-Fernández R, Castillejo M-Á, et al.

Untargeted MS-based metabolomics analysis of the responses to drought stress in *Quercus ilex* L. leaf seedlings and the identification of putative compounds related to tolerance. *Forests*. 2022; 13(4): 551.
76 Jorrin Novo JV. Proteomics and plant biology: contributions to date and a look towards the next decade. *Expert Rev Proteomics*. 2021; 18(2): 93–103.
77 Liu Y, Beyer A, Aebersold R. On the dependency of cellular protein levels on mRNA abundance. *Cell*. 2016; 165(3): 535–550.
78 Rodrigues AM, Ribeiro-Barros AI, António C. Experimental design and sample preparation in forest tree metabolomics. *Metabolites*. 2019; 9(12): 285.
79 Dihazi H, Asif AR, Beißbarth T, Bohrer R, Feussner K, Feussner I, et al. Integrative omics – from data to biology. *Expert Rev Proteomics*. 2018; 15(6): 463–466.
80 Balotf S, Wilson R, Tegg RS, Nichols DS, Wilson CR. Shotgun proteomics as a powerful tool for the study of the proteomes of plants, their pathogens, and plant-pathogen interactions. *Proteomes*. 2022; 10(1): 5.
81 Kushalappa AC, Gunnaiah R. Metabolo-proteomics to discover plant biotic stress resistance genes. *Trends Plant Sci*. 2013; 18(9): 522–531.
82 Fahy E, Sud M, Cotter D, Subramaniam S. LIPID MAPS online tools for lipid research. *Nucleic Acids Res*. 2007; 35: W606–W612.
83 Djoumbou Feunang Y, Eisner R, Knox C, Chepelev L, Hastings J, Owen G, et al. ClassyFire: automated chemical classification with a comprehensive, computable taxonomy. *J Cheminf*. 2016; 8: 61.
84 The Metabolomics Workbench, https://www.metabolomicsworkbench.org/.
85 Tian T, Liu Y, Yan H, You Q, Yi X, Du Z, et al. agriGO v2.0: a GO analysis toolkit for the agricultural community, 2017 update. *Nucleic Acids Res*. 2017; 45(W1): W122–W129.
86 Ge SX, Jung D, Yao R. ShinyGO: a graphical gene-set enrichment tool for animals and plants. *Bioinformatics*. 2020; 36(8): 2628–2629.
87 UniProt Consortium. UniProt: the Universal protein knowledgebase in 2023. *Nucleic Acids Res*. 2023; 51(D1): D523–D531.
88 Bahrman N, Damerval C. Linkage relationships of loci controlling protein amounts in maritime pine (*Pinus pinaster* Ait.). *Heredity*. 1989; 63: 267–274.
89 Abril N, Gion J-M, Kerner R, Müller-Starck G, RMN C, Plomion C, et al. Proteomics research on forest trees, the most recalcitrant and orphan plant species. *Phytochemistry*. 2011; 72(10): 1219–1242.
90 Rey M-D, Castillejo MÁ, Sánchez-Lucas R, Guerrero-Sanchez VM, López-Hidalgo C, Romero-Rodríguez C, et al. Proteomics, Holm oak (*Quercus ilex* L.) and other recalcitrant and orphan forest tree species: How do they see each other? *Int J Mol Sci*. 2019; 20(3): 692.

91 Saiz-Fernández I, Milenković I, Berka M, Černý M, Tomšovský M, Brzobohatý B, et al. Integrated proteomic and metabolomic profiling of *Phytophthora cinnamomi* attack on sweet chestnut (*Castanea sativa*) reveals distinct molecular reprogramming proximal to the infection site and away from it. *Int J Mol Sci.* 2020; 21(22): 8525.

92 Szuba A, Marczak Ł, Ratajczak I, Kasprowicz-Maluśki A, Mucha J. Integrated proteomic and metabolomic analyses revealed molecular adjustments in *Populus × canescens* colonized with the ectomycorrhizal fungus *Paxillus involutus*, which limited plant host growth. *Environ Microbiol.* 2020; 22(9): 3754–3771.

93 Sekiya A, Marques FG, Leite TF, Cataldi TR, de Moraes FE, Pinheiro ALM, et al. Network analysis combining proteomics and metabolomics reveals new insights into early responses of *Eucalyptus grandis* during rust infection. *Front Plant Sci.* 2020; 11: 604849.

94 Pascual J, Cañal MJ, Escandón M, Meijón M, Weckwerth W, Valledor L. Integrated physiological, proteomic, and metabolomic analysis of ultra violet (UV) stress responses and adaptation mechanisms in *Pinus radiata. Mol Cell Proteomics.* 2017; 16(3): 485–501.

95 Budzinski IGF, Moon DH, Lindén P, Moritz T, Labate CA. Seasonal variation of carbon metabolism in the cambial zone of *Eucalyptus grandis. Front Plant Sci.* 2016; 7: 932.

96 Correia B, Valledor L, Hancock RD, Renaut J, Pascual J, Soares AMVM, et al. Integrated proteomics and metabolomics to unlock global and clonal responses of *Eucalyptus globulus* recovery from water deficit. *Metabolomics.* 2016; 12: 141.

97 Valledor L, Guerrero S, García-Campa L, Meijón M. Proteometabolomic characterization of apical bud maturation in *Pinus pinaster. Tree Physiol.* 2021; 41(3): 508–521.

98 Chong J, Wishart DS, Xia J. Using MetaboAnalyst 4.0 for comprehensive and integrative metabolomics data analysis. *Curr Protoc Bioinformatics.* 2019; 68(1): e86.

99 Weckwerth W. Green systems biology – From single genomes, proteomes and metabolomes to ecosystems research and biotechnology. *J Proteomics.* 2011; 75(1): 284–305.

100 Srivastava V, Obudulu O, Bygdell J, Löfstedt T, Rydén P, Nilsson R, et al. OnPLS integration of transcriptomic, proteomic and metabolomic data shows multi-level oxidative stress responses in the cambium of transgenic hipI- superoxide dismutase *Populus* plants. *BMC Genomics.* 2013; 14(1): 893.

101 Wang JP, Matthews ML, Williams CM, Shi R, Yang C, Tunlaya-Anukit S, et al. Improving wood properties for wood utilization through multi-omics integration in lignin biosynthesis. *Nat Commun.* 2018; 9(1).

102 Obudulu O, Mähler N, Skotare T, Bygdell J, Abreu IN, Ahnlund M, et al. A multi-omics approach reveals function of Secretory Carrier-Associated Membrane Proteins in wood formation of *Populus* trees. *BMC Genomics*. 2018; 19(1): 11.

103 Kasper K, Abreu IN, Feussner K, Zienkiewicz K, Herrfurth C, Ischebeck T, et al. Multi-omics analysis of xylem sap uncovers dynamic modulation of poplar defenses by ammonium and nitrate. *Plant J*. 2022; 111(1): 282–303.

104 López-Hidalgo C, Guerrero-Sánchez VM, Gómez-Gálvez I, Sánchez-Lucas R, Castillejo-Sánchez MA, Maldonado-Alconada AM, et al. A multi-omics analysis pipeline for the metabolic pathway reconstruction in the orphan species Quercus ilex. *Front Plant Sci*. 2018; 9: 935.

105 Fernie AR, Schauer N. Metabolomics-assisted breeding: a viable option for crop improvement? *Trends Genet*. 2009; 25(1): 39–48.

106 Movahedi A, Almasi A, Yaghuti Z, Wei H, Rutland P, Sun W, et al. Plant secondary metabolites with an overview of populus. *Int J Mol Sci*. 2021: 1–14.

107 González-Martínez SC, Hamanishi ET, Barchet GL, Dauwe R, et al. Poplar trees reconfigure the transcriptome and metabolome in response to drought in a genotype- and time-of-day-dependent manner. *BMC Genomics*. 2015; 16: 329 Wheeler NC, Ersoz E, Nelson CD, Neale DB. Association genetics in *Pinus taeda* L. I. wood property traits. Genetics. 2007;175(1):399–409.

108 Grattapaglia D, Plomion C, Kirst M, Sederoff RR. Genomics of growth traits in forest trees. *Curr Opin Plant Biol*. 2009; 12(2): 148–156.

109 Kirst M, Myburg A, Sederoff R. Genetic mapping in forest trees: markers, linkage analysis and genomics. *Genet Eng (N Y)*. 2004; 26: 105–141.

110 Kearsey MJ, Farquhar AGL. QTL analysis in plants; where are we now? *Heredity (Edinb)*. 1998; 80(2): 137–142.

111 Zhu C, Gore M, Buckler ES, Yu J. Status and prospects of association mapping in plants. *Plant Genome*. 2008; 1(1): 5–20.

112 Holland JB. Genetic architecture of complex traits in plants. *Curr Opin Plant Biol*. 2007; 10(2): 156–161.

113 Caseys C, Stritt C, Glauser G, Blanchard T, Lexer C. Effects of hybridization and evolutionary constraints on secondary metabolites: the genetic architecture of phenylpropanoids in European *Populus* species. *PLoS One*. 2015; 10(5): 1–23.

114 Teh SL, Rostandy B, Awale M, Luby JJ, Fennell A, Hegeman AD. Genetic analysis of stilbenoid profiles in grapevine stems reveals a major mQTL hotspot on chromosome 18 associated with disease-resistance motifs. *Hortic Res*. 2019; 6(1).

115 Laoué J, Depardieu C, Gérardi S, Lamothe M, Bomal C, Azaiez A, et al. Combining QTL, mapping and transcriptomics to decipher the genetic architecture of phenolic compounds metabolism in the conifer white spruce. *Front Plant Sci.* 2021; 12: 1–27.

116 Brown GR, Bassoni DL, Gill GP, Fontana JR, Wheeler NC, Megraw RA, et al. Identification of quantitative trait loci influencing wood property traits in Loblolly pine (*Pinus taeda* L.). III. QTL verification and candidate gene mapping. *Genetics.* 2003; 164(4): 1537–1546.

117 Novaes E, Osorio L, Drost DR, Miles BL, Boaventura-Novaes CRD, Benedict C, et al. Quantitative genetic analysis of biomass and wood chemistry of *Populus* under different nitrogen levels. *New Phytol.* 2009; 182(4): 878–890.

118 Thumma BR, Nolan MF, Evans R, Moran GF. Polymorphisms in cinnamoyl CoA reductase (CCR) are associated with variation in microfibril angle in *Eucalyptus* spp. *Genetics.* 2005; 171(3): 1257–1265.

119 Muchero W, Guo J, DiFazio SP, Chen JG, Ranjan P, Slavov GT, et al. High-resolution genetic mapping of allelic variants associated with cell wall chemistry in *Populus. BMC Genomics.* 2015; 16(1): 1–14.

120 Muneera Parveen AB, Muthupandi M, Kumar N, Chauhan SS, Vellaichamy P, Senthamilselvam S, et al. Quantitative genetic analysis of wood property traits in biparental population of *Eucalyptus camaldulensis* x *E. tereticornis. J Genet.* 2021; 100(2).

121 Morreel K, Goeminne G, Storme V, Sterck L, Ralph J, Coppieters W, et al. Genetical metabolomics of flavonoid biosynthesis in *Populus*: a case study. *Plant J.* 2006; 47(2): 224–237.

122 Bai B, Wang L, Lee M, Zhang Y, Alfiko Y, Ye BQ, et al. Genome-wide identification of markers for selecting higher oil content in oil palm. *BMC Plant Biol.* 2017; 17(1): 1–11.

123 Rosa JRBF, Mantello CC, Garcia D, de Souza LM, da Silva CC, Gazaffi R, et al. QTL detection for growth and latex production in a full-sib rubber tree population cultivated under suboptimal climate conditions. *BMC Plant Biol.* 2018; 18(1): 1–16.

124 Du Q, Lu W, Quan M, Xiao L, Song F, Li P, et al. Genome-wide association studies to improve wood properties: challenges and prospects. *Front Plant Sci.* 2018; 9: 1912.

125 Fang C, Luo J. Metabolic GWAS-based dissection of genetic bases underlying the diversity of plant metabolism. *Plant J.* 2019; 97(1): 91–100.

126 Xiao Y, Liu H, Wu L, Warburton M, Yan J. Genome-wide association studies in maize: praise and stargaze. *Mol Plant.* 2017; 10(3): 359–374.

127 Eckert AJ, Wegrzyn JL, Cumbie WP, Goldfarb B, Huber DA, Tolstikov V, et al. Association genetics of the loblolly pine (*Pinus taeda, Pinaceae*) metabolome. *New Phytol.* 2012; 193(4): 890–902.

128 Lu M, Seeve CM, Loopstra CA, Krutovsky KV. Exploring the genetic basis of gene transcript abundance and metabolite levels in loblolly pine (*Pinus* *taeda* L.) using association mapping and network construction. *BMC Genet*. 2018; 19(1): 1–13.

129 Zhang J, Yang Y, Zheng K, Xie M, Feng K, Jawdy SS, et al. Genome-wide association studies and expression-based quantitative trait loci analyses reveal roles of HCT2 in caffeoylquinic acid biosynthesis and its regulation by defense-responsive transcription factors in *Populus*. *New Phytol*. 2018; 220(2): 502–516.

130 Guerra FP, Suren H, Holliday J, Richards JH, Fiehn O, Famula R, et al. Exome resequencing and GWAS for growth, ecophysiology, and chemical and metabolomic composition of wood of *Populus trichocarpa*. *BMC Genomics*. 2019; 20(1): 1–14.

131 Wegrzyn JL, Eckert AJ, Choi M, Lee JM, Stanton BJ, Sykes R, et al. Association genetics of traits controlling lignin and cellulose biosynthesis in black cottonwood (*Populus trichocarpa, Salicaceae*) secondary xylem. *New Phytol*. 2010; 188(2): 515–532.

132 Cappa EP, El-Kassaby YA, Garcia MN, Acuña C, Borralho NMG, Grattapaglia D, et al. Impacts of population structure and analytical models in genome-wide association studies of complex traits in forest trees: a case study in *Eucalyptus globulus*. *PLoS One*. 2013; 8(11).

133 Fahrenkrog AM, Neves LG, Resende MFR, Vazquez AI, de los Campos G, Dervinis C, et al. Genome-wide association study reveals putative regulators of bioenergy traits in *Populus deltoides*. *New Phytol*. 2017; 213(2): 799–811.

134 Gong C, Du Q, Xie J, Quan M, Chen B, Zhang D. Dissection of insertion–deletion variants within differentially expressed genes involved in wood formation in *Populus*. *Front Plant Sci*. 2018; 8: 2199.

135 Quan M, Liu X, Du Q, Xiao L, Lu W, Fang Y, et al. Genome-wide association studies reveal the coordinated regulatory networks underlying photosynthesis and wood formation in *Populus*. *J Exp Bot*. 2021; 72(15): 5372–5389.

136 Chanroj V, Rattanawong R, Phumichai T, Tangphatsornruang S, Ukoskit K. Genome-wide association mapping of latex yield and girth in Amazonian accessions of *Hevea brasiliensis* grown in a suboptimal climate zone. *Genomics*. 2017; 109(5–6): 475–484.

137 Grimberg Å, Lager I, Street NR, Robinson KM, Marttila S, Mähler N, et al. Storage lipid accumulation is controlled by photoperiodic signal acting via regulators of growth cessation and dormancy in hybrid aspen. *New Phytol*. 2018; 219(2): 619–630.

138 Xia W, Luo T, Dou Y, Zhang W, Mason AS, Huang D, et al. Identification and validation of candidate genes involved in fatty acid content in oil palm by genome-wide association analysis. *Front Plant Sci*. 2019; 10: 1263.

139 Song Y, Chen P, Xuan A, Bu C, Liu P, Ingvarsson PK, et al. Integration of genome wide association studies and co-expression networks reveal roles of PtoWRKY 42-PtoUGT76C1-1 in trans-zeatin metabolism and cytokinin sensitivity in poplar. *New Phytol.* 2021; 231(4): 1462–1477.

140 Yao L, Li P, Du Q, Quan M, Li L, Xiao L, et al. Genetic architecture underlying the metabolites of chlorogenic acid biosynthesis in *Populus tomentosa. Int J Mol Sci.* 2021; 22(5): 1–21.

141 Ding X, Diao S, Luan Q, Wu HX, Zhang Y, Jiang J. A transcriptome-based association study of growth, wood quality, and oleoresin traits in a slash pine breeding population. *PLos Genet.* 2022; 18(2): 1–24.

142 Desta ZA, Ortiz R. Genomic selection: genome-wide prediction in plant improvement. *Trends Plant Sci.* 2014; 19(9): 592–601.

143 Grattapaglia D, Silva-Junior OB, Resende RT, Cappa EP, Müller BSF, Tan B, et al. Quantitative genetics and genomics converge to accelerate forest tree breeding. *Front Plant Sci.* 2018; 9: 1693.

144 Lebedev VG, Lebedeva TN, Chernodubov AI, Shestibratov KA. Genomic selection for forest tree improvement: methods, achievements and perspectives. *Forests.* 2020; 11(11): 1–36.

145 Ahmar S, Ballesta P, Ali M, Mora-Poblete F. Achievements and challenges of genomics-assisted breeding in forest trees: from marker-assisted selection to genome editing. *Int J Mol Sci.* 2021; 22(19): 10583.

146 Resende MDV, Resende MFR, Sansaloni CP, Petroli CD, Missiaggia AA, Aguiar AM, et al. Genomic selection for growth and wood quality in *Eucalyptus*: capturing the missing heritability and accelerating breeding for complex traits in forest trees. *New Phytol.* 2012; 194(1): 116–128.

147 Cortés AJ, Restrepo-Montoya M, Bedoya-Canas LE. Modern strategies to assess and breed forest tree adaptation to changing climate. *Front Plant Sci.* 2020; 11: 583323.

148 Beaulieu J, Nadeau S, Ding C, Celedon JM, Azaiez A, Ritland C, et al. Genomic selection for resistance to spruce budworm in white spruce and relationships with growth and wood quality traits. *Evol Appl.* 2020; 13(10): 2704–2722.

149 Zapata-Valenzuela J, Isik F, Maltecca C, Wegrzyn J, Neale D, McKeand S, et al. SNP markers trace familial linkages in a cloned population of *Pinus taeda*-prospects for genomic selection. *Tree Genet Genomes.* 2012; 8(6): 1307–1318.

150 de Moraes BFX, dos Santos RF, de Lima BM, Aguiar AM, Missiaggia AA, da Costa DD, et al. Genomic selection prediction models comparing sequence capture and SNP array genotyping methods. *Mol Breed.* 2018; 38: 115.

151 Rambolarimanana T, Ramamonjisoa L, Verhaegen D, Leong Pock Tsy JM, Jacquin L, Cao-Hamadou TV, et al. Performance of multi-trait genomic selection for *Eucalyptus robusta* breeding program. *Tree Genet Genomes.* 2018; 14: 71.

152 Cappa EP, de Lima BM, da Silva-Junior OB, Garcia CC, Mansfield SD, Grattapaglia D. Improving genomic prediction of growth and wood traits in *Eucalyptus* using phenotypes from non-genotyped trees by single-step GBLUP. *Plant Sci.* 2019; 284: 9–15.

153 Bouvet JM, Makouanzi Ekomono CG, Brendel O, Laclau JP, Bouillet JP, Epron D. Selecting for water use efficiency, wood chemical traits and biomass with genomic selection in a *Eucalyptus* breeding program. *For Ecol Manage.* 2020; 465: 118092.

154 Mphahlele MM, Isik F, Mostert-O'Neill MM, Reynolds SM, Hodge GR, Myburg AA. Expected benefits of genomic selection for growth and wood quality traits in *Eucalyptus grandis*. *Tree Genet Genomes.* 2020; 16(4): 1–12.

155 Perez-Riverol Y, Wang R, Hermjakob H, Müller M, Vesada V, Vizcaíno JA. Open source libraries and frameworks for mass spectrometry based proteomics: a developer's perspective. *Biochim Biophys Acta.* 2014; 1844(1 Pt A): 63–76.

156 Jones OAH. Illuminating the dark metabolome to advance the molecular characterisation of biological systems. *Metabolomics.* 2018; 14(8): 101.

157 Desmet S, Brouckaert M, Boerjan W, Morreel K. Seeing the forest for the trees: retrieving plant secondary biochemical pathways from metabolome networks. *Comput Struct Biotechnol J.* 2021; 19: 72–85.

158 Buche G, Colas C, Fougère L, Destandau E. Oak species *Quercus robur* L. and *Quercus petraea* Liebl. Identification based on UHPLC-HRMS/MS molecular networks. *Metabolites.* 2021; 11(10): 684.

159 Dos Santos BM, Zibrandtsen JFS, Gunbilig D, Sørensen M, Cozzi F, Boughton BA, et al. Quantification and localization of formylated phloroglucinol compounds (FPCs) in *Eucalyptus* species. *Front Plant Sci.* 2019; 10: 186.

160 Koo HB, Hwang H-S, Han JY, Cheong EJ, Kwon Y-S, Choi YE. Enhanced production of pinosylvin stilbene with aging of *Pinus strobus* callus and nematicidal activity of callus extracts against pinewood nematodes. *Sci Rep.* 2022; 12(1): 770.

161 Emwas A-HM. The strengths and weaknesses of NMR spectroscopy and mass spectrometry with particular focus on metabolomics research. *Methods Mol Biol.* 2015; 1277: 161–193.

162 Patyra A, Dudek MK, Kiss AK. LC-DAD-ESI-MS/MS and NMR analysis of conifer wood specialized metabolites. *Cells.* 2022; 11(20): 3332.

163 Hedenström M, Wiklund-Lindström S, Oman T, Lu F, Gerber L, Schatz P, et al. Identification of lignin and polysaccharide modifications in *Populus* wood by chemometric analysis of 2D NMR spectra from dissolved cell walls. *Mol Plant.* 2009; 2(5): 933–942.

164 Lourenço A, Rencoret J, Chemetova C, Gominho J, Gutiérrez A, Del Río JC, et al. Lignin composition and structure differs between xylem, phloem and phellem in *Quercus suber* L. *Front Plant Sci.* 2016; 7: 1612.
165 Rencoret J, Gutiérrez A, Nieto L, Jiménez-Barbero J, Faulds CB, Kim H, et al. Lignin composition and structure in young versus adult *Eucalyptus globulus* plants. *Plant Physiol.* 2011; 155(2): 667–682.
166 Kanehisa M, Goto S, Sato Y, Furumichi M, Tanabe M. KEGG for integration and interpretation of large-scale molecular data sets. *Nucleic Acids Res.* 2012; 40(Database issue): D109–D114.
167 Naithani S, Preece J, D'Eustachio P, Gupta P, Amarasinghe V, Dharmawardhana PD, et al. Plant Reactome: a resource for plant pathways and comparative analysis. *Nucleic Acids Res.* 2017; 45(D1): D1029–D1039.
168 Horai H, Arita M, Kanaya S, Nihei Y, Ikeda T, Suwa K, et al. MassBank: a public repository for sharing mass spectral data for life sciences. *J Mass Spectrom.* 2010; 45: 703–714.
169 Smith CA, O'Maille G, Want EJ, Qin C, Trauger SA, Brandon TR, et al. METLIN: a metabolite mass spectral database. *Ther Drug Monit.* 2005; 27(6): 747–750.
170 Aron AT, Gentry EC, McPhail KL, Nothias L-F, Nothias-Esposito M, Bouslimani A, et al. Reproducible molecular networking of untargeted mass spectrometry data using GNPS. *Nat Protoc.* 2020; 15(6): 1954–1991.
171 da Silva RR, Wang M, Nothias L-F, van der Hooft JJJ, Caraballo-Rodríguez AM, Fox E, et al. Propagating annotations of molecular networks using in silico fragmentation. *PLoS Comput Biol.* 2018; 14(4): e1006089.
172 Beauxis Y, Genta-Jouve G. MetWork: a web server for natural products anticipation. *Bioinformatics.* 2019; 35(10): 1795–1796.
173 Dührkop K, Fleischauer M, Ludwig M, Aksenov AA, Melnik AV, Meusel M, et al. SIRIUS 4: a rapid tool for turning tandem mass spectra into metabolite structure information. *Nat Methods.* 2019; 16(4): 299–302.
174 PlantGenIE.org. The Plant Genome Integrative Explorer Resource [Internet]. Available from https://plantgenie.org/, accessed 18 Feb 2023.
175 Sundell D, Mannapperuma C, Netotea S, Delhomme N, Lin Y-C, Sjödin A, et al. The plant genome integrative explorer resource: PlantGenIE.org. *New Phytol.* 2015; 208(4): 1149–1156.
176 Nystedt B, Street NR, Wetterbom A, Zuccolo A, Lin Y-C, Scofield DG, et al. The Norway spruce genome sequence and conifer genome evolution. *Nature.* 2013; 497: 579–584.
177 Sjödin A, Street NR, Sandberg G, Gustafsson P, Jansson S. The Populus Genome Integrative Explorer (PopGenIE): a new resource for exploring the *Populus* genome: Methods. *New Phytol.* 2009; 182(4): 1013–1025.
178 Christie N, Mannapperuma C, Ployet R, van der Merwe K, Mähler N, Delhomme N, et al. qtlXplorer: an online systems genetics browser in the

Eucalyptus Genome Integrative Explorer (EucGenIE). *BMC Bioinf*. 2021; 22(1): 595.

179 Wegrzyn JL, Lee JM, Tearse BR, Neale DB. TreeGenes: a forest tree genome database. *Int J Plant Genomics*. 2008; 2008: 412875.

180 Ramos AM, Usié A, Barbosa P, Barros PM, Capote T, Chaves I, et al. The draft genome sequence of cork oak. *Sci Data*. 2018; 5: 180069.

181 Tarkka MT, Herrmann S, Wubet T, Feldhahn L, Recht S, Kurth F, et al. OakContigDF159.1, a reference library for studying differential gene expression in *Quercus robur* during controlled biotic interactions: use for quantitative transcriptomic profiling of oak roots in ectomycorrhizal symbiosis. *New Phytol*. 2013; 199(2): 529–540.

182 Goodstein DM, Shu S, Howson R, Neupane R, Hayes RD, Fazo J, et al. Phytozome: a comparative platform for green plant genomics. *Nucleic Acids Res*. 2012; 40(Database issue): D1178–D1186.

183 Koonin EV, Fedorova ND, Jackson JD, Jacobs AR, Krylov DM, Makarova KS, et al. A comprehensive evolutionary classification of proteins encoded in complete eukaryotic genomes. *Genome Biol*. 2004; 5(2): R7.

184 Punta M, Coggill PC, Eberhardt RY, Mistry J, Tate J, Boursnell C, et al. The Pfam protein families database. *Nucleic Acids Res*. 2012; 40(Database issue): D290–D301.

185 Mi H, Guo N, Kejariwal A, Thomas PD. PANTHER version 6: protein sequence and function evolution data with expanded representation of biological pathways. *Nucleic Acids Res*. 2007; 35(Database issue): D247–D252.

186 Richelle A, Joshi C, Lewis NE. Assessing key decisions for transcriptomic data integration in biochemical networks. *PLoS Comput Biol*. 2019; 15(7): e1007185.

187 Krassowski M, Das V, Sahu SK, Misra BB. State of the field in multi-omics research: from computational needs to data mining and sharing. *Front Genet*. 2020; 11: 610798.

188 Yang Y, Saand MA, Huang L, Abdelaal WB, Zhang J, Wu Y, et al. Applications of multi-omics technologies for crop improvement. *Front Plant Sci*. 2021; 12: 563953.

5

A Workflow for Metabolomics of Forest Tree Biotic Stress Response and Applications for Management

Anna O. Conrad[1], *Caterina Villari*[2], *and Pierluigi Bonello*[3]

[1] *USDA Forest Service, Northern Research Station, West Lafayette, IN, USA*
[2] *Warnell School of Forestry & Natural Resources, University of Georgia, Athens, USA*
[3] *Department of Plant Pathology, The Ohio State University, Columbus, USA*

5.1 Introduction

Trees are exposed to a variety of stressors over their lifetimes. To survive and thrive under diverse conditions, they have evolved a suite of defenses, which confer resistance or tolerance to most biotic and abiotic stressors. Even though disease (or insect pest infestation) is the exception and not the rule, pests and pathogens have had and continue to have a profound impact on the health of forests globally. This is especially true for invasive and/or non-native pests and pathogens and, more recently, for native pests and pathogens whose range or virulence has changed because of climate change (1–3). Any estimate of the impact of pests and pathogens on forests is likely conservative and an underestimate of the actual impact on forests and the communities that rely on them. The incidence of non-native insects (and pathogens) continues to increase every year (4), and increased tree mortality within just the United States is predicted in 63% of forestland due to non-native insect pests and pathogens (5). Over the last 100 years, invasive pests and pathogens have functionally eradicated or decimated multiple forest tree species in local populations or across species ranges. Examples of impacted forest tree species include, but are not limited to, chestnut (chestnut blight), pines (white pine blister rust and mountain pine beetle), ash (emerald ash borer and ash dieback), and tanoak, oaks, and larch (sudden oak death and Ramorum blight).

The impacts of pests and pathogens are widespread, from economic losses to loss of forest ecosystem services and negative outcomes for human health. For example, emerald ash borer, which was first identified in the early 2000s,

Monitoring Forest Damage with Mass Spectrometry-Based Metabolomics Methods, First Edition.
Edited by Carla António.

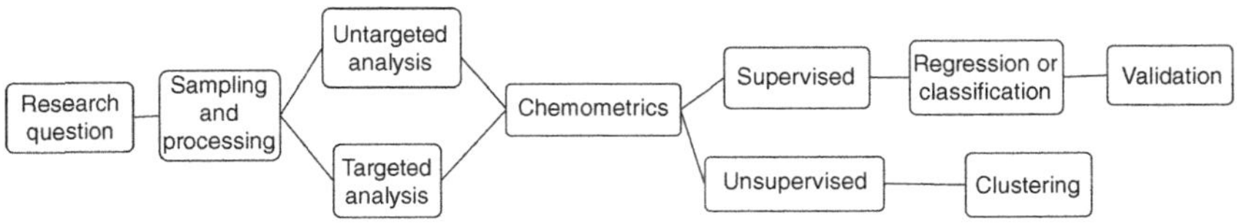

Figure 5.1 A workflow for metabolomics studies of forest tree biotic stress interactions. Typical research questions focus on identifying resistant trees or detecting diseased or pest-infected trees based on metabolomics data.

is estimated to have killed tens of millions of ash trees in the eastern United States (6). In areas of emerald ash borer infestation, increased human mortality from cardiovascular illness has been reported (7). Impacts have also been felt in the housing industry, where emerald ash borer infection negatively impacted home values in one Midwestern United States city (8). These studies do not even include more direct impacts, such as costs of forest management for the treatment, removal, or replacement of susceptible ash trees (9). Wide-reaching impacts are not exclusive to the emerald ash borer. Similar impacts have been observed or estimated for other pathogens and pests of trees across wildlands and urban landscapes (10, 11).

To counter biotic threats, trees possess a suite of defenses that act directly and/or indirectly on the attacking pest or pathogen, and range from physical defenses, including bark and lignin, to chemical defenses, such as phenolics and terpenoids. These defenses can be general or specific for a given organism and can be produced constitutively (pre-infection) or induced following infection. The reason why invasive and non-native pests and pathogens have been particularly problematic is because affected tree species lack a co-evolutionary history with them, so the pests and pathogens have no co-evolved natural enemies. Thus, the suite of defenses that the novel host possesses is unable to prevent infection or minimize adverse impacts to tree health, ultimately, resulting in higher-than-normal levels of mortality of impacted host tree species, often across larger spatial scales.

In this chapter, we focus on the role of metabolite defenses in forest tree responses to biotic stress. Within this context, we describe a workflow (Figure 5.1) for use in metabolomics studies of forest tree-pest/pathogen interactions. Applications for the workflow as well as considerations for applying it across scales, from individual trees to landscapes, are also discussed. Finally, case studies from multiple tree-pathogen and pest systems are presented, including oak (sudden oak death and European oak leafroller), ash (emerald ash borer and ash dieback), and pine (Diplodia tip blight and mountain pine beetle).

5.2 Methods

5.2.1 Research Question

The research question(s) will dictate the experimental approach and statistical methods employed in a study of forest tree biotic stress response. Questions can direct the focus of study to all metabolites produced in a biological system (i.e. the metabolome of a host tree), or to a sub-set (i.e. metabolite profiling). Metabolite profiling is of relevance to studies of forest tree biotic stress response because metabolite profiling aims to identify metabolites correlated with a specific response, such as infection or disease resistance (12, 13). In some situations, a more rapid method of assessing the metabolome of a host tree is needed. This is referred to as metabolic fingerprinting (12). In contrast to metabolite profiling, individual metabolites are not identified or quantified. For this reason, metabolic fingerprints can be used for rapid classification (12). Key features of the metabolome, such as metabolites associated with specific treatments, might not be resolvable using the latter approach, so follow-up studies will be required for more precise identification and quantification.

5.2.2 Sample Selection and Processing

The life cycles of forest trees and of the pathogens and pests that attack them need to be considered when selecting samples, both in determining optimal tissues to sample (e.g. leaves, stem, or roots) and when to sample (i.e. timing corresponding both with host and pathogen/pest cycles). Metabolites are known to change with the season (tree phenology), age, and position on the tree (e.g. sun versus shade leaves) (14–17), so considerations for standardizing sample collection, regardless of whether the focus is on constitutive or induced metabolites, is necessary.

Under optimal conditions, samples are harvested and flash frozen (in liquid nitrogen) and then stored in a low-temperature freezer. This minimizes sample degradation. However, where liquid nitrogen is not available, or its use is unfeasible (e.g. in rugged field conditions), dry ice can be used to preserve samples until returning to the laboratory.

Tissue homogenization is the next step, and the best approach will depend on the tissue. Woody tissues can be difficult to homogenize. Lyophilizing (freeze drying) samples is sometimes useful or needed. Homogenization can be accomplished by hand (with mortar and pestle and liquid nitrogen to maintain frozen tissue) or with a ball mill. When sampling bark or wood from mature trees, using a cordless hand drill to sample tissues is useful for pre-grinding samples and making subsequent homogenization easier. Various extraction methods can be employed depending on the metabolites of interest, although methanol is widely used as a solvent for extraction of biotic stress-related metabolites; for further examples and protocols, see (18–20).

5.2.3 Analytical Methods

The analytical method used will depend on the research question and which metabolites or metabolite classes are of interest to the system. For example, phenolic compounds are well known for their association with biotic stress response in forest trees (21, 22). The analytical approach can be classified into two categories: untargeted and targeted. Sedio et al. (23) provide a protocol for high-throughput, untargeted metabolomics analysis. A targeted approach focuses on either a class of metabolites or specific, individual metabolites.

For studies of metabolites in forest trees subjected to biotic stress, liquid chromatography mass spectrometry (LC–MS) or gas chromatography mass spectrometry (GC–MS)-based approaches are the norm for separating out individual compounds for subsequent identification and quantification; see (24) for a comprehensive review of the topic. GC–MS is suitable for volatile compounds or compounds that can be derivatized, like terpenes (25). MS libraries can aid in compound identification, although libraries are more well-developed for GC–MS versus LC–MS (24). Internal standards are used for quantification and normalization, and to ensure technical accuracy, such as extraction efficiency and run-to-run consistency (26). External standards are also useful for quantifying individual compounds and for confirming the presence of specific metabolites in samples based on retention time and spectral properties (26).

5.2.4 Chemometrics

Chemometrics is the multivariate analysis of chemical data and covers the analysis of different types of metabolomics data from spectral data to MS data sets (27, 28). While univariate approaches (e.g. analysis of variance [ANOVA]), are useful for evaluating treatment effects on individual metabolites, chemometrics is useful for simultaneous analysis of multiple metabolites. Approaches that aim to reduce redundancy in the data, a result of co-linearity, are of particular importance when developing models to predict disease-resistant trees or to discriminate between healthy and infected trees based on metabolite levels. Otherwise, models are at risk of being overfit, which can lead to poor performance when models are applied to new datasets (29).

There are two approaches for chemometric analysis, namely, supervised and unsupervised. Supervised analysis considers a priori groupings and labels the data, while unsupervised analysis does not consider any grouping or treatment. Unsupervised analysis (e.g. principal component analysis [PCA]) can be useful for identifying natural groupings or clustering of samples and metabolites associated with differences between treatments or groups without considering the group or treatment (28). While both approaches can be used to identify metabolites that vary between treatments or groups, supervised analysis is especially well suited for identifying metabolites most important for distinguishing between groups or identifying metabolites most influenced by treatment effects (30). See Table 5.1 for a summary of common chemometric methods applied to forest tree biotic stress studies.

For studies focused on predicting or modeling specific traits, such as those focused on distinguishing between trees that differ in disease severity or identifying resistant trees, availability of robust and reliable phenotypic data (paired with metabolomics data) is essential. Many forest tree responses to biotic stressors are quantitative in nature. So, symptoms fall along a continuum with trees ranging from resistant (no to limited symptoms) to susceptible (severe symptoms often resulting in death), and progression of symptoms occurs over months to years. For these reasons, accurate phenotyping can be a challenge. Therefore, evaluation of multiple disease or pest severity metrics, both qualitative and quantitative, might be needed.

5.2.5 Pre-processing

Once thresholds for detection are specified, metabolite peaks are identified, and raw data is extracted from analytical software (e.g. XCMS (43)), metabolite data needs to be prepared for analysis (pre-processing). For metabolite profiling studies, functional data analysis (FDA) can be used to identify potential outliers (44). FDA considers the depth of the curve, such that the depth is inversely related to the sample outlyingness. Outliers can also be identified from multivariate analyses, for instance clustering analysis (28). Once outliers are identified and trimmed, standardization and normalization of data are needed. Internal standards are useful for normalizing data (26). Mean centering and scaling, and other types of transformation, including log transformation, can also be used.

5.2.6 Classification

Classification of metabolite data is conducted using supervised or unsupervised approaches. Supervised methods are commonly used for examining forest tree biotic stress responses (see Table 5.1), although unsupervised approaches can be useful for examining natural clustering of samples, without influence of the grouping or treatment. Regardless of whether a priori groupings are considered, classification can be accomplished using multivariate statistical methods or machine learning. Machine learning algorithms learn

Table 5.1 Commonly used chemometric methods in forest tree biotic stress studies.

Method	Type	Supervised or unsupervised	Variable selection	Application	Selected references
Principal component analysis	Classification	Unsupervised, supervised	Yes	Infection, resistance	(26, 31–34)
Support vector machine	Classification	Supervised	No	Resistance	(35, 36)
Random forest	Classification	Supervised	Yes	Resistance, infection	(35, 37)
Linear discriminant analysis	Classification	Unsupervised, supervised	Yes	Resistance	(38, 39)
Partial least squares discriminant analysis	Classification	Supervised	Yes	Infection, resistance	(32, 35, 37, 40)
Soft independent modeling of class analogy	Classification	Supervised	Yes	Resistance	(36, 41, 42)
Logistic regression	Regression	N/A	Yes	Resistance	(38, 39)
Partial least squares regression	Regression	N/A	Yes	Resistance	(36)

from experience and aim to generalize trends or patterns in the data (45–47). Examples of classification methods include PCA (can be supervised or unsupervised), partial least squares discriminant analysis (PLS–DA), linear discriminant analysis (LDA), support vector machine (SVM), and random forest (RF) (Table 5.1).

5.2.7 Regression

Regression is useful for identifying associations between data, where the response variable is quantitative, including continuous and binomial response variables. Regression can also be used to make predictions based on known data. Common regression methods include linear regression (continuous response variable), logistic regression (binomial response variable), and partial least squares regression (PLSR) (Table 5.1). Regression approaches, such as backward stepwise logistic regression, can also be used to identify the most important variables (see Section 5.2.8) (38, 39).

5.2.8 Variable Selection

Reducing the number of variables, especially in metabolomics data sets with co-linearity, or a high number of variables relative to sample size can be essential to building accurate and robust models and preventing overfitting (28, 48). In addition, variable selection can be useful for teasing apart biological treatment effects and identifying metabolites most associated with biotic stress response. Various approaches can be used and include classification and regression methods (see Table 5.1). For example, sparse PLS–DA (a variation of PLS–DA) combines classification with variable selection to identify and include only the most important features (49). Even with unsupervised approaches, like PCA, loading values can be used to identify metabolites most important for separating treatment groups. Random forest can also be used for variable selection (e.g. variable selection using random forests [VSURF]) or variable importance score (VIP) (50, 51).

5.2.9 Validation

For classification and regression models, including machine learning approaches, validation is needed to assess model quality and transferability (28). Model performance (e.g. accuracy, precision, and area under the receiver operator characteristic curve) is ideally assessed using external validation (separate data set) or cross-validation. The latter approach is especially useful when sample sizes are limited; see (28) for further information on validation in chemometrics.

5.2.10 Available Tools for Analyzing Metabolomics Data

Various registered software and open-source software packages are available for exploring and analysis of multivariate data, including those tailored specifically to metabolomics data analysis or analysis of complex omics data sets (e.g. Metaboanalyst, mixOmics, mdatools) (30,52–54). Some of these tools can be used directly in a web browser (e.g. Metaboanalyst) (52). In addition, Conrad et al. developed a workflow for the analysis of spectral data for use in disease prediction (35, 55) and resistance screening (36).

5.3 Application

The workflow presented in Section 5.2 (see Figure 5.1 for overview) is customizable depending on the research question. Consequently, the specific methods used within each step will vary depending on the research question and the tree-pathogen/pest system. Two primary applications were tested: (i) identifying infected trees; and (ii) identifying resistant trees. The workflow, however, is broadly applicable and may be useful beyond metabolomics studies. This is because many of the methods used for chemometrics (e.g. PLS–DA) are appropriate for use with other omics datasets (30).

5.3.1 Detection of Diseased and Pest-Infected Trees

Disease and pest management relies on identifying infected trees. This often occurs after symptoms develop, and unfortunately, after the pest or pathogen has become established in an area. Thus, early detection, including identifying infected trees before symptoms develop, is key. This is because the lag time between initial infection and symptom development provides the pathogen or pest time to spread without intervention. Examples include identifying infected but asymptomatic American beech trees with *Litylenchus crenatae mccannii*, a non-native nematode associated with beech leaf disease (35), and distinguishing between symptomatic and asymptomatic Norway spruce infected with the fungus *Herterobasidion* spp. (56).

5.3.2 Identifying Resistant Trees

The workflow can also be used to identify disease or pest-resistant trees. Use of resistant trees is a viable option for disease/pest management, restoration of ecologically important forest tree species, and as part of improvement programs for economically important species (57, 58). Identifying resistant trees using metabolomics data is the primary application of this workflow to date. Examples include identifying coast live oak resistant to sudden oak death (38, 39, 41), European ash resistant to ash dieback (42), and Austrian pine resistant to Diplodia tip blight (36). Constitutive or induced metabolites can be analyzed. If the aim is to develop a non-invasive screen for resistance (without the need for natural or artificial inoculations), then the focus should be on constitutive metabolites (e.g. (38, 41)).

5.3.3 Landscape-Level Applications

Translating metabolomics tools from individual trees or tree populations to the landscape scale is a topic of interest not only for identifying infected or resistant trees, but more broadly within the fields of high-throughput plant phenotyping (59) and digital forestry. High-throughput and landscape-level assessment of forest tree biotic stress responses can be limited with MS-based approaches, unless applied only to a subset of individuals within a species or population. However, more sensitive MS-based approaches can be integrated with more high-throughput remote sensing platforms for application at larger scales, such as aerial hyperspectral sensing, which is capable of picking up on physiological differences in trees (e.g. water stress associated with pathogen infection) as well as foliar biochemical variation (60). Use of proximal and remote sensing (with spectral sensors) has already been shown to be effective at identifying oak wilt-infected trees at multiple scales (61, 62).

5.4 Case Studies

The following case studies demonstrate the use of metabolomics to examine and predict pest and pathogen infection and resistance within forest trees.

5.4.1 Ash

Ash dieback is caused by the invasive fungal pathogen *Hymenoscyphus fraxineus* and is currently impacting European ash (*Fraxinus excelsior*) across Europe (63). It is reported that some trees are resistant to the disease, and genotypes with resistance have been identified (63, 64). Untargeted metabolomics analysis of leaves from high- and low-susceptibility individuals was performed by Sambles et al. (37) to identify metabolites associated with disease resistance. Partial least squares discriminate analysis was used along with variable influence on projection and RF to determine important metabolites. Putative iridium glycosides were found to differ between resistant and susceptible trees. Nemesio-Gorriz et al. (40) also identified metabolites associated with resistance (tolerance) to ash dieback. Using PLS–DA combined with variable selection, the authors identified that higher levels of fraxetin and escalation were associated with lower disease susceptibility. These compounds inhibited the growth of *H. fraxineus in vitro*. Finally, constitutive bark spectral profiles of resistant and susceptible European ash grown across six countries could be distinguished using soft independent modeling of class analogy (SIMCA) (42).

Emerald ash borer (*Agrilus planipennis*) is an invasive, non-native insect pest that has destroyed North American ash populations across the eastern United States and Canada (reviewed in (65)). While resistant (lingering) ash have been identified (66), most studies to date have examined metabolite differences between susceptible North American ash species (e.g. green ash, *Fraxinus pennsylvanica* and white ash, *Fraxinus americana*) and resistant Manchurian ash (*Fraxinus mandshurica*). For example, Qazi et al. (67) used untargeted HPLC–MS/MS to examine differences in constitutive foliar metabolite profiles of ash species with differing susceptibility to the pest. Using PCA, the authors identified metabolites whose levels increased in resistant Manchurian ash. Similarly, Chakraborty et al. (26) compared phloem phenolic profiles of resistant Manchurian ash and susceptible black ash (*Fraxinus nigra*) following infection with emerald ash borer and drought stress. Using a multi-faceted approach of PCA, hierarchical cluster analysis (HCA), and ANOVA, the authors discovered that pinoresinol A was induced by emerald ash borer infection and that levels of six other phenolic compounds decreased following infection.

5.4.2 Oak

Since its discovery in the western United States in the 1990s, sudden oak death, caused by the non-native pathogen *Phytophthora ramorum*, has led to the extensive mortality of tanoak (*Notholithocarpus densiflorus*) and true oaks in the red oak group (*Quercus* section *Lobatae*) (68). The pathogen is also found in the United Kingdom, where impacts are felt in particular on plantation-grown larch (69). In the United States, multiple studies have examined the role of metabolites, particularly phenolic compounds, in coast live oak (*Quercus agrifolia*) defense and resistance to *P. ramorum* (38,39,41,70,71). McPherson et al. (39) and Conrad et al. (38) identified phenolic biomarkers of resistance in coast live oak previously exposed to the pathogen and in naïve trees, respectively. To do this, sparse linear discriminant analysis (SLDA) and logistic regression were used to identify biomarkers and predict resistance based on constitutive levels of the identified compounds. In naïve trees, biomarkers of resistance included myrictrin and three incompletely characterized flavonoids (38).

The European oak leafroller (*Tortrix viridana*) causes severe defoliation of pedunculate oak (*Quercus robur*) in Europe. Differences in *Q. robur* susceptibility to *T. viridana* have been observed and two classes of trees with differing susceptibility have been defined: T-oaks (resistant, less defoliation) and S-oaks (susceptible and greater defoliation) (31). Using multivariate data analysis, specifically PCA and PLSR, Kersten et al. (31) identified metabolic differences between resistant and susceptible trees. The authors found that resistant trees contained higher levels of galloylated flavonol glycosides compared to susceptible trees. Similarly, Bertić et al. (32) identified metabolic biomarkers to distinguish *T. viridana* resistant and susceptible *Q. robur* using PCA and orthogonal partial least squares discriminant analysis (OPLS–DA). One of the goals of this study was to identify biomarkers for use in assisting tree nurseries in seedling selection. For that reason, biomarkers were used to develop prediction models for the selection of resistant trees.

5.4.3 Pine

Diplodia tip blight is caused by the pathogen *Diplodia sapinea* and affects two-needle pine species, including Austrian pine (*Pinus nigra*). The pathogen is widely distributed and can be especially problematic in urban areas or where stressed pine trees are found (i.e. disease symptoms become more pronounced with drought stress) (72). Wallis et al. (73) identified associations between Austrian pine susceptibility to *D. sapinea* and metabolites, including lignin (negatively correlated with susceptibility) and pinosylvin and pinosylvin monomethyl ether (positively correlated) using correlation analysis. Using PLSR, SIMCA, and SVM, constitutive metabolite profiles from the mid-infrared spectrum were used to distinguish between trees that differed in disease susceptibility and to predict differences in pine susceptibility to *D. sapinea* (36).

Mountain pine beetle (*Dendroctonus ponderosae*) is native to western North America. Its range and devastation, particularly of lodgepole pine (*Pinus contorta*), has increased due to climate change (74). Six et al. (33) examined differences in terpenoids among two groups of whitebark pine, *Pinus albicaulis*: survivors (trees that showed no signs of beetle attack) and generals (trees that were mature but below the size threshold for beetle attack). Permutational multivariate analyses of variance (PERMANOVA) and PCA were used to examine differences in terpenoids between groups and across sites. Concentrations of terpenoids differed between generals and survivors and were mostly higher in survivors. A similar approach of PCA and PERMANOVA was used to examine the impacts of mountain pine beetle fungal symbionts on lodgepole pine metabolites (34). Potential biomarkers of resistance were identified using multivariate empirical Bayes (MEBA) time-series analysis and included β-phellandrene, β-myrcene, β-pinene, and 3-carene.

5.5 Conclusions and Future Perspectives

As forest trees continue to be threatened by new and existing pests and pathogens, alone and in combination with climate change and other stressors, new solutions for disease and pest management will be needed. One solution is to use metabolites associated with tree biotic stress response as biomarkers for pathogen and pest infection and/or for differentiating between resistant and susceptible trees. The workflow presented here has been successfully tested in multiple tree-pathogen and pest systems to distinguish between resistant and susceptible trees and for disease detection. However, additional approaches, beyond those highlighted in this chapter, will likely need to be considered and evaluated on a case-by-case basis. Finally, translating results and metabolite-based predictive models from controlled studies to real-world scenarios, either in the forest or in managed environments (e.g. nurseries and tree plantations) remains a challenge. However, advances in digital forestry and high-throughput phenotyping will likely facilitate translation and application at larger spatial scales.

Acknowledgments

The authors thank Carrie Fearer for providing helpful comments on an earlier draft of this chapter. The authors declare no conflicts of interest or competing interests. The use of trade names is for the information and convenience of the reader and does not imply official endorsement or approval by the USDA or the Forest Service of any product to the exclusion of others that may be suitable. This work was supported in part by the US Department of Agriculture, Forest Service.

References

1 Ramsfield TD, Bentz BJ, Faccoli M, Jactel H, Brockerhoff EG. Forest health in a changing world: effects of globalization and climate change on forest insect and pathogen impacts. *Forestry*. 2016; 89: 245–252.
2 Simler-Williamson AB, Rizzo DM, Cobb RC. Interacting effects of global change on forest pest and pathogen dynamics. *Annu Rev Ecol Evol Syst*. 2019; 50: 381–403.
3 Dukes JS, Pontius J, Orwig D, Garnas JR, Rodger VL, Brazee N, et al. Responses of insect pests, pathogens, and invasive plant species to climate change in the forests of northeastern North America: what can we predict? *Can J For Res*. 2009; 39: 231–248.
4 Aukema JE, McCullough DG, Von Holle B, Liebhold AM, Britton K, Frankel SJ. Historical accumulation of nonindigenous forest pests in the continental United States. *Bioscience*. 2010; 60: 886–897.
5 Lovett GM, Weiss M, Liebhold AM, Holmes TP, Leung B, Lambert KF, et al. Nonnative forest insects and pathogens in the United States: impacts and policy options. *Ecol Appl*. 2016; 26: 1437–1455.

6 USDA APHIS. Emerald ash borer beetle [Internet]. Available from (https://www.aphis.usda.gov/aphis/resources/pests-diseases/hungry-pests/the-threat/emerald-ash-borer/emerald-ash-borer-beetle), accessed 30 March 2023.

7 Donovan GH, Butry DT, Michael YL, Prestemon JP, Liebhold AM, Gatziolis D, et al. The relationship between trees and human health: evidence from the spread of the emerald ash borer. *Am J Prev Med.* 2013; 44: 139–145.

8 Li X, Holmes TP, Boyle KJ, Crocker EV, Nelson CD. Hedonic analysis of forest pest invasion: the case of emerald ash borer. *Forests.* 2019; 10: 1–15.

9 Kovacs KF, Haight RG, McCullough DG, Mercader RJ, Siegert NW, Liebhold AM. Cost of potential emerald ash borer damage in U.S. communities, 2009–2019. *Ecol Econ.* 2010; 69: 569–578.

10 Kovacs K, Holmes TP, Englin JE, Alexander J. The dynamic response of housing values to a forest invasive disease: evidence from a sudden oak death infestation. *Environ Resour Econ.* 2011; 49: 445–471.

11 Fei S, Morin RS, Oswalt CM, Liebhold AM. Biomass losses resulting from insect and disease invasions in US forests. *Proc Natl Acad Sci USA.* 2019; 116: 17371–17376.

12 Fiehn O. Combining genomics, metabolome analysis, and biochemical modelling to understand metabolic networks. *Comp Funct Genomics.* 2001; 2: 155–168.

13 Fiehn O. Metabolomics – the link between genotypes and phenotypes. *Plant Mol Biol.* 2002; 48: 155–171.

14 Salminen JP, Roslin T, Karonen M, Sinkkonen J, Pihlaja K, Pulkkinen P. Seasonal variation in the content of hydrolyzable tannins, flavonoid glycosides, and proanthocyanidins in oak leaves. *J Chem Ecol.* 2004; 30: 1693–1711.

15 Yarnes CT, Boecklen WJ, Salminen J-P. No simple sum: seasonal variation in tannin phenotypes and leaf-miners in hybrid oaks. *Chemoecology.* 2008; 18: 39–51.

16 Lichtenthaler HK, Ač A, Marek MV, Kalina J, Urban O. Differences in pigment composition, photosynthetic rates and chlorophyll fluorescence images of sun and shade leaves of four tree species. *Plant Physiol Biochem.* 2007; 45: 577–588.

17 Conrad AO. Metabolomics of *Quercus* spp. to understand and predict resistance to *Phytophthora ramorum.* PhD Thesis. Ohio State University; 2015. http://rave.ohiolink.edu/etdc/view?acc_num=osu1429716653

18 Buche G, Colas C, Fougère L, Giordanengo T, Destandau E. Untargeted UHPLC-Q-TOF-HRMS based determination of discrimating compounds for oak species *Quercus robur* L. and *Quercus petraea* Liebl. identification. *Phytochem Anal.* 2021; 32: 660–671.

19 Bonello P, Blodgett JT. *Pinus nigra-Sphaeropsis sapinea* as a model pathosystem to investigate local and systemic effects of fungal infection of pines. *Physiol Mol Plant Pathol.* 2003; 63: 249–261.

20 Di Masi S, De Benedetto GE, Malitesta C, Saponari M, Citti C, Cannazza G, Ciccarella G. HPLC-MS/MS method applied to an untargeted metabolomics approach for the diagnosis of "olive quick decline syndrome". *Anal Bioanal Chem.* 2022; 414: 465–473.

21 Appel HM. Phenolics in ecological interactions – the importance of oxidation. *J Chem Ecol.* 1993; 19: 1521–1552.

22 Witzell J, Martin JA. Phenolic metabolites in the resistance of northern forest trees to pathogens – past experiences and future prospects. *Can J For Res.* 2008; 38: 2711–2727.

23 Sedio BE, Boya PCA, Rojas Echeverri JC. A protocol for high-throughput, untargeted forest community metabolomics using mass spectrometry molecular networks. *APPS.* 2018; 6(3): e1033.

24 Moritz T, Johansson AI. Plant metabolomics. In: Griffiths WJ, editor. *Metabolomics, Metabonomics and Metabolite Profiling.* Cambridge (UK): Royal Society of Chemistry; 2008. p. 254–272.

25 Lundborg L, Sampedro L, Borg-Karlson AK, Zas R. Effects of methyl jasmonate on the concentration of volatile terpenes in tissues of maritime pine and Monterey pine and its relation to pine weevil feeding. *Trees-Struct Funct.* 2019; 33: 53–62.

26 Chakraborty S, Whitehill JGA, Hill AL, Opiyo SO, Cipollini D, Herms DA, et al. Effects of water availability on emerald ash borer larval performance and phloem phenolics of Manchurian and black ash. *Plant Cell Environ.* 2014; 37: 1009–1021.

27 Paul A, de Boves Harrington P. Chemometric applications in metabolomic studies using chromatography-mass spectrometry. *Trends Anal Chem.* 2021; 135: 116165.

28 Wehrens R. *Chemometrics with R: Multivariate Data Analysis in the Natural Sciences and Life Sciences.* Heidelberg, New York: Springer; 2011.

29 Sperschneider J. Tansley insight machine learning in plant–pathogen interactions: empowering biological predictions from field scale to genome scale. *New Phytol.* 2020; 228: 35–41.

30 Rohart F, Gautier B, Singh A, Lê Cao KA. mixOmics: an R package for 'omics feature selection and multiple data integration. *PLoS Comput Biol.* 2017; 13: 1–19.

31 Kersten B, Ghirardo A, Schnitzler J-P, Kanawati B, Schmitt-Kopplin P, Fladung M, et al. Integrated transcriptomics and metabolomics decipher differences in the resistance of pedunculate oak to the herbivore *Tortrix viridana* L. *BMC Genomics.* 2013; 14: 737.

32 Bertić M, Schroeder H, Kersten B, Fladung M, Orgel F, Buegger F, et al. European oak chemical diversity – from ecotypes to herbivore resistance. *New Phytol.* 2021; 232: 818–834.

33 Six DL, Trowbridge A, Howe M, Perkins D, Berglund E, Brown P, et al. Growth, chemistry, and genetic profiles of whitebark pine forests affected by climate-driven mountain pine beetle outbreaks. *Front For Global Change.* 2021; 4: 1–22.

34 Zaman R, May C, Ullah A, Erbilgin N. Bark beetles utilize ophiostomatoid fungi to circumvent host tree defenses. *Meta.* 2023; 13: 239.

35 Fearer CJ, Conrad AO, Marra RE, Georskey C, Villari C, Slot J, et al. A combined approach for early in-field detection of beech leaf disease using near-infrared spectroscopy and machine learning. *Front For Global Change.* 2022; 5: 1–16.

36 Conrad AO, Villari C, Sherwood P, Bonello P. Phenotyping Austrian pine for resistance using Fourier-transform infrared spectroscopy. *AUF.* 2020; 46: 276–286.

37 Sambles CM, Salmon DL, Florance H, Howard TP, Smirnoff N, Nielsen LR, et al. Data descriptor: ash leaf metabolomes reveal differences between trees tolerant and susceptible to ash dieback disease. *Sci Data.* 2017; 4: 1–13.

38 Conrad AO, McPherson BA, Wood DL, Madden LV, Bonello P. Constitutive phenolic biomarkers identify naïve *Quercus agrifolia* resistant to *Phytophthora ramorum*, the causal agent of sudden oak death. *Tree Physiol.* 2017; 37: 1686–1696.

39 McPherson BA, Mori SR, Opiyo SO, Conrad AO, Wood DL, Bonello P. Association between resistance to an introduced invasive pathogen and phenolic compounds that may serve as biomarkers in native oaks. *For Ecol Manag.* 2014; 312: 154–160.

40 Nemesio-Gorriz M, Menezes RC, Paetz C, Hammerbacher A, Steenackers M, Schamp K, et al. Canditate metabolites for ash dieback tolerance in *Fraxinus excelsior. J Exp Bot.* 2020; 71: 6074–6083.

41 Conrad AO, Rodriguez-Saona LE, McPherson BA, Wood DL, Bonello P. Identification of *Quercus agrifolia* (coast live oak) resistant to the invasive pathogen *Phytophthora ramorum* in native stands using Fourier-transform infrared (FT-IR) spectroscopy. *Front Plant Sci.* 2014; 5: 521.

42 Villari C, Dowkiw A, Enderle R, Ghasemkhani M, Kirisits T, Kjær ED, et al. Advanced spectroscopy-based phenotyping offers a potential solution to the ash dieback epidemic. *Sci Rep.* 2018; 8: 17448.

43 Benton HP, Ivanisevic J, Mahieu NG, Kurczy ME, Johnson CH, Franco L, et al. Autonomous metabolomics for rapid metabolite identification in global profiling. *Anal Chem.* 2015; 87: 884–891.

44 Febrero-Bande M, de la Fuente M. Statistical computing in functional data analysis: the R package fda.usc. *J Stat Softw.* 2012; 51: 1–28.

45 Singh A, Ganapathysubramanian B, Singh AK, Sarkar S. Machine learning for high-throughput stress phenotyping in plants. *Trends Plant Sci.* 2016; 21: 110–124.

46 Jordan MI, Mitchell TM. Machine learning: trends, perspectives, and prospects. *Science.* 2015; 349: 255–260.

47 Liakos KG, Busato P, Moshou D, Pearson S, Bochtis D. Machine learning in agriculture: a review. *Sensors (Switzerland).* 2018; 18: 1–29.

48 Vabalas A, Gowen E, Poliakoff E, Casson AJ. Machine learning algorithm validation with a limited sample size. *PLoS One.* 2019; 14: 1–20.

49 Lê Cao K-A, Boitard S, Besse P. Sparse PLS discriminant analysis: biologically relevant feature selection and graphical displays for multiclass problems. *BMC Bioinf.* 2011; 12: 253.

50 Genuer R, Poggi J-M, Tuleau-Malot C. VSURF: an R package for variable selection using random forests. *R J.* 2015; 7: 19–33.

51 Stock NL, Doran MC, Bonners RF, March RE. Liquid chromatography/mass spectrometry for the detection of ash tree metabolites following emerald ash borer infestation. *Rapid Commun Mass Spectrom.* 2018; 32: 385–392.

52 Xia J, Psychogios N, Young N, Wishart DS. MetaboAnalyst: a web server for metabolomic data analysis and interpretation. *Nucleic Acids Res.* 2009; 37: W652–W660.

53 Xia J, Sinelnikov IV, Han B, Wishart DS. MetaboAnalyst 3.0—making metabolomics more meaningful. *Nucleic Acids Res.* 2015; 43: W251–W257.

54 Kucheryavskiy S. mdatools – R package for chemometrics. *Chemom Intell Lab Syst.* 2020; 198: 103937.

55 Conrad AO, Li W, Lee D-Y, Wang G-L, Rodriguez-Saona L, Bonello P. Machine learning-based presymptomatic detection of rice sheath blight using spectral profiles. *Plant Phenomics.* 2020; 2020: 1–10.

56 Mukrimin M, Conrad AO, Kovalchuk A, Julkunen-Tiitto R, Bonello P, Asiegbu FO. Fourier-transform infrared (FT-IR) spectroscopy analysis discriminates asymptomatic and symptomatic Norway spruce trees. *Plant Sci.* 2019; 289: 110247.

57 Naidoo S, Slippers B, Plett JM, Coles D, Oates CN. The road to resistance in forest trees. *Front Plant Sci.* 2019; 10: 273.

58 Sniezko RA, Koch J. Breeding trees resistant to insects and diseases: putting theory into application. *Biol Invasions.* 2017; 19: 3377–3400.

59 Hall RD, D'Auria JC, Silva Ferreira AC, Gibon Y, Kruszka D, Mishra P, et al. High-throughput plant phenotyping: a role for metabolomics? *Trends Plant Sci.* 2022; 27: 549–563.

60 Chlus A, Townsend PA. Characterizing seasonal variation in foliar biochemistry with airborne imaging spectroscopy. *Remote Sens Environ.* 2022; 275: 113023.

61 Fallon B, Yang A, Lapadat C, Armour I, Juzwik J, Montgomery RA, et al. Spectral differentiation of oak wilt from foliar fungal disease and drought is correlated with physiological changes. *Tree Physiol.* 2020; 40: 377–390.

62 Sapes G, Lapadat C, Schweiger AK, Juzwik J, Montgomery R, Gholizadeh H, et al. Remote sensing of environment canopy spectral reflectance detects oak wilt at the landscape scale using phylogenetic discrimination. *Remote Sens Environ.* 2022; 273: 112961.

63 Enderle R, Stenlid J, Vasaitis R. An overview of ash (*Fraxinus* spp.) and the ash dieback disease in Europe. *CAB Rev Perspect Agric Vet Sci Nutr Nat Resour.* 2019; 14: 025.

64 Cleary MR, Andersson PF, Broberg A, Elfstrand M, Daniel G, Stenlid J. Genotypes of *Fraxinus excelsior* with different susceptibility to the ash dieback pathogen *Hymenoscyphus pseudoalbidus* and their response to the phytotoxin viridiol – a metabolomic and microscopic study. *Phytochemistry.* 2014; 102: 115–125.

65 Villari C, Herms DA, Whitehill JGA, Cipollini D, Bonello P. Progress and gaps in understanding mechanisms of ash tree resistance to emerald ash borer, a model for wood-boring insects that kill angiosperms. *New Phytol.* 2016; 209: 63–79.

66 Koch JL, Carey DW, Mason ME, Poland TM, Knight KS. Intraspecific variation in *Fraxinus pennsylvanica* responses to emerald ash borer (*Agrilus planipennis*). *New For.* 2015; 46: 995–1011.

67 Qazi SS, Lombardo DA, Abou-Zaid MM. A metabolomic and HPLC-MS/MS analysis of the foliar phenolics, flavonoids and coumarins of the *Fraxinus* species resistant and susceptible to emerald ash borer. *Molecules.* 2018; 23: 1–19.

68 Rizzo DM, Garbelotto M, Davidson JM, Slaughter GW, Koike ST. *Phytophthora ramorum* as the cause of extensive mortality of *Quercus* spp. and *Lithocarpus densiflorus* in California. *Plant Dis.* 2002; 86: 205–214.

69 Grunwald NJ, LeBoldus JM, Hamelin RC. Ecology and evolution of the sudden oak death pathogen *Phytophthora ramorum. Annu Rev Phytopathol.* 2019; 57: 301–321.

70 Ockels FS, Eyles A, McPherson BA, Wood DL, Bonello P. Phenolic chemistry of coast live oak response to *Phytophthora ramorum* infection. *J Chem Ecol.* 2007; 33: 1721–1732.

71 Nagle AM, McPherson BA, Wood DL, Garbelotto M, Bonello P. Relationship between field resistance to *Phytophthora ramorum* and constitutive phenolic chemistry of coast live oak. *For Pathol.* 2011; 41: 464–469.

72 Sherwood P, Villari C, Capretti P, Bonello P. Mechanisms of induced susceptibility to *Diplodia* tip blight in drought-stressed Austrian pine. *Tree Physiol.* 2015; 35: 549–562.

73 Wallis C, Eyles A, Chorbadjian RA, Riedl K, Schwartz S, Hansen R, et al. Differential effects of nutrient availability on the secondary metabolism of Austrian pine (*Pinus nigra*) phloem and resistance to *Diplodia pinea*. *For Pathol.* 2011; 41: 52–58.
74 Cudmore TJ, Björklund N, Carroll AL, Staffan Lindgren B. Climate change and range expansion of an aggressive bark beetle: evidence of higher beetle reproduction in naïve host tree populations. *J Appl Ecol.* 2010; 47: 1036–1043.

6

Analysis of Volatile Organic Compounds

Andrea Ghirardo[1] *and Federico Brilli*[2]

[1] *Research Unit Environmental Simulation (EUS), Helmholtz Zentrum München, Neuherberg, Germany*
[2] *Institute for Sustainable Plant Protection (IPSP), National Research Council of Italy (CNR), Sesto Fiorentino (Florence), Italy*

6.1 Plant Volatile Organic Compounds

Plants, particularly trees, are able to emit many volatile organic compounds (VOCs) in relatively large amounts into the atmosphere. VOCs are molecules with low molecular weight and high vapor pressure at ambient temperatures. Chemically, VOCs include alkanes, alkenes, alcohols, aldehydes, ethers, esters, carboxylic acids, and isoprenoids (also called terpenes). Global biogenic VOC emissions into the atmosphere account for about 1100 $Tg\,yr^{-1}$ and mainly comprise isoprene (530 $Tg\,yr^{-1}$) followed by monoterpenes (160 $Tg\,yr^{-1}$), which often dominate the VOC profile of woody plants (1–3). All the other VOCs together comprise around 390 $Tg\,yr^{-1}$ (3).

Plant VOCs are classified according to their biosynthetic origin or chemical compositions (Figure 6.1). Isoprenoids (terpenes) contain one or multiple (C_5H_8) units of isoprene (e.g. $C_{10}H_{16}$, monoterpenes; $C_{15}H_{24}$, sesquiterpenes), their alcohol derivatives contain an additional oxygen (e.g. $C_{10}H_{14}O$, $C_{10}H_{18}O$, monoterpenoids; $C_{15}H_{24}O$, $C_{15}H_{26}O$ sesquiterpenoids) and may be irregular, such as the $C_{11}H_{18}$ homoterpenes (*E*)-4,8-dimethyl-1,3,7-nonatriene (DMNT) and the $C_{16}H_{26}$ 4,8,12-trimethyl-1,3(*E*),7(*E*), 11-tridecatetraene (TMTT). Among the VOCs, terpenes represent the broadest class of natural plant products accounting for over 25,000 structures (4) synthesized by a highly diversified family of terpene synthases (TPS) (5). Despite the wide spread of TPS in the plant kingdom, the high diversification of the TPS gene family results in the ability of each plant species to produce only a small fraction of terpenes (5). Biosynthesis of terpenes is derived from the C_5 building blocks isopentenyl diphosphate (IPP) and its allylic isomer, dimethylallyl diphosphate (DMAPP), the latter being used to produce the hemiterpene isoprene (6).

Monitoring Forest Damage with Mass Spectrometry-Based Metabolomics Methods, First Edition.
Edited by Carla António.

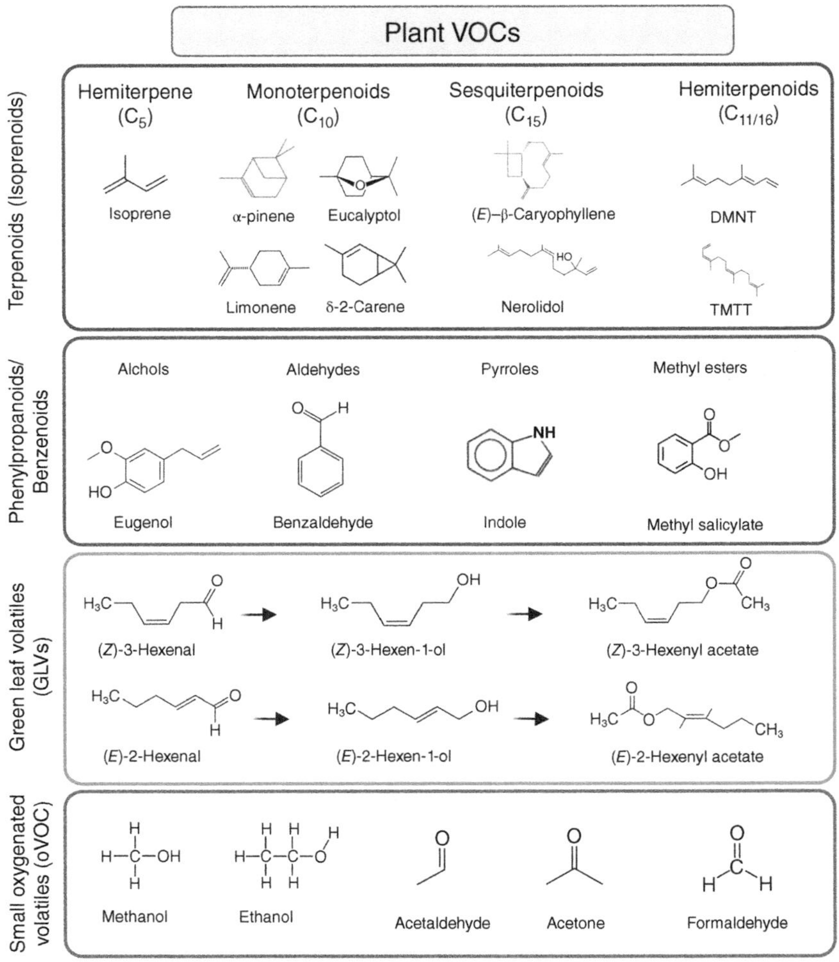

Figure 6.1 Major classes of plant volatile organic compounds (VOCs). Terpene and their alcohol derivatives (terpenoids, syn. isoprenoids) are sub-classified according to the C_5 branched isoprene units. Phenylpropanoids and benzenoids contain a benzene ring in their skeleton. Green leaf volatiles (GLVs) are a mixture of C_6 compounds, including alcohols, aldehydes, and esters originating from the lipoxygenase (LOX) pathway during the oxidation of the unsaturated linoleic and linolenic fatty acids. Oxygenated volatiles (oVOC) are usually referred to as small (C_1–C_6) aliphatic aldehydes, alcohols, acids, esters, and ketones. In the figure, the low abundant amino acids-derived aliphatic VOCs are omitted. Source: Figure created with BioRender.com.

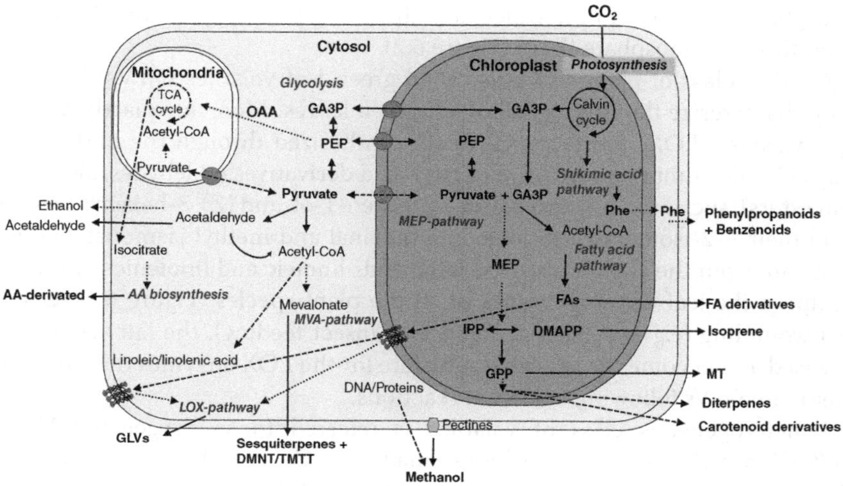

Figure 6.2 Main origins of plant VOCs, their biochemical pathways, and link to photosynthesis. Photosynthetic CO_2 fixation through the Calvin cycle delivers intermediates for the plastidic MEP pathway, the shikimic acid pathway, and the fatty acids pathway. Export from the plastids to the cytosol of the metabolic intermediates GA3P, PEP, and pyruvate supply directly the MVA-pathway and the AA biosynthesis-pathway (or indirectly via the mitochondrial TCA cycle). LOX-pathway occurs when the linoleic/linolenic fatty acids become available from plastidic and cell membranes. Abbreviations: AA, amino acids; CoA, coenzyme A; DMAPP, dimethylallyl pyrophosphate; DMNT, (*E*)-4,8-dimethyl-1,3,7-nonatriene; FA, fatty acid; GA3P, glyceraldehyde 3-phosphate; GLVs, green leaf volatiles; IPP, isopentenyl pyrophosphate; LOX, lipoxygenase; MEP, 2-C-methylerythritol 4-phosphate; MVA, mevalonate; MT, monoterpenes/monoterpenoids; OAA, oxaloacetate; PEP, 2-phosphoenolpyruvate; Phe, phenylalanine; TCA, tricarboxylic acid; TMTT, 4,8,12-trimethyl-1,3(*E*),7(*E*), 11-tridecatetraene.

In plants, terpenes are synthesized by both the cytosolic mevalonic acid (MVA) and the plastidial methylerythritol phosphate (MEP) pathways (Figure 6.2). MVA pathway is responsible for the formation of sesquiterpenes, DMNT and TMTT (6, 7). The MEP-pathway produces isoprene, monoterpenes, diterpenes, and volatile carotenoid derivatives. Although spatially separated, cross talk between the two pathways has been documented (8–10).

The second-largest class of plant VOCs includes phenylpropanoids and benzenoids, which contain a benzene ring in their skeleton (Figure 6.1). Biochemically, they originate from the aromatic amino acid phenylalanine of the shikimate pathway via the chorismate intermediate of the phenylpyruvate pathway (11). Typical examples of VOC belonging to this class are indole, benzaldehyde, phenylethanol, methyl salicylate, eugenol, methyleugenol, and chavicol. The phenylpropanoid pathway shares with the MEP-pathway the same metabolic precursors of the Calvin cycle, glycolysis, and pentose phosphate pathway (PPP), which are the phosphoenolpyruvate (PEP) and D-erythrose 4-phosphate (E4P) (Figure 6.2).

Another class of plant VOCs is called green leaf volatiles (GLVs) because they characterize the fresh smell of cut green leaves. More adequately named lipoxygenase (LOX) products, GLVs are synthesized through the LOX pathway (12) and comprise a mixture of fatty acid derivatives (aldehydes, alcohols, and esters), such as (*Z*)-3-hexenal, (*Z*)-3-hexen-1-ol, and (*Z*)-3-hexenyl acetate (and their *E*-2-isomers), also including nonanal and methyl jasmonate. GLVs originate from the C_{18} unsaturated fatty acids linoleic and linolenic acids that compose the leaf cell membranes of all the plant species (Figure 6.1). Upon leaf wounding (e.g. mechanical damage or insect feeding), the fatty acids are released and become the available substrate for the LOX enzymes that convert them into GLVs following oxidation reactions.

A heterogeneous class of volatiles is referred to as oxygenated VOC (oVOC), which commonly includes small (e.g. C_1–C_6) aliphatic aldehydes, alcohols, acids, esters, and ketones synthesized from different pathways. The most important oVOC emitted from plants is methanol; other oVOCs are formaldehyde, formic acid, ethanol, acetaldehyde, acetone, and acetic acid (13). Another class of VOCs originates from multiple routes linked to the branched-chain amino acids, such as alanine, valine, leucine, isoleucine, and methionine (6). Those amino acids-derived aliphatic VOCs are mainly aldehydes, often contain nitrogen and sulfur, and are commonly found in floral scents and fruit aromas (14). Some examples are 3-methylbutanal from leucine, 2-methylbutanoate from isoleucine, 2-methylpropanal from valine, and ethyl 3-methylthiopropionate from methionine (15).

6.1.1 Ecological and Physiological Functions of Plant VOCs

6.1.1.1 VOCs and Biotic Stresses

VOCs play key roles in plant defense. VOCs can act either directly as chemical weapons to prevent natural enemies (e.g. herbivory and pathogens) or indirectly as chemical cues to alarm neighboring plants of incoming danger, besides attracting beneficial predators (16). Many studies demonstrated that VOCs emitted from leaves could efficiently repel herbivory insects and nematodes and deter insect feeding (17–21). In addition, VOCs act as messages to recruit predators of the herbivores (i.e. indirect plant defense). For instance, the herbivory-induced release of specific VOC can attract predators and parasitoids to contrast insect damage upon feeding. Typical examples of signaling VOCs are the terpenoid (*E*)-β-caryophyllene (22) and the benzenoid indole (23). Furthermore, VOCs are key priming signals in plant systemic immunity to contrast pathogens (24). Once emitted, VOCs can be perceived by distal plant parts and trigger defense responses against herbivores or pathogens (18, 25–29). Plant VOCs have also been demonstrated to be potent antibacterial and antifungal chemicals because they exert an inhibitory action on the microbial pathogens' growth and germination (30, 31), besides having a direct toxic effect on herbivory (23). Overall, VOCs mediate antagonistic and beneficial intra- and interspecies interactions among organisms, thus functioning as crucial infochemicals in above and belowground plant communication, improving the fitness of both emitter and neighboring receiver plants (16, 26, 32–35). Altogether, VOCs are optimal non-invasive biomarkers of plant defense.

6.1.1.2 VOCs and Abiotic Stresses

Plant VOCs, particularly terpenes, help plants to cope with various abiotic stresses (36, 37). Studies initially conducted on isoprene, further extended to monoterpenes, have shown that terpenes can protect the photosynthetic apparatus against severe oxidative stress due to excessive light intensities, high temperature, drought, and ozone exposure (38–45). Furthermore, the production of terpenes can stabilize hydrophobic interactions in membranes by minimizing lipid peroxidation and membrane stiffness, which enhances the integrity of thylakoid membranes and improve the photosystem II photochemical efficiency (40, 46, 47). In addition, terpenes improve plant resistance to oxidative stresses by directly modulating the formation of reactive oxygen and reactive nitrogen species (ROS/RNS) (48, 49) or indirectly by regulating the level of gene expression and proteins involved in plant defense (50–53). While some terpenes can direct scavenge ROS, others form with ROS more reactive volatiles, such as methacrolein (MAC) and the methyl vinyl ketone (MVK), which have to be metabolized within the leaf through detoxification processes (54, 55).

6.1.2 Controls of Plant VOC Emissions

Emissions of VOCs are species-specific (1, 56) and vary with plant organs, developmental stage, and fluctuating environmental conditions (2, 57, 58) (Figure 6.3). The release of VOC from leaves can be constitutive or stress-induced (33, 59). Induction of VOCs following abiotic and biotic stresses is ubiquitous to all plant species (36, 60). Differently, only some species can constitutively produce and emit VOCs from leaves at high rates (1). Constitutive terpene emissions from leaves are controlled by light intensity, temperature, and CO_2 concentration, which affect photosynthesis (2) (Figure 6.3). However, the physicochemical characteristic of VOCs can also affect their emissions by influencing the gas-, aqueous-, and lipid-phase diffusion within the leaves (61). Leaf emission of VOCs, having a high Henry's law volatility constant (i.e. very high partition in the gas phase rather than in the liquid phase, e.g. isoprene and monoterpenes), does not depend on the degree of stomatal opening (61). On the other hand, water-soluble VOCs

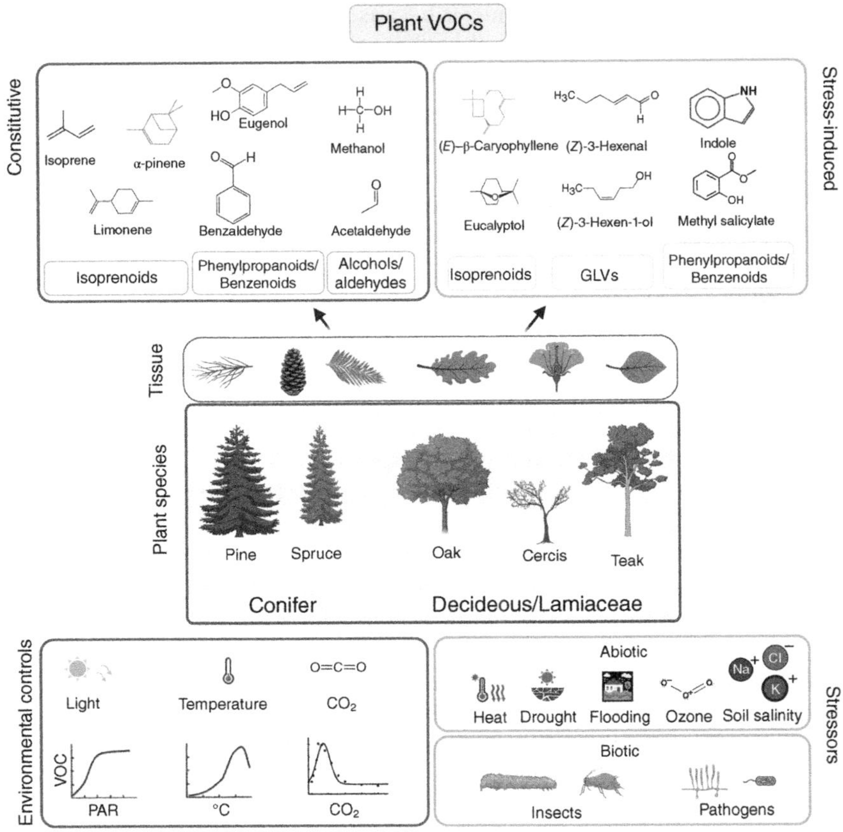

Figure 6.3 Plant VOCs are emitted mainly from foliage, fruits, and flowers. Constitutive emissions of VOC are controlled by light, temperature, and CO_2 concentration and are plant species-specific. Conifers and Lamiaceae species possess specialized structures to store VOCs; broadleaf plant species typically produce VOC by *de novo* emission. Abiotic and biotic stress factors can cause stress-induced VOC emissions. Source: Figure created with BioRender.com.

(e.g. acetone, methanol, and ethanol) with a low Henry's law constant are constrained by leaf stomata (61).

6.1.2.1 Constitutive VOC Emissions

Constitutive VOCs are those that some plant species regularly emit into the environment under both optimal and stressful conditions (59). Terpenes and methanol dominate the constitutive emissions, followed by low emissions of some oVOC (acetaldehyde, acetic acid, and acetone), GLVs, and benzenoids (57, 62–64). Most of the constitutive VOCs are emitted from foliage, although all other plant organs and tissues, such as flowers, fruits, and roots, can synthesize and release VOCs (Figure 6.3).

The constitutive emission rate of terpenes depends on plant genotype and environmental conditions (36). For instance, leaves of plants belonging to the genera *Populus, Salix,* and *Quercus* possess the genetic information that codifies isoprene synthases (ISPS), which produce isoprene (65). Differently, conifers (i.e. pines and spruces) are strong emitters of monoterpenes produced by the same gene family of terpene synthases (*TPS*). Generally, monocotyledons are low emitters of terpenes and high emitters of oVOC, most notably methanol, formaldehyde, formic acid, ethanol, acetaldehyde, acetone, and acetic acid (13).

Constitutive emission of terpenes can immediately follow their biosynthesis (so-called *de novo* emissions) or originate from specialized storage structures (so-called "pool emissions") (36, 66). *De novo* emission of terpenes is strongly light and temperature dependent and varies with atmospheric CO_2 concentrations (66–70). Increasing light intensities stimulate emissions up to a plateau value, while the temperature dependency of terpenes emission follows an Arrhenius-type function that reaches a maximum at a temperature optimum (e.g. 42–47 °C for isoprene), beyond which emission rates decrease quickly (69, 71, 72) (Figure 6.3). Substrate availability is another crucial factor controlling terpene emissions and can affect the temperature dependency of plastidic emissions (73, 74). Pool emissions of terpenes are mainly controlled by temperature (66) through evaporation and occur from those plant species that store large amounts of VOCs in specialized structures, such as resin ducts in conifers and glands in laminaceous species. Moreover, terpene emission capacity varies with plant ontogeny and nutrient availability (57,75).

Methanol, the second most abundant biogenic VOC (76), is ubiquitously released from leaves of all plant species mainly during the demethylation of cell wall pectins by pectin methyl esterases (PMEs) during growth processes and senescence (77, 78). Methanol is also produced by other cell metabolic processes, such as the demethylation of RNA, DNA, and proteins (78). Methanol formation is affected by temperature, light, water status, stomatal conductance, and leaf developmental stage (77,79,80).

6.1.2.2 Stress-Induced VOC Emissions

The occurrence of stressful conditions can induce large VOC emissions in all plant species. Stress-induced emissions from leaves have been demonstrated to depend on the type of stress and to scale with stress intensity (56,59,60,81–84). Biotic stresses, such as herbivory and aphid feedings as well as pathogenic infection, cause a large release of VOCs (Figure 6.4). Similarly, exposure to acute abiotic stress (e.g. heat and ozone) can result in substantial

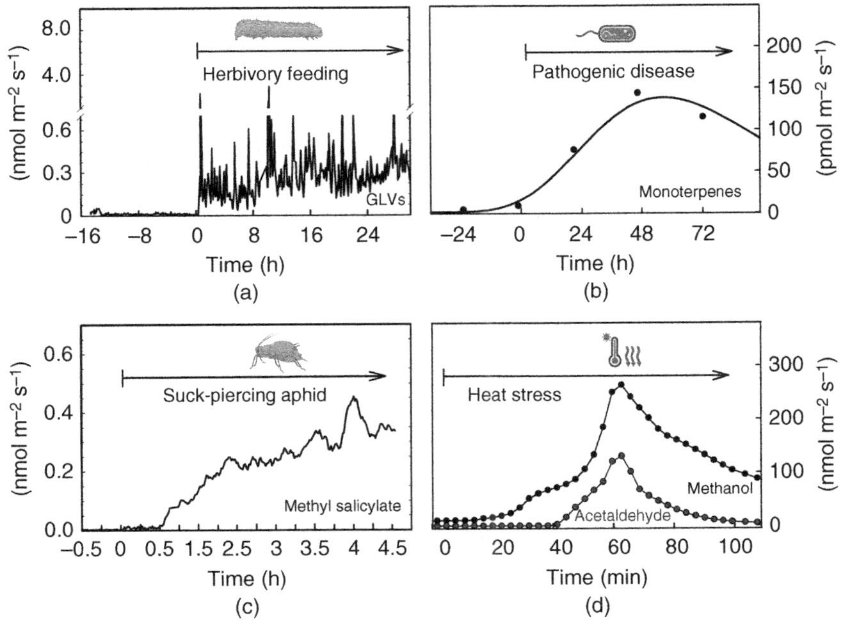

Figure 6.4 Examples of VOCs as biomarkers elicited following specific (a)biotic stress occurrence. (a) GLV emissions from European oaks (*Quercus robur*) upon herbivory (*Tortrix viridana*) feeding. (b) Monoterpene emissions from thale cress (*Arabidopsis thaliana*) after pathogenic infection of the bacteria *Pseudomonas syringae* pathovar *tomato*. (c) Methyl salicylate emissions from poplar (*Populus* × *canescens*) upon a suck-piercing aphid (*Chaitophorus leucomelas*) feeding. The burst of methanol and acetaldehyde emissions from poplar (*Populus* × *canescens*) upon cell membrane collapse as a consequence of heat-shock stress.

VOC emissions (56,85) (Figure 6.4). Pool emissions from specialized structures are very sensitive to mechanical damage, which causes an abrupt release of stored VOCs (86).

The onset of mild and severe stressful conditions differently impacts the induction of VOC emissions (59, 85). For instance, the emission of terpenes (i.e. isoprene and monoterpenes) resulted slightly affected by mild drought stress and fully recovered to pre-stress conditions together with the plants' status when the stress ceased (87–89). However, under severe drought stress, cell damages can abruptly cause a burst of VOC or even hamper photosynthesis along with the associated terpenes biosynthesis, thus preventing a full recovery of the emission rates (71,90,91).

Concomitant to the release of stress-induced VOCs, the occurrence of stressful conditions can decrease the emissions of constitutive VOCs by impairing their biosynthesis. For instance, heat stress lowered isoprene emissions while eliciting the release of oVOCs, GLV, and some monoterpenes in

Eucalyptus plants (92). This is because constitutive isoprene emission is tightly connected to photosynthesis. However, while heat stress conditions impair the photosynthesis-dependent *de novo* emissions of plastidic terpenes, emissions of GLVs (from membrane damage) and the cytosolic sesquiterpenoids (from activations of TPS) can be induced (93, 94).

Heat and drought stress (when occurs either separately or in combination) can dramatically affect VOCs emission, especially when they lead to foliage necrosis or tree mortality (95). In particular, the response of VOC emissions from dying trees is usually faster and may be detected earlier than other physiological parameters, such as photosynthesis and transpiration (95). Typically, early stages of foliage collapse lead to a quick and large burst of methanol, acetaldehyde, and GLVs (e.g. see Figure 6.4) coupled with monoterpenes emissions in coniferous plant species (94, 95).

Ubiquitous to all plant species, stress-induced emissions of the GLVs occur quickly upon leaf wounding (80, 83). This is the result of enzymatic reactions that start at the time of membrane damage when the linoleic/linolenic acids become available substrates for LOX enzymes, which are already present in the cellular sites (Figure 6.2). Moreover, other oVOCs, such as ethanol (96), acetaldehyde (97), acetic acid (98), and acetone (99, 100), are produced by pyruvate decarboxylase, alcohol dehydrogenase, and acetoacetate decarboxylase after leaf cutting and fatty acids degradation as well as under drought, salt stress or during anoxic conditions, such as flooding (94, 96, 99, 101, 102). Because the plant's response to abiotic and biotic stresses involves the production of ROS/RNS, this may trigger a similar change in VOC emission. Indeed, both ozone and herbivory feeding resulted in the activation of the same pathways that produce the phytohormones salicylic acid (SA), jasmonic acid (JA), and ethylene (ET) (49), leading also to the formation of sesquiterpenes and methyl-salicylate (MeSa) (103).

Typically, the release of specific monoterpenes (e.g. (*E*)-*β*-ocimene, linalool, and eucalyptol), sesquiterpenes (e.g. (*E,E*)-*α*-farnesene, (*E*)-*β*-farnesene), homoterpenes (e.g. 4,8-dimethyl-1,3*E*,7-nonatriene (DMNT), and 4,8,12-trimethyl-1,3(*E*),7(*E*), 11-tridecatetraene (TMTT)) and phenylpropanoids/benzenoids (e.g. indole and MeSa) are observed in response to (a)biotic stresses. Importantly, the profiles of stress-induced terpenoids can be specific to herbivory presence (104). Different from GLVs, changes in terpene emissions take longer to occur since they require preliminary enzyme and gene activations triggered by plant defensive mechanisms. Besides monoterpenes, late responses of VOCs following stress comprise MeSa, methyl jasmonate, cis-jasmone, indole, and sesquiterpene emissions.

Since changes in VOC emissions represent an early plants' response occurring immediately or after short-term (hours to days) stress episodes, measuring VOCs might become a useful non-invasive method to monitor the health status of forests under climate change.

6.2 Methodologies for Detecting Plant VOCs

Various techniques are available for detecting and quantifying the complex blends of VOCs emitted by plants (105, 106). Emissions of VOC can be measured at different spatial (leaf, whole plants, and ecosystem) and temporal (minutes to hours) scales (Figures 6.5 and 6.6). At a smaller spatial scale, single plant parts are enclosed in (static or dynamic) gas exchange systems ranging from small cuvettes to large chambers that allow the detection of

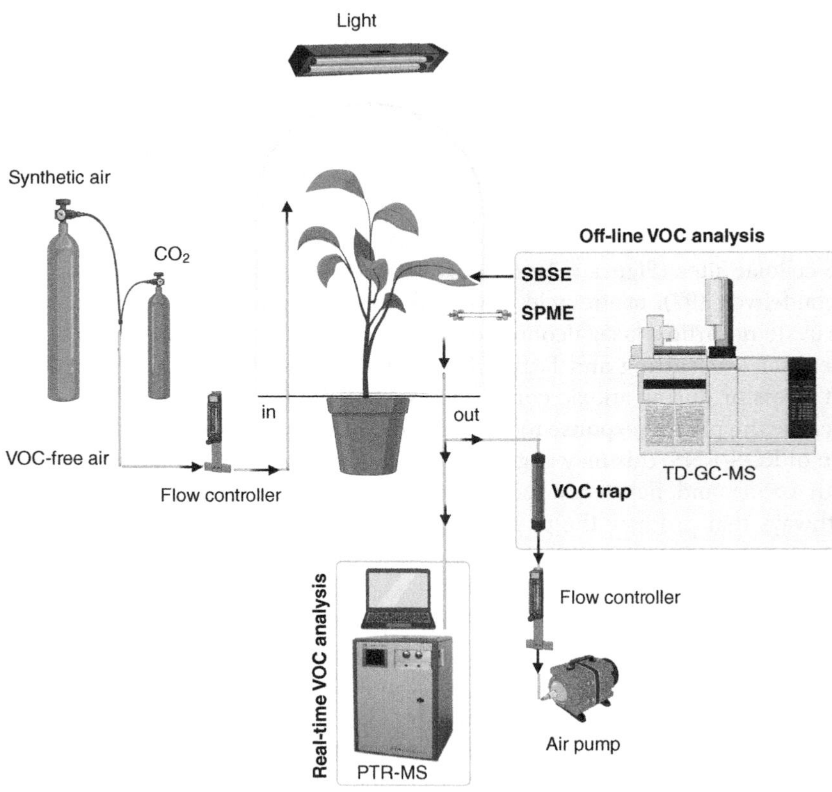

Figure 6.5 Example of a schematic analytical system to analyze plant VOCs. A whole plant is enclosed in a glass cuvette gas-exchange system under environmentally controlled or monitored conditions of light, temperature, air humidity, and CO_2 concentration. In dynamic headspace analysis (depicted in the figure), the cuvette is continuously flushed with a defined rate of VOC-free air. The air exiting the cuvette can be analyzed in real-time by PTR–MS and/or offline by TD–GC–MS techniques. Conversely, in *static* headspace analysis, the cuvette is not actively flushed by VOC-free air, allowing VOCs to accumulate for a certain period. This results in an enrichment of VOCs in the air surrounding the plant, which can then be collected: (i) passively, using SPME or SBSE; or (ii) actively, through VOC trap methods. Source: Figure created with BioRender.com.

Figure 6.6 Examples of different enclosure systems used to measure plant VOCs. (a) Portable, commercial cuvette system (GFS-3000, HeinzWalz, Effeltrich, Germany) designed for gas-exchange measurements of CO_2 and H_2O and adapted for VOC analysis (56, 107). (b) Custom-made portable cuvette system which includes a thermoelectric Peltier element and LED light source connected to an Infra-Red Gas Analyzer (IRGA) (66, 74, 108–110). (c) FEP foil is used to enclose the whole plant (*Quercus ilex*) in greenhouse VOC studies. (d) Simple and cost-effective PET "cooking bag" for field assays on *Acer Platanoides*. (e) Enclosure of branch and stem with PET bags to analyze VOC emissions from different plant tissues of *Parrotia persica*. Source: (Photo courtesy of Agnes Meyer-Brandis). Insert: Glass cuvette used for soil/root VOC emissions measurements using SBSE and purge and trap method (a charcoal filter cleans the inflow air). (f) Glass cuvettes for parallel measurements of immunity-associated VOC emissions from pathogen-infected (*Blumeria graminis* f. sp. *hordei*) versus control barley (*Hordeum vulgare*) plants (27). (g) Automatic VOC-screen system made of 24 cuvettes accommodated in a phytotrone facility for climate simulation experiments (111, 112). Source: (111). Jud et.al., 2018/Springer Nature/CC0 1.0. (h) Dynamic flow-through cuvette based on FEP foil used in the field to measure VOC from shoots of the conifer *P. sylvestris*. Source: (113). Vanhatalo et.al., 2018/European Geosciences Union/CC BY 4.0. (i–j) Chamber system for field ecosystem flux measurements and chemical characterization of VOCs from the biofuel crops maize, oil-seed rape, and ryegrass using PTR–QMS and TD–GC–MS (57, 114).

VOC emissions under monitored environmental conditions. At a larger spatial scale, emission fluxes of VOC are measured above entire ecosystems through micrometeorological eddy covariance methods. Because VOC concentrations in air range from part per trillion volume (pptv) to part per billion volume (ppbv), sophisticated and sensitive instrumentation is required. Mass spectrometry (MS)-based methods are commonly used to reliably measure a broad range of VOCs at trace level, and instrument resolution, sensitivity as well as reproducibility are crucial parameters in VOC analysis. The traditional technique for VOC analysis is thermal desorption gas chromatography mass spectrometry (TD–GC–MS), which involves adsorption and pre-concentration of VOCs during sample collection (Figure 6.5). Later, proton transfer reaction mass spectrometry (PTR–MS) has been developed, thus allowing for fast and real-time measurements of plant VOC emissions. To prevent interactions between the experimental system and VOCs, all equipment, such as enclosures and tubing, must be inert and non-sticky (115). Commonly used materials for VOC analysis are glass, plexiglass, polytetrafluoroethylene (PTFE), perfluoroalkoxy alkanes (PFA), polyvinylfluoride (PVF), fluorinated ethylene propylene (FEP), and polyethylene terephthalate (PET).

6.2.1 Static Headspace Analysis

The simplest procedure to collect VOCs is to enclose an entire plant or part of it (e.g. a leaf) into a gas-tight cuvette and allow VOCs to accumulate into a defined air volume for a specified period of time. This procedure is referred to as static headspace analysis and may be assisted by adsorbent materials that can trap and concentrate VOCs for TD–GC–MS analysis (see Section 6.3.1); alternatively, VOCs contained in the headspace can be directly analyzed in real-time by PTR–MS (see Section 6.3.2). When adsorbents are used, sample collection is either active or passive (Figure 6.5). In active sampling, a defined amount of air is forced through a trap (a tube filled with adsorbents) using a pump. In passive sampling techniques, the (ad)sorbents are left in the headspace for a certain time to passively accumulate VOCs. The most widely used passive techniques are solid-phase microextraction (SPME) and stir bar sorptive extraction (SBSE) (105).

Static headspace analysis can also be used to measure VOC fluxes. In this technique, enclosures are kept mainly open for growing plants within them under natural conditions and shortly closed for measuring the buildup of VOC concentrations (115). The VOC fluxes can be derived through the differences between the VOC concentrations measured before and after closing the enclosures and by assuming a constant emission rate during the sampling. However, this method has some drawbacks. The increase of VOC concentrations and perturbation of the environmental conditions (e.g. increasing temperature, pressure, and humidity) within the enclosures can directly affect VOC emissions, thus leading to inaccurate estimation of VOC fluxes. These limitations can be avoided by using automatic large chamber systems and performing multiple VOC measurements for short time periods (e.g. a few minutes) to ensure a linear increase in VOC concentrations (57). In such semi-static headspace measurements, the slopes of significant linear regressions provide the emission rates of plants or small ecosystems.

6.2.2 Dynamic Headspace Analysis

In dynamic headspace analysis, part or a whole plant is enclosed either in commercially available or customized cuvettes that are continuously flushed with a constant and known flow of VOC-free air, typically ranging from 0.2 to 10 L min^{-1} (see Figure 6.6). Within the dynamic cuvettes, all the environmental conditions (light intensity, air humidity, temperature, and CO_2 concentrations) are controlled (e.g. in laboratory studies) or monitored (e.g. in field studies): light is supplied by an external source (e.g. metal-halide lamp, quartz halogen lamp, and light-emitting diode [LED]) and/or measured by photosynthetic active radiation (PAR) light sensors; CO_2 concentration is monitored by Infra-Red Gas Analyzer (IRGA), and eventually supplied by a gas cylinder in measurements that uses synthetic air (which is a mixture of N_2 and O_2); leaf temperature is monitored by thermocouples and controlled by Peltier resistances; air humidity can be monitored by IRGA and regulated by dew point generators; the airflow through the cuvette is maintained constant by rotameters or mass flow controllers (MFCs).

The use of dynamic, flow-through cuvettes is ideal for quantifying VOC emission rates from the vegetation under constant or monitored environmental conditions based on either the surface area or the amount of biomass enclosed (62, 116, 117). Emission fluxes of VOCs are calculated from the concentration of VOCs measured in the air leaving the cuvette and then background-corrected by measurements of the empty cuvette. Dynamic cuvette measurements enable precise and rapid monitoring of VOC fluxes from a few seconds to hours, in addition to the possibility of recording the gas exchange of CO_2 and H_2O when combined with IRGA. Furthermore, this experimental approach allows studying *in vivo* the relationships between plant physiological processes (photosynthetic CO_2 assimilation, stomatal conductance) and VOCs biosynthesis, which is crucial to investigate the physiological role of VOCs as well as for building mathematical models that can estimate plant VOCs emission from vegetation under fluctuating environmental conditions.

6.2.3 Eddy Covariance

At the ecosystem level, vertical bidirectional (emission and deposition) turbulent fluxes of VOC are measured (e.g. from a tower) through the so-called eddy covariance (EC) method (118). EC method allows estimations of fluxes over a broad surface area whose dimensions depend on air turbulence and wind direction. EC is based on fast and highly frequent (between 5 and 20 Hz) monitoring of changes in VOC concentrations measured by PTR–MS (see Section 6.3.2) and vertical wind speed measured by an ultrasonic anemometer. Mathematically, EC fluxes are computed as the covariance between the instantaneous deviation in vertical wind speed and the instantaneous deviation in VOC mixing ratios from their respective mean values, multiplied by mean air density. Furthermore, computed VOC fluxes undergo several processing steps, including conversion of signals from voltages to physical units, de-spiking, correction for time delays, de-trending, application of frequency response, and other corrections that account for instrumental features and measurement conditions. Finally, processed VOC fluxes are evaluated through a procedure that considers the terrain heterogeneities and changing meteorological conditions to identify high-quality data (119). Eventually, in long-term datasets, missing VOC flux data can be gap-filled following mathematical methods (120). Tower-based estimation of EC fluxes of VOC has, typically, a time resolution of half-hour and it is representative of the footprint area.

6.3 Analytical Systems for Measuring Plant VOCs

6.3.1 Thermal Desorption Gas Chromatography Mass Spectrometry

Plant VOCs have been traditionally analyzed by trapping techniques coupled with gas chromatography mass spectrometry (GC–MS) (121,122), as the use of adsorbents has the advantage of increasing VOC-detection sensitivity. Trapped VOCs can be eluted from the adsorbent either with solvents or by increasing the temperature in a thermal desorption (TD) unit interfaced with the GC–MS (i.e. TD–GC–MS). The disadvantages of using solvents are that the analytic process is laborious, causes sample dilution, and can potentially introduce contaminations leading to artifacts. Therefore, TD–GC–MS represents the most widely used technique for VOC analysis because it facilitates a sensitive and accurate quantitative analysis, while enabling chemical identification of VOCs.

Overall, the analysis of VOCs through TD–GC–MS comprises a three-step procedure (Figure 6.7). In the first step (sample collection), adsorbent materials are used to trap and concentrate VOCs. In the second step (TD–GC–MS analysis), the adsorbents are heated in the TD unit to release the VOCs into the GC–MS, where compound separation is achieved via the GC column and

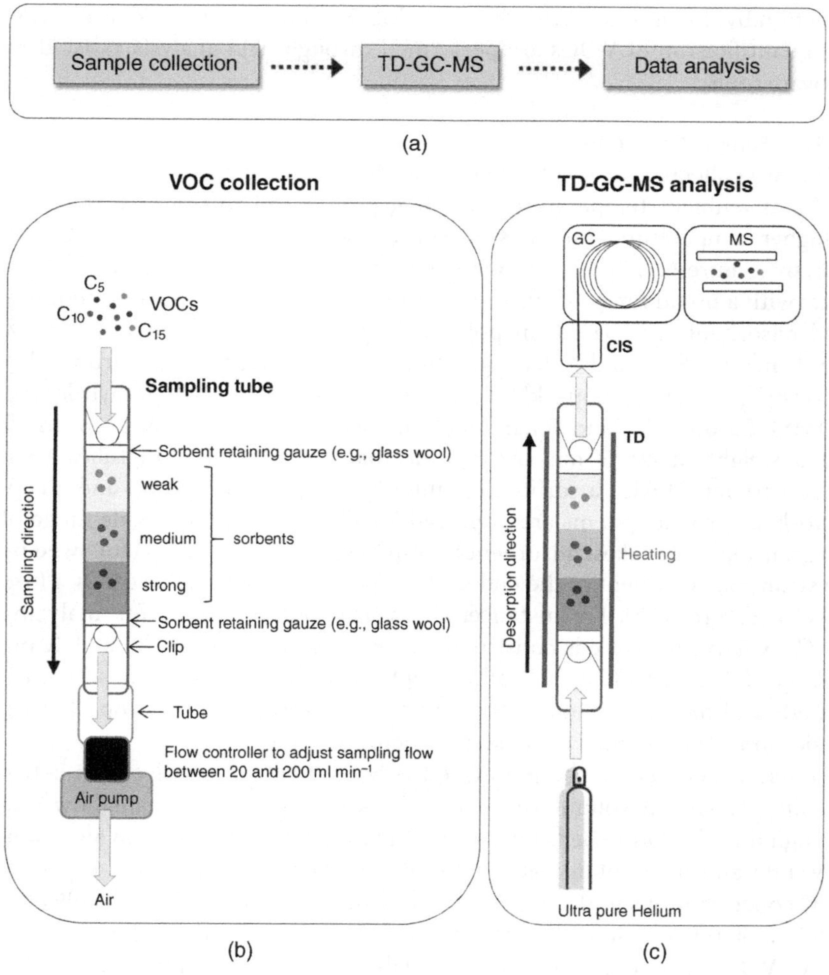

Figure 6.7 (a) Schematic workflow of VOC analysis with TD–GC–MS. (b) A constant airflow (typically between 20 and 200 mL min^{-1}) of sampling air is diverted into a (glass) tube containing adsorbents. The figure depicts a three-bed adsorbent tube, where the first one is the weakest to trap large molecules of VOC (e.g. C_{10}–C_{15}) with high boiling points, then there is the medium (e.g. for C_5–C_{10}), and finally the strongest adsorbent for low boiling point VOC (e.g. C_3–C_5). This procedure allows the detection of a broad range of VOCs and avoids a poor recovery of the analytes. (c) For the analysis of sampled VOCs, the sorbents are heated inside a thermal desorption (TD) unit in a reverse flow of ultrapure carrier gas (e.g. helium) and injected first into a narrow cold trap (cryo injection system, CIS), which is quickly heated up (>100 °C min^{-1}) to increase chromatographic resolution (sharp peaks) and sensitivity (VOC detection) of a wide range of VOCs having different volatility.

detection by the mass selective detector (MSD). In the last step, identification and quantification of VOCs are performed through data analysis assisted by software tools.

6.3.1.1 Sample Collection

VOCs are collected using adsorbent materials, which have the ability to trap VOCs at ambient temperature (sampling phase) and to fully release them at higher temperatures (analysis phase) during desorption prior to GC–MS analysis (Figure 6.7). There are many commercially available types of adsorbents with a broad range of surface densities (SD, 10–300 $m^2\ g^{-1}$). Commonly used adsorbents are based on poly (2,6-diphenyl-*p*-phenylenoxid (TENAX TA; 24 $m^2\ g^{-1}$ SD), and several graphitized carbon black (e.g. Carbopack™ B 100 $m^2\ g^{-1}$ SD, Carbopack™ X, 240 $m^2\ g^{-1}$ SD), or polydimethylsiloxane (PDMS). Carbopack™ and Tenax® materials with a particle size of 40–80 mesh and a weight of 40–60 mg are typically packed into GC–MS tubes (also called cartridges) with a sorbent retaining gauze (e.g. glass wool) and a clip. Multi-bed sorbent systems are preferred for the collection of a wide range of VOCs, using the weakest adsorbent to trap bigger VOCs (e.g. C_{15}) followed by the strongest adsorbent at the end of the tube to trap the smallest VOCs (e.g. C_3–C_5) (Figure 6.7b). For instance, a two-bed sorbent system for analyzing C_5–C_{15} volatile terpenoids consists of 40 mg of Tenax® TA 60–80 and 40 mg Carbopack™ X 40–60, which allows optimal collection of both sesquiterpenoids and monoterpenoids (Tenax®) as well as isoprene (Carbopack™ X), in addition to GLVs and a wide range of benzenoids (62).

Sample collection can be performed either actively or passively. In active sampling, a known volume of air sample is passed at a constant flow rate through the adsorption beds for a specific period of time by using an air pump. Flow rate and final volume are preliminarily chosen based on the expected VOC concentrations in the air. For accurate quantitative analysis, all collected VOCs must remain on the adsorbent trap during the collection. Thus, to avoid loss of VOCs from the adsorbent materials, the flow and volume of sampled air should not exceed half of their breakthrough volume (BV). The BV is the volume of air per gram of adsorbent that purges VOCs out of the adsorbent tube at a specific temperature. A practical approach to check for BV is to recollect and analyze the air at the tube outlet using another tube.

In passive sampling, the adsorbent materials are let in contact with air for a certain amount of time to accumulate VOCs (106). The most used techniques are SPME and SBSE, which rely on the adsorption of VOCs on an inert fiber coated with one or more adsorbents or on a stir bar coated with a thin layer of PDMS, respectively. In SPME and SBSE, VOCs are partitioned between the headspace and the sorbent material. This partition is affected by temperature, volume, and chemical/physical properties of both sorbent and VOCs. Therefore, the concentrations of VOCs can be calculated by knowing the uptake rate of the sorbents and the environmental conditions or through calibration with known amounts of VOCs.

6.3.1.2 TD–GC–MS Measurements

After collection into adsorbent tubes, VOCs are thermally desorbed inside a TD unit by gradually increasing the temperature (e.g. from 35 to 300 °C at a rate of 280 °C min^{-1}) of the tubes under a stream of an inert carrier gas (e.g. helium or nitrogen) (Figure 6.8). All the desorbed VOCs enter a cryo injection system (CIS) where they are cryo-trapped (e.g. −50 °C using liquid nitrogen) onto a small-volume injector glass liner filled with an adsorbent (e.g. Tenax®). In the CIS, the VOCs are positioned at the head of the column and then released from the trap to enter the chromatography in a very narrow band using a fast heating rate (e.g. 12 °C s^{-1}). This procedure, called cryo-focusing, results in very sharp chromatographic peaks with a high signal-to-noise ratio; it can also be helpful in removing unwanted molecules with a different boiling point from those being studied. Once into the GC column, the VOCs are separated through chromatography based on their physical and chemical properties. The GC instrument consists of a temperature-controlled oven capable of increasing the temperature of the column up to over 300 °C, which allows VOCs to be released from the stationary phase to the mobile phase under a carrier gas stream. The temperature-dependent release of VOC leads to the elution of single VOCs at specific times, known as retention times (RT). The stationary phase and the length of the capillary column are chosen depending on the chemical properties and the complexity of the VOC mixture. For instance, a long and thin nonpolar column (60 m × 250 μm, film thickness 0.25 μm) based on 5/95% diphenyl/dimethylpolysiloxane (DB-5) can be used for a comprehensive analysis of nonpolar and moderately polar VOCs (56).

After GC separation, neutral VOCs enter the MS, where they become first ionized by an ion source (Figure 6.8). In the electron impact ionization (EI) mode, electrons ($e^{-\cdot}$) are produced from the cathode of the ion source as a result of heating a metal filament to an incandescent temperature (thermionic emission), which is achieved by passing an electric current through it. On the other side of the cathode, the anode function as an electron trap to collect the unused electrons. The emitted electron beam is accelerated to 70 electron volts (eV) and, under vacuum pressure (10^{-5}–10^{-6} mbar), induces ionization and fragmentation of neutral VOCs. In this process, an electron from the VOC molecule (the so-called precursor or parent molecule, M) is expelled during the collision process; thus, M is converted into a radical cation ($M^{+\cdot}$) (Figure 6.8). This type of hard ionization leads to the fragmentation of the precursor molecules of VOC. The fragmentation pattern of M depends on the chemical structure, representing a specific fingerprint of each VOC, which is then used for identification. Then, the $M^{+\cdot}$ are accelerated by a repeller electrode and focused into an ion lens system to finally enter the MSD

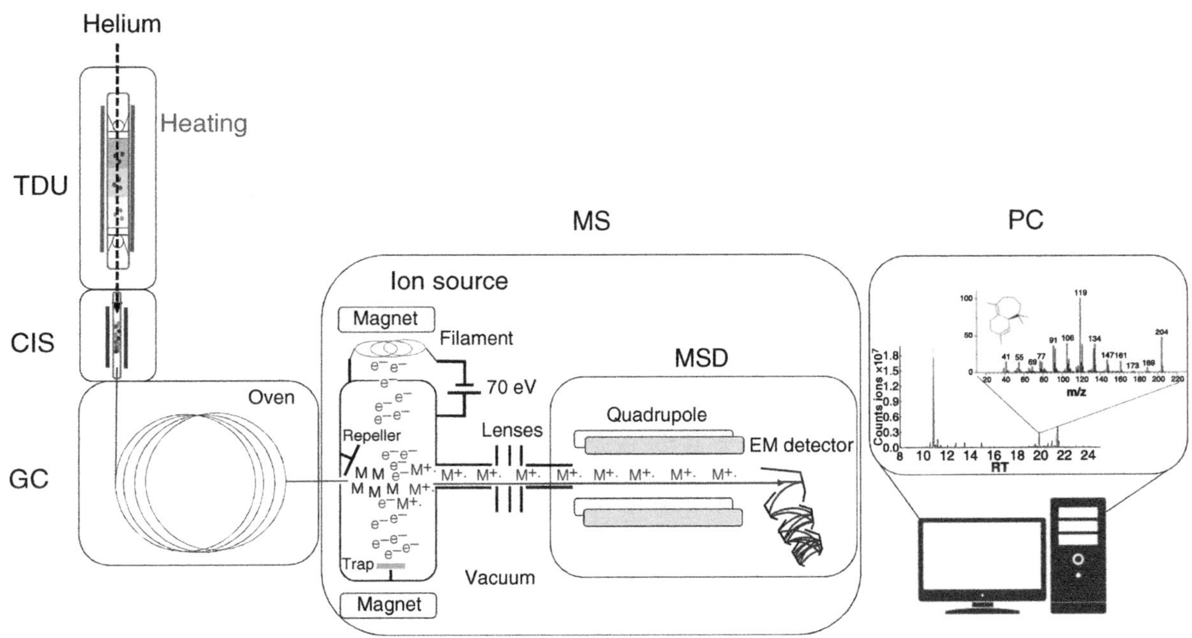

Figure 6.8 Schematic representation of thermal desorption gas chromatography mass spectrometer (TD–GC–MS) using an electron impact (EI) ion source and a quadrupole-based mass selective detector (MSD) for VOC analysis. VOCs trapped in adsorbents are thermally desorbed in the TD unit and moved to the cryo injection system (CIS) for refocusing and purging before being separated by GC and entering the MS. In the MS, the neutral molecules (M) are ionized to radicals $M^{+\cdot}$, separated by the quadrupole according to their mass-to-charge ratios (m/z) and detected by an electron multiplier. Ion signals are recorded in a chromatogram containing a series of mass spectra.

with fixed kinetic energy. The MSD separates the VOCs according to their mass-to-charge ratio (m/z) and detects VOCs by producing signals, which are amplified by an electron multiplier (EM).

There are several types of MSD having different mass resolutions (i.e., the ability to distinguish two peaks of different m/z). The most common type is the more affordable quadrupole, in which the electric current (voltage) can be varied between the four poles in order to select specific m/z to reach the detector. The resolution of a quadrupole-based MSD is typically 0.5–0.7 atomic mass unit (amu), whereas that of both time-of-flight (TOF) and Ion Trap mass analyzers can resolve up to 100,000 $M/\Delta M$ (where M is the nominal mass of a peak related to VOC within the mass spectrum, and the ΔM is the peak width measured at half of the peak height).

Quadrupole-based GC–MS can acquire mass spectra of VOCs either in full scan mode for qualitative/quantitative analysis or in single ion monitoring (SIM) mode for quantitative analysis. In full scan mode, the whole range of m/z is continuously measured (e.g. from 35 to 300 amu) at high speed (two to five times per second), with a dwell time (time that the MSD employs to collect the intensity of a specific m/z) of 0.3–3 ms, thus resulting in a chromatogram composed by summing up all the measured m/z intensities (so-called total ion chromatogram, TIC) which allow wide spectra acquisition and VOC identification. Differently, in SIM mode, only a few ions (those fragments characterizing a VOC) are selected and monitored (e.g. m/z 93, 121, 136 for monoterpenes) with a high dwell time (i.e. 20–100 ms) to enhance sensitivity and more accurate quantitative analysis. In both TIC and SIM modes, 8–15 mass spectra need to be acquired in order to define a single chromatographic peak.

6.3.1.3 Qualitative Analysis

The qualitative analysis aims to identify VOCs by using spectral data matching and RT. Recorded spectra are compared with those found in available databases (e.g. Wiley and the National Institute of Standards and Technology (NIST) libraries, or user-built databases from pure standards). Identification is achieved unambiguously when both spectra and RT match those of pure standards. However, when VOCs are not available as standards or the spectral match is ambiguous, the calculation of the retention index (RI) based on the relative RT to alkane standard mixture (C_5–C_{20}) allows result comparison with existing literature.

A common procedure is the use of the Kováts RI (123) defined as:

$$RI = 100c + 100\frac{\log(t'_R)_x - \log(t'_R)_c}{\log(t'_R)_{c+1} - \log(t'_R)_c} \tag{6.1}$$

where the $(t'_R)_c$ and $(t'_R)_{c+1}$ are the corrected RT values ($t'_R = t_R - t_0$) of the alkane standards eluting before $(t'_R)_c$ and after $(t'_R)_{c+1}$ the RT peak of the VOC $(t'_R)_x$ to be identified. The "c" and "c+1" refer to the numbers of carbon atoms of the two n-alkane standards. t_0 is the void time, also known as dead time, which is the volume of the packed capillary column divided by the flow rate that roughly corresponds to the injection time. This calculation can be applied to an isothermal GC temperature program. For non-isothermal linear temperature programs, the Kováts RI calculation is modified as follows (124):

$$RI = 100c + 100\frac{(t'_R)_x - (t'_R)_c}{(t'_R)_{c+1} - (t'_R)_c} \tag{6.2}$$

The resulting Kováts index is, therefore, independent of the GC–MS used to analyze the VOCs and allows comparison with literature data originated under different analytical conditions.

6.3.1.4 Blank Measurements

The TD–GC–MS system and both the cleaning and storage of tubes needs to be contaminant-free in order to avoid artifacts in VOC analysis. For this reason, quality checks are performed using two types of blank measurements, which are run together with the sample batch: (i) the TD–GC–MS blank; and (ii) the tube blank. The TD–GC–MS blank is performed by using the same analytical procedure but without any adsorbent tube. The tube blank is performed with conditioned (clean) tubes from the same batch of those used for sampling and that have been stored under the same conditions. Typically, 10% of tube blanks are randomly analyzed from the same batch used for sampling. During the conditioning procedure, tubes are cleaned form VOCs using an oven heated at a high temperature (e.g. the adsorption max temperature of 300–350 °C), or 10 °C higher than the desorption temperature set on the TD–GC–MS, under a stream of 80 mL min^{-1} ultrapure He or N_2.

6.3.1.5 Internal Standard

Internal standard (ISD) is a known quantity of a compound, having chemical characteristics similar to the VOC of interest, that is added to the sample before the TD–GC–MS analysis. A labeled (e.g. ^{13}C-, ^{2}H) analog of the targeted analyte is typically used, although, for VOC analysis, it is sufficient to use an isomer of the targeted VOC that does not appear in the collected sample. The use of ISD is required to correct variations in VOC analysis (data normalization) mainly due to changes in MSD sensitivity (i.e. variations in the ionization efficiency and detector aging) but also to take into account the losses in the collected samples. Before VOC collection, each adsorbent tube is spiked with an appropriate ISD dissolved into a low-boiling point solvent, which quickly evaporates from the adsorbent material and does not interfere with VOC analysis. Examples of ISD and solvent used for monoterpene analysis are δ-2-carene (which rarely occurs naturally) and methanol, respectively.

6.3.1.6 Quantitative Analysis

Quantitative analysis of VOCs is performed through calibration, which measures different amounts (moles) of a standard mixture of VOCs in a range consistent with that absorbed onto TD-GC–MS tubes (e.g. 40–800 pmol). Regulatory guidelines recommend using five to six different standard concentrations to define a calibration curve and determine the MSD linearity for each standard. In addition, all standards employed for calibration must contain the same ISD concentration as the VOC sample to perform data normalization.

Overall, the quantitation of VOCs is based on the linear response of the MSD to the peak area resulting from a known concentration of VOC standard analyzed, and defined as follows:

$$y = mx + b \tag{6.3}$$

where y is the TD-GC–MS response (peak area), x is the concentration of the VOC, m is the slope of the calibration curve, and b is the y-intercept indicating the VOC concentration in the background. However, when the response is corrected by the background measurements, the calibration curve can be forced to pass through the axis origin by assuming that $b = 0$.

The background quantification of the experimental system is required to define the limit of detection (LOD) for a specific VOC. The LOD is the smallest amount of a VOC that can be accurately detected and quantified by the TD-GC–MS. The LOD can be calculated by using the standard deviation (σ) of the background measurements divided by the signal-to-noise (S/N) ratio. A common S/N value is 3, but other values can be used depending on the desired level of confidence. To calculate σ, multiple background measurements are needed over a sample size large enough (at least the same number of the sample replicates) to obtain a representative value. Whereas the LOD defines the lowest quantity of VOC that can be detected, the limit of quantification (LOQ) indicates the lowest quantity of VOC that can be quantified according to an acceptable uncertainty (μ) range. The μ of the absolute quantification is calculated by taking into account the uncertainties that occurred during standard preparation, sample collection, and measurements of the calibration curve. Air samples are often replicated three times to calculate the measurement uncertainty.

After measuring the calibration curve, the mole response factor (RF) is calculated as the slope of the regression line between the quantity of the VOC standard analyzed (STD) and the normalized peak area (response) to the ISD (Figure 6.9). Therefore, the RF defines the ratio between the response of the MSD and the quantity of a VOC. The peak area related to a single VOC is thus divided by the RF value, resulting in the amount of VOC trapped in the TD-GC-MS adsorbent tube.

$$\mathrm{RF}(A) = \frac{\mathrm{Peakarea(STD)}}{\mathrm{mole(STD)}} \tag{6.4}$$

$$\mathrm{mole}(A) = \frac{\mathrm{Peakarea}(A)}{\mathrm{RF}(A)} \tag{6.5}$$

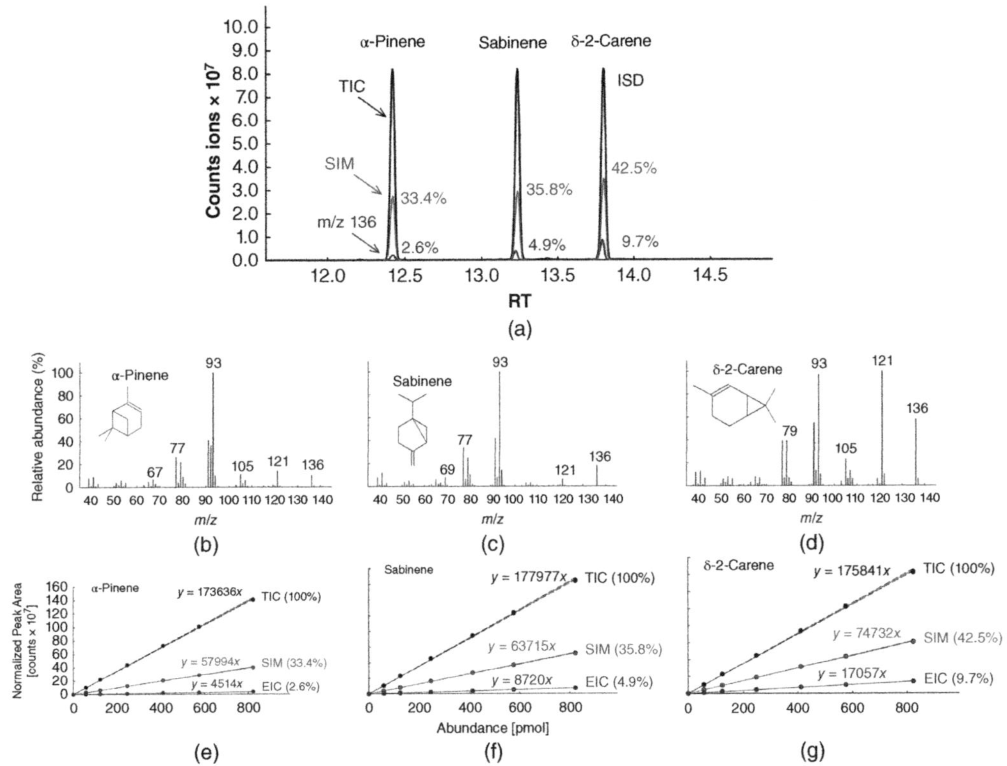

Figure 6.9 Example of the mass selective detector (MSD) response to the same amount of three different monoterpenes and their effect on the resulting response factors (RFs) based on (a) total ion chromatograms (TIC), single ion monitoring (SIM), and extracted ion chromatogram (EIC). Spectra of: (b) α-pinene; (c) sabinene; (d) δ-2-carene showing their different fragmentation patterns. The RFs based on TIC (RF_{TIC}) are similar (± 2%) in (e) α-pinene; (f) sabinene; (g) δ-2-carene. RFs calculated based on SIM and EIC are proportional to RF_{TIC} and specific to VOCs, representing the percentages of their measured ions relative to the sum of ions composing the VOC spectrum and are related to the characteristic fragmentation pattern of a given VOC.

where *A* is the VOC analyte, and the peak area resulting from the TD-GC–MS analysis is normalized for the quantity of the ISD in Eqs. (6.4) and (6.5).

The RF can be calculated from the data acquired in TIC or SIM mode, although the SIM is preferred to increase the sensitivity of the measurements. Quantification based on TIC can be biased by potential contaminants (i.e., co-eluting molecules within the same chromatographic peak). On the other hand, SIM has the disadvantage of requiring *a priori* knowledge in selecting the ions to be measured. However, VOC, which are not monitored in SIM mode but are contaminated by co-eluting molecules, can still be reliably quantified from the TIC by extracting the ion(s) characterizing a specific VOC (that do not appear in the contaminant) (see Section 6.3.1.7). This mode is commonly referred to as extracted ion chromatogram (EIC).

In the case that a specific VOC is not commercially available as standard, it is possible to perform a quantitative analysis using isomers based on the calculated RF. This method requires the determination of the fragmentation pattern and calculation of the percentage of a quantified ion (so-called target ions, Tgt) in the TIC (scan) mode (Figure 6.9). This concept is simplified in Figure 6.9, which depicts the sensitivity of the MSD to three hydrocarbons, namely the $C_{10}H_{16}$ monoterpenes α-pinene, sabinene, and δ-2-carene (used as ISD). Since all three monoterpenes have the same molecular mass and are chemically similar, the MSD will produce the same response, and thus, the resulting RF based on the TIC (scan) should be similar ($\pm$ 0.5–5%) (Figure 6.9). The RF based on SIM and EIC is essentially the proportion between the respective Tgt and TIC.

In addition, the mass dependence sensitivity of the MSD to specific chemical classes needs to be determined for a broad range of VOCs (e.g. 8–10 different VOCs within an m/z 35–250) because the signal intensity is proportional to the VOCs molecular mass, and the nonmatching polarity can lead to a chemical loss in the GC column (e.g. polar compounds separated in nonpolar columns). In order to describe the linear relationship between the MSD signal and the molecular mass of VOCs, it is necessary to consider the m/z range of measurement (e.g. 35–250 m/z), which usually starts after the ^{18}O isotope of oxygen. Once VOC-mass correction is made, the measurement's uncertainty is 2–10% (62).

6.3.1.7 Deconvolution

TD-GC–MS analysis of air samples containing a complex mixture of VOCs can lead to artifacts and erroneous quantification due to the interference of other molecules present in the background. An example is illustrated in Figure 6.10, where the peak of MeSa coelutes with that of decanal.

In order to correctly identify and extract all the quantitative information, each spectrum related to individual VOC needs to be reconstructed based on the complex composite spectrum (Figure 6.10). In GC–MS analysis, the spectrum construction is called deconvolution. The Automated Mass Spectral Deconvolution and Identification System (AMDIS) software (www.amdis.net) allows the separation and identification of overlapping peaks.

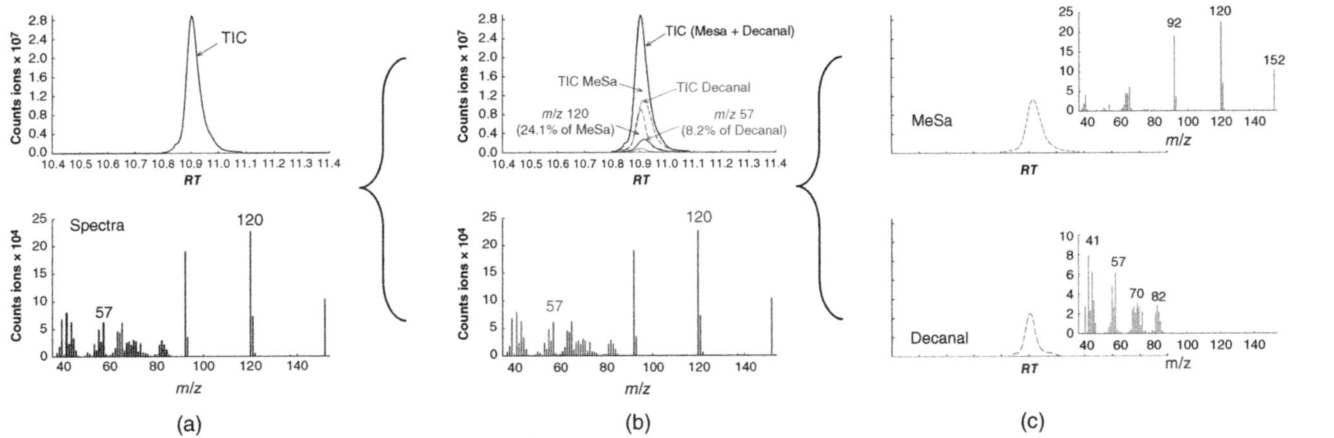

Figure 6.10 Examples of a chromatogram, mass spectra, and deconvolution processes to separate overlapping peaks and spectra. (a) The TIC at RT 10.9 minutes and its spectrum, containing (b) the two volatile MeSa (in blue) and decanal (in red). (c) Deconvolution of mass spectra allows a quantitative analysis of MeSa and decanal based on their characteristic and unique ions *m/z* 120 (MeSa) and m/z 57 (decanal). For a specific VOC, the proportion of the quantitative ions (Qion) to the total of ion fragments composing the VOC must be determined experimentally from data of uncontaminated peak samples or of pure standards.

The deconvolution process provides a well-defined peak that can be further integrated, thus leading to more accurate results. Knowing the proportion of the quantitation ion (i.e. Tgt) with respect to the total ions, a targeted VOC can then be reliably quantified (Figure 6.10). Following deconvolution, detection of peaks related to VOCs in a complex mixture is enhanced.

6.3.1.8 Applications of TD–GC–MS in Plant VOC Analysis

TD–GC–MS (and GC–MS) has been used to analyze biogenic VOCs since the 1960s, when their importance for atmospheric chemistry was first recognized (125). Specifically, TD–GC–MS has been traditionally employed worldwide to study VOC emission fluxes from vegetation: for example in Europe (126–129), the United States (130, 131), Asia (132), Africa (133), and the Arctic (117, 134). All these studies allowed compiling inventories of biogenic VOC emissions (3, 64, 135–138), which is essential for understanding the chemistry of the atmosphere, the climate, and the impact of VOCs on air quality (56, 139–142).

The TD–GC–MS approach has assisted early studies aiming at clarifying the environmental controls of light, temperature, and stomatal conductance on the emissions of terpenes (143–145). In addition, TD-GC–MS analyses have supported studies on the biosynthesis of VOCs by using the ^{13}C-labeling technique (62, 66, 146–150), and to understand their biosynthetic origins (e.g. from subarctic ecosystems) (62). Over the years, TD–GC–MS analyses have revealed to be an essential tool for complementing real-time PTR–MS measurements (57).

In plant ecology, TD–GC–MS analyses have been key to: (i) investigate the role of VOCs in plant-to-plant communication (28, 151–155), and plant-herbivore chemical interaction at the community level (156); (ii) unravel VOCs as chemical cues that guide caterpillar insects to find appropriate host plants for oviposition (157). Moreover, in plant defense (28, 29, 155, 158–161), TD–GC–MS analyses have highlighted how VOC emission act in priming and fortifying the plant immunity system (18, 25, 27).

Therefore, the use of static and dynamic cuvette/chamber systems, combined with passive or active adsorption techniques, followed by TD–GC–MS analysis represents a powerful approach for VOC analysis that nowadays has become a standard procedure (106). TD–GC–MS analysis is still the gold standard for elucidating the chemical composition of the VOC emission blend at leaf, branch, plant, or ecosystem levels (62, 157, 161, 162). In particular, TD–GC–MS analysis is crucial to distinguish different isomers of VOC (having the same number of atoms) that cannot be resolved only by PTR–MS analysis (163). Moreover, TD–GC–MS allows the quantification as well as the analysis of the enantiomeric and non-enantiomeric composition of VOCs (i.e. terpenes), in combination with headspace SPME or purge and trap techniques (164, 165).

Passive trapping techniques, such as SPME and SBSE, have the advantage of being small, cost-effective, and of allowing a large number of samples to be collected in parallel (e.g. optimal for field studies). Therefore, these have been used for high-throughput TD–GC–MS analysis (166). In addition, PDMS-coated stir bars have been employed to collect VOCs directly on the plant surface by placing the trapping adsorbents on one side of the leaves held by magnets on the other side (167, 168), or enclosing adsorbents together with leaves into bags of PTFE (169, 170), PET (171), or in cups (156).

TD–GC–MS is notoriously known as an off-line technique because it requires the adsorption and desorption of VOCs. However, in a few cases, TD–GC–MS has also been used as an real-time technique in which samples have been rapidly trapped and desorbed in sequence to monitor the dynamics of VOC emissions upon (a)biotic stress occurrence (56, 172–174). Custom-build fast TD–GC-MS, based on a cryogen VOC enrichment system, allowed the quantification of VOCs at higher time resolution (two to three minutes) at sub-pptv levels (140).

Another advantage of TD–GC–MS is that sample collection is disconnected from the analytical detection. This allows sampling VOCs in remote areas because the material for VOC collections (i.e. adsorbents, tubes, air pumps) is easy to ship or transport in the field. Recently, adsorption-assisted VOC collection systems have been applied to aerial drones (175, 176) and enabled fast (i.e. 10 minutes intervals) collection of air samples to perform vertical profiles of VOCs in difficult accessible areas, such as the Amazon rainforest or wetlands.

6.3.2 Proton Transfer Reaction Mass Spectrometry

6.3.2.1 General Principles and Analytical Instrument Development

The theoretical study of chemical ionization (CI) goes back to the experiments of Talrose in the early 1950s. This pioneering research led to the idea of exploiting CI for analyzing VOCs (177) through non-dissociative proton transfer reaction (PTR) between primary reactant ions, among which the most suitable are hydronium ions (H_3O^+), and VOC:

$$H_3O^+ + VOC \rightarrow VOC - H^+ + H_2O \tag{6.6}$$

PTR occurs with most of the VOCs that have proton affinities (PA) higher than that of H_2O (= 691 kJ mol^{-1}) (178), and it has an exoergicity small enough to avoid strong fragmentation so that only one (or a few) protonated ion(s) resulted for each detected VOC. In addition, the selectivity and sensitivity of PTR are ensured for VOCs because the PA of the primary gasses in the air (N_2, O_2, CO_2) is lower than that of H_2O, thus avoiding any interference with VOC detection.

Over the years, analytical instruments have been employed to exploit PTR for real-time detection of VOCs: from the pioneering "flowing afterglow" technique (179), where the source producing primary reactant ions was positioned within the drift tube, to the "selected ion flow tube" (SIFT), where the primary ions were produced in a detached source (180), which was further improved into the "selected ion flow drift tube" (SIFDT) (181). The concept of combining a SIFDT with a CI reaction system was the basis for creating a commercial system sensitive enough to allow monitoring mixing ratios of VOCs at pptv level with PTR–MS (182) (Figure 6.11a).

In PTR–MS, the primary ions (H_3O^+) are produced at high density with a purity of around 99.5% using a hollow cathode ion source (185), which has two advantages: (i) a high concentration of H_3O^+ ions does not require pre-selection and, (ii) the only significant impurity observed are O_2^+ ions. However, O_2^+ ions are not converted into another ionic form, and because their concentration is <1–3% than that of H_3O^+, they do not significantly interfere with the resulting protonated spectra of VOCs. After being produced, H_3O^+ ions enter a drift tube flushed continuously with ambient air that contains the VOCs to be analyzed, where an electric field is applied in the axial direction. Because no gas buffer is needed, the whole original mole fraction of VOCs in the air is retained in the drift tube, and PTR between H_3O^+ ions and VOC molecules occur under constant and standard pressure (e.g. 2.2 mbar), temperature (e.g. 50 °C), and electric field (e.g. $600\,V\,cm^{-1}$), resulting in a $E/N = 120\,Td$ ($1\,Td = 10^{-17}\,V\,cm^2$), where E represents the electric field strength and N the buffer gas density in the drift tube. Under these PTR–MS ionization conditions, all VOCs react mainly with H_3O^+ and their fragmentation is still limited. The PTR–MS operating conditions can be modified to change the parameter E/N, and this affects the degree of VOC fragmentation as well as the formation of water clusters ($H_2O_n - H_3O^+$, with $n = 1, 2, 3, \ldots$) because a small amount of water vapor flows through the ion source and reaches the drift tube (186). At higher E/N values (i.e. $E/N > 120$ Td), the water clustering is negligible, but the collision-induced dissociations (CID) and the amount of low-mass VOC fragments increase. At lower E/N values (i.e. $E/N < 120$ Td), the VOC fragmentation pattern decreases but most of the reagent ions associated with water clusters ($H_2O_n - H_3O^+$) have a different proton affinity for VOC than that of H_3O^+, making the sensitivity strongly dependent on the air sample humidity (187).

In the first version of the PTR–QMS instrument (Figure 6.11a), protonated-VOC ions were selected through a single quadrupole (Q) mass filter, detected by a secondary electron multiplier (SEM), and counted within integer [amu] for a specific dwell time, resulting in a mass resolution of one unit. A drawback of the low mass resolution of PTR–QMS is the inability to discriminate between isobaric VOCs having the same molecular weight [amu] but a different chemical formula.

To improve the mass resolution, the quadrupole has been replaced with a time-of-flight (TOF) mass analyzer, which resulted in a new commercially available PTR–TOF–MS (188). This highly performing instrument allows to detect, through multi-channel-plates (MCP) coupled with a digital converter (TDC), the whole mass spectra of VOC-protonated ions in a

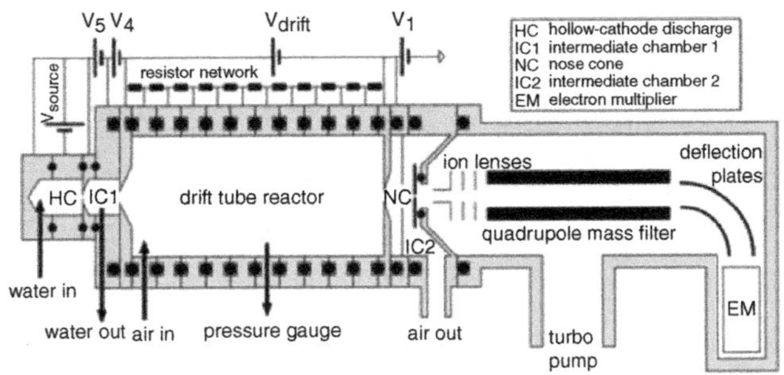

(a)

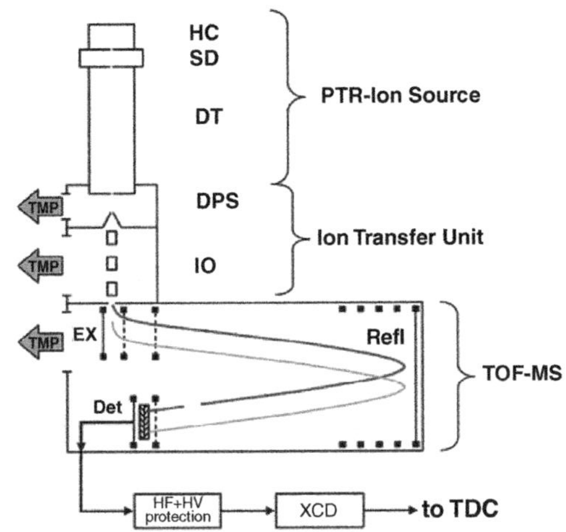

(b)

Figure 6.11 Schemes describing the working parts of PTR–QMS and PTR–TOF–MS instruments. (a) In PTR–QMS, the primary reactants ions H_3O^+ are produced within the hollow cathode (ion source) before entering the drift tube where proton transfer reaction (PTR) occurs with the VOCs; the resulting protonated-VOC are filtered according to their *m/z* ratio by the quadrupole mass analyzer (QMS) and further detected by an electron multiplier (EM). Source: ((183), figure 1, p. 225. Reproduced with permission of John Wiley & Sons). (b) In PTR–TOF–MS, the hollow cathode (HD) and the source drift (SD) (which form the ion source), and the drift tube (DT) are similar to those of the PTR–QMS, and have been interfaced with a differential pumping stage (DPS) and ions optics (IO) which drive the protonated-VOC to the pulsed extraction region (EX) before being accelerated orthogonally toward the reflectron (Refl) and finally refocused by detector (Det) into the time-of-flight mass spectrometer (TOF–MS) region. Then, all the detected signals are processed by an amplifier/discriminator unit (XCD) combined with a time to digital converter (TDC), which determines the *m/z* ratio from the flight times of the protonated VOC measured and each extraction pulse and generates a complete mass spectrum. Source: ((184), figure 2, p. 897. Reproduced with permission of Elsevier).

wide range and in less than 1 s, by maintaining the sensitivity at pptv level (Figure 6.11b). In particular, the higher mass resolution (around $M/\Delta M$ = 4000) provided by the PTR–TOF–MS than PTR–QMS enables to identify the molecular formula and separate between isobaric VOCs (184). However, PTR–TOF–MS analysis generates a large amount of data that contain hundreds of mass peaks related to the different VOCs. Various approaches have been developed for post-processing the raw data recorded with PTR–TOF–MS through programming routines able to perform various operations such as: accurate mass calibration, peak peaking, baseline correction, peak shape determination and fitting, mass discrimination, and dead-time correction.

PTR–QMS and PTR–TOF–MS analytical techniques have been further implemented with a switching reagent ion (SRI) technology, which enables the selection of primary ions among H_3O^+, NH_4^+, NO^+, and O_2^+ (189). The use of SRI results in a wider selectivity and improved detection of VOCs such as aromatics (e.g. toluene and ethylbenzene), aldehydes (e.g. butanal and valeraldehyde), and ketones (e.g. acetone and butanone), besides enhancing the ability to distinguish isomeric compounds. Furthermore, PTR–TOF–MS has been coupled with a gas chromatography (GC) column, where the mixture of VOCs is preliminarily separated according to their retention time before being ionized and detected; this improved the analytical capabilities of PTR–MS by distinguishing between different isomers (monoterpenes) emitted from *Quercus ilex* leaves (190).

6.3.2.2 Quantification of VOC Concentrations: Direct Calibration and Theoretical Calculation

The sensitivity of PTR–MS for VOCs is determined through direct measurements of the ion signal intensities of a calibrated gas. Pure gas standards are accurately diluted by mean of MFCs, with a constant flow of either synthetic VOC-free air or ambient air where VOC have been removed without changing the RH% (i.e., by charcoal filter or catalytic conversion of VOC into CO_2) to create different VOC concentrations (e.g. ranging from 1 to 50 ppbv), as follows:

$$\text{Sensivity}\left(\frac{\text{cps}}{\text{ppbv}}\right) = \frac{[\text{VOC–H}^+]}{[\text{VMR}]} \tag{6.7}$$

where $[\text{VOC–H}^+]$ are the recorded ion signals (in counts s^{-1}) of the protonated VOC and [VMR] is the concentration (volume mixing ratio) of the VOC in the gas standard. This experimental measurement of PTR–MS sensitivity includes k, the collision rate constant, and the detector transmission efficiency (also called the transmission factor) of a given VOC. However, at low E/N, the sensitivity of some VOCs is humidity dependent, thus requiring further calibration for different air humidity (RH%) (183, 186).

Moreover, to account for the effect of humidity on the cluster ion distribution, the protonated VOC ion signals need to be normalized to a standard primary ion signal (e.g. 10^6–10^7 cps) by using the following equation:

$$\text{Normalized ion signal (ncps)} = \frac{10^6 * [\text{VOC} - \text{H}^+]}{([\text{H}_3\text{O}^+] + \text{XR} * [\text{H}_3\text{O} - \text{H}_2\text{O}])} \tag{6.8}$$

where [VOC-H^+], [H_3O^+], and [H_3O^+–H_2O] are the ion signals (in counts s^{-1}) of the protonated VOC, the primary ion and the first water cluster, respectively, and XR is a VOC specific factor which reflects both the difference in mass dependent detection efficiencies and PTR rate coefficients for H_3O^+ and H_3O^+–H_2O (191).

Following direct calibration, limits of detection (LOD) for each VOC are defined (192) as:

$$\text{LOD} = 2\,\frac{a}{b} \tag{6.9}$$

where "*a*" is the standard deviation of the background signal measured in VOC-free air, and "*b*" is the instrument's sensitivity resulting from calibration.

However, the quantification of VOCs can be estimated theoretically from known ion-molecular reaction kinetics (183, 193). This calculation is based on the principles that: (i) H_3O^+ does not decrease significantly, and thus, can be assumed to remain constant; (ii) VOCs are ionized in a very short time into the drift tube, following the reaction (Eq. (6.6)), as follows:

$$[\text{VOC} - \text{H}^+] = [\text{H}_3\text{O}^+](1 - e^{-k[\text{VOC}-\text{H}^+]t}) \approx [\text{H}_3\text{O}^+]\,[\text{VOC} - \text{H}^+]kt \tag{6.10}$$

where [VOC–H^+] and [H_3O^+] are the concentration of the protonated VOC and the primary ions, respectively, "*k*" is the reaction rate constant of the PTR, "*t*" is reaction time (i.e. the time the protonated ions spend in the drift tube (194)). Since the measured count rates of H_3O^+ and VOC–H^+ are proportional to the respective ion concentrations, in PTR–QMS, the concentration of a specific VOC can be calculated from the following equation:

$$[\text{VOC}] = \frac{[\text{VOC} - \text{H}^+] * \text{trans}(\text{H}_3\text{O}^+)}{[\text{H}_3\text{O}^+] * \text{trans}(\text{VOC} - \text{H}^+) * k * t} \tag{6.11}$$

where "trans" is the mass-dependent quadruple transmission for H_3O^+ and VOC that can be directly measured; "*t*" is the residence time of the H_3O^+ in the drift tube, which is typically 100 μs; "*k*" is the PTR rate constant, which corresponds to the ion-molecule capture collisions; [VOC–H^+] and [H_3O^+] are the ion signals of the protonated VOC and the primary ions, respectively, as recorded by the instrument in count s^{-1} (cps). Since the amount of the primary ions [H_3O^+], usually 10^6–10^7, exceeds the saturation range of the SEM, it is more reliable to measure the ^{18}O isotope of the primary ion ($H_3{}^{18}O^+$) at *m/z* 21 which, due to the natural abundance of $^{18}O/^{16}O$ of about 0.20%, must be multiplied by 500 to obtain the concentrations (cps) of the primary ions [$H_3{}^{16}O^+$] (195).

Differently, in PTR–TOF–MS measurements, due to the effect of the TOF detector, which depends on the time of arrival and hence on the molecular weight of the protonated ions, the theoretical formula to calculate VOC concentrations (Eq. (6.6)) should be corrected as follows:

$$[\mathrm{VOC}] = \frac{[\mathrm{VOC}-\mathrm{H}^+] * \sqrt{(m/z\,\mathrm{H_3O^+})}}{[\mathrm{H_3O^+}] * \sqrt{(\mathrm{VOC}-\mathrm{H}^+)} * k * t} \tag{6.12}$$

where (m/z H_3O^+) is the molecular weight of the protonated primary ion m/z = 19.0178 Th (or alternatively the isotope of H_3O^+ at m/z = 21.0221 Th * 500), and (m/z VOC–H^+) is the molecular weight of the protonated VOC (196).

Overall, theoretical quantification of VOCs has a lower accuracy than direct calibration with gas standards due to the approximations of some of the parameters (i.e. t and k) of the formula (Eq. (6.10)). However, this methodology allows VOC quantification when calibration standards are unavailable (183, 193).

6.3.2.3 Application of PTR–MS in Plant VOCs Analysis

In recent years, PTR–MS technology has been applied to detect VOCs across different research fields spanning multiple domains: atmospheric chemistry (197), environmental sciences (198), food chemistry (199), microbiology (200, 201), medical science (202), and plant science (105). With particular regard to the latter, PTR–MS has been employed to: deepen the understanding of the VOCs biosynthesis (see Section 6.3.2.3.1); investigate the physiological mechanisms that regulate VOC emissions upon stresses (see Section 6.3.2.3.2); characterize and quantify the VOC fluxes emitted by different ecosystems (see Section 6.3.2.3.3).

6.3.2.3.1 Labeling Experiments

Pioneering experiments demonstrated that isoprene is synthesized by plant leaves within the chloroplasts via the MEP-pathway, mainly from the CO_2 assimilated through the light-dependent photosynthetic process. Feeding leaves of *Quercus rubra* with $^{13}CO_2$ showed rapid and high incorporation (about 80%) of ^{13}C into the isoprene molecule (150). Consistent with these results, the latest application of PTR–QMS allowed to monitor, in real-time, the kinetic of ^{13}C-labeling of isoprene upon exposure of poplar leaves to $^{13}CO_2$ (203) by distinguishing the transient appearance of individual intermediate molecules of isoprene isotopologues containing one to five ^{13}C atoms (Figure 6.12a), and hence the progressive change in protonated isoprene molecular weight from unlabeled m/z 69 to fully-labeled m/z 74. At the same time, PTR–MS analysis allowed to track the ^{13}C-labeling of a fragment of protonated isoprene (at m/z 41) containing both the two carbon atoms of isoprene derived from cytosolic pyruvate (Figure 6.12a), plus one of the three carbons coming from glyceraldehyde-3-phosphate (204), which suggested that the 20%

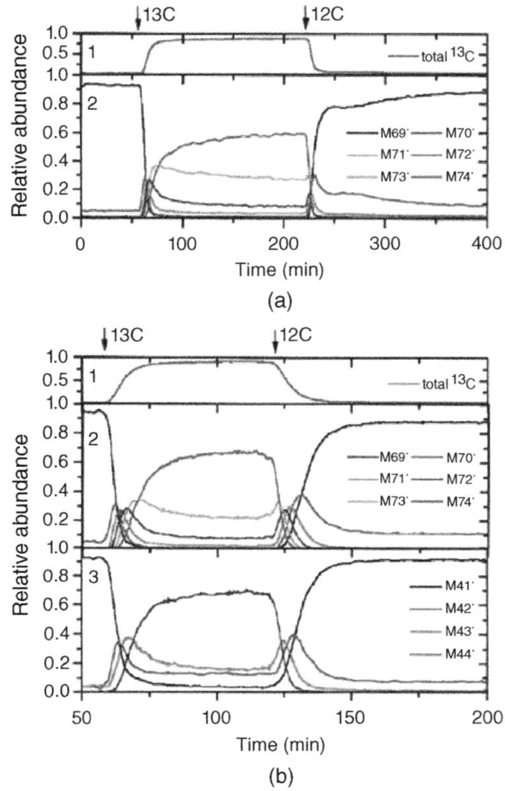

Figure 6.12 (a) Real-time monitoring of ^{13}C-incorporation into isoprene by PTR–QMS following exposure of *Quercus agrifolia* (upper panel) and (b) *Populus deltoides* (lower panel) leaves to $^{13}CO_2$. The measurements of ^{13}C-labeling kinetic showed the progressive appearance of individual intermediate molecules of isoprene containing an increasing number of ^{13}C atoms (from unlabeled *m/z* 69 to fully labeled *m/z* 74) when leaves were fed with $^{13}CO_2$ as well as the lightening of isoprene molecule (from fully labeled *m/z* 74 to unlabeled *m/z* 69) when the $^{13}CO_2$ was replaced by $^{12}CO_2$. Moreover, PTR–MS analysis showed the ^{13}C-labeling/unlabeling of a C_3-isoprene protonated fragment at *m/z* 41 (lower panel), which contains both the two carbon atoms derived from cytosolic pyruvate and one of the three carbons coming from glyceraldehyde-3-phosphate. Source: ((203), figure 1, p. 1032. Reproduced with permission of Springer Nature).

of the unlabeled isoprene is of extrachloroplastic origin. Alternative carbon sources involved in isoprene biosynthesis, in addition to the CO_2 assimilated through photosynthesis, were further investigated by PTR–QMS (108, 205). These analyses revealed, in poplar leaves, that 7–11% of unlabeled carbon used for isoprene formation originates from carbohydrates transported through the xylem, while 8–11% from photosynthetic intermediates have slower turnover rates (e.g. starch).

Moreover, carbon allocation and translocation dynamics within grey poplar plants were more deeply investigated by tracing ^{13}C-incorporation into isoprene emission from different plant parts. PTR–QMS measurements of the isotopic composition of labeled isoprene emission showed that part of the extracholoplastic carbon sources sustaining isoprene emission originated from carbon initially transported to the roots, through the phloem, following rapid translocation and reallocation to the apical plant parts via xylem transported sugars (108). On the other hand, the lack of ^{13}C-incorporation into methyl vinyl ketone (MVK) and methacrolein (MACR) upon $^{13}CO_2$-labeling demonstrated, in red oaks, that these carbonyls do not originate from isoprene oxidation, as could be expected from within leaves reaction of isoprene with ROS. Therefore, the metabolic origin of MVK and MACR remains still unknown (206). However, another study coupling PTR–TOF–MS measurements with global modeling suggested that, after being emitted, isoprene is converted into toxic isoprene oxidation products (i.e. MVK and MACR) and absorbed by leaves, then transformed into methyl ethyl ketone (MEK) before being re-emitted into the atmosphere (55).

Exposure to $^{13}CO_2$ followed by PTR–QMS analysis enabled to dissect specific fractions of monoterpenes emission produced *de novo* with photosynthetically fixed carbon and those synthesized from storage pools among the most common plant of the boreal forest (i.e. *Pinus sylvestris, Picea abies, Larix decidua,* and *Betula pendula*) (66). Moreover, PTR–MS monitoring of the leaves ^{13}C-labeling emission patterns after feeding *Halimium halimifolium* cut branches with position-specific ^{13}C-labeled pyruvate helped to disentangle possible biosynthetic pathways leading to the production of other different VOCs than terpenes. These experiments highlighted: (i) the origin of methyl acetate to be related to that of acetic acid; (ii) the cross-talk between MEP and MEV pathways; (iii) the production of some benzenoids to be linked to the MEP-pathway (207). Furthermore, ^{13}C-pyruvate labeling coupled with PTR–MS measurements has been used to investigate the emission of diterpenes from leaves of Cistaceae shrubs (*H. halimifolium* and *Cistus ladanifer*), pointing out that emission of karuene originates mainly from storage pools rather than be synthesized *de novo* from freshly assimilated carbon (208).

6.3.2.3.2 Real-Time Stress-Induced VOC Emission Changes at Leaf- and Whole-Plant Level

One of the earliest successful applications of PTR–QMS in plant VOC analysis was real-time monitoring of the kinetic of GLV emissions following wounding of aspen, beech, and clover leaves (209). More recently, PTR–TOF–MS provided a high-time resolved analysis by dissecting the rapid and transient release of multiple VOCs after leaf wounding (Figure 6.12a) and, at the same time, confirmed GLVs to be a suitable indicator of the degree of mechanical damage to the vegetation (80).

Other studies have investigated by PTR–QMS (100, 101, 210, 211) and by PTR–TOF–MS (80, 107), the prompt emission of acetaldehyde along with many different oVOC following a sudden switch from light to dark in leaves of herbaceous (i.e. *Dacrtilyis glomerata*) and woody (i.e. *Populus alba, Q. ilex, Prosopis velutina*) plant species. In fact, light-to-dark transitions induce stress by variating the extracellular and intracellular leaf pH, besides affecting the membrane stability. Moreover, PTR–MS has been employed to gain new knowledge on the biosynthesis and emission of VOC in response to high-light and high-temperature fluctuations, revealing that acetaldehyde and GLVs are very sensitive markers of abiotic stress in plants (93). Indeed, analyses of the combined effect of heat, drought stress, and elevated CO_2 concentrations of the VOC released from grey poplar suggested that the bursts of acetaldehyde, ethanol, and GLVs following rapid light-dark transition might be potentially useful phenotypic markers to signal stress occurrence (107) as well as to assess, non-invasively, the sensitivity of different plant genotypes to abiotic stresses (111).

Following the application of drought stress to *P. alba* sapling, measurements of ^{13}C-incorporation into isoprene highlighted the large contribution of extra-chloroplastic alternative carbon sources to support isoprene emission at high rates when the photosynthetic assimilation of CO_2 has hampered (91). In drought-stressed Scots pine seedlings, the carbon partitioning into terpenes and oVOCs emissions have been investigated by monitoring the labeling patterns after either exposure to $^{13}CO_2$ or feeding position-specific ^{13}C-labeled pyruvate. Results of these experiments showed a re-direction of pyruvate that lowered sesquiterpenes production but maintained isoprene and monoterpene emissions when photosynthesis has strongly inhibited by drought (212).

In other labeling studies, PTR–MS analysis has been used to deepen the understanding of the carbon employed for isoprene biosynthesis under changing temperature conditions (213). After fumigating $^{13}CO_2$ or applying [2-^{13}C]-glycine to mango (*Mangifera indica*) leaves, a photorespiration process was suggested to supply the carbon that supported higher isoprene emission under high temperatures. The dynamics of photosynthetic assimilated carbon allocation for VOC synthesis were also investigated during a simulated heat wave in which the temperature increased to 38 °C for 10 days (214). Tracking position-specific ^{13}C-labeled pyruvate into VOC emitted by the Mediterranean shrub *H. halimifolium* L. indicated an enhanced carbon allocation into *de novo* synthesis of oVOC (i.e. acetaldehyde) and terpenes when photosynthesis was impaired and respiration increased following heat stress. Furthermore, other measurements confirmed photosynthesis to uncouple from isoprene emission following heat stress and, at the same time, demonstrated an increased amount of cytosolic pyruvate incorporated into isoprene to sustain higher emission rates in two contrasting tropical species, *Ficus benjamina* and *Pachira aquatica* (215).

In another study, climatic chambers have been used to simulate global warming (+5 °C rise in temperature) on subarctic ecosystems dominated by dwarf shrubs of the genera *Salix* or *Betula*. Analysis of a wide range of VOCs by PTR–MS showed that warming significantly increased emissions of terpenoids, alcohol, and benzenoid compounds and led to changes in the chemical composition of the emission from the subarctic tundra toward more reactive VOCs (62).

Experiments studying the response of plants to herbivory have coupled PTR–QMS and TD-GC–MS with insect behavioral test to analyze how changes in VOC emission profile induced by the green oak leaf roller (*Tortrix viridana* L., Lepidopera:Tortricidae) is involved in the defense strategy of *Quercus robur* phenotypes (21). It was found that the blend of VOC emitted from leaves of resistant oaks (i.e. higher amount of sesquiterpenes) repelled, whereas that of susceptible oaks (i.e. higher amounts of (*E*)-4,8-dimethyl-1,3,7-nonatriene, DMNT and the monoterpene (*E*)-β-ocimene) attracted the female moths. PTR–MS technology was applied to further explore belowground plant-herbivore interactions by monitoring, in real-time, induced changes in VOC emission following the feeding activities of root herbivores, which could be exploited as biomarkers of root damage (216).

Newly developed PTR–TOF–MS technology improved VOC analysis through instantaneous and highly sensitive real-time detection of the whole spectra of VOC with high resolving power. This provides the first known complete and high-throughput measurement of the entire blend of volatile metabolites (the "plant volatilome"). Phenotyping the plant volatilome offers a fast and sensitive measure of biomarkers (217) exploitable in a wide range of practical applications, such as: (i) non-invasive screening of cultivars either resistant to diseases or tolerant to stresses (111); (ii) real-time diagnosis of both crop health status and early stress occurrence (218); and (iii) taxonomic classification of tree species based on resins exudation profile of VOCs (219).

6.3.2.3.3 Ecosystem-Level Measurements of VOC Emissions

In the field, PTR–QMS has been initially deployed for measuring real-time above-canopy fluxes of biogenic VOC fluxes (i.e. methanol, acetaldehyde, acetone, pentenols, 2-methylbutanal, hexenals, hexanal, and hexenols) at a mowed hayfield through the disjunct eddy covariance (DEC) method (220). Although this PTR–QMS application permitted intermitted sampling of a set of preselected protonated ions, it enabled monitoring of many biogenic VOC fluxes from a mixed coniferous subalpine forest site, thus showing daily peak emission fluxes of 2-methyl-3-buten-2-ol (MBO), methanol, acetone and acetaldehyde (221), and emission fluxes of monoterpenes from a Norway spruce forest (222). Recently, direct eddy covariance measurement of VOC fluxes was made possible by the use of the PTR–TOF–MS, which can acquire the full mass spectra at a high frequency (i.e. 10 Hz) (223). In particular, the mass resolving power coupled with high-duty cycles of PTR–TOF–MS allowed measuring the emission and deposition fluxes of a multitude of VOCs from an orange orchard site, thus providing a full mass balance of the VOCs exchanged by this ecosystem (224). Over the years, PTR–MS measurements of bi-directional VOC fluxes have characterized different forest ecosystems worldwide: tropical rain forest in Central America (225) and South-East Asia (226), boreal forest (227), Mediterranean macchia (228), Danish beech forest (229), subarctic peatland (230), and poplar plantation (231).

In addition, changes in VOC fluxes in ecosystems challenged by abiotic stress have been measured by PTR–MS. Higher emission fluxes of methyl-salicylate (MeSa) were observed in a deciduous walnut orchard in response to increasing temperature, as this VOC-like hormone may act as a signaling molecule helping plants to cope with occurring environmental stresses (232). Furthermore, under mild drought conditions, emission fluxes of isoprene and monoterpenes measured above an oak–hickory forests transiently increased before dropping when drought became severe. In contrast, the bidirectional exchange of methanol was less sensitive to decreasing soil water availability (233). The exchange pattern of VOCs was also monitored by PTR–TOF–MS above a managed temperate mountain grassland following mechanical stress during leaves cutting and drying (234). Results of this field campaign showed a wide range of VOC emitted from freshly cut grass comprising mainly methanol, in addition to acetaldehyde, ethanol, GLVs, acetone, acetic acid, monoterpenes, and sesquiterpenes, besides highlighting deposition of MVK, furfural, and terpenes on drying grass (Figure 6.13b). Another study, by combining PTR–MS plant-level laboratory cuvette measurements with field EC fluxes, confirmed that methanol and GLVs substantially represented the VOCs that are released in response to mowing and harvesting of managed grassland. Moreover, the same study revealed species-specific emission of terpenes and other VOCs after cut and drying, that may be further used as proxies for characterizing the plant diversity and functional group composition (i.e. legumes, forbs, and C_3 grasses) of a grassland (235). Recently, continuous PTR–MS monitoring of VOC fluxes in a grain sorghum plantation captured an enhanced emission of acetaldehyde coupled with other oVOC after lodging of plant stems, which represents a specific stress-induced VOC signature distinguishing it from the blend of VOC released after cutting-harvesting of plants (236).

6.3.2.4 Computational Approaches to Analyze PTR–MS Data

Real-time VOC detection by either PTR–QMS or PTR–TOF–MS generates a large amount of data which requires the development of computational approaches to extract usable information. Therefore, different data mining methods have been proposed for exploring the large datasets collected

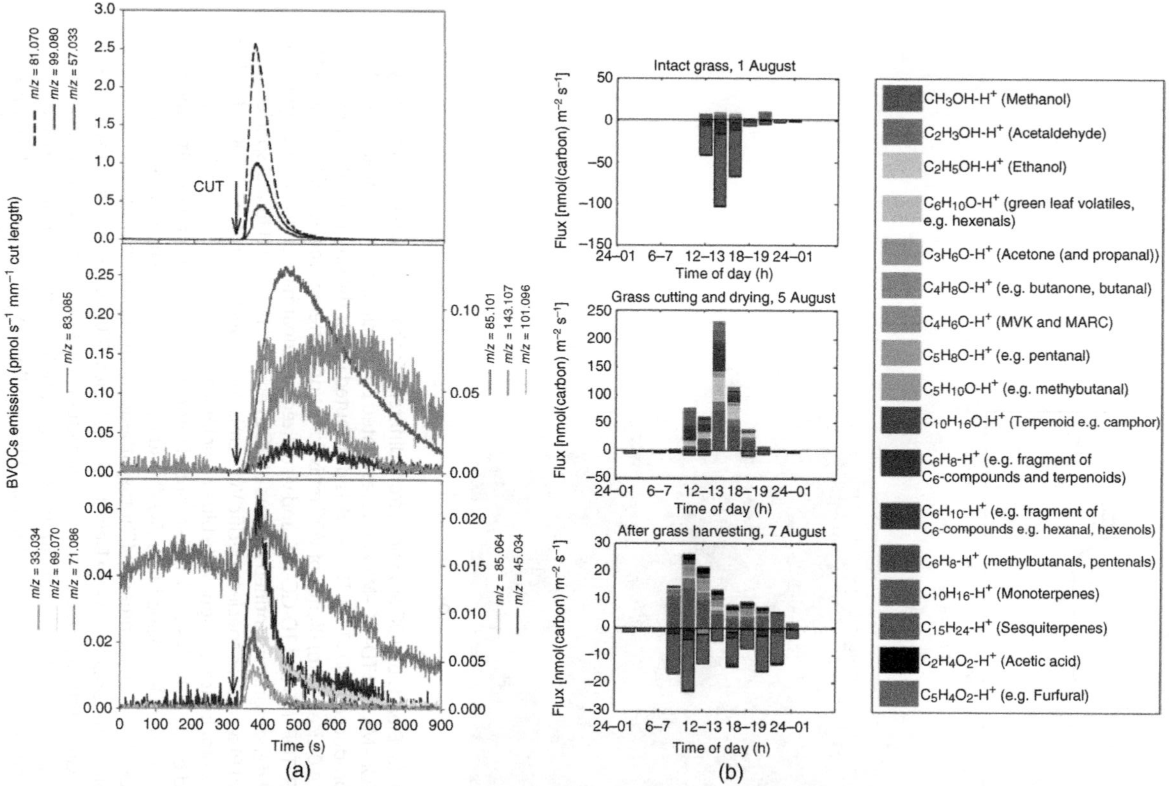

Figure 6.13 (a) Real-time evolution of the VOC emission blend induced by wounding of *D. glomerata* leaves enclosed in a cuvette and measured by PTR–TOF–MS. Source: Figure (80) is reproduced by © under the license agreement Creative Commons Attribution 3.0 License. (b) Monitoring of the mixture of VOC emission fluxes following cut drying of grasses by PTR–TOF–MS at ecosystem level. Source: (Reproduced by © (234). This work is distributed under the Creative Commons Attribution 3.0 License).

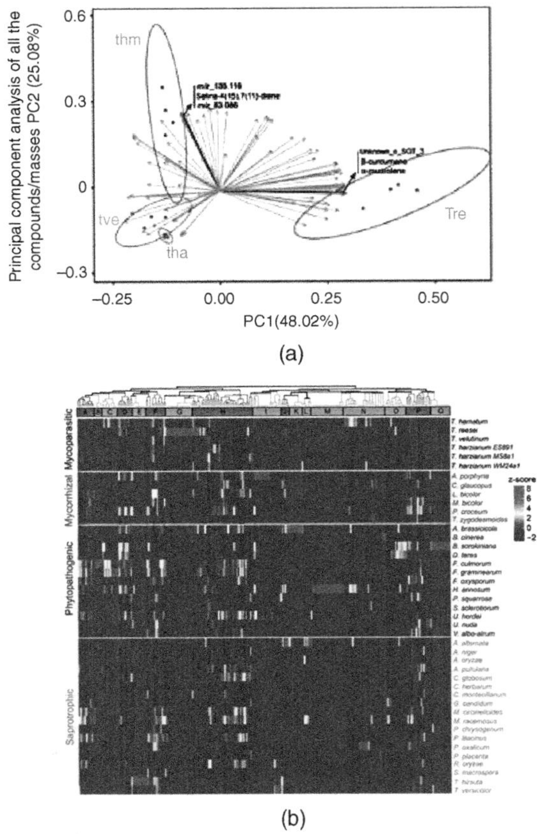

Figure 6.14 (a) Blends of VOC emissions from different *Trichoderma* species detected by both PTR–TOF–MS and TD–GC–MS and investigated by principal component analysis (PCA). Source: Reproduced by © (239). This work is distributed under the Creative Commons Attribution 4.0 License. (b) Blends of VOC emissions from different species of fungi detected by both PTR–TOF–MS and TD-GC–MS, and visualized by a heatmap; on top of the figure, the hierarchical dendrogram identified clusters of VOCs emitted by several fungal species (i.e. A, D, F, H, J, and P) as well as species-specific VOC emission clusters (i.e. B, C, E, G, I, K, L, M. N, O, and Q). Red color indicates high while blue color indicates low emission rates. Source: Reproduced by © (240). This work is distributed under the Creative Commons Attribution 4.0 License). (c) Self-Organizing Maps (SOM) visualization of the multiple correlations between VOC fluxes, environmental (Photosynthetic Active Radiation, water EvapoTraspiration, Vapour Pressure Deficit), physiological (CO_2 Net Ecosystem Exchange, Respiration, Gross Primary Productivity) and structural (Leaf Area Index) parameters recorded in a "coppiced" poplar plantation during a whole growing season. Dark red represents the highest (emission), while dark blue has the lowest (deposition) flux rates. The same position of a colored area in different panels indicates a correlation between the VOC fluxes and other parameters ((241), figure 2, p. 545. Source: Reproduced with permission of John Wiley & Sons).

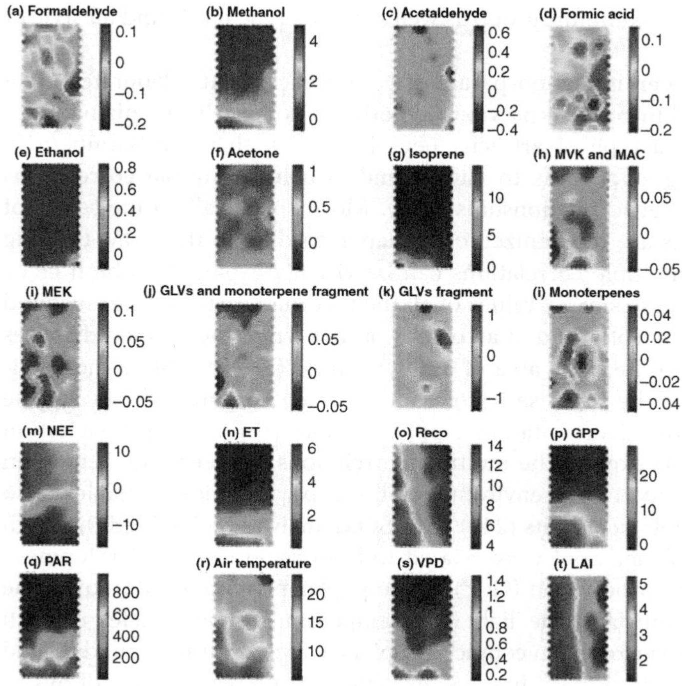

(c)

Figure 6.14 (*Continued*)

with PTR–MS instruments and to characterize VOC emission patterns and dependencies (237). Multivariate analyses aid the separation and organization of data into categories by using unsupervised and supervised methods. Unsupervised exploratory approaches, where classes are not predefined, have been commonly employed as suitable tools for visualizing high-dimensional VOC data (e.g. containing many data points) (238). Indeed, principal component analysis (PCA), by reducing the dataset into a few summary indexes, enabled to show the relations between data in a biplot simultaneously representing the space of both the recorded emission of VOC and the type of variable, thus helping to interpret the results (Figure 6.14a). Moreover, quick and preliminary inspection of complex VOC datasets has been conducted through cluster heatmaps to reveal hierarchical clusters in PTR–MS data matrices (Figure 6.14b). PCA coupled with heatmaps has been successfully applied to define species-specific metabolic profiles of VOC emitted from plant growth-promoting fungal species of *Trichoderma* spp. (242). Unsupervised data mining approach of complex VOC emission patterns measured from forty-three fungal species allowed to identify specific VOCs as biomarkers that discriminated the taxonomic group as well as the trophic mode and lifestyle of the fungi (240).

Another unsupervised non-parametric approach that elaborates data without the need to confirm previous hypotheses is the self-organizing maps (SOM). SOM is a type of artificial neural network that uses competitive machine learning algorithms to cluster and visualize multiple correlations into simple geometric relationships (243). More specifically, time-series of different variables are reorganized on a map according to the SOM training algorithm, and multiple correlations can be visually explored by both color and position because similar values of all the recorded variables are clustered and have the same color and, if a correlation occurred, these colored spaces are located within the same area of different maps (Figure 6.14c). The application of SOM to the time- series of data collected continuously during the growing season of a short rotation coppiced poplar (*Populus* spp.) plantation allowed to visually explore the multiple correlations between VOC emission and deposition fluxes with environmental and physiological variables, also including morphological traits (241). This SOM analysis highlighted that high emission rates of methanol were related to both rapid canopy development and high primary production (GPP) of the poplar plantation. Moreover, the same analysis confirmed the light and temperature dependencies of high isoprene emission fluxes from coppiced poplars, which were also found related to the seasonal trend of GPP, besides revealing that emission fluxes of oVOC (i.e. acetaldehyde, acetone, formic acid, ethanol, and fragments of GLVs) were mainly in relation with low values of both GPP and of leaf area index, which were recorded late in the season in the onset of plant senescence. In a different study, SOM was applied to analyze the big dataset recorded from a sorghum plantation during a whole growing season (236). The interpretation of this SOM output allowed us to successfully identify three distinctive patterns of VOC emission fluxes that correlated with growth indexes and environmental variables and were associated with phonological stages of sorghum plants. Moreover, a different SOM analysis carried out by using a subset of the same data collected only during the occurrence of stressful events of lodging and cutting revealed two distinctive signatures of VOC emission fluxes: while cutting was indicated by a correlation between the highest emission fluxes of many VOC (i.e. acetaldehyde, methanol, acetic acid, formic acid, acetone, and GLVs), lodging was identified by a correlation between peak emission fluxes of only acetaldehyde and formic acid.

In addition, supervised machine learning approaches have been employed for the classification of large VOC emission dataset. These involve the use of "classifiers," which learn from a subset of training data containing both predictors and corresponding predefined classes resulting from other studies, which are applied before testing the data. Among the supervised methods, random forest (RF) has been employed for highly efficient classification of VOC emitted from apples belonging to different cultivars and clones and having different resistance to fungal pathogens, with high accuracy based on multiple decision trees (196). Furthermore, the application of Random Forest-Recursive Feature Elimination (RF–RFE) method was able to identify the VOCs that contribute more to the separation of different classes (244). In addition, support vector machine (SVM) provided a very accurate classification of the raspberry cultivar by maximizing classifiers separation of VOC emission during the training of the algorithms, although it resulted in a prolonged process (245). Partial least squares discriminant analysis (PLS–DA), which combines dimensionality reduction and discriminant analysis, was also used to estimate a classification model that differentiated strawberry varieties based on VOC emission measurements from strawberry fruits (246).

6.4 Concluding Remarks and Future Perspectives

The emission of biogenic VOCs plays a crucial role in plant defense and communication as well as in atmospheric chemistry. Over time, research on VOC analysis has resulted in technical advances in mass spectrometry instrumentation, such as TD–GC–MS and PTR–MS, which enabled the detection of VOCs at very low concentrations and in real time. This resulted in the chemical identification and quantification of a wide array of VOC emissions fluxes from vegetation at different temporal and spatial scales. Today, we have gained comprehensive knowledge of the biosynthesis of plant VOCs and their response to environmental changes. Abiotic factors like drought and heat, or biotic factors like pathogenic diseases and insect outbreaks, impact plants' physiology and cause the production of a specific, stress-induced blend of VOCs. Therefore, measurements of VOC emissions may now be adopted as a non-invasive method for assessing plant health because VOCs are biomarkers that provide specific information on stress occurrence in plants. Monitoring stress-induced emissions of VOCs from plants can be particularly important under climate change, which impacts forest ecosystems by increasing the frequency and intensity of (a)biotic stress and extreme events, including drought, heatwaves, or the spread of herbivores and pathogenic diseases. Early detection of plant physiological changes through data mining and multivariate analyses of plants' VOC emissions could assist scientists and forest managers in identifying the onset of stress and taking appropriate measures to mitigate and prevent further damage. In conclusion, studying VOC emissions from forest ecosystems can provide valuable insights into how trees respond to climate-related stressors. By implementing VOC analysis in forest metabolomics, we might better monitor the health of our forests. Moreover, knowledge of VOC emissions from plants can help identify species-specific or chemotypes (i.e. plants with same genetic information but different chemical compositions) that are particularly stress-resistant, thus contributing to the conservation and maintenance of healthy forests while also mitigating the impact of climate change.

References

1 Kesselmeier J, Staudt M. Biogenic volatile organic compounds (VOC): an overview on emission, physiology and ecology. *J Atmos Chem*. 1999; 33: 23–88.

2 Laothawornkitkul J, Taylor JE, Paul ND, Hewitt CN. Biogenic volatile organic compounds in the Earth system. *New Phytol*. 2009; 183(1): 27–51.

3 Guenther AB, Jiang X, Heald CL, Sakulyanontvittaya T, Duhl T, Emmons LK, et al. The model of emissions of gases and aerosols from nature version 2.1 (MEGAN2.1): an extended and updated framework for modeling biogenic emissions. *Geosci Model Dev*. 2012; 5(6): 1471–1492.

4 Gershenzon J, Dudareva N. The function of terpene natural products in the natural world. *Nat Chem Biol*. 2007; 3(7): 408–414.

5 Chen F, Tholl D, Bohlmann J, Pichersky E. The family of terpene synthases in plants: a mid-size family of genes for specialized metabolism that is highly diversified throughout the kingdom. *Plant J*. 2011; 66(1): 212–229.

6 Dudareva N, Klempien A, Muhlemann JK, Kaplan I. Biosynthesis, function and metabolic engineering of plant volatile organic compounds. *New Phytol*. 2013; 198(1): 16–32.

7 Herde M, Gärtner K, Köllner TG, Fode B, Boland W, Gershenzon J, et al. Identification and regulation of TPS04/GES, an *Arabidopsis* geranyllinalool synthase catalyzing the first step in the formation of the insect-induced volatile C_{16}-homoterpene TMTT. *Plant Cell*. 2008; 20(4): 1152–1168.

8 Dudareva N, Andersson S, Orlova I, Gatto N, Reichelt M, Rhodes D, et al. The nonmevalonate pathway supports both monoterpene and sesquiterpene formation in snapdragon flowers. *Proc Natl Acad Sci USA*. 2005; 102(3): 933–938.

9 Hemmerlin A, Hoeffler J-F, Meyer O, Tritsch D, Kagan IA, Grosdemange-Billiard C, et al. Cross-talk between the cytosolic mevalonate and the plastidial methylerythritol phosphate pathways in tobacco bright yellow-2 cells. *J Biol Chem*. 2003; 278(29): 2666–2676.

10 Gerber E, Hemmerlin A, Hartmann M, Heintz D, Hartmann M-A, Mutterer J, et al. The plastidial 2-C-methyl-D-erythritol 4-phosphate pathway provides the isoprenyl moiety for protein geranylgeranylation in tobacco BY-2 cells. *Plant Cell*. 2009; 21(1): 285–300.

11 Dudareva N, Pichersky E. Metabolic engineering of plant volatiles. *Curr Opin Biotechnol*. 2008; 19: 181–189.

12 Hatanaka A, Kajiwara T, Matsui K. The biogeneration of green odour by green leaves and it's physiological functions – past, present and future. *Zeitschrift fuer Naturforsch*. 1995; (50c): 467–472.

13 Oderbolz DC, Aksoyoglu S, Keller J, Barmpadimos I, Steinbrecher R, Skjøth CA, et al. A comprehensive emission inventory of biogenic volatile

organic compounds in Europe: improved seasonality and land-cover. *Atmos Chem Phys.* 2013; 13(4): 1689–1712.
14 Knudsen JT, Eriksson R, Gershenzon J, Ståhl B. Diversity and distribution of floral scent. *Bot Rev.* 2006; 72(Mar): 1–120.
15 Gonda I, Bar E, Portnoy V, Lev S, Burger J, Schaffer AA, et al. Branched-chain and aromatic amino acid catabolism into aroma volatiles in *Cucumis melo* L. fruit. *J Exp Bot.* 2010; 61(4): 1111–1123.
16 Brosset A, Blande JD. Volatile-mediated plant–plant interactions: volatile organic compounds as modulators of receiver plant defence, growth, and reproduction. *J Exp Bot.* 2022; 73(2): 511–528.
17 Frank L, Wenig M, Ghirardo A, van der Krol A, Vlot AC, Schnitzler JP, et al. Isoprene and β-caryophyllene confer plant resistance via different plant internal signalling pathways. *Plant Cell Environ.* 2021; 44(44): 1151–1164.
18 Wenig M, Ghirardo A, Sales JH, Pabst ES, Breitenbach HH, Antritter F, et al. Systemic acquired resistance networks amplify airborne defense cues. *Nat Commun.* 2019; 10(1): 3813.
19 Gfeller V, Huber M, Förster C, Huang W, Köllner TG, Erb M. Root volatiles in plant-plant interactions I: High root sesquiterpene release is associated with increased germination and growth of plant neighbors. *Plant Cell Environ.* 2019; (Jan): 1–14.
20 Yi H-S, Heil M, Adame-Alvarez RM, Ballhorn DJ, Ryu C-M. Airborne induction and priming of plant defenses against a bacterial pathogen. *Plant Physiol.* 2009; 151: 2152–2161.
21 Ghirardo A, Heller W, Fladung M, Schnitzler J-PP, Schroeder H. Function of defensive volatiles in pedunculate oak (*Quercus robur*) is tricked by the moth *Tortrix* viridana. *Plant cell Environ.* 2012; 35(12): 2192–2207.
22 Huang M, Sanchez-Moreiras AM, Abel C, Sohrabi R, Lee S, Gershenzon J, et al. The major volatile organic compound emitted from *Arabidopsis thaliana* flowers, the sesquiterpene (*E*)-β-caryophyllene, is a defense against a bacterial pathogen. *New Phytol.* 2012; 193(4): 997–1008.
23 Veyrat N, Robert CAM, Turlings TCJ, Erb M. Herbivore intoxication as a potential primary function of an inducible volatile plant signal. *J Ecol.* 2016; 104(2): 591–600.
24 Vlot AC, Sales JH, Lenk M, Bauer K, Brambilla A, Sommer A, et al. Systemic propagation of immunity in plants. *New Phytol.* 2021; 229(3): 1234–1250.
25 Riedlmeier M, Ghirardo A, Wenig M, Knappe C, Koch K, Georgii E, et al. Monoterpenes support systemic acquired resistance within and between plants. *Plant Cell.* 2017; (Jun): 1440–1459.
26 Bouwmeester H, Schuurink RC, Bleeker PM, Schiestl F. The role of volatiles in plant communication. *Plant J.* 2019; 100(5): 892–907.

27 Brambilla A, Sommer A, Ghirardo A, Wenig M, Knappe C, Weber B, et al. Immunity-associated volatile emissions of β-ionone and nonanal propagate defence responses in neighbouring barley plants. *J Exp Bot.* 2022; 73(2): 615–630.

28 Frost CJ, Appel HM, Carlson JE, De Moraes CM, Mescher MC, Schultz JC. Within-plant signalling via volatiles overcomes vascular constraints on systemic signalling and primes responses against herbivores. *Ecol Lett.* 2007; 10(6): 490–498.

29 Arimura G, Ozawa R, Shimoda T, Boland W, Takabayashi J. Herbivory-induced volatiles elicit defence genes in lima bean leaves. *Nature.* 2000; 406(Aug): 1–4.

30 Quintana-Rodriguez E, Morales-Vargas AT, Molina-Torres J, Ádame-Alvarez RM, Acosta-Gallegos JA, Heil M. Plant volatiles cause direct, induced and associational resistance in common bean to the fungal pathogen *Colletotrichum lindemuthianum. J Ecol.* 2015; 103(1): 250–260.

31 Zeringue HJ, Brown RL, Neucere JN, Cleveland TE. Relationships between C_6-C_{12} alkanal and alkenal volatile contents and resistance of maize genotypes to *Aspergillus flavus* and aflatoxin production. *J Agric Food Chem.* 1996; 44(2): 403–407.

32 Ninkovic V, Markovic D, Rensing M. Plant volatiles as cues and signals in plant communication. *Plant Cell Environ.* 2020; (Oct): 1030–1043.

33 Holopainen JK, Gershenzon J. Multiple stress factors and the emission of plant VOCs. *Trends Plant Sci.* 2010; 15(3): 176–184.

34 Turlings TCJ, Erb M. Tritrophic interactions mediated by herbivore-induced plant volatiles: mechanisms, ecological relevance, and application potential. *Annu Rev Entomol.* 2018; 63: 433–452.

35 Yu H, Kivimäenpää M, Blande JD. Volatile-mediated between-plant communication in Scots pine and the effects of elevated ozone. *Proc R Soc B Biol Sci.* 2022; 289(1982).

36 Loreto F, Schnitzler J-P. Abiotic stresses and induced BVOCs. *Trends Plant Sci.* 2010; 15(3): 154–166.

37 Monson RK, Weraduwage SM, Rosenkranz M, Schnitzler J-P, Sharkey TD. Leaf isoprene emission as a trait that mediates the growth-defense tradeoff in the face of climate stress. *Oecologia.* 2021; 197(4): 885–902.

38 Loreto F, Velikova V. Isoprene produced by leaves protects the photosynthetic apparatus against ozone damage, quenches ozone products, and reduces lipid peroxidation of cellular membranes. *Plant Physiol.* 2001; 127(4): 1781–1787.

39 Velikova V, Loreto F. On the relationship between isoprene emission and thermotolerance in *Phragmites australis* leaves exposed to high temperatures and during the recovery from a heat stress. *Plant Cell Environ.* 2005; 28(3): 318–327.

40 Behnke K, Ehlting B, Teuber M, Bauerfeind M, Louis S, Hänsch R, et al. Transgenic, non-isoprene emitting poplars don't like it hot. *Plant J.* 2007; 51(3): 485–499.

41 Vickers CE, Possell M, Cojocariu CI, Velikova VB, Laothawornkitkul J, Ryan A, et al. Isoprene synthesis protects transgenic tobacco plants from oxidative stress. *Plant, Cell Environ.* 2009; 32(5): 520–531.

42 Tattini M, Loreto F, Fini A, Guidi L, Brunetti C, Velikova V, et al. Isoprenoids and phenylpropanoids are part of the antioxidant defense orchestrated daily by drought-stressed *Platanus* × *acerifolia* plants during Mediterranean summers. *New Phytol.* 2015; 207(3): 613–626.

43 Peñuelas J, Llusià J. Linking photorespiration, monoterpenes and thermotolerance in Quercus. *New Phytol.* 2002; 155(2): 227–237.

44 Delfine S, Csiky O, Seufert G, Loreto F. Fumigation with exogenous monoterpenes of a non-isoprenoid-emitting oak (*Quercus suber*): Monoterpene acquisition, translocation, and effect on the photosynthetic properties at high temperatures. *New Phytol.* 2000; 146(1): 27–36.

45 Singsaas EL, Lerdau M, Winter K, Sharkey TD. lsoprene increases thermotolerance of isoprene-emitting species. *Plant Physiol.* 1997; 115: 1413–1420.

46 Pollastri S, Jorba I, Hawkins TJ, Llusià J, Michelozzi M, Navajas D, et al. Leaves of isoprene-emitting tobacco plants maintain PSII stability at high temperatures. *New Phytol.* 2019; 223(3): 1307–1318.

47 Pollastri S, Tsonev T, Loreto F. Isoprene improves photochemical efficiency and enhances heat dissipation in plants at physiological temperatures. *J Exp Bot.* 2014; 65(6): 1565–1570.

48 Vanzo E, Merl-Pham J, Velikova V, Ghirardo A, Lindermayr C, Hauck SM, et al. Modulation of protein S-nitrosylation by isoprene emission in poplar. *Plant Physiol.* 2016; 170(Apr): 1945–1961.

49 Vickers CE, Gershenzon J, Lerdau MT, Loreto F. A unified mechanism of action for volatile isoprenoids in plant abiotic stress. *Nat Chem Biol.* 2009; 5(5): 283–291.

50 Zuo Z, Weraduwage SM, Lantz AT, Sanchez LM, Weise SE, Wang J, et al. Isoprene acts as a signaling molecule in gene networks important for stress responses and plant growth. *Plant Physiol.* 2019; 180(1): 124–152.

51 Xu C, Wang B, Luo Q, Ma Y, Zheng T, Wang Y, et al. The uppermost monoterpenes improving *Cinnamomum camphora* thermotolerance by serving signaling functions. *Front Plant Sci.* 2022; 13: 5189.

52 Lantz AT, Allman J, Weraduwage SM, Sharkey TD. Isoprene: new insights into the control of emission and mediation of stress tolerance by gene expression. *Plant Cell Environ.* 2019; 42(10): 2808–2826.

53 Harvey CM, Sharkey TD. Exogenous isoprene modulates gene expression in unstressed *Arabidopsis thaliana* plants. *Plant Cell Environ.* 2016; 39(6): 1251–1263.

54 Muramoto S, Matsubara Y, Mwenda CM, Koeduka T, Sakami T, Tani A, et al. Glutathionylation and reduction of methacrolein in tomato plants account for its absorption from the vapor phase. *Plant Physiol.* 2015; 169(Nov): 1744–1754.

55 Canaval E, Millet DB, Zimmer I, Nosenko T, Georgii E, Partoll EM, et al. Rapid conversion of isoprene photooxidation products in terrestrial plants. *Commun Earth Environ.* 2020; 1(1): 1–9.
56 Ghirardo A, Xie J, Zheng X, Wang Y, Grote R, Block K, et al. Urban stress-induced biogenic VOC emissions impact secondary aerosol formation in Beijing. *Atmos Chem Phys.* 2016; 16(16): 2901–2920.
57 Wiß F, Ghirardo A, Schnitzler J-P, Nendel C, Augustin J, Hoffmann M, et al. Net ecosystem fluxes and composition of biogenic volatile organic compounds over a maize field-interaction of meteorology and phenological stages. *GCB Bioenergy.* 2017; 9: 1627–1643.
58 Niinemets Ü, Arneth A, Kuhn U, Monson RK, Peñuelas J, Staudt M. The emission factor of volatile isoprenoids: Stress, acclimation, and developmental responses. *Biogeosciences.* 2010; 7(7): 2203–2223.
59 Niinemets U. Mild versus severe stress and BVOCs: thresholds, priming and consequences. *Trends Plant Sci.* 2010; 15(3): 145–153.
60 Niinemets Ü, Kännaste A, Copolovici L. Quantitative patterns between plant volatile emissions induced by biotic stresses and the degree of damage. Front. *Plant Sci.* 2013; 4(Jul): 1–15.
61 Niinemets U, Loreto F, Reichstein M. Physiological and physicochemical controls on foliar volatile organic compound emissions. *Trends Plant Sci.* 2004; 9(4): 180–186.
62 Ghirardo A, Lindstein F, Koch K, Buegger F, Schloter M, Albert A, et al. Origin of volatile organic compound emissions from subarctic tundra under global warming. *Global Change Biol.* 2020; 26: 1908–1925.
63 Portillo-Estrada M, Zenone T, Arriga N, Ceulemans R. Contribution of volatile organic compound fluxes to the ecosystem carbon budget of a poplar short-rotation plantation. *GCB Bioenergy.* 2018; 10(6): 405–414.
64 Sindelarova K, Markova J, Simpson D, Huszar P, Karlicky J, Darras S, et al. High-resolution biogenic global emission inventory for the time period 2000-2019 for air quality modelling. *Earth Syst Sci Data.* 2022; 14(1): 251–270.
65 Monson RK, Jones RT, Rosenstiel TN, Schnitzler J-P. Why only some plants emit isoprene. *Plant Cell Environ.* 2013; 36(3): 503–516.
66 Ghirardo A, Koch K, Taipale R, Zimmer I, Schnitzler J-P, Rinne J. Determination of de novo and pool emissions of terpenes from four common boreal/alpine trees by $^{13}CO_2$ labelling and PTR-MS analysis. *Plant cell Environ.* 2010; 33(5): 781–792.
67 Tingey DT, Manning M, Grothaus L, c., Burns W f. The influence of light and temperature on isoprene emission rates from live oak. *Physiol Plant.* 1979; 47(2): 112–118.
68 Loreto F, Ciccioli P, Cecinato A, Brancaleoni E, Frattoni M, Tricoli D. Influence of environmental factors and air composition on the emission of α-pinene from *Quercus ilex* leaves. *Plant Physiol.* 2016; 110(1): 267–275.

69 Grote R, Monson RK, Niinemets Ü. Leaf-level models of constitutive and stress-driven volatile organic compound emissions. In: Niinemets Ü, & Monson RK, editors. *Biology, Controls and Models of Tree Volatile Organic Compound Emission.* Volume 5. New York (NY): Springer; 2013. p. 315–355.

70 Monson RK, Grote R, Niinemets Ü, Schnitzler J-P. Modeling the isoprene emission rate from leaves. *New Phytol.* 2012; 195(3): 541–559.

71 Ghirardo A, Nosenko T, Kreuzwieser J, Winkler JB, Kruse J, Albert A, et al. Protein expression plasticity contributes to heat and drought tolerance of date palm. *Oecologia.* 2021; 197(4): 903–919.

72 Schnitzler J-P, Zimmer I, Bachl A, Arend M, Fromm J, Fischbach RJ. Biochemical properties of isoprene synthase in poplar (*Populus* x *canescens*). *Planta.* 2005; 222(5): 777–786.

73 Rasulov B, Hüve K, Bichele I, Laisk A, Niinemets Ü. Temperature response of isoprene emission in vivo reflects a combined effect of substrate limitations and isoprene synthase activity: a kinetic analysis. *Plant Physiol.* 2010; 154(3): 1558–1570.

74 Ghirardo A, Wright LP, Bi Z, Rosenkranz M, Pulido P, Rodríguez-Concepción M, et al. Metabolic flux analysis of plastidic isoprenoid biosynthesis in poplar leaves emitting and nonemitting isoprene. *Plant Physiol.* 2014; 165(1): 37–51.

75 Niinemets Ü, Monson RK, Arneth A, Ciccioli P, Kesselmeier J, Kuhn U, et al. The leaf-level emission factor of volatile isoprenoids: Caveats, model algorithms, response shapes and scaling. *Biogeosciences.* 2010; 7(6): 1809–1832.

76 Bates KH, Jacob DJ, Wang S, Hornbrook RS, Apel EC, Kim MJ, et al. The global budget of atmospheric methanol: new constraints on secondary, oceanic, and terrestrial sources. *J Geophys Res Atmos.* 2021; 126(4): 1–23.

77 Nemecek-Marshall M, Macdonald RC, Franzen JJ, Wojciechowski C, Fall R. Methanol emission from leaves. *Plant Physiol.* 1995; 1995(108): 1359–1368.

78 Dorokhov YL, Sheshukova EV, Komarova TV. Methanol in plant. *Life.* 2018; 9(November): 1–6.

79 Hüve K, Christ MM, Kleist E, Uerlings R, Niinemets Ü, Walter A, et al. Simultaneous growth and emission measurements demonstrate an interactive control of methanol release by leaf expansion and stomata. *J Exp Bot.* 2007; 58(7): 1783–1793.

80 Brilli F, Ruuskanen TM, Schnitzhofer R, Müller M, Breitenlechner M, Bittner V, et al. Detection of plant volatiles after leaf wounding and darkening by proton transfer reaction 'time-of-flight' mass spectrometry (ptr-tof). *PLoS One.* 2011; 6(5).

81 Beauchamp J, Wisthaler A, Hansel A, Kleist E, Miebach M, Niinemets Ü, et al. Ozone induced emissions of biogenic VOC from tobacco:

relationships between ozone uptake and emission of LOX products. *Plant Cell Environ.* 2005; 28: 1334–1343.

82 Brilli F, Gioli B, Ciccioli P, Zona D, Loreto F, Janssens I, et al. Proton Transfer Reaction Time-of-Flight Mass Spectrometric (PTR-TOF-MS) determination of volatile organic compounds (VOCs) emitted from a biomass fire developed under stable nocturnal conditions. *Atmos Environ.* 2014; 97: 54–67.

83 Portillo-Estrada M, Kazantsev T, Talts E, Tosens T, Niinemets Ü. Emission timetable and quantitative patterns of wound-induced volatiles across different leaf damage treatments in Aspen (*Populus Tremula*). *J Chem Ecol.* 2015; 41(12): 1105–1117.

84 Leitner M, Boland W, Mithöfer A. Direct and indirect defences induced by piercing-sucking and chewing herbivores in Medicago truncatula. *New Phytol.* 2005; 167(2): 597–606.

85 Kask K, Kännaste A, Talts E, Copolovici L, Niinemets Ü. How specialized volatiles respond to chronic and short-term physiological and shock heat stress in *Brassica nigra. Plant Cell Environ.* 2016; 39(9): 2027–2042.

86 Niinemets Ü. Storage of defense metabolites in the leaves of Myrtaceae: news of the eggs in different baskets. *Tree Physiol.* 2018; 38(10): 1445–1450.

87 Pegoraro E, Rey A, Greenberg J, Harley P, Grace J, Malhi Y, et al. Effect of drought on isoprene emission rates from leaves of *Quercus virginiana* Mill. *Atmos Environ.* 2004; 38(36): 6149–6156.

88 Penuelas J, Staudt M, Peñuelas J, Staudt M. BVOCs and global change. *Trends Plant Sci.* 2010; 15(3): 133–144.

89 Sharkey TD, Loreto F. Water stress, temperature, and light effects on the capacity for isoprene emission and photosynthesis of kudzu leaves. *Oecologia.* 1993; 95(3): 328–333.

90 Brilli F, Barta C, Fortunati A, Lerdau M, Loreto F, Centritto M. Response of isoprene emission and carbon metabolism to drought in white poplar (*Populus alba*) saplings. *New Phytol.* 2007; 175(2): 244–254.

91 Fortunati A, Barta C, Brilli F, Centritto M, Zimmer I, Schnitzler JP, et al. Isoprene emission is not temperature-dependent during and after severe drought-stress: a physiological and biochemical analysis. *Plant J.* 2008; 55(4): 687–697.

92 Guidolotti G, Pallozzi E, Gavrichkova O, Scartazza A, Mattioni M, Loreto F, et al. Emission of constitutive isoprene, induced monoterpenes, and other volatiles under high temperatures in *Eucalyptus camaldulensis* : a ^{13}C labelling study. *Plant Cell Environ.* 2018; 2019(42): 1–10.

93 Loreto F, Barta C, Brilli F, Nogues I. On the induction of volatile organic compound emissions by plants as consequence of wounding or fluctuations of light and temperature. *Plant Cell Environ.* 2006; 29(9): 1820–1828.

94 Behnke K, Ghirardo A, Janz D, Kanawati B, Esperschütz J, Zimmer I, et al. Isoprene function in two contrasting poplars under salt and sunflecks. *Tree Physiol.* 2013; 33(6): 562–578.

95 Birami B, Bamberger I, Ghirardo A, Grote R, Arneth A, Gaona-Colmán E, et al. Heatwave frequency and seedling death alter stress-specific emissions of volatile organic compounds in Aleppo pine. *Oecologia.* 2021; 197(4): 939–956.

96 Kreuzwieser J, Scheerer U, Rennenberg H. Metabolic origin of acetaldehyde emitted by poplar (*Populus tremula* x *P. alba*) trees. *J Exp Bot.* 1999; 50(335): 757–765.

97 Jardine K, Karl T, Lerdau M, Harley P, Guenther A, Mak JE. Carbon isotope analysis of acetaldehyde emitted from leaves following mechanical stress and anoxia. *Plant Biol.* 2009; 11(4): 591–597.

98 Oikawa PY, Lerdau MT. Catabolism of volatile organic compounds influences plant survival. *Trends Plant Sci.* 2013; 18(12): 695–703.

99 Warneke C, Luxembourg SL, De Gouw JA, Rinne HJI, Guenther AB, Fall R. Disjunct eddy covariance measurements of oxygenated volatile organic compounds fluxes from an alfalfa field before and after cutting. *J Geophys Res Atmos.* 2002: 107(7–8).

100 Jardine K, Barron-Gafford GA, Norman JP, Abrell L, Monson RK, Meyers KT, et al. Green leaf volatiles and oxygenated metabolite emission bursts from mesquite branches following light-dark transitions. *Photosynth Res.* 2012; 113(1–3): 321–333.

101 Graus M, Schnitzler J-P, Hansel A, Cojocariu C, Rennenberg H, Wisthaler A, et al. Transient release of oxygenated volatile organic compounds during light-dark transitions in Grey poplar leaves. *Plant Physiol.* 2004; 135(4): 1967–1975.

102 de Gouw JA, Howard CJ, Custer TG, Baker BM, Fall R. Proton-transfer chemical-ionization mass spectrometry allows real-time analysis of volatile organic compounds released from cutting and drying of crops. *Environ Sci Technol.* 2000; 34(12): 2640–2648.

103 Mentel TF, Kleist E, Andres S, Dal Maso M, Hohaus T, Kiendler-Scharr A, et al. Secondary aerosol formation from stress-induced biogenic emissions and possible climate feedbacks. *Atmos Chem Phys.* 2013; 13(17): 8755–8770.

104 Heil M. Herbivore-induced plant volatiles: targets, perception and unanswered questions. *J Physiol.* 2014; 204(2): 297–306.

105 Tholl D, Boland W, Hansel A, Loreto F, Röse USRR, Schnitzler J-PP. Practical approaches to plant volatile analysis. *Plant J.* 2006; 45(4): 540–560.

106 Tholl D, Hossain O, Weinhold A, Röse USR, Wei Q. Trends and applications in plant volatile sampling and analysis. *Plant J.* 2021; 106(2): 314–325.

107 Jud W, Vanzo E, Li Z, Ghirardo A, Zimmer I, Sharkey TD, et al. Effects of heat and drought stress on post-illumination bursts of volatile organic compounds in isoprene-emitting and non-emitting poplar. *Plant Cell Environ.* 2016; 39(6): 1204–1215.
108 Ghirardo A, Gutknecht J, Zimmer I, Brüggemann N, Schnitzler J-P. Biogenic volatile organic compound and respiratory CO_2 emissions after ^{13}C-labeling: online tracing of C translocation dynamics in poplar plants. *PLoS One.* 2011; 6(2): e17393.
109 Wright LP, Rohwer JM, Ghirardo A, Hammerbacher A, Ortiz-Alcaide M, Raguschke B, et al. Deoxyxylulose 5-phosphate synthase controls flux through the methylerythritol 4-phosphate pathway in Arabidopsis. *Plant Physiol.* 2014; 165(4): 1488–1504.
110 Loivamäki M, Gilmer F, Fischbach RJ, Sörgel C, Bachl A, Walter A, et al. Arabidopsis, a model to study biological functions of isoprene emission? *Plant Physiol.* 2007; 144(2): 1066–1078.
111 Jud W, Winkler JB, Niederbacher B, Niederbacher S, Schnitzler JP. Volatilomics: a non-invasive technique for screening plant phenotypic traits. *Plant Methods.* 2018; 14(1): 1–18.
112 Roy J, Rineau F, De Boeck HJ, Nijs I, Pütz T, Abiven S, et al. Ecotrons: powerful and versatile ecosystem analysers for ecology, agronomy and environmental science. *Global Change Biol.* 2020; 27(27): 1387–1407.
113 Vanhatalo A, Ghirardo A, Juurola E, Schnitzler JP, Zimmer I, Hellén H, et al. Long-term dynamics of monoterpene synthase activities, monoterpene storage pools and emissions in boreal Scots pine. *Biogeosciences.* 2018; 15(16): 5047–5060.
114 Havermann F, Ghirardo A, Schnitzler JP, Nendel C, Hoffmann M, Kraus D, et al. Modeling intra- and interannual variability of BVOC emissions from maize, oil-seed rape, and ryegrass. *J Adv Model Earth Syst.* 2022; 14(3).
115 Niinemets Ü, Kuhn U, Harley PC, Staudt M, Arneth A, Cescatti A, et al. Estimations of isoprenoid emission capacity from enclosure studies: measurements, data processing, quality and standardized measurement protocols. *Biogeosciences.* 2011; 8(8): 2209–2246.
116 Ortega J, Helmig D. Approaches for quantifying reactive and low-volatility biogenic organic compound emissions by vegetation enclosure techniques – Part A. *Chemosphere.* 2007; 72: 343–364.
117 Faubert P, Tiiva P, Rinnan A, Michelsen A, Holopainen JK, Rinnan R. Doubled volatile organic compound emissions from subarctic tundra under simulated climate warming. *New Phytol.* 2010; 187(1): 199–208.
118 Aubinet M, Grelle A, Ibrom A, Rannik Ü, Moncrieff J, Foken T, et al. Estimates of the annual net carbon and water exchange of forests: the EUROFLUX methodology. *Adv Ecol Res.* 1999; 30(C): 113–175.
119 Göckede M, Rebmann C, Foken T. A combination of quality assessment tools for eddy covariance measurements with footprint modelling for

the characterisation of complex sites. *Agric For Meteorol.* 2004; 127(3-4): 175–188.

120 Bamberger I, Hörtnagl L, Walser M, Hansel A, Wohlfahrt G. Gap-filling strategies for annual VOC flux data sets. *Biogeosciences.* 2014; 11(8): 2429–2442.

121 Ciccioli P. VOCs and air pollution. In: Bloeme HJT, & Burn J, editors. *Chemistry and Analysis of Volatile Organic Compounds in the Environment.* Blackie Academic & Professional; 1992. p. 23–174.

122 Ciccioli P, Cecinato A, Brancaleoni E, Frattoni M, Liberti A. Use of carbon adsorption traps combined with high resolution gas chromatography – mass spectrometry for the analysis of polar and non-polar C_4-C_{14} hydrocarbons involved in photochemical smog formation. *J High Resolut Chromatogr.* 1992; 15(2): 75–84.

123 Ettre LS. The Kováts retention index system. *Anal Chem.* 1964; 36(8): 31A–41A.

124 Van den Dool H, Kratz PD. A generalization of the retention index system including linear temperature programmed gas-liquid partition chromatography. *J Chromatogr.* 1963; 11(3): 463–471.

125 Went FW. Blue hazes in the atmosphere. *Nature.* 1960; 187(4738): 641–643.

126 Kesselmeier J, Schäfer L, Ciccioli P, Brancaleoni E, Cecinato A, Frattoni M, et al. Emission of monoterpenes and isoprene from a Mediterranean oak species *Quercus ilex* L. measured within the BEMA (Biogenic Emissions in the Mediterranean Area) project. *Atmos Environ.* 1996; 30(10–11): 1841–1850.

127 Ciccioli P, Brancaleoni E, Frattoni M, Di Palo V, Valentini R, Tirone G, et al. Emission of reactive terpene compounds from orange orchards and their removal by within-canopy processes. *J Geophys Res.* 1999; 104: 8077.

128 Hakola H, Tarvainen V, Laurila T, Hiltunen V, Hellén H, Keronen P. Seasonal variation of VOC concentrations above a boreal coniferous forest. *Atmos Environ.* 2003; 37(12): 1623–1634.

129 Steinbrecher R, Ziegler H, Eichstädter G, Fehsenfeld U, Gabriel R, Kolb C, et al. Monoterpene and isoprene emission in Norway spruce forests. In: Slanina S, editor. *Biosphere-Atmosphere Exchange of Pollutants and Trace Substances. Transport and Chemical Transformation of Pollutants in the Troposphere.* Volume 4. Berlin, Heidelberg: Springer; 1997. p. 352–365.

130 Geron C, Guenther A, Greenberg J, Karl T, Rasmussen R. Biogenic volatile organic compound emissions from desert vegetation of the southwestern US. *Atmos Environ.* 2006; 40(9): 1645–1660.

131 Constable JVH, Litvak ME, Greenberg JP, Monson RK. Monoterpene emission from coniferous trees in response to elevated CO_2 concentration and climate warming. *Global Change Biol.* 1999; 5(3): 252–267.

132 Baker B, Bai JH, Johnson C, Cai ZT, Li QJ, Wang YF, et al. Wet and dry season ecosystem level fluxes of isoprene and monoterpenes from a southeast Asian secondary forest and rubber tree plantation. *Atmos Environ.* 2005; 39(2): 381–390.

133 Serça D, Guenther A, Klinger L, Vierling L, Harley P, Druilhet A, et al. EXPRESSO flux measurements at upland and lowland Congo tropical forest site. *Tellus B Chem Phys Meteorol.* 2001; 53(3): 220–234.

134 Tiiva P, Faubert P, Michelsen A, Holopainen T, Holopainen JK, Rinnan R. Climatic warming increases isoprene emission from a subarctic heath. *New Phytol.* 2008; 180(4): 853–863.

135 Simpson D, Guenther A, Hewitt CN, Steinbrecher R. Biogenic emissions in Europe 1. Estimates and uncertainties. *J Geophys Res.* 1995; 100(D11): 22–890.

136 Guenther A. A global model of natural volatile organic compound emissions. *J Geophys Res.* 1995; 100(D5): 8873–8892.

137 Sindelarova K, Granier C, Bouarar I, Guenther A, Tilmes S, Stavrakou T, et al. Global data set of biogenic VOC emissions calculated by the MEGAN model over the last 30 years. *Atmos Chem Phys.* 2014; 14(17): 9317–9341.

138 Zhihui W, Yuhua B, Shuyu Z. A biogenic volatile organic compounds emission inventory for Beijing. *Atmos Environ.* 2003; 37(27): 3771–3782.

139 Bouvier-Brown NC, Goldstein AH, Gilman JB, Kuster WC, de Gouw JA. In-situ ambient quantification of monoterpenes, sesquiterpenes, and related oxygenated compounds during BEARPEX 2007: implications for gas- and particle-phase chemistry. *Atmos Chem Phys.* 2009; 9(15): 5505–5518.

140 Bourtsoukidis E, Helleis F, Tomsche L, Fischer H, Hofmann R, Lelieveld J, et al. An aircraft gas chromatograph-mass spectrometer System for Organic Fast Identification Analysis (SOFIA): Design, performance and a case study of Asian monsoon pollution outflow. *Atmos Meas Tech.* 2017; 10(12): 5089–5105.

141 Fuzzi S, Baltensperger U, Carslaw K, Decesari S, Denier Van Der Gon H, Facchini MC, et al. Particulate matter, air quality and climate: lessons learned and future needs. *Atmos Chem Phys.* 2015; 15(14): 8217–8299.

142 Elsasser M, Crippa M, Orasche J, Decarlo PF, Oster M, Pitz M, et al. Organic molecular markers and signature from wood combustion particles in winter ambient aerosols: aerosol mass spectrometer (AMS) and high time-resolved GC-MS measurements in Augsburg, Germany. *Atmos Chem Phys.* 2012; 12(14): 6113–6128.

143 Harley P, Eller A, Guenther A, Monson RK. Observations and models of emissions of volatile terpenoid compounds from needles of ponderosa pine trees growing in situ: control by light, temperature and stomatal conductance. *Oecologia.* 2014; 176: 35–55.

144 Staudt M, Seufert G. Light-dependent emission of monoterpenes by Holm oak (*Quercus ilex* L.). *Naturwissenschaften*. 1995; 82(2): 89–92.

145 Monson RK, Fall R. Isoprene emission from aspen leaves. The influence of environment and relation to photosynthesis and photorespiration. *Plant Physiol*. 1989; 90(1): 267–274.

146 Shao M, Czapiewski KV, Heiden AC, Kobel K, Komenda M, Koppmann R, et al. Volatile organic compound emissions from Scots pine: mechanisms and description by algorithms. *J Geophys Res*. 2001; 106(D17): 20483.

147 Wu C, Pullinen I, Andres S, Kiendler-Scharr A, Kleist E, Wahner A, et al. ^{13}C labelling study of constitutive and stress-induced terpenoide missions from Norway spruce and Scots pine. *Biogeosci Discuss*. 2017; (Jul): 1–29.

148 Loreto F, Pinelli P, Brancaleoni E, Ciccioli P. ^{13}C labeling reveals chloroplastic and extrachloroplastic pools of dimethylallyl pyrophosphate and their contribution to isoprene formation. *Plant Physiol*. 2004; 135(Aug): 1903–1907.

149 Loreto F, Ciccioli P, Cecinato A, Brancaleoni E, Frattoni M, Fabozzi C, et al. Evidence of the photosynthetic origin of monoterpenes emitted by *Quercus ilex* L. leaves by ^{13}C labeling. *Plant Physiol*. 1996; 110(4): 1317–1322.

150 Delwiche CF, Sharkey TD. Rapid appearance of ^{13}C in biogenic isoprene when $^{13}CO_2$ is fed to intact leaves. *Plant Cell Environ*. 1993; 650(16): 587–591.

151 McCall PJ, Turlings TCJ, Loughrin J, Proveaux AT, Tumlinson JH. Herbivore-induced volatile emissions from cotton (*Gossypium hirsutum* L.) seedlings. *J Chem Ecol*. 1994; 20(12): 3039–3050.

152 Engelberth J, Alborn HT, Schmelz E, a, Tumlinson JH. Airborne signals prime plants against insect herbivore attack. *Proc Natl Acad Sci USA*. 2004; 101(6): 1781–1785.

153 Heil M. Indirect defence via tritrophic interactions. *New Phytol*. 2008; 178(1): 41–61.

154 Heil M, Silva Bueno JC. Within-plant signaling by volatiles leads to induction and priming of an indirect plant defense in nature. *Proc Natl Acad Sci USA*. 2007; 104(13): 5467–5472.

155 Kalske A, Shiojiri K, Uesugi A, Sakata Y, Morrell K, Kessler A. Insect herbivory selects for volatile-mediated plant-plant communication. *Curr Biol*. 2019; 29(18): 3128–3133.e3.

156 Zu P, Boege K, del-Val E, Schuman MC, Stevenson PC, Zaldivar-Riverón A, et al. Information arms race explains plant-herbivore chemical communication in ecological communities. *Science (80-)*. 2020; 368(6497): 1377–1381.

157 De Moraes CM, Mescher MC, Tumlinson JH. Caterpillar-induced nocturnal plant volatiles repel conspecific females. *Nature*. 2001; 410(6828): 577–580.

158 Arimura G, Kost C, Boland W. Herbivore-induced, indirect plant defences. *Biochim Biophys Acta BBA Mol Cell Biol Lipids*. 2005; 1734: 91–111.

159 Erb M. Plant biology: evolution of volatile-mediated plant–plant interactions. *Curr Biol*. 2019; 29(18): R873–R875.

160 Karban R, Yang LH, Edwards KF. Volatile communication between plants that affects herbivory: a meta-analysis. *Ecol Lett*. 2014; 17(1): 44–52.

161 Erb M, Veyrat N, Robert CAMM, Xu H, Frey M, Ton J, et al. Indole is an essential herbivore-induced volatile priming signal in maize. *Nat Commun*. 2015; 6: 1–10.

162 Turlings TCJ, Tumlinson JH, Lewis WJ. Exploitation of herbivore-induced plant odors by host-seeking parasitic wasps. *Science*. 1990; 250(4985): 1251–1253.

163 Misztal PK, Heal MR, Nemitz E, Cape JN. Development of PTR-MS selectivity for structural isomers: monoterpenes as a case study. *Int J Mass spectrom*. 2012; 310: 10–19.

164 Yassaa N, Custer T, Song W, Pech F, Kesselmeier J, Williams J. Quantitative and enantioselective analysis of monoterpenes from plant chambers and in ambient air using SPME. *Atmos Meas Tech*. 2010; 3(6): 1615–1627.

165 Ribeiro C, Gonçalves R, Tiritan ME. Separation of enantiomers using gas chromatography: application in forensic toxicology, food and environmental analysis. *Crit Rev Anal Chem*. 2021; 51(8): 787–811.

166 Kallenbach M, Oh Y, Eilers EJ, Veit D, Baldwin IT, Schuman MC. A robust, simple, high-throughput technique for time-resolved plant volatile analysis in field experiments. *Plant J*. 2014; 78(6): 1060–1072.

167 Kfoury N, Scott E, Orians C, Robbat A. Direct contact sorptive extraction: a robust method for sampling plant volatiles in the field. *J Agric Food Chem*. 2017; 65(38): 8501–8509.

168 Aksenov AA, Pasamontes A, Peirano DJ, Zhao W, Dandekar AM, Fiehn O, et al. Detection of huanglongbing disease using differential mobility spectrometry. *Anal Chem*. 2014; 86: 2481–2488.

169 Padmanaban PBS, Rosenkranz M, Zhu P, Kaling M, Schmidt A, Schmitt-Kopplin P, et al. Mycorrhiza-tree-herbivore interactions: alterations in poplar metabolome and volatilome. *Metabolites*. 2022; 12(2).

170 Lemaitre-Guillier C, Dufresne C, Chartier A, Cluzet S, Valls J, Jacquens L, et al. Vocs are relevant biomarkers of elicitor-induced defences in grapevine. *Molecules*. 2021; 26(14): 1–18.

171 Clancy MV, Zytynska SE, Senft M, Weisser WW, Schnitzler JP. Chemotypic variation in terpenes emitted from storage pools influences early aphid colonisation on tansy. *Sci Rep*. 2016; 6: 38087.

172 Vercammen J, Pham-Tuan H, Sandra P. Automated dynamic sampling system for the on-line monitoring of biogenic emissions from living organisms. *J Chromatogr A*. 2001; 930(1–2): 39–51.

173 Aharoni A, Giri AP, Deuerlein S, Griepink F, De Kogel WJ, Verstappen FWA, et al. Terpenoid metabolism in wild-type and transgenic arabidopsis plants. *Plant Cell.* 2003; 15(12): 2866–2884.
174 Behnke K, Kleist E, Uerlings R, Wildt JJ, Rennenberg H, Schnitzler JPJ-P. RNAi-mediated suppression of isoprene biosynthesis in hybrid poplar impacts ozone tolerance. *Tree Physiol.* 2009; 29(5): 725–736.
175 Mckinney KA, Wang D, Ye J, De FJB, Guimarães PC, Batista CE, et al. A sampler for atmospheric volatile organic compounds by copter unmanned aerial vehicles. *Atmos Meas Tech.* 2019; 12(6): 3123–3135.
176 Ruiz-Jimenez J, Zanca N, Lan H, Jussila M, Hartonen K, Riekkola ML. Aerial drone as a carrier for miniaturized air sampling systems. *J Chromatogr A.* 2019; 1597(2): 202–208.
177 Munson B. Chemical ionization mass spectrometry. *Anal Chem.* 1971; 43(13): 28A–43A.
178 Blake RS, Monks PS, Ellis AM. Proton-transfer reaction mass spectrometry. *Chem Rev.* 2009; 109(3): 861–896.
179 Ferguson EE, Fehsenfeld FC, Schmeltekopf AL. Flowing afterglow measurements of ion-neutral reactions. *Adv At Mol Opt Phy.* 1969; 5(C):1–56.
180 Adams NG, Smith D. The selected ion flow tube (SIFT); a technique for studying ion-neutral reactions. *Int J Mass Spectrom Ion Phys.* 1976; 21(3–4): 349–359.
181 Futrell JH. *editor.* United States: Gaseous Ion Chemistry and Mass Spectrometry; 1986.
182 Lindinger W, Hansel A, Jordan A. On-line monitoring of volatile organic compounds at pptv levels by means of proton-transfer-reaction mass spectrometry (PTR-MS) medical applications, food control and environmental research. *Int J Mass Spectrom Ion Processes.* 1998; 173(3): 191–241.
183 De Gouw J, Warneke C. Measurements of volatile organic compounds in the earth's atmosphere using proton-transfer-reaction mass spectrometry. *Mass Spectrom Rev.* 2007; 26(2): 223–257.
184 Graus M, Müller M, Hansel A. High resolution PTR-TOF: quantification and formula confirmation of VOC in real time. *J Am Soc Mass Spectrom.* 2010; 21(6): 1037–1044.
185 Hansel A, Jordan A, Holzinger R, Prazeller P, Vogel W, Lindinger W. Proton transfer reaction mass spectrometry: on-line trace gas analysis at the ppb level. *Int J Mass Spectrom Ion Processes.* 1995; (C): 149, 609–150, 619.
186 de Gouw J, Warneke C, Karl T, Eerdekens G, van der Veen C, Fall R. Sensitivity and specificity of atmospheric trace gas detection by proton-transfer-reaction mass spectrometry. *Int J Mass spectrom.* 2003; 223–224: 365–382.
187 Warneke C, van der Veen C, Luxembourg S, de Gouw J, Kok a. Measurements of benzene and toluene in ambient air using proton-

transfer-reaction mass spectrometry: calibration, humidity dependence, and field intercomparison. *Int J Mass spectrom.* 2001; 207(3): 167–182.

188 Jordan A, Haidacher S, Hanel G, Hartungen E, Märk L, Seehauser H, et al. A high resolution and high sensitivity proton-transfer-reaction time-of-flight mass spectrometer (PTR-TOF-MS). *Int J Mass spectrom.* 2009; 286(2–3): 122–128.

189 Jordan A, Haidacher S, Hanel G, Hartungen E, Herbig J, Märk L, et al. An online ultra-high sensitivity proton-transfer-reaction mass-spectrometer combined with switchable reagent ion capability (PTR + SRI - MS). *Int J Mass spectrom.* 2009; 286(1): 32–38.

190 Bsaibes S, Piel F, Gros V, Truong F, Lafouge F, Ciuraru R, et al. Monoterpene chemical speciation with high time resolution using a fastGC/PTR-MS: results from the COV3ER experiment on *Quercus ilex. Atmosphere (Basel).* 2020; 11(7).

191 de Gouw JA, Goldan PD, Warneke C, Kuster WC, Roberts JM, Marchewka M, et al. Validation of proton transfer reaction-mass spectrometry (PTR-MS) measurements of gas-phase organic compounds in the atmosphere during the New England Air Quality Study (NEAQS) in 2002. *J Geophys Res D Atmos.* 2003; 108(21): 1–18.

192 Karl TG, Christian TJ, Yokelson RJ, Artaxo P, Hao WM, Guenther A. The tropical forest and fire emissions experiment: method evaluation of volatile organic compound emissions measured by PTR-MS, FTIR, and GC from tropical biomass burning. *Atmos Chem Phys.* 2007; 7(22): 5883–5897.

193 Taipale R, Ruuskanen TM, Rinne J, Kajos MK, Hakola H, Pohja T, et al. Technical note: quantitative long-term measurements of VOC concentrations by PTR-MS – measurement, calibration, and volume mixing ratio calculation methods. *Atmos Chem Phys.* 2008; 8(3): 9435–9475.

194 Lindinger W, Hansel A, Jordan A, Hansel A. Proton-transfer-reaction mass spectrometry (PTR – MS): on-line monitoring of volatile organic compounds at pptv levels. *Chem Soc Rev.* 1998; 27: 347–375.

195 Steinbacher M, Dommen J, Ammann C, Spirig C, Neftel A, Prevot ASH. Performance characteristics of a proton-transfer-reaction mass spectrometer (PTR-MS) derived from laboratory and field measurements. *Int J Mass spectrom.* 2004; 239(2–3): 117–128.

196 Cappellin L, Karl T, Probst M, Ismailova O, Winkler PM, Soukoulis C, et al. On quantitative determination of volatile organic compound concentrations using proton transfer reaction time-of-flight mass spectrometry. *Environ Sci Technol.* 2012; 46(4): 2283–2290.

197 Yuan B, Koss AR, Warneke C, Coggon M, Sekimoto K, De Gouw JA. Proton-transfer-reaction mass spectrometry: applications in atmospheric sciences. *Chem Rev.* 2017; 117(21): 13187–13229.

198 Hewitt CN, Hayward S, Tani A. The application of proton transfer reaction-mass spectrometry (PTR-MS) to the monitoring and analysis of volatile organic compounds in the atmosphere. *J Environ Monit.* 2003; 5(1): 1–7.

199 Majchrzak T, Wojnowski W, Rutkowska M, Wasik A. Real-time volatilomics: a novel approach for analyzing biological samples. *Trends Plant Sci.* 2020; 25(3): 302–312.

200 Bunge M, Araghipour N, Mikoviny T, Dunkl J, Schnitzhofer R, Hansel A, et al. On-line monitoring of microbial volatile metabolites by proton transfer reaction-mass spectrometry. *Appl Environ Microbiol.* 2008; 74(7): 2179–2186.

201 Khomenko I, Stefanini I, Cappellin L, Cappelletti V, Franceschi P, Cavalieri D, et al. Non-invasive real time monitoring of yeast volatilome by PTR-ToF-MS. *Metabolomics.* 2017; 13(10): 1–13.

202 Herbig J, Müller M, Schallhart S, Titzmann T, Graus M, Hansel A. On-line breath analysis with PTR-TOF. *J Breath Res.* 2009; 3(2).

203 Karl T, Fall R, Rosenstiel TN, Prazeller P, Larsen B, Seufert G, et al. On-line analysis of the $^{13}CO_2$ labeling of leaf isoprene suggests multiple subcellular origins of isoprene precursors. *Planta.* 2002; 215(6): 894–905.

204 Lichtenthaler HK. The 1-deoxy-D-xylulose-5-phosphate pathway of isoprenoid biosynthesis in plants. *Annu Rev Plant Physiol Plant Mol Biol.* 1999; 50: 47–65.

205 Schnitzler J-P, Graus M, Kreuzwieser J, Heizmann U, Rennenberg H, Wisthaler A, et al. Contribution of different carbon sources to isoprene biosynthesis in poplar leaves. *Plant Physiol.* 2004; 135(1): 152–160.

206 Cappellin L, Loreto F, Biasioli F, Pastore P, McKinney K. A mechanism for biogenic production and emission of MEK from MVK decoupled from isoprene biosynthesis. *Atmos Chem Phys.* 2019; 19(5): 3125–3135.

207 Fasbender L, Yáñez-Serrano AM, Kreuzwieser J, Dubbert D, Werner C. Real-time carbon allocation into biogenic volatile organic compounds (BVOCs) and respiratory carbon dioxide (CO_2) traced by PTR-TOF-MS, 13CO_2 laser spectroscopy and ^{13}C-pyruvate labelling. *PLoS One.* 2018; 13(9): 1–22.

208 Yáñez-Serrano AM, Fasbender L, Kreuzwieser J, Dubbert D, Haberstroh S, Lobo-Do-Vale R, et al. Volatile diterpene emission by two Mediterranean Cistaceae shrubs. *Sci Rep.* 2018; 8(1): 1–13.

209 Fall R, Karl T, Hansel A, Jordan A, Lindinger W. Volatile organic compounds emitted after leaf wounding: on-line analysis by proton-transfer-reaction mass spectrometry. *J Geophys Res Atmos.* 1999; 104(D13): 15963–15974.

210 Holzinger R, Sandoval-Soto L, Rottenberger S, Crutzen PJ, Kesselmeier J. Emissions of volatile organic compounds from *Quercus ilex* L. measured by Proton Transfer Reaction Mass Spectrometry under different

environmental conditions. *J Geophys Res Atmos.* 2000; 105(D16): 20573–20579.

211 Rasulov B, Copolovici L, Laisk A, Niinemets U. Postillumination isoprene emission: in vivo measurements of dimethylallyldiphosphate pool size and isoprene synthase kinetics in aspen leaves. *Plant Physiol.* 2009; 149(3): 1609–1618.

212 Kreuzwieser J, Meischner M, Grün M, Yáñez-Serrano AM, Fasbender L, Werner C. Drought affects carbon partitioning into volatile organic compound biosynthesis in Scots pine needles. *New Phytol.* 2021; 232(5): 1930–1943.

213 Jardine K, Chambers J, Alves EG, Teixeira A, Garcia S, Holm J, et al. Dynamic balancing of isoprene carbon sources reflects photosynthetic and photorespiratory responses to temperature stress. *Plant Physiol.* 2014; 166(4): 2051–2064.

214 Werner C, Fasbender L, Romek KM, Yáñez-Serrano AM, Kreuzwieser J. Heat waves change plant carbon allocation among primary and secondary metabolism altering CO_2 assimilation, respiration, and VOC emissions. Front. *Plant Sci.* 2020; 11(Aug): 1–17.

215 Yáñez-Serrano AM, Mahlau L, Fasbender L, Byron J, Williams J, Kreuzwieser J, et al. Heat stress increases the use of cytosolic pyruvate for isoprene biosynthesis. *J Exp Bot.* 2019; 70(20): 5827–5838.

216 Danner H, Samudrala D, Cristescu SM, van Dam NM. Tracing hidden herbivores: time-resolved non-invasive analysis of belowground volatiles by proton-transfer-reaction mass spectrometry (PTR-MS). *J Chem Ecol.* 2012; 38(6): 785–794.

217 Niederbacher B, Winkler JB, Schnitzler JP. Volatile organic compounds as non-invasive markers for plant phenotyping. *J Exp Bot.* 1909; 2015.

218 Brilli F, Loreto F, Baccelli I. Exploiting plant volatile organic compounds (VOCs) in agriculture to improve sustainable defense strategies and productivity of crops. *Front Plant Sci.* 2019; 10: 264.

219 Taiti C, Costa C, Figorilli S, Billi M, Caparrotta S, Comparini D, et al. Volatome analysis approach for the taxonomic classification of tree exudate collection using Proton Transfer Reaction Time of Flight Mass Spectrometry. *Flavour Fragr J.* 2018; 33(3): 245–262.

220 Karl T, Guenther A, Jordan A, Fall R, Lindinger W. Eddy covariance measurement of biogenic oxygenated VOC emissions from hay harvesting. *Atmos Environ.* 2001; 35(3): 491–495.

221 Karl TG, Spirig C, Rinne J, Stroud C, Prevost P, Greenberg J, et al. Virtual disjunct eddy covariance measurements of organic compound fluxes from a subalpine forest using proton transfer reaction mass spectrometry. *Atmos Chem Phys.* 2002; 2(4): 279–291.

222 Grabmer W, Graus M, Lindinger C, Wisthaler A, Rappenglück B, Steinbrecher R, et al. Disjunct eddy covariance measurements of

monoterpene fluxes from a Norway spruce forest using PTR-MS. *Int J Mass spectrom.* 2004; 239(2–3): 111–115.

223 Müller M, Graus M, Ruuskanen TM, Schnitzhofer R, Bamberger I, Kaser L, et al. First eddy covariance flux measurements by PTR-TOF. *Atmos Meas Tech.* 2010; 3(2): 387–395.

224 Park JH, Goldstein AH, Timkovsky J, Fares S, Weber R, Karlik J, et al. Active atmosphere-ecosystem exchange of the vast majority of detected volatile organic compounds. *Science.* 2013; 341(6146): 643–647.

225 Karl T, Potosnak M, Guenther A, Clark D, Walker J, Herrick JD, et al. Exchange processes of volatile organic compounds above a tropical rain forest: Implications for modeling tropospheric chemistry above dense vegetation. *J Geophys Res Atmos.* 2004; 109(18).

226 Langford B, Misztal PK, Nemitz E, Davison B, Helfter C, Pugh TAM, et al. Fluxes and concentrations of volatile organic compounds from a South-East Asian tropical rainforest. *Atmos Chem Phys.* 2010; 10(17): 8391–8412.

227 Schallhart S, Rantala P, Kajos MK, Aalto J, Mammarella I, Ruuskanen TM, et al. Temporal variation of VOC fluxes measured with PTR-TOF above a boreal forest. *Atmos Chem Phys.* 2018; 18(2): 815–832.

228 Davison B, Taipale R, Langford B, Misztal P, Fares S, Matteucci G, et al. Concentrations and fluxes of biogenic volatile organic compounds above a Mediterranean macchia ecosystem in western Italy. *Biogeosciences.* 2009; 6(8): 1655–1670.

229 Schade GW, Solomon SJ, Dellwik E, Pilegaard K, Ladstätter-Weissenmayer A. Methanol and other VOC fluxes from a Danish beech forest during late springtime. *Biogeochemistry.* 2011; 106(3): 337–355.

230 Seco R, Holst T, Sillesen Matzen M, Westergaard-Nielsen A, Li T, Simin T, et al. Volatile organic compound fluxes in a subarctic peatland and lake. *Atmos Chem Phys.* 2020; 20(21): 13399–13416.

231 Brilli F, Gioli B, Zona D, Pallozzi E, Zenone T, Fratini G, et al. Simultaneous leaf- and ecosystem-level fluxes of volatile organic compounds from a poplar-based SRC plantation. *Agric For Meteorol.* 2014; 187: 22–35.

232 Karl T, Guenther A, Turnipseed A, Patton EG, Jardine K. Chemical sensing of plant stress at the ecosystem scale. *Biogeosciences.* 2008; 5(5): 1287–1294.

233 Seco R, Karl T, Guenther A, Hosman KP, Pallardy SG, Gu L, et al. Ecosystem-scale volatile organic compound fluxes during an extreme drought in a broadleaf temperate forest of the Missouri Ozarks (central USA). *Global Change Biol.* 2015; 21(10): 3657–3674.

234 Ruuskanen TM, Müller M, Schnitzhofer R, Karl T, Graus M, Bamberger I, et al. Eddy covariance VOC emission and deposition fluxes above grassland using PTR-TOF. *Atmos Chem Phys.* 2011; 11(2): 611–625.

235 Brilli F, Hörtnagl L, Bamberger I, Schnitzhofer R, Ruuskanen TM, Hansel A, et al. Qualitative and quantitative characterization of volatile organic compound emissions from cut grass. *Environ Sci Technol.* 2012; 46(7): 3859–3865.

236 Manco A, Brilli F, Famulari D, Gasbarra D, Gioli B, Vitale L, et al. Cross-correlations of biogenic volatile organic compounds (BVOC) emissions typify different phenological stages and stressful events in a Mediterranean Sorghum plantation. *Agric For Meteorol.* 2020; 2021(303): 108380.

237 Cappellin L, Biasioli F, Granitto PM, Schuhfried E, Soukoulis C, Costa F, et al. On data analysis in PTR-TOF-MS: from raw spectra to data mining. *Sens Actuators, B.* 2011; 155(1): 183–190.

238 Lubes G, Goodarzi M. Analysis of volatile compounds by advanced analytical techniques and multivariate chemometrics. *Chem Rev.* 2017; 117(9): 6399–6422.

239 Guo Y, Jud W, Ghirardo A, Antritter F, Benz JP, Schnitzler JP, et al. Sniffing fungi – phenotyping of volatile chemical diversity in *Trichoderma* species. *New Phytol.* 2020; 227(1): 244–259.

240 Guo Y, Jud W, Weikl F, Ghirardo A, Junker RR, Polle A, et al. Volatile organic compound patterns predict fungal trophic mode and lifestyle. *Commun Biol.* 2021; 4(1): 1–12.

241 Brilli F, Gioli B, Fares S, Terenzio Z, Zona D, Gielen B, et al. Rapid leaf development drives the seasonal pattern of volatile organic compound (VOC) fluxes in a 'coppiced' bioenergy poplar plantation. *Plant, Cell Environ.* 2016; 39: 539–555.

242 Guo Y, Ghirardo A, Weber B, Schnitzler J-P, Benz JP, Rosenkranz M. *Trichoderma* species differ in their volatile profiles and in antagonism toward ectomycorrhiza *Laccaria bicolor*. *Front Microbiol.* 2019; 10: 1–15.

243 Kohonen T. The basic SOM. In: Kohonen T, editor. *Self-Organizing Maps.* Volume 30. Heidelberg: Springer Berlin; 2001. p. 105–176.

244 Granitto PM, Gasperi F, Biasioli F, Trainotti E, Furlanello C. Modern data mining tools in descriptive sensory analysis: a case study with a Random forest approach. *Food Qual Preference.* 2007; 18(4): 681–689.

245 Cappellin L, Aprea E, Granitto P, Romano A, Gasperi F, Biasioli F. Multiclass methods in the analysis of metabolomic datasets: the example of raspberry cultivar volatile compounds detected by GC-MS and PTR-MS. *Food Res Int.* 2013; 54(1): 1313–1320.

246 Biasioli F, Gasperi F, Aprea E, Mott D, Boscaini E, Mayr D, et al. Coupling proton transfer reaction-mass spectrometry with linear discriminant analysis: a case study. *J Agric Food Chem.* 2003; 51(25): 7227–7233.

7

Assessing Specialized Metabolites in Tree Bark Using Wide-Targeted LC–MS Analysis

Maria Kenosis Emmanuelle Galingay Lachica, Mutsumi Watanabe, and Takayuki Tohge

Division of Biological Sciences, Nara Institute of Science and Technology (NAIST) – Graduate School of Science and Technology, Ikoma, Nara, Japan

7.1 Introduction

As sessile organisms, plants have evolved the capacity to synthesize diverse secondary (specialized) metabolites that function physiologically for defense against biotic and abiotic stresses (1, 2) and, in some cases, facilitate reproductive processes (3). Metabolomics, or the analytical study of the entire metabolite profile of a living system, is considered an indispensable tool that addresses the need for the comprehensive analysis of chemical diversifications and metabolic polymorphisms demonstrated by inter/intra-plant species. Metabolomics approaches often have been employed for the analysis of plant systemic responses toward genetic and environmental changes, such as genetic modification studies (4) and responses to specific environmental stresses (5). Coupled chromatographic and mass spectrometric techniques are the most common approaches used in studying the plant metabolome (6, 7). Plant metabolomics using liquid chromatography mass spectrometry (LC–MS) provides a comprehensive type of approach optimized with a wide variety of extraction methods, separation columns, and mass detections (5, 6, 8, 9), allowing the detection of a large number of compound families. Environmental stressors, such as extreme temperature, salinity, water availability, and high light intensity (7), can trigger metabolic changes in the specialized metabolism of a plant to maintain homeostasis. Model organisms, such as *Arabidopsis thaliana*, common agricultural crops, and medicinal plants (4, 6), have been prime targets for large-scale metabolomics studies due to their direct impact on human health and lifestyle. In contrast, studies involving forest ecosystems, which are equally vulnerable in response to environmental changes (10), are relatively lacking (5).

Monitoring Forest Damage with Mass Spectrometry-Based Metabolomics Methods, First Edition.
Edited by Carla António.

Forest trees are the most productive organisms in forest ecosystems and are highly valued for industrial products, such as lumber, plywood, paper, and as benign alternatives to fossil fuels in the form of biofuels (11). However, forest trees are possibly sensitive to a changing climate, specifically temperature stress, drought, and climate-driven disease and pest outbreaks during their long life span (10, 11). To face the current increasing demand for wood and wood-derived products, it is imperative to focus on the improvement of the production, health, and performance of commercially valued forest tree species (7, 10). The bark is comprised of all the tissues found outside the vascular cambium of a tree (12) and contributes to approximately 20% of its above-ground biomass (13). Older trees possess an outer bark, or a cork cambium, composed of the dead primary phloem and periderm, and an inner bark containing the living secondary phloem (14). As the outermost layer, the bark protects plant stems against mechanical damage, herbivory, infectious agents (15), and other environmental stressors, such as frost, drought, and fires (16), resulting in overall wood quality and plant longevity (17).

Polyphenolic compounds, such as flavonoids, anthocyanins, and lignins, have been highlighted as important specialized metabolites in response to pathogenic infection (15), UV-B irradiation, drought, and high-temperature stress, functioning mainly by suppressing stress-induced oxidation in plant cells (7, 18, 19). In poplar, the model for woody plant research (5), phenolic glycosides, such as salicin, salicortin, tremuloidin and tremulacin, hydroxycinnamate derivatives, and flavonoid-derived condensed tannins have been found to constitute a significant part of its specialized metabolites (20). Salicortin and tremulacin have been previously studied as potential herbivore toxins (21). In *Populus tremuloides*, a species belonging to the same genus, leaves and bark have been reported to synthesize pyrocatechol, benzoic acid, and salicylic acid, leading to the repression of the fungus *Hypoxylon pruinatum* (22). Phenolic glycosides and tannins produced by the same species have also been observed to confer resistance against bacterial (23) and fungal infections and the latter case, to inhibit spore germination (24). Ten percent of the total extracted compounds from the cambium tissue of poplar are specialized metabolites; of which, catechol (1,2-dihydroxybenzene) is the most notable bioactive compound used to defend the plant against herbivores (5). Crucial metabolites found in the bark of forest trees allow us to glean information as to their roles in assessing wood quality and improving tree survival, providing a physiological status "snapshot" (7) of the plant in response to environmental stress factors. Such information is proven valuable to plant breeders and forest managers in sustaining healthy and productive forest ecosystems and complements current knowledge of the bioactive phytochemicals found in bark tissues of select forest tree species.

This chapter presents a general methodology for wide-targeted metabolomics using LC–MS, emphasizing on plant specialized metabolites from the bark tissue of common forest trees, such as those from the genera *Populus, Pinus,* and *Eucalyptus,* along with useful strategies for specialized metabolite peak identification and annotation.

7.2 Materials and Methods

This section provides an optimized protocol that can be used to evaluate specialized metabolites obtained from the bark of selected forest trees, such as *Eucalyptus globulus, Pinus thunbergii,* and *Populus deltoides.* Due to the differences in terms of chemical composition between plant tissues and species, sample preparation and extraction buffer volumes might have to be adjusted.

7.2.1 Plant Cultivation and Sampling

It is ideal that the plant of interest is considerably healthy with a well-developed bark along the main trunk and branches. Harvesting was done by cutting a fully-developed branch using a handheld saw. If the bark was dry, shallow cuts were created on the bark surface using a sharp blade, and clean tweezers were used to remove them from the surface of the branch, and scissors to portion them into smaller pieces. A similar approach can be done in the case that samples are collected from the main trunk, but only remove bark from one side. Girdling, or the complete removal of a strip of bark around a branch or trunk of a woody plant (25), inhibits the flow of nutrients from the roots to the rest of the plant (26), and can leave the specimen open to desiccation and infection. Before metabolite extraction, samples can be stored at room temperature in a 50 mL conical tube with the lid partly open to allow moisture evaporation.

Prior to powdering the bark samples, 50-mL conical tubes used for storage were quenched in liquid nitrogen, after which the frozen plant tissues were transferred in a pre-cooled grinding jar with a metal ball, and ground with a ball mill for two minutes at 25 Hz. This step must be performed quickly to prevent samples from thawing. If samples are not completely fine powdered after processing with the ball mill, or if a ball mill is unavailable, further powdering can be done manually, using a pre-cooled mortar and pestle. Briefly, sterilize the surface of the mortar and pestle that will be in contact with the sample using 70% ethanol, then wipe the excess ethanol with a clean tissue. Add liquid nitrogen into the mortar to initially cool the surface, then fill approximately a quarter of the mortar with liquid nitrogen. Add the tissue sample and grind it down with the pre-cooled pestle. All powdered frozen samples were collected with a pre-cooled metal spatula and placed in a pre-chilled 50 mL conical tube. These can be stored at −80°C until further use.

7.2.2 Sample Aliquoting

A 2.0-mL microcentrifuge tube was pre-cooled by submerging it in liquid nitrogen for five seconds, after which it was weighed on an electric analytical balance. Approximately 30.0 ± 5.0 mg of the frozen powdered sample was aliquoted into the 2.0-mL microcentrifuge tube, and the corresponding sample weight was noted. All samples must be kept in liquid nitrogen (to avoid thawing), but none should enter the sample tubes as this can forcibly open the tubes upon any increase in temperature, and powdered samples might be lost in the process. This situation can be prevented by placing a metal or cardboard holding rack into the liquid nitrogen to keep the tubes in place. All sample aliquots can be stored at −80 °C until metabolite extraction.

7.2.3 Extraction of Specialized Metabolites from Tree Bark Tissue

Numerous methods can be considered for the extraction of specialized metabolites. Two-phase extraction utilizes a single solvent of a specific polarity range, but it excludes compounds of lower or higher polarities. Three-phase extraction methods, which are demonstrated below, permit the collection of both polar and non-polar metabolites, expanding the compound scope for analysis. In addition, polar and nonpolar fractions can be used for other metabolomics analyses, such as gas chromatography mass spectrometry (GC–MS) primary metabolite profiling, and LC–MS lipidomics and wax measurement, respectively. The extraction solvent was prepared by adding a designated internal standard to 100% methanol at a 1 : 100 dilution. Internal standards are compounds that are used to correct or normalize ionization inefficiencies and matrix ion suppression effects to obtain the accurate concentration of the plant-specialized metabolite being evaluated (27–29). Compounds that are not endogenous in the biological sample can be used as internal standards, which ensures that they can be distinguished from the analyte by their mass differences (29), thereby preventing cases of cross-contribution caused by insufficient chromatographical separation of the standard from a sample peak (30). Examples of commonly used internal standards are: (i) for dicot samples, apigenin-6-*C*-glucoside (isovitexin), a glycoflavone only present in monocots; and (ii) for monocot samples, hesperetin-7-*O*-rutinoside (hesperidin), a flavanone glycoside exclusive to dicots (9).

A pre-cooled zirconia bead was placed inside the sample tubes (for 2.0-mL microcentrifuge tubes, 3 mm diameter zirconia beads will suffice). Note that the lid should not be forcibly opened as it can lead to lid breakage; the lid hinge can be gradually warmed for a few seconds before the lid is opened entirely. Ten microliters of the pre-prepared extraction solvent were added per 1 mg of powdered sample, and immediately vortex-mixed for approximately 15 seconds. To further homogenize the mixture, sample tubes can be placed in vial adapters, and the extraction is further performed using a ball mill for 60 seconds at 20 Hz. After homogenization, 200 µL of chloroform was added, followed by vortex-mixing for 15 seconds. Bark tissue contains nonpolar compounds, such as suberin, an insoluble polymer composed primarily of glycerol, long-chain α,ω-di-acid, and long-chain ω-hydroxy acid monomers (31). The addition of chloroform allows the separation of non-polar from polar compounds in the extract. Subsequently, 400 µL of water (LC–MS grade) were added, and again vortex-mixed for 15 seconds. Tubes were centrifuged at 4 °C and 15,000 g for 10 minutes. After centrifugation, three phases were visible in the tube: (i) the upper polar phase; (ii) the interphase; and (iii) the lower nonpolar phase. Approximately 200 and 100 µL of the upper and lower phases, respectively, were aliquoted into new separate 1.5-mL microcentrifuge tubes. The aliquots were dried in a vacuum concentrator overnight at 25° C. The dried aliquots can be stored at −30 °C prior to LC–MS analysis.

7.2.4 LC–MS Analysis of Polar Fractions

The obtained polar fractions were found to be more soluble in lower methanol concentrations; hence, 200 µL of 50% methanol was used to resuspend dried aliquots. This can, however, vary depending on the specialized metabolites that constitute the bark tissue from the selected plant species. The sample tube was initially centrifuged at 4 °C and 15,000 g for 10 minutes. Subsequently, the liquid sample was transferred into a new 1.5-mL microcentrifuge tube. Any debris present in the liquid was excluded in the final sample aliquot, after which, 100 µL of the diluted sample was aliquoted into vials suitable for LC–MS analysis. Samples at higher concentrations can leave artefacts along the column; to avoid shifts in the retention time and/or matrix ion suppression effects, it is advisable to wash the column after use. Another preventive measure is to perform a preliminary check using samples at a lower concentration, which can be gradually increased depending on the peak quality after analysis. One large contributing factor to saturated sample aliquots is the use of dry material (in contrast to fresh material) of the tissue sample upon sample preparation. In most cases, using dry material yields a higher concentration of metabolites due to the lack of water, but this might depend on the plant species and tissue sampled. However, in the case that there is an observed retention time shift in the collected peak data, it can still be validated using the internal standard. Lastly, the samples were placed in the autosampler tray with a separate vial containing the extraction buffer blank. It is recommended to perform at least five injections of the extraction buffer blank to equilibrate the LC–MS system prior to analyzing the first sample (32).

Liquid chromatography (LC) was performed on a Vanquish (Thermofisher Scientific), and a Luna C_{18} column (100 Å, 150 × 2.00 mm, 3 µm, Phenomenex), with the column oven and sample trays set to 28 and 4 °C, respectively. Table 7.1

Table 7.1 Liquid chromatography gradient for specialized secondary metabolite profiling.

Time (min)	Flow rate (μL min^{-1})	Buffer A[a] (%)	Buffer B[b] (%)
0.0	200	100	0
1.0	200	100	0
2.0	200	90	10
16.0	200	60	40
22.0	200	0	100
25.0	200	0	100
25.5	200	100	0
30.0	200	100	0

a) Buffer A: Formic acid in water at a final concentration of 0.1% formic acid (v/v)
b) Buffer B: Formic acid in acetonitrile at a final concentration of 0.1% formic acid (v/v)

presents the pump gradient applied during LC. Electrospray ionization mass spectrometry (ESI–MS) was performed on a LTQ XL (Thermofisher Scientific), with the following parameters: scan mode at positive/negative ESI, mass range from *m/z* 150–1500, capillary temperature of 275 °C, and spray voltage set to 3.50 kV.

7.3 Data Analysis

7.3.1 Peak Evaluation

Various software can be configured to be used in the analysis of binary. RAW files from the LC–MS operating system. Licensed software is available from the manufacturer (e.g. Xcalibur by Thermofisher Scientific, Analyst by SCIEX, MassHunter by Agilent), while Metalign, MSFact, XCMS2, BINBASE, MZmine2, and TagFinder are available online to download for free (9, 32).

Peaks are evaluated and picked on the basis of peak area, if not peak height, with consideration of the peak shape and scan points or modes (32). Several of the aforementioned software are capable of selecting peaks and their peak values by means of peak intensity. Manual peak picking can also be done by visually checking the sample chromatogram, identifying the peaks of interest, and noting their corresponding retention times (Rt), *m/z* values at a given scan mode, and fragments; the latter of which can be obtained from the sample spectrum. Subsequently, the signal intensity data of the peaks of interest are exported into a whole data matrix, followed by data normalization using the aliquoted sample weight (prior to metabolite extraction), and the peak intensity of the internal standard used. This can be supplemented by dedicated data analysis tools, such as MetTailor, Normalyzer, IntCor, and EigenMS, which further eliminate biased trends in the data (9, 32, 33). Note that peak evaluation is dependent on the type of analysis to be conducted; in untargeted metabolomics, all detected metabolite signals are picked to provide a metabolic overview of the plant tissue being examined. In contrast, targeted metabolomics is a more focused approach, and only defined groups of chemically characterized and annotated metabolites are selected (9).

7.3.2 Peak Annotation

In targeted metabolomics studies, the most accurate methods used for peak identification involve nuclear magnetic resonance (NMR)-based approaches, in which the compounds that correspond to the selected peaks are purified and analyzed, and coelution analysis is performed using authentic standards or related reference compounds (9, 32). Generally, due to the diversity of specialized metabolites in plants, standard compounds for each target metabolite might be commercially unavailable. To circumvent such limitations, peak annotation is performed to characterize selected peaks, and potentially predict how the said compounds participate in specific biosynthetic pathways. Nonetheless, metabolite annotation remains to be the major bottleneck for metabolomics studies of plant-specialized metabolism (34), and is often performed through a combination of several approaches:

7.3.2.1 Tandem Mass Spectrometry (MS/MS, MS^n)

Different fragmentation patterns are obtained using several collision energy intensities and charges, which provide key fragments for the identification of parental ions, aglycones, and various combinations of the parental ion with detached moieties (9). Figure 7.1 shows the fragmentation patterns from a single metabolite peak under different charges and collision energies. The said patterns can then be compared to data from available public MS/MS databases (e.g. MASSBANK) (35), and plant-specific databases, such as Spektraris (36) and the RIKEN MS^n Spectral Database for Phytochemicals (ReSpect) (37).

At a given collision energy intensity and charge, a metabolite peak in the chromatogram will have a corresponding fragmentation pattern. However, the identification of the molecular parent ion peak from the fragments might be challenging at first glance. For example, the ionization process can yield adducts, or charged parental ions due to the addition or loss of atoms, and are present as fragments that can be mistaken as the parental ion peak (38). Spectra comparison of negative and positive ion modes is one way to check a suspected parental ion peak as it should be present in both, except if compounds are produced due to neutral loss.

Aglycones, which are basic structures with acceptors of sugar decorations replaced with an H atom, can be detected as individual peaks. However, since

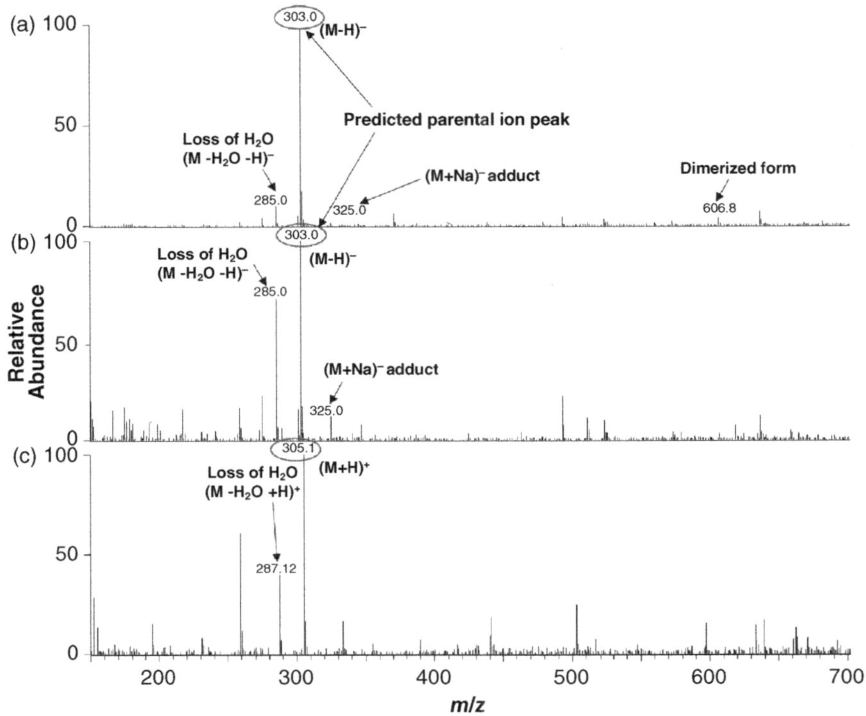

Figure 7.1 Fragmentation pattern of a metabolite peak (Rt 12.17 minutes) from *Pinus thunbergii* bark tissue. Compounds are detected by negative scan mode with a collision energy of (a) 5 V and (b) 30 V and (c) positive scan mode with a collision energy of 20 V. A 2 *m/z* difference is observed between the molecular parent ion peak in the negative and positive scan modes due to the loss and gain of H^+, respectively. Based on the fragments in the spectrum, the parental ion peak (encircled) was putatively determined to be taxifolin or dihydroquercetin (MW = 304).

aglycones are often decorated or glycosylated *in planta* because these forms are more functionally stable against environmental stresses (39), it is best to confirm whether it is an accumulation of the undecorated compound or a fragment of its decorated form. In most cases, there are standards available for purchase, and can be co-eluted with the plant extract. Retention times of aglycones can also differ depending on the attached moieties (Figure 7.2). Figure 7.2 illustrates how flavonoid aglycones, and their common sugar moieties, elute relative to each other in reverse-phase columns in liquid chromatography.

Annotating aglycones detected in plant extracts is particularly useful. For instance, by viewing the chromatogram using the mass range plot type, and delimiting it to include the *m/z* value of the annotated aglycone, it shows the peaks of related compounds, such as their glycosylated forms and derivatives. The related peaks can be further annotated and used in biosynthetic pathway

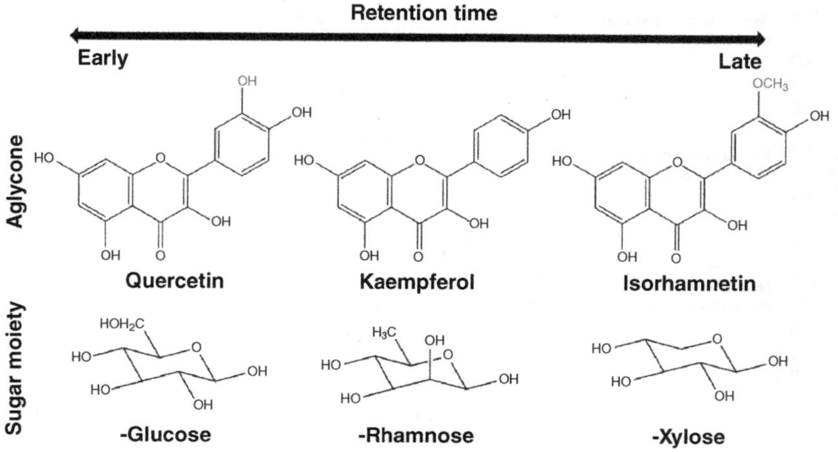

Figure 7.2 Relative retention times of common flavonoid aglycones and attached sugar moieties. Reverse-phase columns in liquid chromatography are used for measurement.

construction. In Figure 7.3, the putative peak of taxifolin with the actual compound standard has similar retention times and *m/z* values upon comparison. After delimiting the mass range to a certain *m/z* tolerance (e.g. ±0.2 to 0.5), chromatogram peaks that possess a fragment of the same *m/z* value are shown (Figure 7.3b,c).

7.3.2.2 Reference Plant Compounds and Extracts with Literature Data

Previously reported compounds that have different *m/z* values but similar fragmentation patterns to the peaks of interest can have similar modification forms. Additionally, literatures that studied extracts similar or related to the target sample with identified and/or annotated compounds can be used to narrowing down candidate metabolites, or to provide a glimpse as to the type of metabolites that can be obtained from the same tissue sample through the comparison of *m/z* values and retention time indexes. However, for this approach, it is ideal that samples are prepared using the same tissue type, growth conditions, extraction and analytical methods, and internal standard/s (9, 32).

7.3.2.3 Database Search

Plant-specific databases provide abundant and curated information on metabolite composition and distribution across various plant species. The largest of these databases is KNApSAcK (40), comprising more than 62,600 metabolites across over 24,700 species, and the Universal Natural Products Database (UNPD) with over 220,000 metabolites from 1,687 species (41). In addition, KNApSAcK allows plant species-specific searches, with approximately 157,100 entries. More specialized databases include, but are not limited to, Dr. Duke's

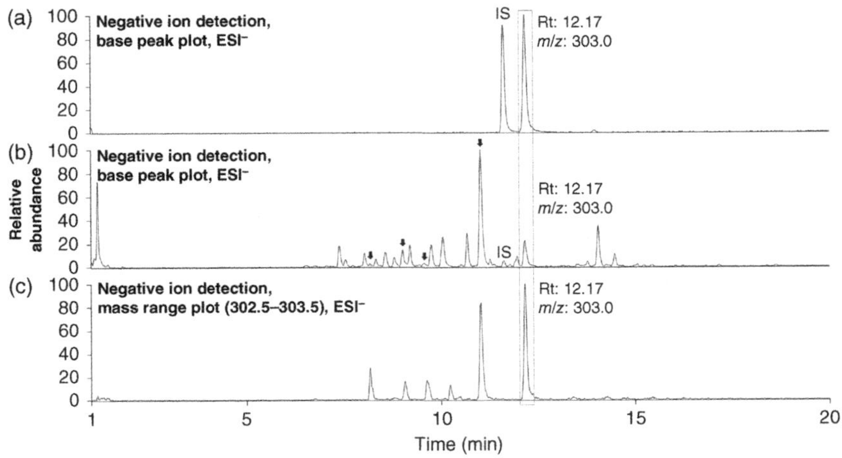

Figure 7.3 Comparison of standard and sample chromatograms using base peak and mass range plots. Extracted ion chromatogram (EIC) in negative ion mode of the (a) internal standard (isovitexin, Rt 11.61 minutes, *m/z* 431.1) and external standard (taxifolin, Rt 12.17 minutes, *m/z* 303.0); (b, c) extract from wild-type *P. thunbergii* bark tissue. The peaks corresponding to taxifolin (red box) and its putatively related compounds (black arrows) are visualized.

Phytochemical and Ethnobotanical Databases for medicinal plants (42), and the Essential Oil Database (EssOilDB) (43). Through the comparison of the predicted molecular weights of the detected metabolites with information from established databases, candidate compounds are narrowed, and additional information, such as possible modifications or decorations of the target metabolites, are deduced (9, 34).

7.3.2.4 Mutant Analysis

Target pathway metabolite changes through knock-out (KO) and overexpression (OX) studies can aid in finding putatively related metabolites from the sample (32). Mutations of key enzymatic genes can lead to the decrease or increase, or in some cases, the lack or sudden production, of downstream metabolites. *F3H* (flavanone 3-hydroxylase), for example, is a key enzyme that channels carbon flow toward the production of 3-hydroxylated flavonoids (44) from naringenin. Studying *F3H* KO mutants of the plant sample and checking the resulting metabolite profile in comparison to its wild-type counterpart can demonstrate which putative peaks represent downstream metabolites from the said enzymatic step.

7.3.2.5 Peak Annotation by the Prediction of Pathway

In the case that a peak was identified using annotation methods mentioned previously, such data are proven useful in the annotation of related peaks by

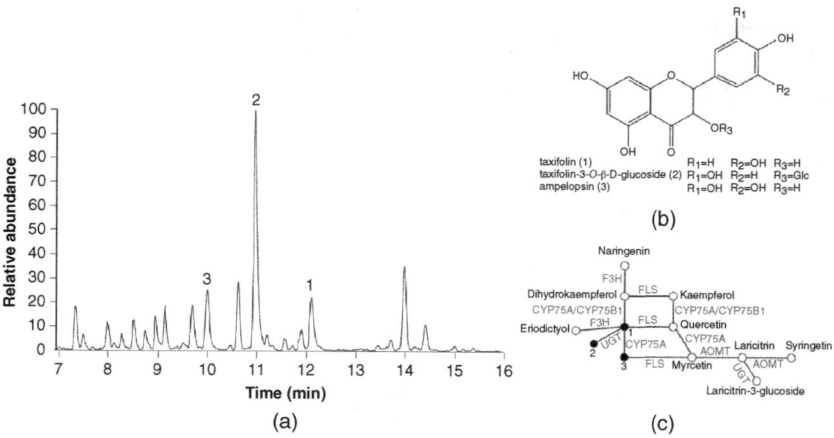

Figure 7.4 Metabolite profiling of *P. thunbergii* bark tissue by LC–MS. (a) Extracted ion chromatogram (EIC) in negative ion mode of the extract from wild-type *P. thunbergii* bark tissue, (b) structures of putatively annotated compounds, and (c) a constructed pathway the annotated peaks (black circle) participate in, with several upstream and downstream compounds that might be found in the sample (white), and the enzymatic genes involved in their biosynthesis. Abbreviations: *F3H*, flavanone 3-hydroxylase; *FLS*, flavonol synthase; *CYP75A*, flavonoid 3′,5′-hydroxylase; *CYP75B1*, flavonoid 3′-hydroxylase; *UGT*, glucosyltransferase; *AOMT*, flavonoid 3′,5′-methyltransferase.

predicting the pathway involved in its biosynthesis. As seen in Figure 7.4, peak **1** detected from *P. thunbergii* bark tissue was putatively identified as the flavonoid aglycone taxifolin (Rt 12.17 minutes, *m/z* 303), under the negative scan mode. Using biosynthetic pathway databases, such as the Kyoto Encyclopedia of Genes and Genomes (KEGG) (45), the biosynthetic pathway involving taxifolin production can be constructed (Figure 7.4c). It was found that taxifolin can also be decorated through an enzymatic step to form glycoside derivatives, such as taxifolin-3-*O*-glucoside, or modified to other hydroxylated forms, like ampelopsin. This implies that the compounds mentioned can be present in the bark tissue sample and were later putatively identified as **2** (Rt 11.02 minutes, *m/z* 465) and **3** (Rt 10.05 minutes, *m/z* 319), respectively. Note that as an aglycone, taxifolin is expected to be eluted after its glycosides and glycoside derivatives, therefore having a relatively longer retention time.

Other examples are seen in Figure 7.5. Peaks **1** and **2** (Figure 7.5a) detected from *E. globulus* bark tissue were annotated as the flavonoid quercetin (Rt 12.79 minutes, *m/z* 301) and isorhamnetin (Rt 13.87 minutes, *m/z* 315), and both 3-*O*-arabinoside (**3**, Rt 11.89 minutes, *m/z* 433; **4**, Rt 15.23 minutes, *m/z* 447) decorated forms under the negative scan mode. Postbiosynthetic pathway construction allowed the annotation of other flavonoid peaks, specifically **D**-catechol or (+)-3′,4′,5,7-tetrahydroxy-2,3-*trans*-flavan-3-ol (**5**, Rt 9.59 minutes, *m/z* 289). Using the pathway as a guide, other metabolite

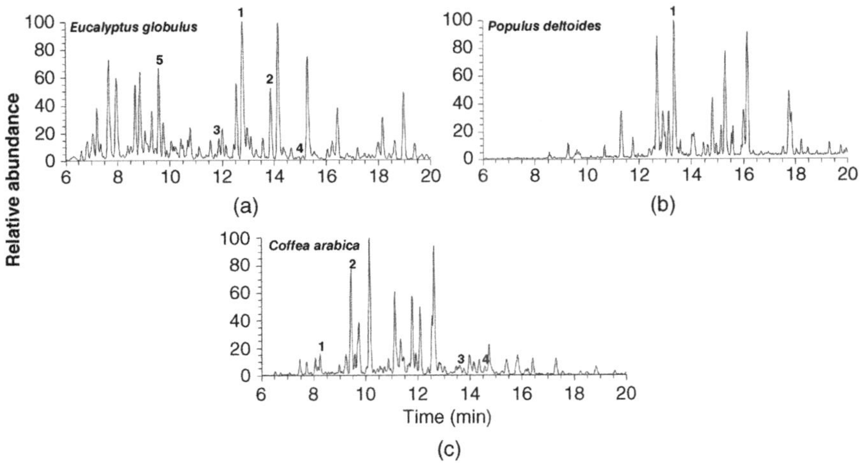

Figure 7.5 Metabolite profiling of forest and commercial crop tree bark tissues by LC–MS. Total ion chromatogram (TIC) in negative ion mode of the extract from (a) wild-type *Eucalyptus globulus*, (b) *Populus deltoides*, and (c) *Coffea arabica* bark tissues.

intermediates, such as dihydrotricetin, dihydroquercetin, and leucocyanidin, which might be present in the analyzed sample, can be identified and annotated as well. Importantly, the aglycone form of flavonoids is highly detectable in bark samples. On the other hand, the bark of *P. deltoides* (Figure 7.5b) contains common flavonol glycosides such as myricetin-hexoside (**1**, Rt 13.35 minutes, *m/z* 463) but without high levels of the myricetin aglycone. Several major peaks annotated in forest tree bark samples were flavonoids and their derivatives. Flavonoids are structurally diverse compounds that have evolved to aid in plant pigmentation, plant-microbe interactions, and UV protection (46). Conversely, bark tissues of tree crop species, such as *Coffea arabica*, present a different spectrum of metabolites. For example, major polyphenolics found in *C. arabica* bark are chlorogenic acid derivatives (Figure 7.5c) (**1**, Rt 8.23 minutes, *m/z* 353; **2**, Rt 9.43 minutes, *m/z* 353; Rt 9.69 minutes, *m/z* 353; **3**, Rt 13.60 minutes, *m/z* 515; Rt 14.04 minutes, *m/z* 515; **4**, Rt 14.57 minutes, *m/z* 515). Both flavonoid and chlorogenic acid biosynthesis diverged from the lignin pathway, but the accumulation mechanisms are slightly different among plant species.

7.4 Data Interpretation

Omics integration allows us to bridge the lack of linear relationship between the genome and the metabolome (34) by relating genomics, transcriptomics, proteomics, and metabolomics data to provide greater insight into metabolite biosynthesis and regulation. Using obtained metabolite data, biosynthetic frameworks are generated and/or refined based on characterized pathways from other plant species. In Figure 7.4c, for example, it is predicted that two enzymatic steps are needed in the production of taxifolin, beginning with naringenin, the central metabolite involved in flavonoid biosynthesis. The identification of the metabolite intermediates and the enzymatic steps involved in their subsequent biosynthesis aids in the prediction of gene expression of possible candidate genes, in conjunction with gene expression databases and web-based tools like the Poplar electronic Fluorescent Pictograph (eFP) browser (47). The information can then be used to perform cross-species analyses of orthologs in other species of interest using genome and transcriptome data in available databases, such as the National Center for Biotechnology Information (NCBI) (48), PLAZA (49), Phytozome (50), and other published literature. Furthermore, coexpression analysis with other genes is possible using coexpression databases, including ATTED-II (51) and coexpressed biological processes (COP) (32, 52).

The presence of stress-defense compounds in forest tree bark, and the identification of their respective biosynthetic and regulatory genes achieved by navigating across other omics platforms, unveils latent applications toward future metabolomics studies in their potential usefulness as reference extracts for other forest tree-related studies and for tree crop species that are highly vulnerable to biotic and abiotic stressors, such as *C. arabica*.

Omics studies is a complex endeavor due to the amount and complexity of data to be handled, from the mapping of metabolite pathways, understanding the roles of these metabolites in a biological context, and integrating data from other omics fields (33, 34) to determine the causes of metabolite variations, be it due to changes in gene and protein expression and regulation and/or enzyme kinetics. Amidst these challenges, omics integration is a useful tool to comprehend, monitor, and control plant cellular responses to genetic and environmental changes (33).

7.5 Conclusions and Future Perspectives

Plant metabolomics, alongside other omics fields, is at the forefront of high-throughput studies in biology, particularly in advancing our understanding of the role of metabolites and their correlation to gene function as well as in biotic and abiotic interactions. Despite this, substantial limitations to the field are present and require continued addressing through protocol standardization and optimization. In tissue-specific metabolomics, leaves are the most commonly evaluated tissues because they are the most accessible and display easily detectable phenotypic changes in stress conditions. To holistically grasp the systemic response of plants to such stressors, it is necessary to extend metabolite evaluation to other tissues. Bark tissue is the first physical line of defense of woody plants, and contains specific flavonoid and flavonoid-derived metabolites, several in undecorated forms. In aglycone forms, compounds are less soluble and amphiphilic, and more importantly, more toxic against parasites and herbivores (53). However, the presence of compounds of lower polarities in bark can create unwanted ion suppression effects in the MS system and lead to unresolvable peaks in the chromatogram. Similarly, seed tissues that are known to accumulate highly decorated flavonoids (e.g. proanthocyanidins) store an abundance of nonpolar oils and might need different solvents (e.g. 80% acetone) to extract target compounds more efficiently. To encompass a wider range of the plant metabolome, lipidomics studies could be explored in tandem with wide-targeted metabolomics, with the former targeting compounds found in lipophilic barriers (e.g. suberin lamellae); nonpolar layers in specific plant tissues and organs prevent water loss and function as mechanical protectants against pathogenic infection and herbivory (54). Another key point is to extend extraction methods from the liquid phase (LPE) to the solid phase (SPE). LPE is more common than the latter due to ease of use, but only covers a portion of the entire metabolite profile of a plant sample restricted by the organic solvent used during extract preparation. The increase of available columns with suitable stationary phases for SPE (55) makes this extraction method more feasible in the research setting. Lastly, because resolving the identities of unknown and unclassified compounds remains a huge challenge in metabolomics studies, it is equally imperative to keep oneself updated on available metabolomics tools that can assist in peak annotation and identification.

References

1 Alseekh S, Fernie AR. Metabolomics 20 years on: what have we learned and what hurdles remain? *Plant J.* 2018; 94: 933–942.

2 Wang S, Alseekh S, Fernie AR, Luo J. The structure and function of major plant metabolite modifications. *Mol Plant.* 2019; 12: 899–919.

3 Vaishnav P, Demain AL. Unexpected applications of secondary metabolites. *Biotechnol.* 2010; 29: 223–229.

4 Saito K, Matsuda F. Metabolomics for functional genomics, systems biology, and biotechnology. *Annu Rev Plant Biol.* 2010; 61: 463–489.

5 Ning K, Ding C, Zhu W, Zhang W, Dong Y, Shen Y, et al. Comparative metabolomic analysis of the cambium tissue of non-transgenic and multi-gene transgenic Poplar (*Populus* x *euramericana* 'Guariento'). *Front Plant Sci.* 2018; 9: 1201.

6 Fernie AR, Tohge T. The genetics of plant metabolism. *Annu. Rev. Genet.* 2017; 51: 287–310.

7 Escandón M, Meijón M, Valledor L, Pascual J, Pinto G, Cañal MJ. Metabolome integrated analysis of high-temperature response in *Pinus radiata*. *Front Plant Sci.* 2018; 9: 485.
8 Doefler H, Lyon D, Nägele T, Sun XL, Fragner L, Hadacek F, et al. Granger causality in integrated GC–MS and LC–MS metabolomics data reveals the interface of primary and secondary metabolism. *Metabolomics.* 2013; 9: 564–574.
9 Shimizu T, Watanabe M, Fernie AR, Tohge T. Targeted LC-MS analysis for plant secondary metabolites. In: António C, editor. *Plant Metabolomics: Methods and Protocols. Methods in Molecular Biology* Volume 1778. New York (NY, USA): Humana Press; 2018. p171–181.
10 Martinich J, Crimmins A, Beach RH, Thomson A, McFarland J. Focus on agriculture and forestry benefits of reducing climate change impacts. *Environ Res Lett.* 2017; 12: 060301.
11 FAO. Forestry and climate change: Roles of forests in climate change [Internet]. Available from (https://www.fao.org/forestry/climatechange/53459/en/), accessed 10 Jan 2023.
12 Poorter L, McNeil A, Hurtado VH, Prins HHT, Putz FE. Bark traits and life-history strategies of tropical dry- and moist forest trees. *Funct Ecol.* 2014; 28: 232–242.
13 Neumann M, Lawes MJ. Quantifying carbon in tree bark: the importance of bark morphology and tree size. *Methods Ecol Evol.* 2020; 12: 646–654.
14 Rosell JA. Bark thickness across the angiosperms: more than just fire. *New Phytol.* 2016; 211(1): 90–102.
15 Abedini A, Colin M, Hubert J, Charpentier E, Angelis A, Bounasri H, et al. Abundant extractable metabolites from temperate tree barks: the specific antimicrobial activity of *Prunus avium* extracts. *Antibiotics.* 2020; 9(3): 111.
16 Pérez-Harguindeguy N, Díaz S, Garnier E, Lavorel D, Poorter H, Jaureguiberry P, et al. New handbook for standardised measurement of plant functional traits worldwide. *Aust J Bot.* 2013; 61: 167–234.
17 Pásztory Z, Mohácsiné IR, Gorbacheva G, Börcsök Z. The utilization of tree bark. *BioRes.* 2016; 11(3): 7859–7888.
18 Hale BK, Herms DA, Hansen RC, Clausen TP, Arnold D. Effects of drought stress and nutrient availability on dry matter allocation, phenolic glycosides, and rapid induced resistance of poplar to two lymantriid defoliators. *J Chem Ecol.* 2005; 31: 2601–2620.
19 Sullivan J, Gitz DC, Peek MS, McElrone AJ. Response of three eastern tree species to supplemental UV-B radiation: leaf chemistry and gas exchange. *Agric.* 2003; 120(1-4): 219–228.
20 Tsai CJ, Harding SA, Tschaplinski TJ, Lindroth RL, Yuan Y. Genome-wide analysis of the structural genes regulating defense phenylpropanoid metabolism in *Populus*. *New Phytol.* 2006; 172(1): 47–62.

21 Osier TL, Lindroth RL. Long-term effects of defoliation on quaking aspen in relation to genotype and nutrient availability: plant growth, phytochemistry and insect performance. *Oecologia.* 2004; 139: 55–65.
22 Hubbes M. Inhibition of *Hypoxylon pruinatum* by pyrocatechol isolated from bark of aspen. *Science.* 1962; 136(3511): 156.
23 Tegelberg R, Julkunen-Tiitto R. Quantitative changes in secondary metabolites of dark-leaved willow (*Salix myrsinifolia*) exposed to enhanced ultraviolet-B radiation. *Physiol Plant.* 2001; 113: 541–547.
24 Flores G, Hubbes M. The nature and role of phytoalexin produced by aspen (*Populus tremuloides* Michx.). *Forest Pathol.* 1980; 10(2-3): 95–103.
25 De Schepper V, Steppe K. Tree girdling responses simulated by a water and carbon transport model. *Ann Bot.* 2011; 108: 1147–1154.
26 Binkley D, Stape JL, Takahashi EN, Ryan MG. Tree-girdling to separate root and heterotrophic respiration in two *Eucalyptus* stands in Brazil. *Oecologia.* 2006; 148(3): 447–454.
27 Perez CJ, Ifa DR. Internal standard application strategies in mass spectrometry imaging by desorption electrospray ionization mass spectrometry. *Rapid Commun Mass Spectrom.* 2021; 35: e9053.
28 Rodrigues AM, António C. Standard key steps in mass spectrometry-based plant metabolomics experiments: Instrument performance and analytical method validation. In: António C, editor. *Plant Metabolomics: Methods and Protocols. Methods in Molecular Biology.* Volume 1778. New York (NY, USA): Humana Press; 2018. p. 19–31.
29 Thomas SN, French D, Jannetto PJ, Rappold BA, Clarke WA. Liquid chromatography–tandem mass spectrometry for clinical diagnostics. *Nat Rev Methods Primers.* 2022; 2: 96.
30 Redestig H, Fukushima A, Stenlund H, Moritz T, Arita M, Saito K, et al. Compensation for systematic cross-contribution improves normalization of mass spectrometry based metabolomics data. *Anal Chem.* 2009; 81: 7974–7980.
31 Graça J, Santos S. Suberin: a biopolyester of plants' skin. *Macromol Biosci.* 2007; 7: 128–135.
32 Tohge T, Fernie A. Combining genetic diversity, informatics and metabolomics to facilitate annotation of plant gene function. *Nat Protoc.* 2010; 5(6): 1210–1227.
33 Fukushima A, Kusano M, Redestig H, Arita M, Saito K. Integrated omics approaches in plant systems biology. *Curr Opin Chem Biol.* 2009; 13(5-6): 532–538.
34 De Souza LP, Naake T, Tohge T, Fernie A. From chromatogram to analyte to metabolite. How to pick horses for courses from the massive web resources for mass spectral plant metabolomics. *GigaScience.* 2017; 6: 1–20.

35 Horai H, Arita M, Kanaya S, Nihei Y, Ikeda T, Suwa K, et al. MassBank: a public repository for sharing mass spectral data for life sciences. *J Mass Spectrom.* 2010; 45(7): 703–714.

36 Cuthbertson DJ, Johnson SR, Piljac-Žegarac J, Kappel J, Schäfer S, Wüst M, et al. Accurate mass-time tag library for LC/MS-based metabolite profiling of medicinal plants. *Phytochemistry.* 2013; 91: 187–197.

37 Sawada Y, Nakabayashi R, Yamada Y, Suzuki M, Sato M, Sakata A, et al. RIKEN tandem mass spectral database (ReSpect) for phytochemicals: a plant specific MS/MS-based data resource and database. *Phytochemistry.* 2012; 82: 38–45.

38 Blumer MR, Chang CH, Brayfindley E, Nunez JR, Colby SM, Renslow RS, et al. Mass spectrometry adduct calculator. *J Chem Inf Model.* 2021; 61: 5721–5725.

39 Saigo T, Wang T, Watanabe M, Tohge T. Diversity of anthocyanin and proanthocyanin biosynthesis in land plants. *Curr Opin Plant Biol.* 2020; 55: 93–99.

40 Afendi FM, Okada T, Yamazaki M, Morita AH, Nakamura Y, Nakamura K, et al. KNApSAcK family databases: integrated metabolite-plant species databases for multifaceted plant research. *Plant Cell Physiol.* 2012; 53(2):e1(1-12).

41 Gu J, Gui Y, Chen L, Yuan G, Lu H, Xu X. Use of natural products as chemical library for drug discovery and network pharmacology. *PLoS One.* 2013; 8(4): e62839.

42 USDA. U.S. Department of Agriculture. Dr. Duke's phytochemical and ethnobotanical databases [Internet]. Available from (http://phytochem.nal.usda.gov/phytochem/search), accessed 29 Jan 2023.

43 Kumari S, Pundhir S, Priya P, Jeena G, Punetha A, Chawla K, et al. EssOilDB: a database of essential oils reflecting terpene composition and variability in the plant kingdom. *Database (Oxford).* 2014; 2014: bau120.

44 Wang L, Liu ACW, Lam PY, Liu G, Godwin ID, Lo C. Transgenic expression of flavanone 3-hydroxylase redirects flavonoid biosynthesis and alleviates anthracnose susceptibility in sorghum. *Plant Biotechnol J.* 2020; 18: 2170–2172.

45 Kanehisa M, Goto S. KEGG: kyoto encyclopedia of genes and genomes. *Nucleic Acids Res.* 2000; 28(1): 27–30.

46 Winkel-Shirley B, Flavonoid biosynthesis. A colourful model for genetics, biochemistry, cell biology, and biotechnology. *Plant Physiol.* 2001; 126(2): 485–493.

47 Wilkins O, Nahal H, Foong J, Provart NJ, Campbell MM. Expansion and diversification of the *Populus* R2R3-MYB family of transcription factors. *Plant Physiol.* 2009; 149(2): 981–993.

48 NCBI. National Center for Biotechnology Information [Internet]. Available from (https://www.ncbi.nlm.nih.gov/), accessed 10 Feb 2023.

49 Bel MV, Silvestri F, Weitz EM, Kreft L, Botzki A, Coppens F, et al. PLAZA 5.0: extending the scope and power of comparative and functional genomics in plants. *Nucleic Acids Res.* 2022; 50: D1468–D1474.
50 Goodstein DM, Shu S, Howson R, Neupane R, Hayes RD, Fazo J, et al. Phytozome: a comparative platform for green plant genomics. *Nucleic Acids Res.* 2012; 40(Database issue): D1178–D1186.
51 Obayashi T, Hibara H, Kagaya Y, Aoki Y, Kinoshita K. 2022. ATTED-II v11: a plant gene coexpression database ysing a sample balancing technique by subagging of principal components. *Plant Cell Physiol.* 2022; 63(6): 869–881.
52 Kazusa DNA Research Institute. Co-expressed biological processes [Internet]. Available from (http://webs2.kazusa.or.jp/kagiana/cop0911/), accessed 10 Feb 2023.
53 Kytidou K, Artola M, Overkleeft HS, Aerts JMFG. Plant glycosides and glycosidases: a treasure trove for therapeutics. *Front Plant Sci.* 2020; 11: 357.
54 Cohen H, Szymanski J, Aharoni A. Assimilation of 'omics' strategies to study the cuticle layer and suberin lamellae in plants. *J Exp Bot.* 2017; 68(19): 5389–5400.
55 Juhaschik MP, Jenkins AJ. Comparison of liquid/liquid and solid-phase extraction for alkaline drugs. *J Chromatogr Sci.* 2009; 47(7): 553–557.

8

Plant Hormone Analysis in Forest Tree Species

Eva Cañizares, Juan Manuel Acién, Miguel González-Guzmán, and Vicent Arbona

Departament de Biologia, Bioquímica i Ciències Naturals, Universitat Jaume I, Castelló de la Plana, Spain

8.1 Importance of Forest Tree Species

Forests provide multiple vital services to the global human population, including food, timber, fuel, fiber, shelter, CO_2 sequestration, flood control and soil protection from erosion, and diversity of forest genetic resources. Moreover, forests are an important source of opportunities for recreation, education, and cultural enrichment (1). According to the FAO Global Forest Resources Assessment 2020 (FAO-FRA 2020), forests occupy 31% of the world's land area, with South America accounting for 48%, followed by Europe (45%) and North and Central America (35%) (2). Owing to their ecological and biotechnological importance, the mapping and assessment of the condition of forest ecosystems and forest ecosystem services have been regarded as an essential part of the EU's Biodiversity Strategy for 2030, a core part of the European Green Deal (EGD) (3). In the EU zone, for example, roundwood production (i.e. wood removed from the forest and other wooded land, or other tree-felling site) was estimated at 490 million m^3 in 2018 accounting for a 12.3% world share, ahead of the United States with 11% (4).

The most important forest tree species worldwide are Norway spruce (*Picea abies*), Scots pine (*Pinus sylvestris*), maritime pine (*Pinus pinaster*), stone pine (*Pinus pinea*), poplars (*Populus nigra*), Douglas-fir (*Pseudotsuga menziesii*), eucalyptus (*Eucalyptus globulus* and others), and common ash (*Fraxinus excelsior*). Other tree species are of capital importance locally, such as oaks from the *Quercus* genus (e.g. cork oak and holm oak) and the flowering ash (*Fraxinus ornus*), or are a source of bioactive compounds, such as sarsaparilla (*Smilax aspera*) that contains furostanol (a saponin that has shown a powerful

Monitoring Forest Damage with Mass Spectrometry-Based Metabolomics Methods, First Edition.
Edited by Carla António.

Table 8.1 Important forest trees and shrubs.

Common name	Scientific name	Distribution	Uses
Norway spruce	*Picea abies* (L.) H. Karst.	Northeast and central Europe (Norway, Poland, Latvia, Estonia, Lithuania, Russia, Belarus, Switzerland, Austria, regions in East Germany, and North of Italy)	Used for timber/pulpwood production and as an ornamental (e.g. Christmas tree); source of Spruce beer (with high vitamin C content).
Scots pine	*Pinus sylvestris* L.	Widely and densely distributed in North and Northeast Europe and China, North Britain, and sparsely found in Northern and central Spain, North Turkey, North Romania, Bosnia, and Bulgaria)	Used for pulpwood/sawnwood production and as an ornamental (e.g. Christmas tree).
Maritime pine	*Pinus pinaster Aiton*	Mostly found in Southern Europe (Spain, Portugal, North Italy, and West France) and sparsely in North Africa (Morocco and Algeria).	Planted for timber in its native area; source of flavonoids, catechins, proanthocyanidins, and phenolic acids.
Poplar	*Populus nigra* L.	Widely distributed throughout Europe, Russia and China, South England, and Ireland,	Primarily used for timber production.
Douglas-fir	*Pseudotsuga menziesii* (Mirbel) Franco	Pacific coastal regions of West-central British Columbia to central California	Extensively used in forestry management; used for softwood timber production and as an ornamental; buds used as odorants in beverages; needles used to prepare tea.
Eucalyptus	*Eucalyptus globulus* Labill (primarily in Europe, but there are many other species)	Primarily Australia, but it is widely present as naturalized non-species in Spain and Portugal, Cyprus, Southern Africa, California, Hawai, and Macaronesian islands.	Used for timber/pulpwood production; source of essential oils (cineole), phenolics (quinic, dihydroxyphenylacetic, and caffeic acids), and flavonoids.
Stone pine	*Pinus pinea* L.	Native areas include Spain, and the Northwest coast of Italy, but it is found as plantation in North Africa, Corsica, Sardinia, and East Coast of France.	Used as food (edible pine nuts) and as an ornamental; resin/bark used for tannin extraction; empty cone shells used as fuel; important ecological service in soil conservation and consolidation of coastal dunes.

(Continued)

Table 8.1 (Continued)

Common name	Scientific name	Distribution	Uses
Common ash	*Fraxinus excelsior* L.	Northern Spain and widely distributed throughout the United Kingdom, France, Germany, Italy, Russia, Northern Greece, Southern Fennoscandia, Turkey, and considered native in Southwestern Asia.	Timber used for fuel and woodworking; forestry management (shelter and forage for grazing animals); firewood; bark and leaves used in modern herbal medicine for their laxative and astringent properties.
Cork oak	*Quercus suber* L.	Portugal and Western and Southwestern Spain, North Africa, Corsica, Sardinia, and Western Italy, but it is found cultivated in Asia, Crimea peninsula in Ukraine,	Used for cork production in Mediterranean countries (Southern Portugal and Spain); bark used for tannin extraction; acorns used for swine feeding.
Holm oak	*Quercus ilex* L. and *Quercus rotundifolia* L.	Spain, Italy, Sardinia, Corsica, Sicily, Greece, Mediterranean, and Atlantic coasts of France, North of Africa (scattered)	Used for charcoal production; acorns used for animal and human consumption; more importantly, used as a host for black truffle (*Tuber melanosporum*) production.
Flowering ash	*Fraxinus ornus* L.	Spain (South Valencia and North Alacant provinces), Italy, Corsica and Sardinia, Greece, Serbia, Romania, Croatia, Hungary, Bulgaria, Macedonia, Bosnia, Albania, Slovenia, and Slovakia. Also, in the Mediterranean Northern coast of Turkey.	Used to obtain sugary syrup from sap and as an ornamental.
Butcher's broom	*Ruscus aculeatus* L.	Widely distributed in Southern Europe (Spain, Portugal, France, Italy, Corsica, and Sardinia), Switzerland, North Africa (Morocco, Tunisia, Algeria, and Libya), England, Turkey, Greece, Serbia, Romania, Croatia, Hungary, Bulgaria, Macedonia, Bosnia, Albania, Slovenia and Slovakia, Georgia, Armenia, Azerbaijan, Syria, Cyprus, Lebanon, Israel, and Jordan.	Source of several naturally occurring steroidal saponins (e.g. ruscogenins) that have anti-inflammatory properties (8).

(Continued)

Table 8.1 (Continued)

Common name	Scientific name	Distribution	Uses
Sarsaparilla	*Smilax aspera* L.	Widespread in central Africa (Democratic Republic of Congo, Kenya, and Ethiopia), Southern Mediterranean Europe (Cyprus, Albania, Croatia, Greece, Italy, Malta, Montenegro, France, Portugal, and Spain), temperate Asia (Israel, Jordan, Lebanon, Syria, and Turkey) and tropical Asia (India, Bhutan, and Nepal).	Roots were traditionally collected to prepare beverages with different medicinal properties; roots contain furostanol, a saponin with powerful antiproliferative properties against cancer cells (5,6).
Arbutus	*Arbutus unedo* L.	It is native to the Mediterranean region: Portugal, Spain, Southeastern France, and North of Algeria, Morocco, Libya, and Tunisia. Coastal regions of Italy, Corsica and Sardinia, Greece, Slovenia, Croatia, Montenegro, Turkey, Israel, Palestine, Jordan, Lebanon, and Syria.	Primarily used as food due to its tasty berries, with high sugar content; leaves are rich in anti-inflammatory flavonoids (e.g. quercetin); provides several ecological services (e.g. nectar for bumblebees, wind barrier for lands close to the sea); resistant to fire making it useful as pioneer plant species in restoration programs; dense foliage used as shelter for insects and small animals during winter; dense root system can help in the stabilization of soils; used as ornamental plant in gardens and other public landscapes; source of several bioactive compounds with anti-tumor potential (7).

antiproliferative activity against cancer cells *in vitro*) (5, 6) and arbutus (*Arbutus unedo*) that produces arbusterin, tannins, and gallic acid, which have been shown to have antiproliferative effect against tumor cells (7) (see Table 8.1 for more details).

8.2 Plant Hormones and Their Roles in Plant Physiology, Biochemistry, and Development

As sessile organisms, plants need to adapt quickly in response to the type and intensity of a given exogenous stimulus. Exogenous cues are abiotic- (e.g. drought, salinity, high irradiance, floods, cold or heat) or biotic-stress factors (e.g. arthropod phytophagous pests, nematodes, bacteria, fungi, or viruses) that under natural conditions, often occur in combination, thereby, constituting a single stimulus with specific physiological, genetic, and metabolic outcomes (9). Plant hormones are phytochemicals of varied chemical origins (see Section 2.1 for more details) that play an important role in integrating the responses of plants with several of these exogenous (environmental) or endogenous (developmental) cues.

8.2.1 Main Plant Hormones Involved in the Interaction of Plants with Their Environment

Ethylene (ET) is a gaseous plant hormone that has been identified as the key signal that triggers organ senescence and shedding, including processes related to fruit ripening in climacteric fruits (10–12). It is produced from methionine through a reaction catalyzed by S-adenosyl methionine synthase to generate S-adenosyl methionine (SAM), which is subsequently converted to 1-aminocyclopropane 1-caboxylate (ACC) by ACC synthase. This compound is further oxidized to ethylene by ACC oxidase (ACO). ET has also been involved: (i) in plant responses to different stress conditions, including drought, salinity, pathogen attack, and hypoxia/anoxia) (10); (ii) in the expression of defense genes, such as chitinase (13–15); and (iii) in the development of specific hypoxia responses, such as aerenchyma or aerial roots (16, 17).

Auxins, predominantly represented by indole-3-acetic acid (IAA), are a group of indole-containing compounds mainly synthesized from the amino acid tryptophan (Trp) via two chemical steps: (i) Trp is converted to indole-3-pyruvic acid (IPA), a reaction catalyzed by tryptophan aminotransferase; and (ii) IPA undergoes oxidative decarboxylation to produce IAA, a reaction catalyzed by the YUC-type flavin monooxygenases (18). The role of auxin has been primarily established as a growth-promoting plant hormone, being the main driver of acid growth, inducing the destabilization of plant cell walls through the activation of hydrolases (19). The polar transport of auxin from the apical shoot meristem is known to regulate the activity of lateral shoot meristems in a mechanism known as apical dominance (20). In forest tree species (e.g. poplar), it has been shown that heavy metal contamination interferes with auxin signaling and hampers lignification (21). Also in this plant species, auxin interacts with other plant hormones in the response to abiotic stress (22).

Cytokinins (CKs) are a group of plant hormones derived from the terpenoid pathway (mevalonate and methyl-erythritol phosphate pathways) that is subsequently linked to an adenosyl ribose phosphate moiety by an isopentenyl phosphate transferase (IPT) (23). The two active CKs are isopentenyl adenine and its hydroxylated form *trans*-zeatin. Both forms can be glycosylated rendering inactive forms; *trans*-zeatin can also be isomerized (zeatin isomerase), reduced (zeatin reductase) into dehydrozeatin, and degraded into adenine and an aldehyde by cytokinin dehydrogenase (CKX). This group of plant hormones has been primarily associated with cell division processes, morphogenesis, shoot, and root branching as well as prevention of chlorophyll degradation due to their role in light-regulated processes (e.g. chloroplast differentiation and de-etiolation).

Gibberellins (GAs) are a wide group of chemically-related compounds, most of them physiologically inactive. These compounds are synthesized from terpenoids and show a characteristic diterpenoid core structure which is modified by different hydroxylations. These plant hormones are also important in the regulation of growth and developmental processes, such as photo- and skotomorphogenesis (24), seed germination (25), bud transition from vegetative to floral developmental stage (26), and regreening of chromoplasts in ripening citrus fruits (27). Active gibberellins are known to interact with other plant hormones to induce different responses to environmental cues, such as shade due to neighbor proximity (28). Other studies revealed that these hormones are also responsible for the development of morphological adaptations to hypoxia (29), while copper toxicity was shown to alter the levels of gibberellins in the root apexes in maize (30).

Jasmonates (JAs) are lipid-derived plant hormones, which are chemically oxygenated fatty acids (or oxylipins) arising from the α-linolenic acid released from membrane phospholipids by a phospholipase. Afterward, a lipoxygenase catalyzes the first rate-limiting step in jasmonate biosynthesis. The biosynthesis of JAs has two major pathways: (i) in the thylakoid membranes of chloroplasts; and (ii) in the peroxisome (31). This group of plant hormones has been typically associated with wound responses (e.g. performed with tools, plant tissue piercing, and chewing by phytophagous arthropods) (32–34). Moreover, JAs have been involved in the maturation of pollen cells (35) and have also been identified as primary signaling compounds in plant abiotic stress responses (36, 37). The JA biosynthetic pathway also shares steps with the biosynthesis of fatty acid-derived volatiles (38), which are important signals in triggering defense responses in neighboring tissues and plants (39).

Salicylic acid (SA) is a plant hormone related to pathogen defense, first identified as a bioactive compound from the bark tissue of the shrub *Salix* spp. (or willow) with keratolytic, comedolytic, and bacteriostatic activities. SA was later confirmed to have a role in mammal cyclooxygenase-1 (COX1) enzyme inhibitor, reducing the production of pro-inflammatory prostaglandins (also oxylipins) (40), and therefore reducing the inflammatory response and the associated pain.

In plants, SA is synthesized from chorismate either through direct conversion to isochorismate via the Isochorismate Synthase 1 (ICS1) pathway or through its conversion to phenylalanine by Chorismate Mutase (CM). Phenylalanine is then converted to *trans*-cinnamic acid in a reaction catalyzed by Phenylalanine

Ammonia-Lyase Lyase (PAL), which in turn is converted to benzoic acid, and finally to SA, catalyzed by a putative benzoic acid 2-hydroxylase (BA2H) (41).

SA is primarily involved in biotrophic pathogen responses by activating the expression of Pathogenesis Related 1 (PR1) proteins that contribute to systemic acquired resistance (SAR); SAR is thus an SA-dependent response that helps to provide a long-lasting broad-spectrum resistance to pathogens (42). Besides its well-known role in defense responses, SA also participates in the regulation of cell growth, stomatal aperture respiration, seed germination, nodulation, expression of senescence-related genes, and tolerance to thermal stress, alone or in combination with other abiotic stresses (43, 44).

Abscisic acid (ABA) is another lipid-derived plant hormone (from terpenoids in this case). Structurally, ABA is a sesquiterpenoid resulting from xanthophylls (9-*cis*-neoxanthin and 9-*cis*-violaxanthin) through oxidative cleavage catalyzed by 9-*cis*-epoxycarotenoid dioxygenase (NCED) into xanthoxin, which is subsequently oxidized into ABA aldehyde by ABA2 and to ABA by ABA Aldehyde Oxidase 3 (AAO3) (45, 46). The activity of this plant hormone has been mainly related to the regulation of abiotic stress responses, such as stomatal closure, biosynthesis of compatible osmolytes (proline), adaptation to extreme desiccation (e.g. during seed ontogeny), and induction of dormancy (47). A particular trait of this plant hormone is that it can be reversibly inactivated by glycosylation into its conjugated form ABA glucose ester (ABAGE) and irreversibly inactivated by oxidization into phaseic acid (PA) (48). Moreover, mechanisms not related to active ABA build-up have been suggested in response to soil flooding (49, 50), therefore making ABA an extraordinarily versatile plant hormone.

Brassinosteroids (BRs) are a large family of steroid class molecules varying on the alkyl group substitution of the side chain: C27, C28, or C29. So far, about 70 naturally occurring compounds from this group have been identified including free forms and their conjugates. At the biosynthetic level, all BRs derive from squalene, which is subsequently converted to cycloartenol. From this point, three pathways arise that conform the three main types of BR synthesized from different precursors: (i) C27 BRs without substituents on C24 are generally synthesized from cholesterol; (ii) C28 BRs with α-methyl, β-methyl, or methylene substituent derive from campesterol, 24-epicampesterol or 24-methylenecholesterol, respectively; and (iii) C29 BRs with an α-ethyl substituent are produced from sitosterol (51). The most active BR is brassinolide, a C28 BR that is also the most widely used BR for exogenous treatments and, for this reason, this is the best characterized pathway (51, 52). Brassinosteroids are present in algae, angiosperms, ferns, and gymnosperms, with important roles in different biological processes, namely, growth, vascular differentiation, fertility, or regulation of responses to environmental cues or stress. BRs have been shown to play a crucial role in the regulation of trade-offs between growth and triggering immune responses (53–56), which is a key aspect in forest and agricultural productivity.

Table 8.2 List of the most important plant hormones in plant physiology.

Name	Chemical class	Site of biosynthesis	Action
Ethylene	volatile alkene	All tissues	Organ abscission, senescence, triggers morphological and biochemical adaptations to hypoxia, controls root radial expansion, and fruit ripening.
Auxin	Alkylated indole	Plant meristems, primarily shoot and root apices	Cell elongation, regulation of lateral shoot meristems, different plant abiotic stress responses.
Cytokinins	Alkylated nitrogen base	All plant tissues	Cell division, shoot and root morphogenesis, along with auxin-regulated axillary bud growth, slowdown of leaf senescence.
Gibberellins	Hydroxylated diterpenes	In all actively growing organs (e.g. ovaries, seed embryo in cereals, plant meristems)	Breaking of dormancy (antagonizes abscisic acid), triggers mobilization of endosperm sugar reserves in cereal seeds, reverts transition of chloroplasts to chromoplasts in ripening citrus fruits, bud transition from vegetative to floral, hypocotyl elongation in response to shade.
Jasmonates	Oxygenated fatty acids (oxylipins)	All plant tissues	Regulation of wound and necrotrophic pathogen response, pollen cell maturation, and stomatal opening.
Abscisic acid	Isoprenoid	Nearly in all plant tissues containing chloroplasts or amyloplasts and in phloem cells.	Seed and bud dormancy, stomatal closure, biosynthesis of compatible osmolytes in vegetative tissues, and flavonoids in certain fruit species.
Salicylic acid	Phenolic acid	Nearly in all plant tissues	Defense against biotrophic pathogens, stomatal closure, senescence, fruit ripening, cell growth, and thermotolerance.
Brassino-steroids	Sterols	Nearly in all plant tissues, but particularly in cambial tissues	Vascular differentiation, growth, response to environmental cues, and balance between growth/defense responses.
Strigolactones	Substituted butenolide	Root exudates and roots	Germination of parasitic plant seeds (*Orobanche* spp., and *Striga* spp.), hyphal growth of AM fungi, inhibition of plant branching, regulation of phosphate, and nutrient starvation and/or deficiency.

Finally, strigolactones (SLs) are a group of plant hormones related to strigol, a molecule constituted by two lactone rings with an enol ether compound. In plants, these compounds are synthesized from carotenoids, specifically *trans*-β-carotene, which is isomerized to *cis*-β-carotene, cleaved by Carotene Cleavage Dioxygenase 7 or CCD7 (MAX3), and subsequently, by CCD8 (MAX4) in the plastids generating carlactone (CL), which is further oxidized in the cytoplasm by cytochrome P450 (MAX1) to form the active 5-deoxystrigol (57). These molecules were first identified as the germination stimulants of parasitic plants of the *Orobanche* and *Striga* genera, as components of host-plant root exudates (58). In recent years, it was discovered that this family of compounds regulates the induction of the hyphal branching of arbuscular mycorrhizal (AM) fungi, present in 80% of plant species. Endogenous SLs have been found to be branching inhibitors, thereby becoming a major player in the regulation of plant growth and development. SLs are responsible for the response to phosphate and other nutrient deficiency or starvation (57,58), integrating root-to-shoot balance, nutrient acquisition, and resource allocation (59). This group of molecules shares structural similarities with smoke-derived compounds karrikins, first identified in 1994, particularly the presence of the butenolide group (60). Karrikins are not plant endogenous compounds because they derive from the combustion of carbohydrates (sugars and polysaccharides from cell walls) during forest wildfires. These compounds are perceived by the seeds of pyrophytic forest tree species and stimulate their germination, constituting a competitive advantage over other species in the colonization of land after deforestation by wildfires (60, 61). Their structural similarity has raised questions on the evolution of the signaling mechanisms of karrikins and SLs, which might have diverged between 65 and 126 million years ago (62) (see Table 8.2 for more details).

8.3 Forest Tree Sampling

There are a number of issues that need to be taken into account when sampling trees from natural environments, including aspects related to the genetic structure of the population and its geographical distribution. Natural populations are likely constituted by several different genotypes, each with different physiological attributes, while isolated (or patched populations) are likely to be constituted by different genotypes of the same plant species (see Section 8.3.1 for more details). In addition, some relevant tree species in forestry management have long life cycles (>20 years), making the selection of the accurate period for sampling difficult (e.g. to study flowering and budbreak) or even its replication under controlled conditions (63). Some forest tree species are also endemic to certain world regions, and sampling them might debilitate their population dynamics, and thus, compromise their survival. Altogether, these concerns make the collection of sufficient and representative samples for research in forest trees challenging.

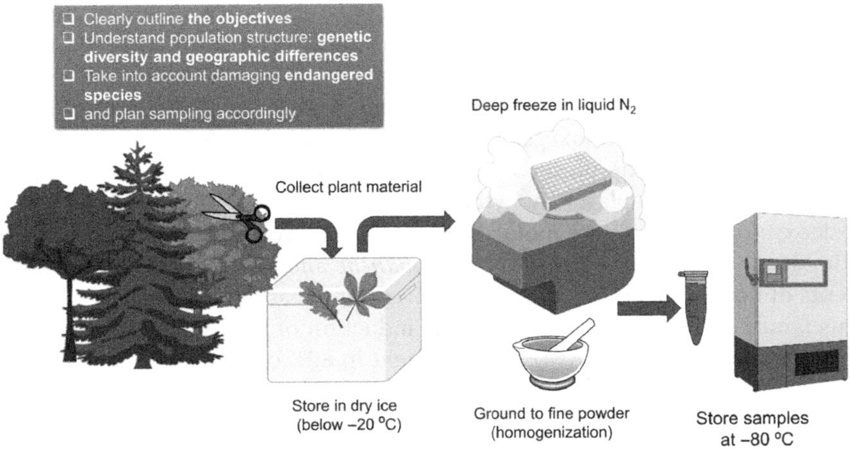

Figure 8.1 Important considerations when planning a field sampling of forest tree species and typical workflow for sample collection and preservation. Source: Created using Biorender.

8.3.1 Issues Related to Forest Tree Population Structure and Distribution

When a study pursues capturing the population variability of a certain plant species or there are biological reasons that hinder the replication of plants under controlled conditions, certain aspects need to be taken into account: (i) the assessment of the genetic variability of natural populations; a simple phenotype assessment might help, but it could also be necessary to perform genetic analyses to identify how many genotypes are present and design a sampling procedure that captures the existing genetic variability; (ii) if a given tree species is distributed in geographically distant clusters, it is important to consider that environmental cues, and potentially genetic differences, might affect sample variability; (iii) the biological question is always on top of any experimental design, for example what tissue(s) have to be sampled and at what developmental stage?; and (iv) when dealing with endemic or endangered species, it is crucial to consider the minimal sample needed to prevent further damage to the population (Figure 8.1).

8.3.2 Practical Aspects Related to Forest Tree Tissue Harvesting and Preservation

Sampling in field conditions holds several practical issues, mostly related to sample integrity. This is particularly important because plant hormones respond rapidly not only to environmental and developmental changes but also to tissue manipulation. Sampling should be performed in order to obtain samples that best represent a snapshot of the plant's metabolic and regulatory status at that particular moment; therefore, any manipulation that potentially alters that condition constitutes an artifact. For these reasons, it is highly recommended to rapidly quench metabolism by flash-freezing in liquid nitrogen immediately after harvesting. However, this might not be possible in natural environments (63) and other methods need to be envisaged. Although it might not be feasible to carry liquid nitrogen for field sampling, it could be possible to carry dry ice in a polystyrene box (e.g. Styrofoam or similar) to keep samples at a very low temperature until being processed in the laboratory. In this regard, it is widely acknowledged that snap-frozen fresh samples are better for plant hormone and metabolite extraction than freeze-dried, despite the latter being a much better form of preservation than cold storage (64). In any case, prior to metabolite extraction and subsequent metabolite analysis, samples need to be properly homogenized to prevent any bias due to different tissue composition (63) (Figure 8.1).

8.3.3 Important Considerations Regarding the Extraction of Plant Hormones

Plant hormone extraction protocols are chosen primarily depending on the chemical nature of the hormone(s) to be analyzed. Therefore, gaseous hormones are extracted from intact or processed tissues through natural or forced release followed by capture in an adequate adsorbent (e.g. Tenax® porous polymers and solid-phase microextraction [SPME] fibers). However, some gaseous plant hormones might be easier to extract, capture, and analyze than others. For instance, ET has been traditionally extracted by incubating tissues at room temperature in an enclosed atmosphere for several hours, after which an air sample is withdrawn with a syringe and analyzed, while other volatiles require a preconcentration step, typically using an adsorbent. In this case, volatiles are forced to migrate to the gaseous layer by extracting the tissue with poorly volatile solvents and slightly heating the mixture.

Non-volatile plant hormones are extracted from processed tissues using different strategies. The most common strategy is to use aqueous mixtures of polar organic solvents followed by concentration and purification steps of the samples using liquid or solid phase extraction and the addition of protective compounds (e.g. hydroxybutyl toluene or diethyldithiocarbamate) to prevent oxidization of analytes (65). In the past, analytical methods required high recovery ratios of the analytes to perform well; however, with the advent of more sensitive and specific mass spectrometers, the tendency is to reduce sample manipulation and use more exhaustive extraction protocols to cover as many chemical classes as possible (66). Most methods based on aqueous solvent extraction are similar and only vary in the composition of the extraction mixture and the presence of purification/preconcentration steps. For example, Chiwocha et al. (67) used isopropanol:glacial acetic acid (99 : 1, v/v), Müller et al. (68) used a mixture of methanol:isopropanol:glacial acetic acid (20 : 79 : 1, v/v/v), while Simura et al. (69) used 50% aqueous acetonitrile. Subsequently, solid phase extraction (SPE) can be used to remove interferents from sample extracts by retention on a solid sorbent (e.g. silica or silica-alkylated) (63), such as the commercially available C18 SPE columns Sep-Pak® and Oasis® HLB (67,69). All these methods reported the recovery and analysis of different plant hormone classes (e.g. ABA, JAs, CKs, and GAs). Gómez-Cadenas et al. (70) reported a method based on mechanical extraction of plant tissues with water. Initially developed to extract and analyze ABA without further purification, it was subsequently refined and included a solvent extraction step to remove interferents and expanded it to the analysis of JA and IAA (71). This method was compared with those based on solvent extraction of plant tissues and rendered higher performance for GAs analysis in citrus fruit tissues (72). Interestingly, as an improvement of the purification step, some authors included immunoaffinity chromatography columns to specifically capture certain analytes (e.g. ABA) (73). A summary of all currently available quantitative methods is provided in Figure 8.2.

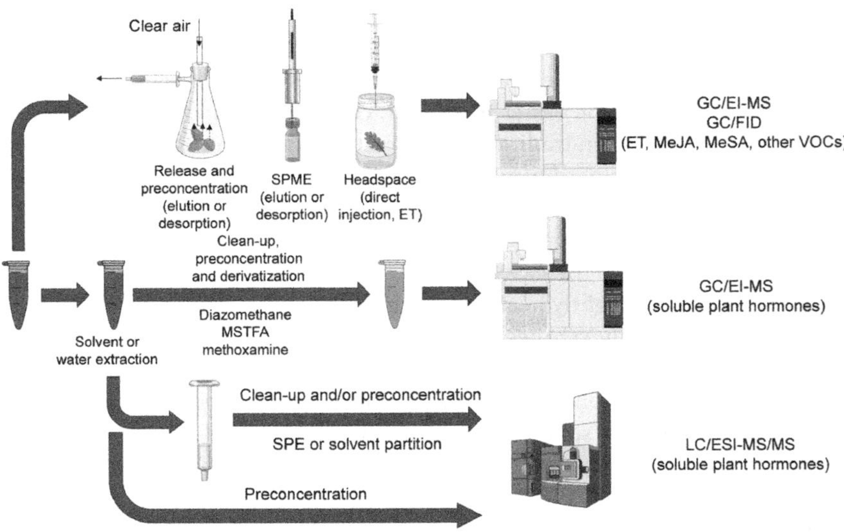

Figure 8.2 Typical analytical methods for plant hormone extraction and analysis according to the chemical nature of the analyte and the available instrumentation. Source: Created using Biorender.

8.4 Analytical Methods for Plant Hormone Analysis and Profiling

Analysis of plant hormones has changed over the years, a long way from activity-based analyses to modern analytical methods, which allow the identification of isomers with different biological activities, derivatives that may constitute inactive or storage forms, or even investigate alternate metabolic pathways leading to hormonal regulators at the organ, tissue, cell, and even subcellular levels. In this regard, the chemical nature of the plant hormone strongly influences the analytical technique to be used, as gaseous hormones (e.g. ET or MeJA) can hardly be analyzed with methods involving liquid handling; however, in some cases, the analysis of precursors of these molecules has provided an alternative to the analysis of the gas phase, such as the analysis of ACC, the precursor of ET, or soluble forms of JAs. As a rule, it must be noted that no exhaustive analytical method exists, and there are always methods better suited for certain compounds.

Early studies of modern plant physiology used the activity of growth regulators to analyze the amount of the regulator present in an extract or isolated fraction. An example is auxins, the first group of plant hormones described in studies dating from the second half of XIX century on the phototropic and gravitropic responses of plants. It was not until 1934 that Kögl and Kostermans isolated IAA, which was later identified as auxin by Berger and Avery in 1944. In 1928, Went described the first quantitative assay of auxin based on the induction of curvature in decapitated *Avena* epicotyls, where the curvature degree was found to be proportional to the amount of IAA present in the sample, which was generally embedded in a gelatin or agar cube and placed on top of the decapitated epicotyl (74). Since the early studies on plant hormone analysis, chromatography has played an important role in the separation of heterogeneous plant extracts to isolate different plant growth regulators. The use of a separation technique prevented researchers from using massive amounts of starting material to isolate compounds by means of more conventional, solvent partitioning-based techniques. Around mid-twentieth century, paper chromatography that was soon replaced by thin-layer chromatography, became available for plant hormone isolation and analysis (74). The technique itself proved to be highly versatile and compatible with different solvent mixtures as well as with colorimetric and bioassay-based analytical methods.

Chromatographic techniques, such as gas and liquid chromatography (GC and LC), are now well-established methods for phytochemical analysis. GC provides a more robust and sensitive analysis to that of LC, being perfectly suited for polar and semipolar compounds, among which plant hormones are found. The GC analytical platform consists of a stationary phase; that is a packed column (generally short and filled with porous silica) or a capillary column (30–60 m long and hollow, the filling coats the inner surface of the capillary). The polarity of compounds (and their interactions with the stationary phase) highly influences the separation process and will also contribute to the differences in the elution profile.

Non-volatile compounds cannot be analyzed by GC without prior derivatization or chemical modification to change their properties. This derivatization step usually involves alkylation reactions with diazomethane to generate methyl esters or with methoxyamine hydrochloride to generate oximes from carbonyls, followed by silylation to produce more thermostable and volatile trimethylsilyl (TMS) derivatives from compounds containing functional groups with active hydrogens (e.g. -OH, -COOH, -NH, and -SH) (75). Reagents used to carry out these reactions are highly reactive and toxic; therefore, should be handled with caution and appropriate protection. In addition, if the conditions for the reaction are not well controlled, each derivatization reaction might generate different products for each endogenous starting molecule, leading to a different compound in the chromatographic separation, which particularly hinders untargeted analyses. For these reasons, solvents used to perform these reactions must be water-free and kept in extremely controlled conditions to prevent rehydration.

Different detectors have been used for plant hormone analysis with GC. Flame ionization detection (FID) has proved to be an efficient and relatively easy ionization technique to implement for the analysis of gaseous hydrocarbons, and GC–FID has been the basis for the analysis of released ET from vegetative tissues and fruits (76). GC–FID also proved useful to analyze ACC after *in vitro* oxidization to ET using mercury chloride ($HgCl_2$) as a catalyzer under extreme alkaline conditions (77). On the other hand, GC with electron capture detection (ECD) has been extensively used for the analysis of non-volatile plant hormones (e.g. ABA and cytokinins) from derivatized extracts. In fact, GC–ECD has proved useful in analyzing ABA and derivatives up to ng/g levels, being more sensitive than GC–FID, but also more sensitive to contaminations that raise the background noise (e.g. electrophilic substances). Unfortunately, there is no post-processing method to remove (or filter out) this noise from FID or ECD data.

For this reason, the analysis of non-volatile plant hormones has found a good ally in LC because these compounds are highly compatible with the liquid mobile phase. Most known plant hormones have semipolar chemical properties (e.g. diterpenes, sesquiterpenes, oxygenated fatty acids, indolic compounds, phenylpropanoids, and hydroxymethyl butenolides/strigolactones), which make modified silica-based stationary phases (e.g. C18) excellent for the separation of such compounds from plant matrices. Extracts can be efficiently separated, and compounds eluted using mixtures of water with different semipolar organic solvents (e.g. methanol or acetonitrile) as mobile phases. Additionally, certain mobile phase modifiers (e.g. formic or acetic acid, ammonium salts) or different buffer mixtures (e.g. acetate, phosphate, and citrate) can be added to improve the retention of compounds and, in some cases, detection.

As a general tool, UV absorption is the most widespread detection method used as single wavelength monitoring with or without acquisition of full spectra on each compound, that provides important structural information (e.g. flavonoids, specific UV spectral band absorption provides information on the position of core flavonoid structure substituents and whether it is a C-bond or an O-bond) (78). However, UV detection is not suitable for all molecules as different plant hormones do not absorb within this range. To improve detection, analytes can be modified through derivatization to add UV-absorbing moieties that improve detectability (e.g. *o*-phthalaldehyde) and also retention.

Fluorescence detection constitutes an improvement of this methodology as UV-absorbing moieties also emit fluorescence in another wavelength, which helps remove interfering wavelengths and increases signal-to-noise ratio. Chloroformates are derivatization agents that target amines at relatively high pH (e.g. 9-Fluorenylmethoxycarbonyl [FMOC] chloride, dansyl chloride, and dabsyl chloride) that have been frequently used for the analysis of, for example polyamines, which show poor UV absorption as well as poor retention on a C18 column. The derivatization process renders less polar compounds with fluorescence properties, which helps to retain these compounds on the column. Nevertheless, LC-column technology has advanced dramatically over the years, and currently, there are several alternative column chemistries (e.g. hydrophilic interaction liquid chromatography [HILIC]) that allow the retention of more polar compounds with the additional advantage of using typical mass spectrometry (MS)-friendly mobile phases for improved sensitivity and efficient online coupling with electrospray ionization (ESI)–MS (79).

MS, either as a stand-alone technique (e.g. direct infusion of complex metabolite mixtures or pure compounds) or coupled to powerful chromatographic techniques (e.g. GC and LC), allows a more insightful analysis of the plant metabolome (in targeted or untargeted approaches) due to its high sensitivity and possibility to obtain structural information from mass spectra; particular important in plant hormone analysis due to their diverse chemical structures and low *in vivo* concentrations.

The first MS studies involved only gas phase samples. This gas phase had to be ionized, usually through very aggressive ionization techniques (e.g. electron impact [EI]) to generate ions and radicals that can be analyzed in the mass spectrometer. This ionization method, still used nowadays, provides a *cloud* of fragments from each compound, useful as *fingerprint* to perform database-based identification of compounds through spectral matching. In 1968, Malcolm Dole first used electrospray ionization (ESI) with MS, which was further developed by John Bennet Fenn who won the Nobel Prize in Chemistry in 2002. This ionization method is based on the dispersion of the liquid containing the analytes (eluates from LC) into a fine aerosol. This is propelling the liquid with a continuous stream of N_2 or nebulization. In this aerosol, ion formation involves extensive solvent evaporation or desolvation as the typical LC solvents include water and organic solvents, as mentioned above. In addition, droplet size is also a parameter to consider regarding ionization of analytes. Therefore, acetic or formic acids are typically added to the mobile phase to increase conductivity and act as sources of protons to facilitate the ionization process. Heating nebulization gas and ionization source help in the desolvation process, which is key in determining the ionization of compounds. Under these conditions, solvent droplets carrying the analytes "explode" generating gas phase ions, which is the most widely accepted model that explains ionization of small molecules by ESI. An advantage of ESI is that it generates singly-charged molecules, ideally as pseudomolecular ions ($[M+H]^+$ in the positive ionization mode and $[M-H]^-$ in the negative ionization mode), which allows the identification of compounds. However, these compounds can also be present as adducts with typical metal ions, such as Na^+ or K^+, NH_4^+, Cl^-, formic acid or acetonitrile, depending on the mobile phase composition. In addition, this soft ionization method has the limitation of requiring the presence of water in the mobile phase and primarily ionizes molecules with heteroatoms present in their structure (O, N, S, P, Cl, Br, or I), which excludes carotenoids and other lipophilic molecules.

Atmospheric pressure chemical ionization (APCI) is another ionization method used in some LC and GC applications. This device is essentially similar to ESI, including the spray needle and the corona discharge electrode where the chemical ionization process takes place. As in ESI, the spray generates droplets that are vaporized by contact with heated parts of the source (usually 350–500 °C). The gas phase eluates are directed to the reaction region, between the corona electrode, the spray, and the mass analyzer entrance, where ions are generated by ion-molecule reactions that go through the transfer region to the analyzer. APCI is considered also a soft ionization method that generates predominantly pseudomolecular ions and causes few fragments to occur. Moreover, it is a gas phase reaction which, unlike ESI, does not require the presence of water as a source of protons; therefore, highly non-polar molecules (i.e. molecules eluted at high organic solvent proportions) can be ionized with APCI (e.g. carotenoids and other lipophilic substances).

Different mass analyzer types are suitable for metabolomics applications. From the configuration perspective, at present, the simplest type is the single quadrupole, which encompasses a quadrupole section after the ionization source and right before the detector. This section can be electronically configured to allow ions with a specific mass-to-charge ratio (m/z) to pass through the mass filter and reach the detector. This method is known as single-ion monitoring (SIM). Despite being highly sensitive to well-established methodologies (i.e. compounds with high m/z) with purified extracts, it is likely that several unrelated plant molecules have the same *m/z* value; therefore, to attain analysis of several analytes with high sensitivity and reliability, other mass analyzer types are required.

An improvement to the single quadrupole is, for example, the triple quadrupole instrument (QqQ). This mass analyzer has two quadrupoles separated by an hexapole section or collision cell, which allows precursor-to-product ion monitoring and multiple reaction monitoring (MRM). In the first quadrupole section, individual precursor ions are selected, which are subsequently fragmented in the hexapole section or collision cell. Afterward, in the second quadrupole section, product ions are selected. This precursor-to-product ion monitoring enhances sensitivity and specificity of the method, resulting in a dramatic reduction of limit of detection (LOD) and limit of quantification (LOQ) values. These are the most widely used instruments to attain sensitive quantitative analysis of soluble molecules in plant extracts. In addition to MRM, and depending on the configuration of the two quadrupole sections, these instruments are capable of performing other different analyses, such as product-to-parent ion scanning, which allow the identification of precursor molecules with a specific transition that might indicate the presence of a particular chemical moiety, neutral loss scanning by which molecules showing a specific neutral loss are identified which might be related to the presence of specific functional groups, among others that have been implemented by the different manufacturers.

Another popular mass analyzer is the time-of-flight (TOF). The structure of this mass analyzer differs from the single quadrupole or QqQ as it is not linear and an important part of the instrument is the flight tube, the section that allows providing exact mass values for each analyte. The principle of the TOF mass analyzer involves the separation of ions based on the time it takes for the ions to travel through the flight tube and reach the detector. The trajectory of the ions depends on their momentum and kinetic energy due to an applied pulsed acceleration voltage and *m/z* ratios of the ions. This is based on the fact that, for a fixed kinetic energy (E, typically around 2000 eV), mass (m), and velocity (v) of ions are correlated ($m = 2 \times E/v^2$). The time that ions take to travel a fixed distance (usually 1000–2000 mm) is measured, allowing calculation of velocity; therefore, by applying the above-mentioned equation, it is possible to calculate the mass of the ions. These mass analyzers are usually commercialized as hybrid quadrupole time-of-flight mass spectrometers (QTOF–MS), which are the gold standard for untargeted analyses. These can also be used in targeted approaches but are usually less sensitive than triple quadrupoles.

Other, less common, configurations but with some advantages over the previously-mentioned designs are the ion trap (IT–MS) and Orbitraps. Both mass analyzers are based on combinations of magnetic fields specially designed to trap ions. The first group provides nominal mass values as single or triple quads, whereas the second group provides exact mass values as TOFs although these mass analyzers are more precise and have a higher resolution (200,000; a typical TOF is around 5000) and dynamic range (over 5000). Both instruments have been used also to attain MRM analyses and, due to their specific configuration, allow multiple-stage ion fragmentation (MS^n), an invaluable tool for unknown metabolite identification (80).

Other powerful instrument configurations have emerged in recent years, either specifically designed for particular applications or as the focus of outstanding technological breakthroughs. One such example is mass spectrometry imaging (MSI), a technique that can use different ionization techniques to measure the spatio-temporal distribution of biomolecules in plant tissues. MSI has been applied in a wide range of areas, from plant defense and plant responses to (a)biotic stress to the growing field of spatial metabolomics. In MSI-based plant science applications, matrix-assisted laser desorption ionization (MALDI) is by far the most used ionization method, typically on a QTOF mass spectrometer (81, 82). Using this technique, Shroff et al. (83) investigated the distribution of glucosinolates in *Arabidopsis thaliana* and found that this distribution controls the feeding pattern of the larvae of the lepidopteran *Helicoverpa armigera*, while Soares et al. (84) investigated the distribution and presence of certain specialized metabolites in citrus at infection sites with *Xylella fastidiosa*. Other studies applied desorption electrospray ionization (DESI)–MSI to the spatial analysis of plant hormones in *Phaseolus vulgaris* seeds (85) or nanoparticle-assisted laser desorption ionization (Nano–PALDI)–MSI to visualize multiple plant hormones in the roots of *Oryza sativa* (86).

Overall, MSI techniques have the added value of being able to spatially resolve different compounds to provide important information on cell/tissue compartmentalization of plant metabolites, local signaling events and transport between cell compartments or tissues. Additionally, MSI sample preparation is not laborious because analyses are typically performed *in situ* (81). Nevertheless, LC–MS provides a vast array of possibilities for the targeted and untargeted analysis of almost all classes of non-volatile metabolites, while GC–MS is an extremely robust tool for the analysis of the volatile fraction.

8.5 Applications of Plant Hormone Profiling to Understand Forest Tree Physiology

It has already been established that climate change is affecting the interrelationship between living organisms and their environment, modifying plant responses to environmental conditions, which are controlled by the coordinated action of plant hormones (87). Researchers have started programs for the identification of more tolerant genotypes to the new climate change conditions, which include biotic and abiotic stresses (88). Hence, researchers are focused not only on the hormones steering plant stress responses but also on the metabolites as end-products of cellular regulatory processes since they represent the final response of plants to environmental-stress factors. For instance, the variation in the plant specialized metabolism is a key factor in the complex biochemical interactions between plants and insects or pathogens (89). Hence, most studies have been focused on the identification of key specific metabolites that define a metabolic phenotype caused by an environmental stress, allowing the characterization of metabolite functionality in different biological processes. Finally, the integration of this information with gene expression data will allow the establishment of gene-metabolite correlations attaining a deeper understanding of plant responses and the identification of specific stress-dependent molecular features (90).

The most common abiotic stress conditions plants face during their life cycles in natural environments are drought, salinity, and reduced or elevated temperatures, which effects are often observed combined. As a result, plants initiate several layers of adaptive responses and tolerance mechanisms controlled by the interplay of several hormones, such as ABA, JAs, ET, or BRs leading to the acclimation to the new environmental conditions (91). Many metabolites with a putative function as osmolytes have been described and used as molecular stress markers, including the amino acids proline and γ-aminobutyric acid (GABA), the soluble sugars trehalose and raffinose, the polyols mannitol and *myo*-inositol, and the polyamines putrescine and spermidine (89). In the case of tree species, the development of new technologies and protocols for metabolomics analysis has increased the number of published works describing specific applications (reviewed in (92)). For example, the tissue-dependent hormonal dynamics analysis of two genotypes of *E. globulus* drought-stressed plants showed that observed tolerance differences were genotype-specific and stress recovery-dependent, decreasing the ABA and ABA-GE content in both genotypes and all plant parts but at different time scales (93). Interestingly, a follow-up work of the same authors showed that the combined analysis and subsequent integration of plant proteome and metabolome of a water-tolerant clone of *E. globulus* revealed new enzymatic activities involved in plant tolerance to water scarcity (94). De Diego et al. (95, 96) focused on the effects of drought stress on *Pinus radiata* populations from different climatic areas of Spain, trying to identify the best genetic background to improve Pinus breeding programs. The authors concluded that the most tolerant genetic backgrounds of *P. radiata* showed high levels of the hormone IAA and metabolites of the glutamate metabolism, such as proline, GABA, and arginine, thus constituting representative metabolic features of drought tolerance in *P. radiata* (95, 96).

However, abiotic stresses very often occur in combination under natural field conditions which provoke a unique and specific plant response according to the stress combination (97). In order to analyze this complex response in *E. globulus*, a drought-tolerant clone was subjected to heat and drought stress alone or in combination. As expected, ABA levels increased when trees were exposed to drought stress alone, accompanied by increases in malondialdehyde (MDA) and a reduction in JAs levels. At the metabolic level, levels of free amino acids, starch, and non-structural carbohydrates increased while sugar phosphates, α-glycerophosphate, and shikimate levels reduced. However, *E. globulus* plants subject to both stresses exhibited a sharp decrease in JA, no alterations in MDA, a universal marker of oxidative damage, or starch contents, and an accumulation of a cinnamate metabolite, which was suggested to be a specific feature of plant metabolism in response to both stresses (9). In another work, the hormonal control of the interaction between host and pathogen in drought conditions was also analyzed by studying the infection caused by the ascomycete fungus *Cenangium ferruginosum* in water-stressed *Pinus koraiensis* plants (98). *Pinus koraiensis* seedlings were inoculated with *C. ferruginosum* showed the accumulation of ABA and a corresponding decline in the defense-associated JA levels, which resulted in a decrease in the expression of some defense-related genes and metabolites (e.g. terpenoids, flavonoids, and phenolic acids). Moreover, the authors concluded that drought does not directly induce pathogenicity of *C. ferruginosum* but weakens pine immunity, favoring *C. ferruginosum* growth (98). In a recent work, Modesto et al. (99) found that resistant *P. pinaster* genotypes to Pine Wilt Disease, caused by the parasitic nematode *Bursaphelenchus xylophilus*, showed upregulation of JA-related defenses along with secondary metabolism including lignin synthesis, whereas susceptible genotypes showed an activation of the SA-defense pathway at later infestation stages. These results suggested that the induction of the jasmonate pathway is crucial for the resistant response.

Current forest productivity includes some economically valuable products, such as cellulose for paper production, cork, plant biomass for wood, coal or pellet production, pine resin, latex, and maple syrup. Traditional breeding of forest species aimed to select plants with increased plant growth which will produce more of these valuable products. However, it is a long and complex process due to tree's life cycle, which normally includes an extensive juvenile phase that lasts for several years (100). Hence, the determination of rapid stem growth has been traditionally assessed by profiling of active GAs, IAA, CKs, and ABA, hormones that modulate the longitudinal and radial stem growth in poplar (101, 102), conifers (103), and *Quercus acutissima* (104), among others. In fact, Kang et al. (104) successfully integrated the hormonal and metabolite profiling of *Q. acutissima* Carruth, identifying a positive correlation between some metabolites (e.g. phosphoric acid, succinic acid, malic acid, and butane-1,3-diol) and high-yield characteristics (e.g. tree height, crown volume, and acorn production) (104). In addition, woody perennial species of northern latitudes have acquired seasonal adaptations which allow them to arrest plant growth and start dormancy in the meristems in a short-day–dependent manner to survive winter conditions and resume growth the following spring, adaptations that are affected by the climate change scenario. For instance, the analysis of the development of an apical bud on white spruce (*Picea glauca*) has identified genes involved in ABA, CKs, and IAA biosynthesis and signaling as well as distinct accumulation profiles of these hormones, which can be related to morphological changes in the bud (105). Moreover, specific auxin-dependent responses in the hybrid aspen (*Populus tremula* L. × *Populus tremuloides* Michx.) have been identified to be responsible for the activity–dormancy transition in the cambial meristem (106). More recently, plant tree architecture has become an important trait because of the ornamental value of some beautiful trees in our city gardens like *Lagerstroemia species* (107) and *Sophora davidii* (108), suggesting an interaction between auxin and GA in regulating the internode length of *Lagerstroemia* as well the putative role of phenylpropanoid and flavonoid biosynthesis in the *S. davidii* plant-height.

8.6 Future Prospects in Plant Hormone Analysis

The analysis of plant metabolites at a very small spatial resolution (e.g. single cell metabolomics) is now possible due to important advances in the field of cell sorting and, more importantly, significant improvements in the sensitivity and mass resolution of mass spectrometers. This is achieved through the use of mechanical probes (109) or focused UV-laser beams (110), which showed to have higher output (111). Nevertheless, when compared to genomics or transcriptomics, metabolomics data cannot be amplified, and therefore, metabolite detectability still relies on the sensitivity of mass spectrometers and chromatographic separation. This could be partially overcome by chemical derivatization, but with added complexity given the enormous chemical diversity and the problem of structural modifications associated with derivatization processes (112). Isolation of single cells from tissues constitutes an important challenge because manipulation always generates variability as an inherent drawback. Laser cell microdissection (LCM) requires ultrathin tissue slices, which result in very low material recovery, thereby, hindering the detectability of analytes (112). MSI constitutes an interesting and attractive alternative because it provides semi-quantitative data on plant hormones and other metabolites at the tissue level (82, 113). MSI techniques provide means to investigate plant hormone accumulation as well as dynamics and transport, which are essential biological processes to understand how plant responses to environmental stimuli are integrated and regulated.

Acknowledgments

All authors acknowledge the financial support provided through projects OPTIMUS PRIME (grant PCI2021-121920 funded by MCIN/AEI/10.13039/501100011033 and by the "European Union NextGenerationEU/PRTR"), EXTREMO (grant PID2020-118126RB-I00 funded by MCIN/AEI/10.13039/501100011033), and UJI-B2019-24 funded by Universitat Jaume I. M.G-G. particularly acknowledges the support received from the Ramón y Cajal program, grant RYC-2016-19325 funded by MCIN/AEI/10.13039/501100011033 and by "ESF Investing in your future."

References

1 US Department of Agriculture, Forest Service, Climate Change Resource Center. Ecosystem Services and Climate Change [Internet]. Available from (www.fs.usda.gov/ccrc/topics/ecosystem-services), accessed 13 February 2022.
2 FAO. 2020. Global Forest Resources Assessment 2020: Main report. Rome.
3 EU. A European Green Deal Striving to be the first climate-neutral continent [Internet]. Available from (https://commission.europa.eu/strategy-and-policy/priorities-2019-2024/european-green-deal_en), accessed 13 February 2022.
4 Eurostat. The EU in the world – agriculture, forestry, and fisheries [Internet]. Available from (https://ec.europa.eu/eurostat/statistics-explained/index.php?title=The_EU_in_the_world_-_agriculture,_forestry_and_fisheries#Forestry), accessed 13 February 2022.
5 Ivanova A, Mikhova B, Batsalova T, Dzhambazov B, Kostova I. New furostanol saponins from *Smilax aspera* L. and their in vitro cytotoxicity. *Fitoterapia*. 2011; 82(2): 282–287.
6 Dalkılıç S, Korkmaz İ, Dalkılıç LK, Akay G, Fidan S. In vitro cytotoxic effects of *Smilax aspera* L. roots on cancer cell lines. *Food Biosci*. 2022; 46: 101505.
7 Cappadone C, Mandrone M, Chiocchio I, Sanna C, Malucelli E, Bassi V, et al. Antitumor potential and phytochemical profile of plants from sardinia (Italy), a hotspot for biodiversity in the mediterranean basin. *Plants*. 2019; 9(1): 26.
8 Helleboid S, Haug C, Lamottke K, Zhou Y, Wei J, Daix S, et al. The identification of naturally occurring neoruscogenin as a bioavailable, potent, and high-affinity agonist of the nuclear receptor RORα (NR1F1). *J Biomol Screening*. 2014; 19: 399–406.
9 Correia B, Hancock RD, Amaral J, Gomez-Cadenas A, Valledor L, Pinto G. Combined drought and heat activates protective responses in eucalyptus

globulus that are not activated when subjected to drought or heat stress alone. *Front Plant Sci.* 2018; 9: 819.

10 Wang F, Cui X, Sun Y, Dong C-H. Ethylene signaling and regulation in plant growth and stress responses. *Plant Cell Rep.* 2013; 32: 1099–1109.

11 Gómez-Cadenas A, Tadeo FR, Talon M, Primo-Millo E. Leaf abscission induced by ethylene in water-stressed intact seedlings of *Cleopatra mandarin* requires previous abscisic acid accumulation in roots. *Plant Physiol.* 1996; 112: 401–408.

12 Merelo P, Agustí J, Arbona V, Costa ML, Estornell LH, Gómez-Cadenas A, et al. Cell wall remodeling in abscission zone cells during ethylene-promoted fruit abscission in citrus. *Front Plant Sci.* 2017; 8: 126.

13 Ponzio C, Gols R, Pieterse CMJ, Dicke M. Ecological and phytohormonal aspects of plant volatile emission in response to single and dual infestations with herbivores and phytopathogens. *Funct Ecol.* 2013; 27(3): 587–598.

14 Gutierrez OA, Wubben MJ, Howard M, Roberts B, Hanlon E, Wilkinson JR. The role of phytohormones ethylene and auxin in plant-nematode interactions. *Russ J Plant Physiol.* 2009; 56(1): 1–5.

15 Lorenzo O, Piqueras R, Sánchez-Serrano JJ, Solano R. ETHYLENE RESPONSE FACTOR1 Integrates signals from ethylene and jasmonate pathways in plant defense. *Plant Cell.* 2003; 15(1): 165–178.

16 Vidoz ML, Loreti E, Mensuali A, Alpi A, Perata P. Hormonal interplay during adventitious root formation in flooded tomato plants. *Plant J.* 2010; 63(4): 551–562.

17 González-Guzmán M, Gómez-Cadenas A, Arbona V. Abscisic acid as an emerging modulator of the responses of plants to low oxygen conditions. *Front Plant Sci.* 2021, 2021; 12: 661789.

18 Zhao Y. Auxin biosynthesis. *The Arabidopsis Book.* 2014; 12(12): e0173.

19 Mockaitis K, Estelle M. Auxin receptors and plant development: a new signaling paradigm. *Annu Rev Cell Dev Biol.* 2008; 24: 55–80.

20 Booker J, Chatfield S, Leyser O. Auxin acts in xylem-associated or medullary cells to mediate apical dominance. *Plant Cell.* 2003; 15(2): 495–507.

21 Elobeid M, Göbel C, Feussner I, Polle A. Cadmium interferes with auxin physiology and lignification in poplar. *J Exp Bot.* 2012; 63(3): 1413–1421.

22 Popko J, Hänsch R, Mendel R-R, Polle A, Teichmann T. The role of abscisic acid and auxin in the response of poplar to abiotic stress. *Plant Biol.* 2010; 12(2): 242–258.

23 Frébort I, Kowalska M, Hluska T, Frébortová J, Galuszka P. Evolution of cytokinin biosynthesis and degradation. *J Exp Bot.* 2011; 62(8): 2431–2452.

24 de Lucas M, Davière J-M, Rodríguez-Falcón M, Pontin M, Iglesias-Pedraz JM, Lorrain S. A molecular framework for light and gibberellin control of cell elongation. *Nature.* 2008; 451(7177): 480–484.

25 Piskurewicz U, Lopez-Molina L. The GA-signaling repressor RGL3 represses testa rupture in response to changes in GA and ABA levels. *Plant Signaling Behav*. 2009; 4(1): 63–65.

26 Valledor L, Guerrero S, García-Campa L, Meijón M. Proteometabolomic characterization of apical bud maturation in *Pinus pinaster*. *Tree Physiol.* 2021; 41(3): 508–521.

27 Keawmanee N, Ma G, Zhang L, Yahata M, Murakami K, Yamamoto M. Exogenous gibberellin induced regreening through the regulation of chlorophyll and carotenoid metabolism in Valencia oranges. *Plant Physiol Biochem*. 2022; 173: 14–24.

28 Pierik R, Djakovic-Petrovic T, Keuskamp DH, de Wit M, Voesenek LACJ. Auxin and ethylene regulate elongation responses to neighbor proximity signals independent of gibberellin and della proteins in Arabidopsis. *Plant Physiol*. 2009; 149(4): 1701–1712.

29 Mustroph A, Steffens B, Sasidharan R. Signalling interactions in flooding tolerance. *Ann Plant Rev Online*. 2018; 1(2).

30 Matayoshi CL, Pena LB, Arbona V, Gómez-Cadenas A, Gallego SM. Early responses of maize seedlings to Cu stress include sharp decreases in gibberellins and jasmonates in the root apex. *Protoplasma*. 2020; 257(4): 1243–1256.

31 Wasternack C. Jasmonates: an update on biosynthesis, signal transduction and action in plant stress response, growth and development. *Ann Bot*. 2007; 100(4): 681–697.

32 Chung HS, Koo AJK, Gao X, Jayanty S, Thines B, Jones AD, et al. Regulation and function of Arabidopsis JASMONATE ZIM-domain genes in response to wounding and herbivory. *Plant Physiol*. 2008; 146(3): 952–964.

33 Zhurov V, Navarro M, Bruinsma KA, Arbona V, Santamaria ME, Cazaux M, et al. Reciprocal responses in the interaction between arabidopsis and the cell-content-feeding chelicerate herbivore spider mite. *Plant Physiol*. 2014; 164(1): 384–399.

34 Salehipourshirazi G, Bruinsma K, Ratlamwala H, Dixit S, Arbona V, Widemann E, et al. Rapid specialization of counter defenses enables two-spotted spider mite to adapt to novel plant hosts. *Plant Physiol*. 2021; 187(4): 2608–2622.

35 Stintzi A, Browse J. The Arabidopsis male-sterile mutant, opr3, lacks the 12-oxophytodienoic acid reductase required for jasmonate synthesis. *Proc Natl Acad Sci USA*. 2000; 97(19): 10625–10630.

36 De Ollas C, Hernando B, Arbona V, Gómez-Cadenas A. Jasmonic acid transient accumulation is needed for abscisic acid increase in citrus roots under drought stress conditions. *Physiol Plant*. 2013; 147: 296–306.

37 De Ollas C, Arbona V, Gómez-Cadenas A. Jasmonoyl isoleucine accumulation is needed for abscisic acid build-up in roots of Arabidopsis under water stress conditions. *Plant Cell Environ.* 2015; 38(10): 2157–2170.

38 Farag MA, Paré PW. C6-green leaf volatiles trigger local and systemic VOC emissions in tomato. *Phytochemistry.* 2002; 61(5): 545–554.

39 Pérez-Hedo M, Alonso-Valiente M, Vacas S, Gallego C, Pons C, Arbona V, et al. Plant exposure to herbivore induced plant volatiles: a sustainable approach through eliciting plant defenses. *J Pest Sci.* 2021; 94: 1221–1235.

40 Pan Z, Camara B, Gardner HW, Backhaus R. Aspirin inhibition and acetylation of the plant cytochrome P450, allene oxide synthase, resembles that of animal prostaglandin endoperoxide H synthase. *J Biol Chem.* 1998; 273(29): 18139–18145.

41 Lefevere H, Bauters L, Gheysen G. Salicylic acid biosynthesis in plants. *Front Plant Sci.* 2020, 2020; (11): 338.

42 Boatwright JL, Pajerowska-Mukhtar K. Salicylic acid: an old hormone up to new tricks. *Mol Plant Pathol.* 2013; 14(6): 623–634.

43 Clarke SM, Mur LAJ, Wood JE, Scott IM. Salicylic acid dependent signaling promotes basal thermotolerance but is not essential for acquired thermotolerance in Arabidopsis thaliana. *Plant J.* 2004; 38(3): 432–447.

44 Zandalinas SI, Rivero RM, Martínez V, Gómez-Cadenas A, Arbona V. Tolerance of citrus plants to the combination of high temperatures and drought is associated to the increase in transpiration modulated by a reduction in abscisic acid levels. *BMC Plant Biol.* 2016; 16(1): 1–16.

45 Nambara E, Marion-Poll A. Abscisic acid biosynthesis and catabolism. *Annu Rev Plant Biol.* 2005; 56: 165–185.

46 González-Guzmán M, Apostolova N, Bellés JM, Barrero JM, Piqueras P, Ponce MR, et al. The short-chain alcohol dehydrogenase ABA2 catalyzes the conversion of xanthoxin to abscisic aldehyde. *Plant Cell.* 2022; 14(8): 1833–1846.

47 Yoshida T, Christmann A, Yamaguchi-Shinozaki K, Grill E, Fernie AR. Revisiting the basal role of ABA – roles outside of stress. *Trends Plant Sci.* 2019; 24(7): 625–635.

48 Seiler C, Harshavardhan VT, Rajesh K, Reddy PS, Strickert M, Rolletschek H, et al. ABA biosynthesis and degradation contributing to ABA homeostasis during barley seed development under control and terminal drought-stress conditions. *J Exp Bot.* 2011; 62(8): 2615–2632.

49 Arbona V, Zandalinas SI, Manzi M, Gonzalez-Guzman M, Rodriguez PL, Gomez-Cadenas A. Depletion of abscisic acid levels in roots of flooded Carrizo citrange (*Poncirus trifoliata* L. Raf. x *Citrus sinensis* L. Osb.) plants is a stress-specific response associated to the differential expression of PYR/PYL/RCAR receptors. *Plant Mol Biol.* 2017; 93(6): 623–640.

50 De Ollas C, Gonzalez-Guzman M, Pitarch Z, Matus JT, Candela H, Rambla JL, et al. Identification of ABA-mediated genetic and metabolic

responses to soil flooding in tomato (*Solanum lycopersicum* L. Mill). *Front Plant Sci.* 2021; 12: 613059.
51 Bajguz A, Chmur M, Gruszka D. Comprehensive overview of the brassinosteroid biosynthesis pathways: substrates, products, inhibitors, and connections. *Front Plant Sci.* 2020; 11: 1034.
52 Chung Y, Choe S. The regulation of brassinosteroid biosynthesis in arabidopsis. *Crit Rev Plant Sci.* 2013; 32(6): 396–410.
53 De Bruyne L, Höfte M, De Vleesschauwer D. Connecting growth and defense: the emerging roles of brassinosteroids and gibberellins in plant innate immunity. *Mol Plant.* 2014; 7(6): 943–959.
54 Lozano-Durán R, Zipfel C. Trade-off between growth and immunity: role of brassinosteroids. *Trends Plant Sci.* 2015; 20(1): 12–19.
55 Trevisan S, Forestan C, Brojanigo S, Quaggiotti S, Varotto S. Brassinosteroid application affects the growth and gravitropic response of maize by regulating gene expression in the roots, shoots and leaves. *Plant Growth Regul.* 2020; 92(1): 117–130.
56 Bajguz A, Tretyn A. The chemical characteristic and distribution of brassinosteroids in plants. *Phytochemistry.* 2003; 62: 1027–1046.
57 Marzec M, Muszynska A, Gruszka D. The role of strigolactones in nutrient-stress responses in plants. *Int J Mol Sci.* 2013; 14(5): 9286–9304.
58 López-Ráez JA, Charnikhova T, Gómez-Roldán V, Matusova R, Kohlen W, De Vos R, et al. Tomato strigolactones are derived from carotenoids and their biosynthesis is promoted by phosphate starvation. *New Phytol.* 2008; 178(4): 863–874.
59 Foo E, Reid JB. Strigolactones: new physiological roles for an ancient signal. *J Plant Growth Regul.* 2012; 32(2): 429–442.
60 Nelson DC, Flematti GR, Ghisalberti EL, Dixon KW, Smith SM. Regulation of seed germination and seedling growth by chemical signals from burning vegetation. *Annu Rev Plant Biol.* 2012; 63(1): 107–130.
61 Morffy N, Faure L, Nelson DC. Smoke and hormone mirrors: action and evolution of karrikin and strigolactone signaling. *Trends Genet.* 2016; 32(3): 176–188.
62 Waters MT, Scaffidi A, Flematti GR, Smith SM. The origins and mechanisms of karrikin signalling. *Curr Opin Plant Biol.* 2013; 16(5): 667–673.
63 Rodrigues AM, Ribeiro-Barros AI, António C. Experimental design and sample preparation in forest tree metabolomics. *Metabolites.* 2019; 9(12): 5–7.
64 Delatorre C, Rodríguez A, Rodríguez L, Majada JP, Ordás RJ, Feito I. Hormonal profiling: development of a simple method to extract and quantify phytohormones in complex matrices by UHPLC–MS/MS. *J Chromatogr B Analyt Technol Biomed Life Sci.* 2017; 1040: 239–249.

65 Ljung K, Sandberg G, Moritz T. Hormone analysis. In: Davies PJ, editor. *Plant Hormones: Biosynthesis, Signal Transduction, Action!* Third Edition. Dordrecht (NL): Springer; 2007. p. 671–694.

66 Wang L, Zou Y, Kaw HY, Wang G, Sun H, Cai L, et al. Recent developments and emerging trends of mass spectrometric methods in plant hormone analysis: a review. *Plant Methods*. 2020; 16(1): 1–17.

67 Chiwocha SDS, Abrams SR, Ambrose SJ, Cutler AJ, Loewen M, Ross ARS, et al. A method for profiling classes of plant hormones and their metabolites using liquid chromatography-electrospray ionization tandem mass spectrometry: an analysis of hormone regulation of thermodormancy of lettuce (*Lactuca sativa* L.) seeds. *Plant J*. 2003; 35(3): 405–417.

68 Müller M, Munné-Bosch S. Rapid and sensitive hormonal profiling of complex plant samples by liquid chromatography coupled to electrospray ionization tandem mass spectrometry. *Plant Methods*. 2011; 7: 37.

69 Simura J, Antoniadi I, Siroka J, Tarkowská D, Strnad M, Ljung K, et al. Plant hormonomics: multiple phytohormone profiling by targeted metabolomics. *Plant Physiol*. 2018; 177: 476–489.

70 Gómez-Cadenas A, Pozo OJ, García-Augustín P, Sancho JV. Direct analysis of abscisic acid in crude plant extracts by liquid chromatography-electrospray/tandem mass spectrometry. *Phytochem Anal*. 2002; 13(4): 228–234.

71 Durgbanshi A, Arbona V, Pozo O, Miersch O, Sancho JV, Gómez-Cadenas A. Simultaneous determination of multiple phytohormones in plant extracts by liquid chromatography-electrospray tandem mass spectrometry. *J Agric Food Chem*. 2005; 53(22): 8437–8442.

72 Manzi M, Gómez-Cadenas A, Arbona V. Rapid and reproducible determination of active gibberellins in Citrus tissues by UPLC/ESI-MS/MS. *Plant Physiol Biochem*. 2015; 94: 1–9.

73 Hradecká V, Novák O, Havlícek L, Strnad M. Immunoaffinity chromatography of abscisic acid combined with electrospray liquid chromatography-mass spectrometry. *J Chromatogr B*. 2007; 847(2): 162–173.

74 Tivendale ND, Cohen JD. Analytical history of auxin. *J Plant Growth Regul*. 2015; 34(4): 708–722.

75 Moldoveanu SC, David V. Derivatization methods in GC and GC/MS. *Intech*. 2016; 11: 13.

76 Yoon GM, Chen YC. Gas chromatography-based ethylene measurement of arabidopsis seedlings. *Methods Mol Biol*. 2017; 1573: 3–10.

77 Lizada C, Yang SF. A simple and sensitive assay for 1-aminocyclopropane-1-carboxylic acid. *Anal Biochem*. 1979; 100(1): 140–145.

78 Djoukeng JD, Arbona V, Argamasilla R, Gomez-Cadenas A. Flavonoid profiling in leaves of citrus genotypes under different environmental situations. *J Agric Food Chem*. 2008; 56(23): 11087–11097.

79 Jorge TF, Rodrigues JA, Caldana C, Schmidt R, van Dongen JT, Thomas-Oates J, António C. Mass spectrometry-based plant metabolomics: metabolite responses to abiotic stress. *Mass Spectrom Rev*. 2016; 35(5): 620–649.
80 Stettin D, Poulin RX, Pohnert G. Metabolomics benefits from orbitrap GC–MS—comparison of low- and high-resolution GC–MS. *Metabolites*. 2020; 10(4): 143.
81 Boughton BA, Thinagaran D, Sarabia D, Bacic A, Roessner U. Mass spectrometry imaging for plant biology: a review. *Phytochem Rev*. 2016; 15: 445–488.
82 Feldberg L, Dong Y, Heinig U, Rogachev I, Aharoni A. DLEMMA-MS-imaging for identification of spatially localized metabolites and metabolic network map reconstruction. *Anal Chem*. 2018; 90(17): 10231–10238.
83 Shroff R, Vergara F, Muck A, Svatos A, Gershenzon J. Nonuniform distribution of glucosinolates in Arabidopsis thaliana leaves has important consequences for plant defense. *Proc Natl Acad Sci USA*. 2008; 105(16): 6196–6201.
84 Soares MS, da Silva DF, Forim MR, da Silva MFDGF, Fernandes JB, Vieira PC, et al. Quantification and localization of hesperidin and rutin in *Citrus sinensis* grafted on *C. limonia* after *Xylella fastidiosa* infection by HPLC-UV and MALDI imaging mass spectrometry. *Phytochemistry*. 2015; 115: 161–170.
85 Enomoto H, Sensu T, Sato K, Sato F, Paxton T, Yumoto E, et al. Visualisation of abscisic acid and 12-oxo-phytodienoic acid in immature *Phaseolus vulgaris* L. seeds using desorption electrospray ionisation-imaging mass spectrometry. *Sci Rep*. 2017; 7: 42977.
86 Shiono K, Taira S. Imaging of multiple plant hormones in roots of rice (*Oryza sativa*) using nanoparticle-assisted laser desorption/ionization mass spectrometry. *J Agric Food Chem*. 2020; 68(24): 6770–6775.
87 González Guzmán M, Cellini F, Fotopoulos V, Balestrini R, Arbona V. New approaches to improve crop tolerance to biotic and abiotic stresses. *Physiol Plant*. 2022; 174: e13547.
88 Liu J, Fernie AR, Yan J. Crop breeding – from experience-based selection to precision design. *J Plant Physiol*. 2021; 256: 153313.
89 Alseekh S, Fernie AR. Metabolomics 20 years on: what have we learned and what hurdles remain? *Plant J*. 2018; 94(6): 933–942.
90 Fernie AR, Tohge T. The genetics of plant metabolism. *Annu Rev Genet*. 2017; 51: 287–310.
91 Gupta A, Rico-Medina A, Caño-Delgado AI. The physiology of plant responses to drought. *Science*. 2020; 368(6488): 266–269.
92 Rodrigues AM, Miguel C, Chaves I, António C. Mass spectrometry-based forest tree metabolomics. *Mass Spectrom Rev*. 2021; 40(2): 126–157.

93 Correia B, Pintó-Marijuan M, Castro BB, Brossa R, López-Carbonell M, Pinto G. Hormonal dynamics during recovery from drought in two Eucalyptus globulus genotypes: from root to leaf. *Plant Physiol Biochem.* 2014; 82: 151–160.

94 Correia B, Valledor L, Hancock RD, Renaut J, Pascual J, Soares AMVM, et al. Integrated proteomics and metabolomics to unlock global and clonal responses of Eucalyptus globulus recovery from water deficit. *Metabolomics.* 2016; 12(8): 141.

95 De Diego N, Pérez-Alfocea F, Cantero E, Lacuesta M, Moncaleán P. Physiological response to drought in radiata pine: phytohormone implication at leaf level. *Tree Physiol.* 2012; 32(4): 435–449.

96 De Diego N, Saiz-Fernández I, Rodríguez JL, Pérez-Alfocea P, Sampedro MC, Barrio RJ, et al. Metabolites and hormones are involved in the intraspecific variability of drought hardening in radiata pine. *J Plant Physiol.* 2015; 188: 64–71.

97 Nguyen D, Rieu I, Mariani C, van Dam NM. How plants handle multiple stresses: hormonal interactions underlying responses to abiotic stress and insect herbivory. *Plant Mol Biol.* 2016; 91(6): 727–740.

98 Ryu M, Mishra RC, Jeon J, Lee SK, Bae H. Drought-induced susceptibility for *Cenangium ferruginosum* leads to progression of Cenangium-dieback disease in *Pinus koraiensis. Sci Rep.* 2018; 8(1): 16368.

99 Modesto I, Sterck L, Arbona V, Gomez-Cadenas A, Carrasquinho I, Van de Peer Y, Miguel CM. Insights into the mechanisms implicated in *Pinus pinaster* resistance to pinewood nematode. *Front Plant Sci.* 2021; 12: 690857.

100 Sork VL, Aitken SN, Dyer RJ, Eckert AJ, Legendre P, Neale DB. Putting the landscape into the genomics of trees: approaches for understanding local adaptation and population responses to changing climate. *Tree Genet Genomes.* 2013; 9(4): 901–911.

101 Eriksson ME, Israelsson M, Olsson O, Moritz T. Increased gibberellin biosynthesis in transgenic trees promotes growth, biomass production and xylem fiber length. *Nat Biotechnol.* 2000; 18(7): 784–788.

102 Tuominen H, Puech L, Fink S, Sundberg B. A radial concentration gradient of indole-3-acetic acid is related to secondary xylem development in hybrid aspen. *Plant Physiol.* 1997; 115(2): 577–585.

103 Park EJ, Lee WY, Kurepin LV, Zhang R, Janzen L, Pharis RP. Plant hormone-assisted early family selection in *Pinus densiflora* via a retrospective approach. *Tree Physiol.* 2014; 35(1): 86–94.

104 Kang J, Lee H, Lim H, Lee W. Identification of potential metabolic markers for the selection of a high-yield clone of *Quercus acutissima* in clonal seed orchard. *Forests.* 2018; 9(3): 116.

105 WEl K, CCG A, CJT J, Adams E, King-Jones S, Zaharia LI, et al. Molecular events of apical bud formation in white spruce, *Picea glauca*. *Plant Cell Environ*. 2011; 34(3): 480–500.

106 Baba K, Karlberg A, Schmidt J, Schrader J, Hvidsten TR, Bako L, et al. Activity-dormancy transition in the cambial meristem involves stage-specific modulation of auxin response in hybrid aspen. *Proc Natl Acad Sci USA*. 2011; 108(8): 3418–3423.

107 Ju Y, Feng L, Wu J, Ye Y, Zheng T, Cai M, et al. Transcriptome analysis of the genes regulating phytohormone and cellular patterning in *Lagerstroemia* plant architecture. *Sci Rep*. 2018; 8(1): 15162.

108 Zhao X, Sun XF, Zhao LL, Huang LJ, Wang PC. Morphological, transcriptomic and metabolomic analyses of *Sophora davidii* mutants for plant height. *BMC Plant Biol*. 2022; 22(1): 144.

109 Sun M, Yang Z, Wawrik B. Metabolomic fingerprints of individual algal cells using the single-probe mass spectrometry technique. *Front Plant Sci*. 2018, 2018; 9: 571.

110 Baumeister TUH, Vallet M, Kaftan F, Svatoš A, Pohnert G. Live single-cell metabolomics with matrix-free laser/desorption ionization mass spectrometry to address microalgal physiology. *Front Plant Sci*. 2019; 10: 172.

111 Svatoš A, Lee YJ, Yang Z. Editorial: single plant cell metabolomics. *Front Plant Sci*. 2020, 2020; 11: 10–11.

112 De Souza LP, Borghi M, Fernie A. Plant single-cell metabolomics — challenges and perspectives. *Int J Mol Sci*. 2020; 21(23): 8987.

113 Zhang C, Žukauskaitė A, Petřík I, Pěnčík A, Hönig M, Grúz J, et al. In situ characterisation of phytohormones from wounded *Arabidopsis* leaves using desorption electrospray ionisation mass spectrometry imaging. *Analyst*. 2021; 146(8): 2653–2663.

9

Metabolomics of Nutrient-Deprived Forest Trees

Sara Adrián Lopez de Andrade, Vinícius Henrique de Oliveira, and Paulo Mazzafera

Department of Plant Biology, Institute of Biology, University of Campinas, Campinas, Brazil

9.1 Introduction

In 2020, the trade of products derived from the exploitation of forests was around 244 billion dollars (1). The products were: roundwood (wood fuel and industrial roundwood), wood pellets and other agglomerates, sawnwood and wood-based panels (plywood, particle board, fibre board, and oriented strand board; OSB), wood pulp, pulp from fibers other than wood, recovered paper, paper, and paperboard. The United States of America (USA) and Canada are the biggest traders of timber and industry-derived wood products due to vast areas of forests in both countries. USA production is three times greater than Canadian production. Forests cover about 66% of the land area of the USA, most of which are native forests. Despite this, there is variation among the largest exporters concerning the products listed above. In 2018, the largest exporters of these products were: roundwood, New Zealand (16%); wood pellets, USA (25%); sawnwood, Russian Federation (20%); Wood-based panels, China (16%); pulp for paper, Brazil (24%); recovered paper, USA (34%); paper and paperboard, Germany (12%) (2). After China, Brazil has the largest planted area of forests in the world. Nine million hectares are occupied mainly by eucalypts, followed by pines, and other species (e.g. acacia, araucaria, paricá and teak), which are primarily used in the cellulose and paper industry (36%), and steel and charcoal industries (12%) (3).

In tropical forests, where the cutting age is lower than in temperate forests, the successive removal of nutrients with harvested wood over the years may increase the possibility of nutrient deficiencies (4). Thus, the removal of nutrients may be significant depending on the plant density and the stem biomass at cutting. Actual global environmental changes can cause nutritional imbalances in trees and create higher nutrient demands that the soil is not able

Monitoring Forest Damage with Mass Spectrometry-Based Metabolomics Methods, First Edition.
Edited by Carla António.

to supply adequately (5, 6). For example, the CO_2 fertilization effect caused by increasing levels of atmospheric CO_2 and the unbalanced deposition of N and P leads to limited availability of nutrients (e.g. Ca, Mg, S, and K) in forest ecosystems (6). In Europe, foliar analysis of different species between 1992 and 2009 showed a reduction of nutrients, which was related to an increased nutrient demand due to global enrichment in atmospheric CO_2 and high atmospheric N deposition, increasing productivity and causing an unbalanced nutritional status (6). The authors warned that if the soil nutrient is not replaced, wood productivity may decrease, mainly in areas where growth is faster.

In Brazil, eucalypt production highly depends on adequate fertilization since the planted genotypes have been bred for rapid growth and wood production (7). The genetic gain of the new eucalypt clones and the improvement in agronomic management increased productivity from around 12–50 m^3 of wood/ha per year. *Eucalyptus grandis* and *Eucalyptus urophylla* and their hybrids are the most planted genetic material when the objective is the pulp and paper industry. The cutting age in this material is around six to seven years. *E. urophylla, Eucalyptus camaldulensis, Eucalyptus cloeziana,* and *Eucalyptus pellita* are usually used in the charcoal industry, where a relative apparent density, lignin content, and fixed carbon yield are important parameters. Other species are used for essential oil production (*Corymbia citriodora*) and lumber (*E. grandis, Eucalyptus saligna,* and *Eucalyptus pirulalis*) (7).

Our knowledge of the nutrition of planted forests is still deficient because, for many years, there was the concept that trees should be planted in areas other than those used for planting crops. Carbon, hydrogen, and oxygen, which are supplied by photosynthesis and water, contributed to 90–95% of plant dry weight. The remaining 5–10% are represented mostly by 13 elements known as macro (N, P, K, Ca, Mg, S) and micronutrients (Mn, Fe, Zn, Cu, B, Cl, Mo), which concentrations vary according to the plant tissue (8).

The nutrient content in the wood varies according to the species, but woody tissues generally accumulate N, Ca, and K (9). In teak and eucalypts, the highest concentrations of N, P, K, and Mg are found in the foliage, but they decline in the bark, branch wood, and stem wood, while Ca concentration in the bark and leaves is similar. The highest accumulation of nutrients is in the foliage in the early years and with aging at the bole wood (9).

The basic physiology of plant nutrition and cell growth in trees are the same as in other plants. However, the wood has different components, which may be influenced differently depending on the nutrient; i.e. the cell wall formation and its quality may be affected by changes in one of its components: cellulose, hemicellulose, lignin, and pectin (10–12).

Compared with herbaceous plants used as genomic models, little is known in trees concerning the wood formation, and poplar has been used mostly as the tree model system (13, 14). Vastly less is known about the changes in the metabolome that occur during wood formation. Metabolomics can be defined as an omics technology able to capture a metabolic snapshot of a cell, tissue, or organ regarding the diversity of low molecular weight molecules in a developmental and physiological stage. The information obtained can be used to understand biological processes and how they change according to a stimulus, such as an environmental condition. Because of the complexity of wood as a biological material, specific procedures related to sample collection and preparation have been suggested for metabolomics studies in woody plants (15), and most of them were focused on leaves (16). In a list of 49 published studies concerning the plant families Fagaceae, Myrtaceae, Pinaceae, and Salicilaceae, only 11 focused partially or entirely on different aspects of the stem (stem, wood, bark) (16). Among all listed references, only one studied nutrient deficiency (phosphorus, P) and stem metabolome (17). The other work reported on the leaf metabolome of eucalypt plants exposed to low P (18). Recently, P influence on the metabolome was studied in the root, leaves, and stem of five eucalyptus species (19).

Here, we intend to focus on how metabolomics contributes to our knowledge of wood formation in trees deprived of nutrients. Initially, we explored the available information on each nutrient and how they impact tree growth and development and wood formation. Then, we explored how metabolomics helped build this knowledge, including references containing general information on wood formation. Because of the lack of information dedicated to the wood metabolome in nutrient-deprived trees, we gathered the little available information in the same section with references related to metabolome and tree nutrition.

9.2 Macronutrient Deficiency and Wood Production

Tree stems are important pools of nutrients, showing highly complex dynamics due to differences among cambium, xylem, and phloem tissues, mobilization processes, and wood (20). However, nutrient dynamics and cycling in living wood are understudied, limiting understanding of how nutrient availability impacts tree functioning and wood formation (21). In trees, nutrient supply significantly influences water transport by regulating vascular traits (e.g. vessel diameter and structure) and impacting wood properties. Nutrient imbalances in wood and foliage can contribute to tree dieback during water deficit stress or higher sensitivity to drought (22, 23).

The wood formation, including lignification, is controlled developmentally and responds to numerous environmental factors; namely, water and nutrient availability. In poplar, for example, nitrogen fertilization caused shorter fibers and thicker cell walls, with higher cellulose deposition than non-fertilizer poplars, profoundly impacting secondary xylem formation (24). Cations, such as K and Na, from parenchymatic cells, may also have a role in the repair of xylem embolism during water stress. Under these conditions, apoplastic pH lowers, activating metal ion transporters and proton pumps (25,26).

9.2.1 Nitrogen

Nitrogen (N) is the main component in nucleotides and proteins vital for living organisms and the mineral nutrient required in the greatest amount by plants (27,28). The major sources of inorganic N taken up by roots are nitrate (NO_3^-) and ammonium (NH_4^+), and its deficiency typically causes chlorosis, stunted growth, lower canopy height, narrow leaves, and premature shedding, which affects the overall plant development and productivity (29, 30). In *Eucalyptus* and *Melaleuca* seedlings, N deficiency reduced growth and photosynthetic activity and decreased the export of photosynthate compounds from leaves (31), further contributing to lower carbon allocation towards wood development.

In forest tree plantations, such as in poplar or eucalypt sites, N availability is often a limiting factor for plant productivity. It is applied to enhance early growth and build the tree crown (32–34). N is the most common nutrient in fertilizers used in boreal conifer forests to improve stem quality and tree diameter, increasing profitability (35). A study on the long-term effects of N (from 14 to 28 years after fertilization) in *Pinus sylvestris* and *Picea abies* sites suggests that tree height and volume improvements occur in the first five years, returning to basal values if not re-applied (36).

A meta-analysis of several fertilization experiments showed that N is likely to limit above-ground net primary production in temperate ($n = 22$) and tropical ($n = 16$) forests alike (37). In *E. grandis*, N fertilization improved above-ground biomass production by 50% in the first year, with little effect on leaf mortality (38). However, several edaphic factors will influence the trees' response to N application, as observed in another *E. grandis* plantation, which showed no significant decrease in the stemwood production due to the lack of N fertilization over 40 months, despite a reduction of 17% in leaf biomass (39). These variations may be linked to the N already present in the soil in control plots, or the influence of N fertilization on soil microorganisms, such as ectomycorrhizal (ECM) fungi, that also contribute to N uptake in trees and can access organic N pools in the soil, being suppressed under high N inputs (35,40,41).

Besides biomass production, N influences wood density and fiber properties that may affect its commercial value. In poplars, N supply decreased wood density, altered cell wall thickness and fiber length (42), and in some cases promoted thinner cell walls and less lignification (43). N application improved growth in both Douglas-fir and Norway spruce plantations but increased by 40% the ligneous dry matter on the former, while only 22% on the latter (44). Moreover, N additions were shown to decrease hydraulic conductivity in pine trees, while the opposite has been reported for eucalypts, which may be related to inherent differences between conifers and angiosperms (45).

9.2.2 Phosphorus

Phosphorus (P) is involved in photosynthesis and enzymatic functions and is a major component of essential molecules in cellular structures, energy metabolism, and genetic processes, namely the phospholipids in membranes, ATP and ADP, and nucleic acids. Thus, its deficiency strongly affects plant development by decreasing shoot growth, leaf expansion, and cell division (30, 46). In this sense, low P availability can affect wood development by disrupting the photosynthetic machinery and electron transport chain, shifting carbon allocation towards root development, and branching and/or maintaining a symbiotic relationship with arbuscular mycorrhizal (AM) fungi (47–49).

Much of the research on P-deficiency tends to focus on tropical environments, where low P availability is a frequent limitation due to the low mobility of phosphate ions and the formation of insoluble complexes with iron (Fe) and aluminum (Al) hydroxides (50). Experiments with eucalypt seedlings and mature trees have shown that P additions improve stem height and biomass (51, 52). In one year, Crous et al. (51) observed 54% increase in the basal area of *Eucalyptus tereticornis* trees from P-fertilized plots compared to controls, and a 109% increase by the end of the third year, despite the same photosynthetic capacity. Similarly, a 30% increase in wood production was reported for *Eucalyptus pauciflora* after P fertilization, which was linked to reduced carbon allocation to roots rather than improvement of canopy biomass (53).

Wood traits are also influenced by P limitation. For instance, in *E. grandis*, the lack of P fertilization for 20 months resulted in a 50% decrease in wood volume (39). Also, in *E. grandis* seedlings, it was observed that low P availability promoted secondary wall thickening, which led to higher wood density and inhibited stem cambial activity (54).

For *P. sylvestris* plantations, a positive correlation has been shown between forest/wood productivity and soil P availability, mainly the highly and moderately labile P fractions (55). High P content in *Pinus radiata* enhanced stem volume but reduced wood density for the first three years of development (56). In pot experiments, P deficiency decreased stem weight, chlorophyll content, and photosynthetic activity of P (57), which may show lower RuBisCO carboxylation rates in such conditions (58). Biomass decrease in different poplar species due to low P availability has been demonstrated in other works (59, 60), albeit not focused on wood productivity.

As for natural environments, in a tropical forest reserve in Sierra Leone, P soil concentration was one of the main determinants of above-ground wood production (61), associated with better nutrition and higher plant diversity at higher soil P. A meta-analysis of 173 natural terrestrial environments detected increased plant growth due to P enrichment and a synergistic effect from N and P additions. However, wood production was not specifically investigated (62). In lowland forest areas in Panama, P supply did not provide a significant increase in tree growth rates (63), yet when considering above-ground wood production in Amazon and Borneo forests it was found to be positively correlated to soil total P (64).

9.2.3 Potassium

Potassium (K) is the most abundant cation in plant cells and is considered as limiting as N and P in terrestrial environments (65). K has essential roles in plants, such as enzyme activation, photosynthesis, ionic homeostasis, cell expansion, photosynthates translocation to sink organs, and detoxification of reactive oxygen species (ROS). But, most notably, its major role pertains to water relations, osmotic balance, and stomatal opening (23, 66–68). Lower wood production under K limitation can be associated with a lower cambial cell expansion, which decreases vessel size (20). Moreover, low K can affect the mesophyll structures and decrease leaf lifespan, hydraulic and stomatal conductance, and transpiration, which impact photosynthetic activity, carbon assimilation, and translocation towards stems (69–72).

Potassium deficiency has been reported to affect wood growth and productivity. After four years of growth, low K fertilization decreased the wood relative productivity of *E. grandis* by 70%, a stronger effect than low N or low P conditions (39), a response associated with lower water use efficiency, and a decrease of leaf lifespan. Battie-Laclau et al. (71) observed similar results in *E. grandis* after three years of K fertilization, which led to higher stemwood biomass (173%), compared to non-fertilized plants, and increased partitioning towards stemwood instead of foliage production. Reduced leaf area and water use efficiency for wood production (stemwood biomass/water transpired) have also been verified *in E. grandis* without receiving a K supply for six years (73). In some cases, about 50% decrease in yield can occur in eucalypt plantations after the first harvest if K is not re-applied (74).

Besides total production, K deficiency can also alter wood characteristics, such as wood density and sapwood/heartwood ratios, which may influence its economic value (75). Poorer wood formation due to low K availability has been reported for tree species of more temperate climates, such as *P. radiata*, *P. abies*, and *Populus* spp. (20). Additionally, wood formation is strongly influenced by seasonal changes when cambial K concentrations vary, decreasing even further during winter months (76).

9.2.4 Calcium

Calcium (Ca) is a nutrient necessary in plant signaling, membrane stability, and cell wall structure (30). Ca ions reach the root mainly by mass flow and, after being taken up by the roots, enter the xylem vessels and follow long-distance transport pushed by the transpiration stream (30). Ca is immobile in the phloem and, after leaving the xylem, is deposited in the cell walls where multiple binding sites are present, mainly due to the carboxyl groups of constituents (e.g. pectin). Thus, Ca is particularly relevant in the middle lamella, where it links pectin chains (20), and during xylem differentiation, with a structural role conferring strength to the cell walls (77). However, once bound to cell wall components, Ca is relatively unavailable to other cellular metabolic processes.

Ca-oxalate crystals have been found in the wood and bark of several tree species (78–80) and, in some cases (e.g. *Populus tremula* bark), they occupy the whole lumen of fibers adjacent to the phloem (80). Seasonal variations of the cambium reactivation in trees have been related to changes in Ca concentrations in both cambium and phloem cells but maintained unchanged in xylem cells (81). This Ca was suggested to arise from those Ca-oxalate crystals in the bark. In a study exploring the relationships between wood density and sulphur (S), K, Ca and manganese (Mn) contents in 12 tropical tree species, it was shown that wood density and Ca and Mn concentrations had intra- and inter-annual patterns that allowed to define growth boundaries of the rings (82). The presence of Ca-oxalate crystals have been related to species with ring boundaries in hardwood species, such as *Cedrela fissilis, Hymenaea courbaril* and *Peltogyne paniculate,* marked by marginal parenchyma (82). Using synchrotron micro-X-ray fluorescence (μ-XRF) spectroscopy in the assessment of inter and intra-ring chemical contents of the mangrove species *Avicennia schauriana,* Ca was the most abundant element in tree-rings, differing in its speciation between heartwood (Ca sulfate and chloride) and sapwood (Ca oxalate and carbonate), that may be related with its physiological roles: metabolic activity of the sapwood and water transport of the heartwood (83).

Ca has essential roles during wood formation, and its deficiency has been shown to strongly modify wood structure and chemistry (84). Ca deficiency alters xylem structure and cambial width (20), as shown under Ca deprivation in hydroponically grown *P. tremula* L. × *Populus tremuloides* Michx. plants, which decreased wood production and the length and size of fibers and vessels, respectively (84). Low Ca contents may also reduce lignification, causing lower lignin proportion and alterations in the hardness and elasticity of the secondary cell wall (84). The reduced cambial growth under Ca deficiency in poplar may be caused by a decrease of carbon allocated to the cambium, as reduced photoassimilate contents loading from the leaves leads to reduced unloading rates at the cambium, also lowering sugar contents in the bark (85). However, glucose, fructose, and sucrose levels increased in the young wood under Ca deficiency (86). The lack of Ca also inhibits compression wood formation by reducing lignification during the differentiation of tracheid cells (87). And in poplar grown under Ca starvation, for example, reduced S-lignin concentrations in the xylem cells were observed, suggesting the dependency of lignin biosynthesis on Ca supply (84).

9.2.5 Sulphur

Sulphur (S) is a constituent of cysteine and methionine amino acids. These amino acids are related to the mechanism of action of several plant proteins and molecules, such as metallothioneins, phytochelatins, and glutathione, among several enzymes and secondary metabolites (30). Even though S can be taken up by plant leaves in the form of atmospheric sulfur dioxides (SO_2), most of it is taken up by roots in the form of sulphate anion (SO_4^{2-}) (88). S deficiency leads to chlorosis, lower chlorophyll, number of leaves, CO_2 assimilation, defense against pathogens, root hydraulic conductivity, N uptake efficiency, and overall yield (30, 89).

Under S limitation, the element may be remobilized from mature leaves towards roots instead of young leaves (90), which affects the development of above-ground parts. In a study with seedlings of *P. tremula* × *Populus alba* under sulphate limitation, hormone contents (indole-3-acetic acid [IAA] and abscisic acid [ABA]), glutathione and phosphate were decreased in young leaves, but an increase in phloem sucrose was observed, indicating the root as a carbon sink (91).

In general, homeostasis of S in trees has been considered to be more robust in comparison to herbaceous plants, mostly due to a larger S stock in stems and roots (deciduous trees) and needles (conifers), and the seasonal variations that increase S storage in the wood (92). Therefore, S is not often reported as a limiting factor in forest productivity (93), although it has been suggested that regions remote from atmospheric pollution inputs may be susceptible to S deficiency (88). Indeed, in a *Pinus contorta* forest in British Colombia, low S levels were reported, where supplementation with N and S for six years promoted larger basal areas and tree height (94). While in New Zealand, *P. radiata* trees suffered from a fungal dieback disease, which was linked to S and boron (B) deficiencies (95).

9.2.6 Magnesium

Magnesium (Mg) regulates carbohydrates synthesis and accumulation, with a central role in sucrose loading into the phloem, influencing the long-distance transport of carbon in the phloem (96). These significant roles lead to adverse effects on biomass production in Mg-deficient plants. In old-growth stands of *P. abies*, fertilization with K, Ca, and Mg positively influenced wood formation (97). Low Mg contents are known to impair vascular lignification, as Mg-dependent ATPases are needed to transport lignin precursors to the apoplast in lignifying xylem tissues (98). Mg is also an allosteric activator and cofactor of many enzymes involved in sugar and polysaccharide biosynthesis and transport (30).

In some tree species, darkwood has much higher Ca and Mg concentrations than lightwood. Such differences in chemical composition have been related to the age of the wood and the redistribution after wood formation or their reallocation from perennial tissues during stand development (99). Remembering that the terms "lightwood" and "darkwood" refer to the color of the wood and do not correlate with the physiological roles of sapwood and heartwood (100).

Waste products from the forestry industry have been used as fertilizers in forest soils to counteract whole-tree harvesting (101). Adding wood ash as fertilizer has been recognized to increase the availability of Ca, Mg, K, and micronutrients and heavy metals in forest soils (102). In a *P. sylvestris* forest, the addition of wood ash as fertilizer changed soil chemical properties in the first 30 cm layer, increasing pH and the availability of P, K, and Ca, and shoot lignification, with a positive correlation between Klason-lignin and K, Ca and Mg in soil, leading to increase stem biomass, diameter, and height (103).

9.2.7 Micronutrients

Micronutrients generally have a narrow range of plant concentrations between deficiency and toxicity, and despite their importance in plant metabolism, the study of micronutrient imbalances in woody plants is usually neglected. The micronutrient content is known to be usually highest in stem bark and tree crowns (leaves) and lowest in stem wood. Forestry management is key to maintaining nutrient availability in soils (104). In commercial tree plantations, removing biomass (wood) may significantly reduce soil fertility, requiring management of fertilization that includes micronutrient replenishment (101). For example, in *E. grandis* × *E. urophylla* clones, the harvest of whole-tree or only the wood, with the reincorporation of leaves and branches into the soil, showed strong potential for the cycling of micronutrients such as Mn, B, Zn, and Na to the planting site (105). For Scots pine stands, micronutrient accumulation was in the following order Mn > Fe > Zn > Cu, with higher contents in branches and needles than in the stems, and the contents of micronutrients have no correlation with the stand density (106).

Micronutrient deficiency also affects the tolerance of woody plants to pests (104, 107), decreasing tree growth, stem diameter, and wood production (104, 108, 109). In *E. grandis* stands, foliar Cu diminished the severity of eucalypt rust (*Austropuccinia psidii*) (104). Cu content was increased when Zn was omitted in the fertilization, highlighting the importance of micronutrient interaction and fertilization management to control fungal diseases and reduce yield losses in eucalypt plantations (104).

One of the most studied micronutrients in forest tree research is boron (B). Its deficiency in forest trees has been confirmed in several regions worldwide (110), especially in Southeast Asia, Eastern Australia, North and South America, and Northern Europe (111), mainly in alkaline and highly weathered soils. B deficiency in trees can be triggered by fertilization with macronutrients (111).

As perennial plants, trees commonly suffer from long-term B deficiency, and B-deficient trees may show stem dieback and growth retardation (112). Initially considered an immobile phloem nutrient, boron (B) is now recognized as a relatively mobile nutrient. Its remobilization capacity can widely vary within species, and it is dependent on B status and sugar alcohols availability (111). Boron concentrations in the bark and wood of adult trees have been reported to be relatively high, up to 520 and 1140 mg kg^{-1}, respectively (112). For trees, adequate foliar B concentrations may be close to 30 mg kg^{-1}, being closely related to wood quality (112). One of the main physiological roles of B, besides sugar synthesis and translocation, is the formation of primary cell walls. B participates in the pectin structure, acting in the crosslinking of rhamnogalacturonans II (RGII), while Ca is linked to pectin-homogalacturonans (113).

Additionally, B may be directly involved in lignin biosynthesis (114), with B-deficient plants showing increased lignification and thickening of the cell walls (115, 116). In the woody species *Neolamarckia cadamba*, B deficiency impacted cell wall lignification, induced changes in pectin structure, and up-regulated cellulose genes (117). In this species, pectin concentration decreased while genes encoding polygalacturonases, beta-galactosidases, and cellulose synthases were increased, suggesting that B deficiency impacts pectin and cellulose synthesis (117). In forest trees, B deficiency affected lignin and phenol content, with negative effects on the quality and quantity of wood.

9.3 General Use of Mass Spectrometry-Based Metabolomics to Study Wood

Among several applications of mass spectrometry (MS)-based metabolomics to study wood, direct analysis in real time time-of-fight mass spectrometry (DART–TOF–MS) (118) showed high potential to discriminate timber species (119). Metabolomics was used to discern biological variation among ten full-sib families of a Douglas-fir tree breeding population replicated on two sites (120). Metabolome data from developing xylem obtained with gas chromatography mass spectrometry (GC–MS) were used together with family, site, tree growth, and quantitative phenotypic wood traits (wood density, microfibril angle, wood chemistry, and fiber morphology) in statistical analyses. Wood sampling was made during the summer, and among 139 metabolites, 78 showed significant family variation, 108 had significant site variation, and 37 had significant family × site interaction. Broad-sense heritability revealed that metabolic and phenotypic traits alike were strongly related to site, while similar associations related to the genetic structure (family) were weak in comparison. Calculated broad-sense heritability showed that metabolic traits were more influenced by the environment, and there was a coherent relationship between genetics, metabolism, environmental and phenotypic expression in wood-forming tissue. The association between xylem-forming metabolites and tree phenotype supports metabolomics as an important tool for understanding secondary cell wall biosynthesis.

Wood formation responds to gravity, forming what is known as tension wood (TW). Gravity provokes an opposite reaction in the stem or branches, altering them biochemically and anatomically the wood (121). Secondary wall biosynthesis is markedly changed in TW. Typically, lignin accumulation is decreased in the upper side (TW) and increased in the lower side of the branch, known as opposite wood (OW) (122). In general, there is also an increase in cellulose content and a decrease in hemicelluloses, but with variations depending on the species (123). A strategy to study cell wall formation is bending the stem or branches and inducing TW (124).

TW was used in *P. tremula* in two growing seasons to understand the regulation of cell wall formation. Transcript profiling and metabolite analysis showed intense remodeling of the cell wall-related genes and a marked change in the carbon flow into various cell wall components (124). About 25% of 350 peaks detected with gas chromatography time-of-fight mass spectrometry (GC–TOF–MS) were identified, and among them, 26 detected metabolites were significantly changed, and 16 of them were identified, mostly related to carbohydrate metabolism. While sucrose, arabinose, and inositol decreased in TW, xylose and xylitol increased. Shikimate, which is an early precursor of phenolics, decreased. Combined transcriptomics and metabolomics analysis allowed the authors to conclude that TW causes a redirecting of carbon metabolism, when there is an increased metabolic activity towards cellulose, lipid and glucosamine biosynthesis and pectin degradation, and a decreased activity of carbon channeling to mannans, the pentose phosphate pathway, lignin biosynthesis, and biosynthesis of cell wall matrix carbohydrates.

Ultra-performance liquid chromatography quadrupole time-of-flight tandem mass spectrometry (UPLC–QTOF–MS/MS) was used to study the metabolome of heart and sapwood regarding their color in *Cunninghamia lanceolata* (Chinese fir) (125). Heartwood is darker than sapwood, where sap still flows. The authors concluded that 21 phenolic compounds (flavonoids, coumarins and derivatives, and stilbenes) in 850 identified peaks in a total of 12,581 detected peaks might be responsible for the darker color of the heartwood because of their higher accumulation compared to sapwood. Independently of the wood, older trees accumulated more metabolites than young trees.

Druart et al. (126) used transcriptomics and metabolomics (GC–TOF–MS) to study the changes in wood formation over the transition of activity–dormancy cycle in wild-grown *P. tremula* trees (6 m tall and 4 cm stem diameter) in North Sweden. A massive reprogramming of the cambial transcriptome and metabolome was observed in the cambium, coordinated by stage-specific transcriptional activators and repressors. It was also suggested a control by chromatin. More than 1,000 peaks were detected, of which 227 significantly changed in the activity–dormancy cycle. Interestingly, in the autumn, during the transition to the winter, when the temperature had not dropped yet, the level of several metabolites changed, suggesting that short days can induce cold acclimation in trees independently of temperature. In general terms, metabolomics analysis helped to conclude that during the transition to the winter, there was an increase in the levels of ABA in the cambium, which was correlated with the late stage of development of cold hardiness, when starch was broken down for supplying substrates, energy generation and biosynthesis of cryoprotectants and fatty acids. During the reactivation stage, spring/summer, sucrose breaking down and induction of the glyoxylate cycle were significantly enhanced, indicating an increase in respiration as growth was resumed. Additionally, stored amino acids were also broken down to generate substrates for the tricarboxylic acid cycle (TCA), and cambial activity was regained and controlled by gibberellin. Overall, the metabolome and transcriptome data revealed complex networks orchestrating the regulation of the activity–dormancy cycle in cambial meristem of *P. tremula*, and how they are coordinately influenced by environmental signals.

A similar approach was used to study changes in the primary metabolism of *E. grandis* bark in response to seasonal variation (127). Barks collected during the summer (wet season) and winter (dry season in Brazil) were subjected to transcriptomics (RT–qPCR), proteomics (2–DE gels), and metabolomics (GC–MS) analysis. There was accumulation of glucose, fructose, and sucrose in the bark collected in the winter but a decrease in the summer. Presumably, stem growth would be a sink of carbohydrates to sustain wood formation during the summer. Other metabolites, including organic acids and carbohydrates, were also significantly changed in the two seasons but without the consistency observed for glucose, fructose, and sucrose. Malate, galactinol, gluconate, and fumarate increase in the summer and erythritol, shikimate, dehydroascorbate, and taxifolin increase in the winter. Budzinski et al. (127) also observed that among 24 genes analyzed, all involved in carbon metabolism, 11 were up-regulated in the summer and three in the winter. Proteomics analysis showed an increased abundance of 77 peptides, with 38 up-regulated in the summer and 37 in the winter. In a complementary study, the same research group analyzed the molecular changes induced by seasonal (summer and winter) variations in the cambial zone of the stem of *E. grandis* trees, using transcripts, proteomics and metabolomics (GC–TOF–MS), emphasizing carbon metabolism (128). Glucose, sucrose, and fructose were analyzed with high-performance liquid chromatography with pulsed amperometric detection (HPLC–PAD). Relative expression of genes related to glycolysis, alcoholic fermentation, TCA cycle, and photosynthesis were obtained with RT–qPCR. A total of 53 metabolites were detected in the barks of summer and winter samples, which were clearly separated with multivariate and univariate statistical methods. Nineteen had their identity established. Contrary to their previous work, there was no variation in fructose, glucose, and sucrose between seasons. On the other hand, other sugars showed variation (raffinose, glycerate, galactinol were abundant in the winter, and xylose and xylitol in the summer). The majority of the other compounds identified in the metabolomics analysis (amino acids, lipids, and organic acids) increased in the summer. Interestingly the authors suggested that fermentation may be an important pathway operating in the bark of eucalypts, as shown by gene expression and metabolite analysis.

Seasonal variation and its influence on wood formation was also explored in maritime pine (*Pinus pinaster*), aiming to understand the effect of edapho-climatic conditions (129). Differentiating xylem samples were collected from the beginning to the end of the growing season. Among other analyses, the metabolic profiles of different tissues were obtained with GC–MS. As shown in other works, the environment significantly influenced secondary wall formation. Among 3512 genes, 19% had their expression changed because of a seasonal effect. The major extracted compounds were hexapyranoses (26.6%), fructose-type compounds (24.3%), non-identified compounds (21.7%), and sucrose (14.8%). The detected compounds were highly variable among samples. The data allowed the authors to conclude that in non-limiting environmental conditions, pine trees direct carbon and energy toward growth, increasing protein and hemicellulose concentrations. Under water deficit, a common environmental constraint for plant growth, carbon allocation to cellulose and lignin biosynthesis is increased, thus leading to thicker cell walls, which is related to hydraulic protection to avoid embolism (130, 131).

Seasonal metabolome variations were also studied in *Fagus sylvatica* and *Populus* × *canescens*. The first species is a temperate climax tree species, and the second is a floodplain tree species with contrasting P nutrition strategies. *F. sylvatica* stores and mobilizes P (17), which is not observed in *P. canescens* (132). To better understand these biological differences, C, N, S, and P-related metabolome and lipidome profiles were determined in poplar twig bud/leaves, bark, and wood during dormancy (winter), spring, summer, and autumn (133). GC–TOF–MS was used for low molecular weight metabolite profiling, and liquid chromatography electrospray ionization mass spectrometry (LC–ESI–MS) for lipids profiling. The lack of P storage capacity was confirmed in *P. canescens*, but not for C and N, in which seasonal variations indicated a storage strategy. Metabolome statistical analysis aimed to determine specific differences between buds/leaves, bark, and wood as well as between seasons. Wood and bark and buds/leaves formed distinct groups, and in each one, seasonal variation was also discriminated in a similar order. Hierarchical clustering analysis of metabolite changes in the analyzed tissues during poplar annual growth showed the formation of distinct groups, and in each one characteristic metabolite could be identified. Coherent variations could also be identified during the annual growth in compounds related to carbon metabolism. N-storage amino acids (arginine and ornithine) were highly abundant in bark and wood during dormancy. In general, it was found that variation in triacylglycerols, sugar-phosphates, metabolites of the TCA cycle, and amino acid abundance in the analyzed tissues was linked to C, N, and S storage and mobilization processes during the annual growth of *P. canescens*. During the winter, the high concentrations of phospholipids in bark and wood were related to cold hardiness. The increase of sugar-phosphate and the decrease of phospholipids indicated a higher metabolic activity linked to the beginning of the growing annual phase.

In most economically explored trees, branching is an undesirable biological event because it diverts energy and nutrients from the main stem and decreases the growth rate. Hormones play a key role in branching (134). They are also responsible for the formation of knots, which affect the technical properties of the wood (135). Aiming to understand branching in trees, Yang et al. (136) studied the transcriptome and hormone profile related to branch development in two genotypes of *E. urophylla*. Samples were taken from the shoot apex, lateral shoot apex, and from a stem segment at the fifth axillary bud from the shoot apex, where the concentrations of IAA, trans-zeatin riboside (TZR), ABA, isopentenyladenine (IP), isopentenyl adenosine (IPA), salicylic acid (SA), methyl jasmonate (MEJA) and jasmonate (JA) were determined with HPLC–MS/MS. Transcriptomics analysis allowed the authors to identify genes which expression was significantly changed when comparing shoot apex and lateral shot apex with stem segment (taken as a control). The authors also built a hormone signal transduction pathway explaining branching development and concluded that ABA and SA promote a larger branch angle and smaller branch number, while IAA has an opposite function (136).

9.4 Tree Nutrition and Metabolome

9.4.1 Nitrogen

N is the most demanded nutrient by plants, and its limitation causes severe negative effects on plant growth. Curiously, *Populus cathayana* female plants greatly differ from male plants regarding N nutrition, as they are more affected by deficiency, as shown by leaf phosphoproteomics and GC–MS metabolomics analysis (137). Amino acids, sugars, and organic acids were found as sex-specific metabolites, but the co-modulation of metabolites was both driven by sex and N deficiency. However, the data showed that N deficiency negatively affects female poplar plants more than male poplar plants. Still related to the gender difference of *P. cathayana* plants, gender-specific metabolites could be identified in response to Pi (inorganic phosphorus) limitation during 60 days (138). Marked changes were observed in metabolites related to carbon flow, and amino acid contents increased in male plants. The authors suggested that these changes in male plants were related to a strategy to save energy during Pi deficiency.

N nutrition may also define wood composition in *Populus trichocarpa* from different provenances. Because of the potential of using wood as feedstock for second-generation bioethanol production, Guerra et al. (139) carried out a comprehensive traits analysis related to growth (height, diameter), phenology (bud flush), ecophysiology (leaf C and N content and isotopic composition), wood composition (contents of sugars and lignin) and metabolomics of wood in 460 clones of *P. trichocarpa* from 101 provenances in the USA. The analyses were carried out on three-year-old trees, all growing in California, and GC–TOF–MS was used for metabolomics analysis. Among 632 metabolites quantified, 161 were identified and distributed mainly in the galactose metabolism, aminoacyl-tRNA biosynthesis, alanine, aspartate, and glutamate metabolism, citric acid cycle, and cyanoamino acid metabolism. The data showed a significant clonal variation and inheritance for the assessed traits. It was possible to identify that variation of 28 metabolites (amino acids, fatty acids, sugars, organic acids, and phosphates) was explained by different experimental factors. The variation in metabolite concentrations was related to three factors: genotypic variation, environmental variation, and the interaction between genetic and environmental variation. In general, the calculated broad-sense heritability for the metabolome data was lower than 0.45, indicating low genetic control and a significant effect of the environment, thus representing the diversity of the samples according to their provenances. The authors related this variation to partially nutritional plant status. Although cultivated in the same place, with the same soil (silt loam soil), the clones may have present genetic variation for nutrient absorption and use efficiency. The authors observed that amino acids such as proline, aspartic acid, and suberylglycine were significantly modulated by the environment and argued that these compounds could be associated with N nutrition.

Metabolomics and transcriptomics were used to study the early response of *Populus tomentosa* to low N availability (140). N shortage significantly impacts plant growth as many biological processes rely on molecules containing this nutrient, such as nucleic acids, amino acids, proteins, chlorophyll, and lipids. The analyses were carried out in 60-day-old plants, which were kept in hydroponic solution with N level for five days, and then transferred to solutions with or without sufficient N. Metabolites were analyzed in the aerial parts using GC–MS. Hormones were also analyzed with HPLC–MS/MS. Additionally, transcriptomics analysis was carried out in the treated plants. The authors identified 1131 metabolites, which were assigned to 18 metabolic pathways, including amino acid and sugar metabolisms. Among all identified metabolites, only 30 varied significantly between N treatments. Most of these metabolites (70%) were significantly suppressed in the low N treatment. A general negative effect of low N in the metabolism was evidenced by metabolomics and transcriptomics analysis, including N absorption and assimilation, photosynthesis, glycolysis, and the TCA cycle. Genes related to gibberellin and IAA pathways were responsive to low-N supply, and their expressions were validated by the determinations of the concentrations of the hormones.

9.4.2 Phosphorus

Gas and liquid chromatography mass spectrometry (GC–MS and LC–MS) and inductively coupled plasma optical emission spectroscopy (ICP–OES) were used to contrast the leaf metabolome and ionome (N, P, S, K, Ca, Mg, Fe, Mn, Zn, Cu, and Na) of two *Quercus variabilis* populations grown at a P-rich and a P-deficient sites (141). Soils in subtropical areas are usually deficient not only in P but also in Ca and Mg, while rich in Fe and Al because of intense weathering and leaching (8). The two *Q. variabilis* populations could be distinguished by 25 identified metabolites, P, N, S, Fe, and Zn. Twenty-one metabolites had higher concentrations in the leaves of the P-rich population. Orthophosphate had a significant contribution to discriminating the populations. Six metabolic pathways could be identified, and amino sugar and nucleotide sugar pathway accommodate the largest number of compounds (UDP-D-galactose, UDP-D-glucose, UDP-*N*-acetylglucosamine, D-mannose, and D-mannose-1-phosphate). The authors could establish positive correlations among nutrients and certain metabolites in the rich and poor P sites. For instance, D-mannose, D-mannose-1-phosphate, and UDP-*N*-acetylglucosamine were most affected due to the P supply in the soil of the P-rich site. On the other hand, N-containing compounds, such as betaine, γ-aminobutyric acid, D-proline, and UDP-*N*-acetyl-glucosamine, were related to leaf N at both sites, independently of P availability. This certainly indicates the development of edaphic ecotypes as a consequence of an adaption to the soil type regarding P availability.

Metabolomics was also used to study the response of *Eucalyptus globulus* plants grown for 2.5 months with four different nutrient solutions regarding P concentration; i.e. 0.1, 0.02, 0.005, and 0.0 mM P (18). Metabolites were extracted from leaves and analyzed by GC–MS. Photosynthesis, P, N, $\delta^{13}C$, and plant growth were also measured in these plants. P nutrition significantly affected metabolites identified by the GC–MS analysis. Ninety peaks were detected, and 70 were identified. P-sugars, such as glucose-6-P and fructose-6-phosphate, were decreased with low P supply, indicating a reallocation of P pools, thus increasing the use efficiency of the nutrient. P had little effect on other metabolites identified as carbohydrates, organic acids of TCA cycle, and amino acids. Arginine and glutamine were an exception and increased dramatically in low P. These amino acids link N metabolic branches and are the major amino acids transported in the phloem because of their high N:C ratio (142, 143), supporting thus the changes in N homeostasis in the P deprived plants. Low P supply also drew down the levels of compounds related to phenol metabolism, such as phenylalanine, coumaroylquinic acid, and catechin. The authors argued that such a decrease might be related to a higher lignin synthesis, as P deficiency increases lignification in proteoid roots of lupin (144) and leaves of young cotton plants (145).

In a metabolomics study dedicated to P deficiency and tree growth, lipid and primary metabolites profiling were obtained from eucalyptus species grown under low P availability using ultra-performance liquid chromatography Fourier transform ion cyclotron resonance mass spectrometry (UPLC–FTICR–MS) and GC–MS, respectively. Leaves, roots, and stems of seedlings of five eucalyptus species (*Eucalyptus acmenoides*, *E. grandis*, *E. globulus*, *E. tereticornis*, and *Corymbia maculata*) were analyzed (19). Growth, gas exchange data, and nutrient concentration were also analyzed. 66, 75, and 66 primary metabolites were identified in the leaves, roots and stems, respectively. In total, 172 lipid species with several chemical variants for the same main compound were identified, which were grouped into eleven lipid classes. Several metabolites were common among the analyzed organs, but their concentrations differed. The concentrations of sugars, organic acids, amino acids, and lipids showed wide variability among eucalypt species under low P supply. Depending on the species, there was a reduction in P-sugars and an increase in sulfur-lipids, indicating different strategies to cope with low P by increasing P use efficiency. Roots of *C. maculata* showed the largest significant changes in lipids. Primary metabolites changed in all organs and in stems, and the largest decreases were observed in *C. maculata* and the largest increases in *E. grandis*. Several metabolites were common among these species, most of them carbohydrates, but also some organic acids and amino acids. Curiously, independent of the species and P treatment, stem accumulated more P than leaves and roots.

9.4.3 Potassium

An analysis of *E. grandis* plants submitted to a combination of K fertilization and water stress in the field showed the great integration between the processes related to the K level and the water status of the plants (146). Through a systems biology perspective, the co-regulation network of genes was obtained, in which it was possible to identify genes highly correlated with cell wall formation. Combined transcriptomics and metabolomics analysis revealed in the applied treatments a total of 5573 differentially expressed genes (DEGs) and 516 differentially accumulated mass signatures (DAMs). Interestingly, a significant part of the DEGs was related to fertilization with K

(>60%) and almost half (42%) with DAMs. On the other hand, only 4% of the DEGs and 12% of the DAMs were exclusively related to water deficit. In the treatment in which K fertilization and water stress were applied, the values were intermediate to the single treatments (DEGs – 34.7%, DAMs – 45%). Most DEGs and DAMs associated with water stress were also associated with K, 90% and 79%, respectively. On the other hand, the opposite was observed concerning DEGs and DAMs regulated both by K and water, 36% and 52%, respectively. Wood phenotypic evaluations (stem biomass, average gross primary productivity allocated to stem growth, saccharification yield, wood density, vessel size, and vessel density) could be correlated with 36 DAMs and 2148 DEGs. Looking specifically in transcription factors highly co-regulated with secondary cell wall biosynthesis in stress conditions, 14 DAMs were correlated with saccharification yield, vessel density, and vessel size. Several of the DEGs identified were related to water deficiency, many of them with the ABA signaling pathway. These genes were highly up-regulated in treatments with K in the presence or absence of water stress, showing that K-fertilized trees were stressed independently of water availability. Probably, this would be related to the greater growth of these plants, increasing water consumption, and decreasing the water available in the soil, mainly during dry periods.

9.4.4 Mycorrhization

The symbiosis between plants and AM or ECM fungi is known to improve nutrient uptake and accumulation in several tree species, especially regarding P (AM) and N (ECM) (41). However, these effects are mostly associated with the fungal ability to intercept nutrients along the soil profile, with little investigation into the changes they induce in the host metabolic profile that could enhance nutrient uptake and affect wood development. Recently, metabolomics was used to study the effect of mycorrhization in *P. canescens* (147). LC–MS/MS analysis showed that wood was little affected by inoculation with the ectomycorrhizal fungus *Paxillus involutus*. Although root colonization was low (2.1%), there was a significant measurable impact on the foliar metabolome. P absorption was increased. The concentrations of phenolic compounds were significantly increased in inoculated as well as in the total soluble carbohydrates and starch, along with the increased abundance of the majority of glycerophosphocholines detected in the roots. Despite the low colonization, significant growth was observed even though leaf proteome analysis did not show alterations in photosynthesis-related proteins. The authors discussed that growth might be related to small changes that together may have had a positive effect. In poplar, the positive effects of mycorrhization on plant nutritional status were suggested to be related to enhanced synthesis of different defense metabolites, such as volatile organic compounds (148) that indirectly can improve biomass production and wood yields.

9.5 Final Remarks

The lack of studies on nutrient deficiency in trees and its effect on wood formation is evident. Understanding the related metabolic changes may help us improve fertilization strategies to keep productivity and quality, considering that trees are perennial plants and are subject to an annual sequence of climatic events, as well as the impact of long-term exposure to climate changes due to global warming. The N and CO_2 atmosphere enrichment caused by anthropogenic activity may cause serious unbalances in tree growth (4) as carbon and N metabolism are tightly related (149, 150). C and N are major plant constituents and may trigger responses that will diminish growth. Additionally, commercial tree plantations remove large amounts of nutrients from the soil each year, and their replenishment is essential to maintain growth, productivity, and wood quality, in several meanings, from timber to second-generation ethanol production.

We can take K as an example of the need for metabolic studies. In the stem, K highest concentration along the stem coincides with greater cambial activity (20, 85, 151–153). This position matches the highest content of IAA, which controls cell proliferation in the cambium (154, 155). However, K can be partially replaced by Na in eucalyptus, without penalty for growth and wood formation (156–158). Therefore, in addition to the individual role of each nutrient, as thoroughly discussed here, nutrient interactions might exist and affect wood formation. Metabolomics studies are scarce compared to the information available about the adverse effects of nutrient deficiency on wood formation. Allied with other omics technologies, metabolic studies certainly would help to better understand the complex participation of nutrients in wood formation.

Acknowledgments

P.M. thanks Conselho Nacional de Desenvolvimento Científico e Tecnológico (CNPq-Brazil) for a research fellowship. V.H.O. thanks Fundação de Amparo à Pesquisa do Estado de São Paulo (FAPESP) for a post-doctoral fellowship.

References

1 FAOSTAT. Forestry production and trade [Internet]. Available from (https://www.fao.org/faostat/en/#data/FO), accessed 13 June 2023.
2 Global Timber and Wood Products Industry Factsheet. (2020). Largest Wood Producing Countries, Largest Wood Exporters and Importers [Internet]. Available from (https://blog.bizvibe.com/blog/largest-wood-producing-countries), accessed 13 June 2023.

3 IBÁ. Indústria Brasileira de Árvores [Internet]. Available from (https://www.iba.org/dados-estatisticos), accessed 13 June 2023.
4 de Moraes Gonçalves JL, Stape JL, Laclau JP, Smethurst P, Gava JL. Silvicultural effects on the productivity and wood quality of eucalypt plantations. *For Ecol Manage.* 2004; 193(1, 2): 45–61.
5 Gojon A, Cassan O, Bach L, Lejay L, Martin A. The decline of plant mineral nutrition under rising CO_2: physiological and molecular aspects of a bad deal. *Trends Plant Sci.* 2023; 28(2): 185–198.
6 Jonard M, Fürst A, Verstraeten A, Thimonier A, Timmermann V, Potočić N, et al. Tree mineral nutrition is deteriorating in Europe. *Global Change Biol.* 2015; 21(1): 418–430.
7 de O Castro CA, Resende RT, Bhering LL, Cruz CD. Brief history of *Eucalyptus* breeding in Brazil under perspective of biometric advances. *Ciência Rural.* 2016; 46(9): 1585–1593.
8 Marschner H. *Mineral Nutrition of Higher Plants.* Second Edition. Academic Press; 1995.
9 Alvarado A. Plant nutrition in tropical forestry. In: Pancel L, & Köhl M, editors. *Tropical Forestry Handbook.* Berlin Heidelberg: Springer-Verlag; 2015. p. 1–91.
10 Ogden M, Hoefgen R, Roessner U, Persson S, Khan G. Feeding the walls: how does nutrient availability regulate cell wall composition? *Int J Mol Sci.* 2018; 19(9): 2691.
11 Moura JCMS, Bonine CAV, de Oliveira Fernandes Viana J, Dornelas MC, Mazzafera P. Abiotic and biotic stresses and changes in the lignin content and composition in plants. *J Integr Plant Biol.* 2010; 52(4): 360–376.
12 Cabane M, Afif D, Hawkins S. Lignins and abiotic stresses. *Adv Bot Res.* 2012; 61: 219–262.
13 Groover AT, Nieminen K, Helariutta Y, Mansfield SD. Wood formation in *Populus.* In: Jansson S, Bhalerao R, & Groover A, editors. *Genetics and Genomics of* Populus. New York (NY): Springer; 2010. p. 201–224.
14 Bradshaw HD, Ceulemans R, Davis J, Stettler R. Emerging model systems in plant biology: Poplar (*Populus*) as a model forest tree. *J Plant Growth Regul.* 2000; 19(3): 306–313.
15 Rodrigues AM, Ribeiro-Barros AI, António C. Experimental design and sample preparation in forest tree metabolomics. *Metabolites.* 2019; 9(12): 285.
16 Rodrigues AM, Miguel C, Chaves I, António C. Mass spectrometry-based forest tree metabolomics. *Mass Spectrom Rev.* 2021; 40(2): 126–157.
17 Netzer F, Schmid C, Herschbach C, Rennenberg H. Phosphorus-nutrition of European beech (*Fagus sylvatica* L.) during annual growth depends on tree age and P-availability in the soil. *Environ Exp Bot.* 2017; 137: 194–207.
18 Warren CR. How does P affect photosynthesis and metabolite profiles of *Eucalyptus globulus*? *Tree Physiol.* 2011; 31(7): 727–739.

19 de O Silva FM, Bulgarelli RG, Mubeen U, Caldana C, SAL A, Mazzafera P. Low phosphorus induces differential metabolic responses in eucalyptus species improving nutrient use efficiency. *Front Plant Sci.* 2022; 13: 989827.

20 Fromm J. Wood formation of trees in relation to potassium and calcium nutrition. *Tree Physiol.* 2010; 30(9): 1140–1147.

21 Zuidema PA, Poulter B, Frank DC. A wood biology agenda to support global vegetation modelling. *Trends Plant Sci.* 2018; 23(11): 1006–1015.

22 Lira-Martins D, Quesada CA, Strekopytov S, Humphreys-Williams E, Herault B, Lloyd J. Wood nutrient-water-density linkages are influenced by both species and environment. *Front Plant Sci.* 2022; 13: 528.

23 Wang M, Zheng Q, Shen Q, Guo S. The critical role of potassium in plant stress response. *Int J Mol Sci.* 2013; 14(4): 7370–7390.

24 Pitre FE, Cooke JEK, Mackay JJ. Short-term effects of nitrogen availability on wood formation and fibre properties in hybrid poplar. *Trees.* 2007; 21(2): 249–259.

25 Brodersen CR, McElrone AJ. Maintenance of xylem network transport capacity: a review of embolism repair in vascular plants. *Front Plant Sci.* 2013; 4: 108.

26 Secchi F, Pagliarani C, Zwieniecki MA. The functional role of xylem parenchyma cells and aquaporins during recovery from severe water stress. *Plant Cell Environ.* 2017; 40(6): 858–871.

27 Kraiser T, Gras DE, Gutiérrez AG, González B, Gutiérrez RA. A holistic view of nitrogen acquisition in plants. *J Exp Bot.* 2011; 62(4): 1455–1466.

28 Xu G, Fan X, Miller AJ. Plant nitrogen assimilation and use efficiency. *Annu Rev Plant Biol.* 2012; 63: 153–182.

29 Will GM. Nutrient deficiencies in *Pinus radiata* in New Zealand. *N Z J For Sci.* 1978; 8(1): 4–14.

30 Hawkesford M, Horst W, Kichey T, Lambers H, Schjoerring J, Møller IS, et al. Functions of macronutrients. In: Marschner P, editor. *Marschner's Mineral Nutrition of Higher Plants.* Third Edition. Amsterdam (NL): Academic Press; 2011. p. 135–189.

31 Nguyen NT, Mohapatra PK, Fujita K, Nakabayashi K, Thompson J. Effect of nitrogen deficiency on biomass production, photosynthesis, carbon partitioning, and nitrogen nutrition status of *Melaleuca* and *Eucalyptus* species. *Soil Sci Plant Nutr.* 2003; 49(1): 99–109.

32 Heilman PE, Hinckley TM, Roberts DA, Ceulemans R. Production physiology. In: Stettler RF, Bradshaw HD, Jr, Heilman PE, & Hinckley TM, editors. *Biology of* Populus *and Its Implications for Management and Conservation.* Ottawa: NRC Research Press; 1996. p. 459–489.

33 Smethurst P, Holz G, Moroni M, Baillie C. Nitrogen management in *Eucalyptus nitens* plantations. *For Ecol Manage.* 2004; 193(1, 2): 63–80.

34 Oliveira IR, Bordron B, Laclau JP, Paula RR, Ferraz AV, JLM G, et al. Nutrient deficiency enhances the rate of short-term belowground transfer

of nitrogen from *Acacia mangium* to *Eucalyptus* trees in mixed-species plantations. *For Ecol Manage.* 2021; 491: 119192.
35 Pukkala T. Optimal nitrogen fertilization of boreal conifer forest. *For Ecosystems.* 2017; 4: 3.
36 Pettersson F, Högbom L. Long-term growth effects following forest nitrogen fertilization in *Pinus sylvestris* and *Picea abies* stands in Sweden. *Scand J For Res.* 2004; 19(4): 339–347.
37 LeBauer DS, Treseder KK. Nitrogen limitation of net primary productivity in terrestrial ecosystems is globally distributed. *Ecology.* 2008; 89(2): 371–379.
38 Laclau JP, Almeida JCR, Gonalves JLM, Saint-Andr L, Ventura M, Ranger J, et al. Influence of nitrogen and potassium fertilization on leaf lifespan and allocation of above-ground growth in *Eucalyptus* plantations. *Tree Physiol.* 2009; 29(1): 111–124.
39 Rocha JHT, de Moraes Gonçalves JL, de Vicente Ferraz A, Poiati DA, Arthur Junior JC, Hubner A. Growth dynamics and productivity of an *Eucalyptus grandis* plantation under omission of N, P, K Ca and Mg over two crop rotation. *For Ecol Manage.* 2019; 447: 158–168.
40 Lambers H, Raven JA, Shaver GR, Smith SE. Plant nutrient-acquisition strategies change with soil age. *Trends Ecol Evol.* 2008; 23(2): 95–103.
41 Smith SE, Read DJ. *Mycorrhizal Symbiosis.* Third Edition. Academic Press; 2008.
42 Pitre FE, Lafarguette F, Boyle B, Pavy N, Caron S, Dallaire N, et al. High nitrogen fertilization and stem leaning have overlapping effects on wood formation in poplar but invoke largely distinct molecular pathways. *Tree Physiol.* 2010; 30(10): 1273–1289.
43 Euring D, Bai H, Janz D, Polle A. Nitrogen-driven stem elongation in poplar is linked with wood modification and gene clusters for stress, photosynthesis and cell wall formation. *BMC Plant Biol.* 2014; 14: 391.
44 Sicard C, Saint-Andre L, Gelhaye D, Ranger J. Effect of initial fertilisation on biomass and nutrient content of Norway spruce and Douglas-fir plantations at the same site. *Trees – Struct Funct.* 2006; 20(2): 229–246.
45 Hacke UG, Plavcová L, Almeida-Rodriguez A, King-Jones S, Zhou W, Cooke JEK. Influence of nitrogen fertilization on xylem traits and aquaporin expression in stems of hybrid poplar. *Tree Physiol.* 2010; 30(8): 1016–1025.
46 Vance CP, Uhde-Stone C, Allan DL. Phosphorus acquisition and use: critical adaptations by plants for securing a non-renewable resource. *New Phytol.* 2003; 157(3): 423–447.
47 Lambers H, Shane MW, Cramer MD, Pearse SJ, Veneklaas EJ. Root structure and functioning for efficient acquisition of phosphorus: matching morphological and physiological traits. *Ann Bot.* 2006; 98: 693–713.

48 Rich MK, Nouri E, Courty PE, Reinhardt D. Diet of arbuscular mycorrhizal fungi: bread and butter? *Trends Plant Sci.* 2017; 22(8): 652–660.

49 Carstensen A, Herdean A, Schmidt SB, Sharma A, Spetea C, Pribil M, et al. The impacts of phosphorus deficiency on the photosynthetic electron transport chain1. *Plant Physiol.* 2018; 177(1): 271–284.

50 Faucon MP, Houben D, Reynoird JP, Mercadal-Dulaurent AM, Armand R, Lambers H. Advances and perspectives to improve the phosphorus availability in cropping systems for agroecological phosphorus management. *Adv Agron.* 2015; 134: 51–79.

51 Crous KY, Ósvaldsson A, Ellsworth DS. Is phosphorus limiting in a mature *Eucalyptus* woodland? Phosphorus fertilization stimulates stem growth. *Plant Soil.* 2015; 391(1, 2): 293–305.

52 Bulgarelli R, Magnum F, Silva O, Adrián S, Andrade L, Mazzafera P. Eucalypts and low phosphorus availability: between responsiveness and efficiency. *Plant Soil.* 2019; 445: 349–368.

53 Keith H, Raison RJ, Jacobsen KL. Allocation of carbon in a mature eucalypt forest and some effects of soil phosphorus availability. *Plant Soil.* 1997; 196(1): 81–99.

54 Thomas DS, Montagu KD, Conroy JP. Why does phosphorus limitation increase wood density in *Eucalyptus grandis* seedlings? *Tree Physiol.* 2005; 26(1): 35–42.

55 Bueis T, Bravo F, Pando V, Kissi YA, Turrión MB. Phosphorus availability in relation to soil properties and forest productivity in *Pinus sylvestris* L. plantations. *Ann For Sci.* 2019; 76(4): 1–13.

56 Nyakuengama JG, Downes GM, Ng J. Growth and wood density responses to later-age fertilizer application in *Pinus radiata. IAWA J.* 2002; 23: 431–448.

57 Zhang S, Jiang H, Zhao H, Korpelainen H, Li C. Sexually different physiological responses of *Populus cathayana* to nitrogen and phosphorus deficiencies. *Tree Physiol.* 2014; 34(4): 343–534.

58 Song H, Lei Y, Zhang S. Differences in resistance to nitrogen and phosphorus deficiencies explain male-biased populations of poplar in nutrient-deficient habitats. *J Proteomics.* 2018; 178: 123–127.

59 Desai S, Naik D, Cumming JR. The influence of phosphorus availability and Laccaria bicolor symbiosis on phosphate acquisition, antioxidant enzyme activity, and rhizospheric carbon flux in *Populus tremuloides. Mycorrhiza.* 2014; 24(5): 369–382.

60 Kavka M, Polle A. Phosphate uptake kinetics and tissue-specific transporter expression profiles in poplar (*Populus × canescens*) at different phosphorus availabilities. *BMC Plant Biol.* 2016; 16: 206.

61 Jucker T, Sanchez AC, Lindsell JA, Allen HD, Amable GS, Coomes DA. Drivers of above-ground wood production in a lowland tropical forest

of West Africa: teasing apart the roles of tree density, tree diversity, soil phosphorus, and historical logging. *Ecol Evol*. 2016; 6(12): 4004–4017.
62 Elser JJ, Bracken MES, Cleland EE, Gruner DS, Harpole WS, Hillebrand H, et al. Global analysis of nitrogen and phosphorus limitation of primary producers in freshwater, marine and terrestrial ecosystems. *Ecol Lett*. 2007; 10(12): 1135–1142.
63 Wright SJ, Yavitt JB, Wurzburger N, Turner BI, Tanner EVJ, Sayer EJ, et al. Potassium, phosphorus, or nitrogen limit root allocation, tree growth, or litter production in a lowland tropical forest. *Ecology*. 2011; 92(8): 1616–1625.
64 Banin L, Lewis SL, Lopez-Gonzalez G, Baker TR, Quesada CA, Chao KJ, et al. Tropical forest wood production: a cross-continental comparison. *J Ecol*. 2014; 102(4): 1025–1037.
65 Sardans J, Peñuelas J. Potassium: a neglected nutrient in global change. *Global Ecol Biogeogr*. 2015; 24(3): 261–275.
66 Sardans J, Peñuelas J. Potassium control of plant functions: ecological and agricultural implications. *Plants*. 2021; 10(2): 419.
67 Oddo E, Inzerillo S, la Bella F, Grisafi F, Salleo S, Nardini A. Short-term effects of potassium fertilization on the hydraulic conductance of *Laurus nobilis* L. *Tree Physiol*. 2011; 31(2): 131–138.
68 Cakmak I. The role of potassium in alleviating detrimental effects of abiotic stresses in plants. *J Plant Nutr Soil Sci*. 2005; 168(4): 521–530.
69 Cornut I, le Maire G, Laclau JP, Guillemot J, Mareschal L, Nouvellon Y, et al. Potassium limitation of wood productivity: a review of elementary processes and ways forward to modelling illustrated by *Eucalyptus* plantations. *For Ecol Manage*. 2021; 494: 119275.
70 Hu W, Ren T, Meng F, Cong R, Li X, White PJ, et al. Leaf photosynthetic capacity is regulated by the interaction of nitrogen and potassium through coordination of CO_2 diffusion and carboxylation. *Physiol Plant*. 2019; 167(3): 418–432.
71 Battie-Laclau P, Delgado-Rojas JS, Christina M, Nouvellon Y, Bouillet JP, de Cassia Piccolo M, et al. Potassium fertilization increases water-use efficiency for stem biomass production without affecting intrinsic water-use efficiency in *Eucalyptus grandis* plantations. *For Ecol Manage*. 2016; 364: 77–89.
72 Epron D, Cabral OMR, Laclau JP, Dannoura M, Packer AP, Plain C, et al. In situ $^{13}CO_2$ pulse labelling of field-grown eucalypt trees revealed the effects of potassium nutrition and throughfall exclusion on phloem transport of photosynthetic carbon. *Tree Physiol*. 2016; 36(1): 6–21.
73 Asensio V, Domec JC, Nouvellon Y, Laclau JP, Bouillet JP, Jordan-Meille L, et al. Potassium fertilization increases hydraulic redistribution and water use efficiency for stemwood production in *Eucalyptus grandis* plantations. *Environ Exp Bot*. 2020; 176: 104085.

74 Silva NF, de Barros NF, Lima Neves JC, Schulthais F, de Novais RF, Mattiello EM. Yield and nutrient demand and efficiency of eucalyptus under coppicing regime. *Forests.* 2020; 11(8): 852.
75 Chambi-Legoas R, Chaix G, Castro VR, Franco MP, Tomazello-Filho M. Inter-annual effects of potassium/sodium fertilization and water deficit on wood quality of *Eucalyptus grandis* trees over a full rotation. *For Ecol Manage.* 2021; 496: 119415.
76 Ache P, Fromm J, Hedrich R. Potassium-dependent wood formation in poplar: seasonal aspects and environmental limitations. *Plant Biol.* 2010; 12(2): 259–267.
77 Brett CT, Hillman JR. *Biochemistry of Plant Cell Walls.* Cambridge University Press; 2009.
78 Deshpande B, Journal AVI. Calcium oxalate crystals in the fusiform cells of the cambium of *Gmelina arborea. IAWA Bull (Netherlands).* 1992, 1992; 13(3): 297–300.
79 Scurfield G, Michell AJ, Silva S. Crystals in woody stems. *Botanical J Linnean Soc.* 1973; 66: 277–289.
80 Trockenbrodt M. Calcium oxalate crystals in the bark of *Quercus robur, Ulmus glabra, Populus tremula* and *Betula pendula. Ann Bot.* 1995; 75(3): 281–284.
81 Follet-Gueye ML, Verdus MC, Demarty M, Thellier M, Ripoll C. Cambium pre-activation in beech correlates with a strong temporary increase of calcium in cambium and phloem but not in xylem cells. *Cell Calcium.* 1998; 24(3): 205–211.
82 Ortega Rodriguez DR, Hevia A, Sánchez-Salguero R, Santini L, de Carvalho HWP, Roig FA, et al. Exploring wood anatomy, density and chemistry profiles to understand the tree-ring formation in Amazonian tree species. *Dendrochronologia (Verona).* 2022; 71: 125915.
83 Alves EEN, Ortega Rodriguez DR, de Azevedo Rocha P, Vergütz L, Santini Junior L, Hesterberg D, et al. Synchrotron-based X-ray microscopy for assessing elements distribution and speciation in mangrove tree-rings. *Results Chem.* 2021; 3: 100121.
84 Lautner S, Fromm J. Calcium-dependent physiological processes in trees. *Plant Biol.* 2009; 12(2): 268–274.
85 Langer K, Levchenko V, Fromm J, Geiger D, Steinmeyer R, Lautner S, et al. The poplar K^+ channel KPT1 is associated with K^+ uptake during stomatal opening and bud development. *Plant J.* 2004; 37(6): 828–838.
86 Schulte-baukloh C, Fromm J. The effect of calcium starvation on assimilate partitioning and mineral distribution of the phloem. *J Exp Bot.* 1993; 44(11): 1703–1707.
87 Du S, Yamamoto F. A study on the role of calcium in xylem development and compression wood formation in *Taxodium distichum* seedlings. *IAWA J.* 2003; 24(1): 75–85.

88 Johnson DW. Sulfur cycling in forests. *Biogeochemistry*. 1984; 1(1): 29–43.
89 Hawkesford MJ, de Kok LJ, Hawkesford MJ, de Kok LJ. Managing sulphur metabolism in plants. *Plant Cell Environ*. 2006; 29(3): 382–395.
90 Abdallah M, Dubousset L, Meuriot F, Etienne P, Avice JC, Ourry A. Effect of mineral sulphur availability on nitrogen and sulphur uptake and remobilization during the vegetative growth of *Brassica napus* L. *J Exp Bot*. 2010; 61(10): 2635–2646.
91 Honsel A, Kojima M, Haas R, Frank W, Sakakibara H, Herschbach C, et al. Sulphur limitation and early sulphur deficiency responses in poplar: significance of gene expression, metabolites, and plant hormones. *J Exp Bot*. 2012; 63(5): 1873–1893.
92 Rennenberg H, Herschbach C, Haberer K, Kopriva S. Sulfur metabolism in plants: are trees different? *Plant Biol*. 2007; 9(5): 620–637.
93 Quilchano C, Haneklaus S, Gallardo JF, Schnug E, Moreno G. Sulphur balance in a broadleaf, non-polluted, forest ecosystem (central-western Spain). *For Ecol Manage*. 2002; 161(1–3): 205–214.
94 Brockley RP. Effects of different sources and rates of sulphur on the growth and foliar nutrition of nitrogen-fertilized lodgepole pine. *Can J For Res*. 2004; 34(3): 728–743.
95 Lambert MJ, Turner J. Dieback in high site quality *Pinus radiata* stands – the role of sulphur and boron deficiencies. *N Z J For Sci*. 1977; 7(3): 333–348.
96 Guo W, Nazim H, Liang Z, Yang D. Magnesium deficiency in plants: an urgent problem. *Crop J*. 2016; 4(2): 83–91.
97 Dünisch O, Bauch J. Influence of mineral elements on wood formation of old growth spruce (*Picea abies* [L.] karst.). *Holzforschung*. 1994; 48(s1): 5–14.
98 Tsuyama T, Kawai R, Shitan N, Matoh T, Sugiyama J, Yoshinaga A, et al. Proton-dependent coniferin transport, a common major transport event in differentiating xylem tissue of woody plants. *Plant Physiol*. 2013; 162(2): 918–926.
99 Arthur MA, Siccama TG, Yanai RD. Calcium and magnesium in wood of northern hardwood forest species: relations to site characteristics. *Can J For Res*. 1999; 29(3): 339–346.
100 Whittaker RH, Likens GE, Bormann FH, Easton JS, Siccama TG. The Hubbard Brook ecosystem study: forest nutrient cycling and element behavior. *Ecology*. 1979; 60(1): 203–220.
101 Österås AH, Greger M. Interactions between calcium and copper or cadmium in Norway spruce. *Biol Plant*. 2006; 50(4): 647–652.
102 Arvidsson H, Lundkvist H. Effects of crushed wood ash on soil chemistry in young Norway spruce stands. *For Ecol Manage*. 2003; 176(1–3): 121–132.

103 Mandre M, Pärn H, Ots K. Short-term effects of wood ash on the soil and the lignin concentration and growth of *Pinus sylvestris* L. *For Ecol Manage.* 2006; 223(1–3): 349–357.

104 Masullo LS, de Vicente Ferraz A, de Moraes Gonçalves JL, LEA C, de Ávila PA, Drago JR, et al. Relationship between forest residue management and micronutrient fertilization with eucalyptus rust severity in *Eucalyptus grandis* plantations. *For Ecol Manage.* 2020; 475: 118443.

105 de Medeiros PL, Pimenta AS, da Silva GGC, de Oliveira EMM, da Silva Júnior DN, de Souza GLF. Efficiency of micronutrients and sodium use of a *Eucalyptus* clone as a function of planting density in short-rotation cropping. *Aust For.* 2021; 84(2): 73–81.

106 Węgiel A, Węgiel A, Polowy K, Bielinis E. The stock and content of micronutrients in above-ground biomass of Scots pine stands of different densities. *J Elem.* 2019; 24: 615–628.

107 Huber D, Römheld V, Weinmann M. Relationship between nutrition, plant diseases and pests. In: *Marschner's Mineral Nutrition of Higher Plants.* Third Edition. San Diego (CA): Academic Press; 2012. p. 283–298.

108 Dell B, Malajczuk N, Xu D, Grove TS. *Nutrient Disorders in Plantation Eucalypts.* Australian Centre for International Agricultural Research; 2001. p. 188.

109 Konsaeng S, Sritharathikhun N, Lordkaew S, Dell B, Rerkasem B. Genotypic variation in response to low boron in eucalypt clones. *South For A J For Sci.* 2012; 74(3): 159–166.

110 Shorrocks VM. The occurrence and correction of boron deficiency. *Plant Soil.* 1997; 193(2): 121–148.

111 Lehto T, Ruuhola T, Dell B. Boron in forest trees and forest ecosystems. *For Ecol Manage.* 2010; 260(12): 2053–2069.

112 Turner J, Knott J, Green P, Turner S. Boron nutritional management in Australian forest plantations. *Trees Forests People.* 2021; 5: 100120.

113 Brown PH, Bellaloui N, Wimmer MA, Bassil ES, Ruiz J, Hu H, et al. Boron in plant biology. *Plant Biol.* 2002; 4(2): 205–223.

114 Kutschera U, Niklas KJ. Boron and the evolutionary development of roots. *Plant Signal Behav.* 2017; 12(7): e1320631.

115 Slack CR, Whittington WJ. The role of boron in plant growth. *J Exp Bot.* 1964; 15(3): 495–514.

116 Dutta TR, McIlrath WJ. Effects of boron on growth and lignification in sunflower tissue and organ cultures. *Bot Gaz.* 1964; 125(2): 89–96.

117 Yin Q, Kang L, Liu Y, Qaseem MF, Qin W, Liu T, et al. Boron deficiency disorders the cell wall in *Neolamarckia cadamba. Ind Crops Prod.* 2022; 176: 114332.

118 Cody RB, Laramée JA, Durst HD. Versatile new ion source for the analysis of materials in open air under ambient conditions. *Anal Chem.* 2005; 77(8): 2297–2302.

119 Deklerck V, Mortier T, Goeders N, Cody RB, Waegeman W, Espinoza E, et al. A protocol for automated timber species identification using metabolome profiling. *Wood Sci Technol.* 2019; 53(4): 953–965.
120 Robinson AR, Ukrainetz NK, Kang K, Mansfield SD. Metabolite profiling of Douglas-fir (*Pseudotsuga menziesii*) field trials reveals strong environmental and weak genetic variation. *New Phytol.* 2007; 174(4): 762–773.
121 Mellerowicz E, Sundberg B. Wood cell walls: biosynthesis, developmental dynamics and their implications for wood properties. *Curr Opin Plant Biol.* 2008; 11(3): 293–300.
122 Aguayo MG, Quintupill L, Castillo R, Baeza J, Freer J, Mendonça RT. Determination of differences in anatomical and chemical characteristics of tension and opposite wood of 8-year old *Eucalyptus globulus. Maderas Ciencia y Tecnología.* 2010; 12(3): 41–251.
123 Felten J, Sundberg B. Biology, chemistry and structure of tension wood. In: Fromm J, editor. *Cellular Aspects of Wood Formation Plant Cell Monographs.* Berlin Heidelberg: Springer; 2013. p. 203–224.
124 Andersson-Gunnerås S, Mellerowicz EJ, Love J, Segerman B, Ohmiya Y, Coutinho PM, et al. Biosynthesis of cellulose-enriched tension wood in *Populus*: global analysis of transcripts and metabolites identifies biochemical and developmental regulators in secondary wall biosynthesis. *Plant J.* 2006; 45(2): 144–165.
125 Yang X, Yu X, Liu Y, Shi Z, Li L, Xie S, et al. Comparative metabolomics analysis reveals the color variation between heartwood and sapwood of Chinese fir *Cunninghamia lanceolata* (Lamb.) Hook. *Ind Crops Prod.* 2021; 169: 113656.
126 Druart N, Johansson A, Baba K, Schrader J, Sjödin A, Bhalerao RR, et al. Environmental and hormonal regulation of the activity-dormancy cycle in the cambial meristem involves stage-specific modulation of transcriptional and metabolic networks. *Plant J.* 2007; 50(4): 557–573.
127 Budzinski IGF, Moon DH, Morosini JS, Lindén P, Bragatto J, Moritz T, et al. Integrated analysis of gene expression from carbon metabolism, proteome and metabolome, reveals altered primary metabolism in *Eucalyptus grandis* bark, in response to seasonal variation. *BMC Plant Biol.* 2016; 16(1): 149.
128 Budzinski IGF, Moon DH, Lindén P, Moritz T, Labate CA. Seasonal variation of carbon metabolism in the cambial zone of *Eucalyptus grandis. Front Plant Sci.* 2016; 7: 932.
129 Paiva JAP, Garnier-Géré PH, Rodrigues JC, Alves A, Santos S, Graça J, et al. Plasticity of maritime pine (*Pinus pinaster*) wood-forming tissues during a growing season. *New Phytol.* 2008; 179(4): 1180–1194.

130 Pereira L, Domingues-Junior AP, Jansen S, Choat B, Mazzafera P. Is embolism resistance in plant xylem associated with quantity and characteristics of lignin? *Trees.* 2018; 32(2): 349–358.

131 Pereira L, Bittencourt PRL, Oliveira RS, Junior MBM, Barros FV, Ribeiro RV, et al. Plant pneumatics: stem air flow is related to embolism - new perspectives on methods in plant hydraulics. *New Phytol.* 2016; 211(1): 357–370.

132 Netzer F, Mueller CW, Scheerer U, Grüner J, Kögel-Knabner I, Herschbach C, et al. Phosphorus nutrition of *Populus* × *canescens* reflects adaptation to high P-availability in the soil. *Tree Physiol.* 2018; 38(1): 6–24.

133 Watanabe M, Netzer F, Tohge T, Orf I, Brotman Y, Dubbert D, et al. Metabolome and lipidome profiles of *Populus* × *canescens* twig tissues during annual growth show phospholipid-linked storage and mobilization of C, N, and S. *Front Plant Sci.* 2018; 9: 1292.

134 Wang J, Tian Y, Li J, Yang K, Xing S, Han X, et al. Transcriptome sequencing of active buds from *Populus deltoides* CL. and *Populus* × *zhaiguanheibaiyang* reveals phytohormones involved in branching. *Genomics.* 2019; 111(4): 700–709.

135 Rocha MFV, Costa LR, Costa LJ, de ACC A, Soares BCD, Hein PRG. Wood knots influence the modulus of elasticity and resistance to compression. *Floresta e Ambiente.* 2018; 25(4): e20170906.

136 Yang H, Xu F, Liao H, Pan W, Zhang W, Xu B, et al. Transcriptome and metabolite analysis related to branch development in two genotypes of *Eucalyptus urophylla. Mol Genet Genom.* 2021; 296(5): 1071–1083.

137 Song H, Cai Z, Liao J, Zhang S. Phosphoproteomic and metabolomic analyses reveal sexually differential regulatory mechanisms in poplar to nitrogen deficiency. *J Proteome Res.* 2020; 19(3): 1073–1084.

138 Zhang S, Tang D, Korpelainen H, Li C. Metabolic and physiological analyses reveal that *Populus cathayana* males adopt an energy-saving strategy to cope with phosphorus deficiency. *Tree Physiol.* 2019; 39(9): 1630–1645.

139 Guerra FP, Richards JH, Fiehn O, Famula R, Stanton BJ, Shuren R, et al. Analysis of the genetic variation in growth, ecophysiology, and chemical and metabolomic composition of wood of *Populus trichocarpa* provenances. *Tree Genet Genomes.* 2016; 12(1): 6.

140 Chen M, Yin Y, Zhang L, Yang X, Fu T, Huo X, et al. Metabolomics and transcriptomics integration of early response of *Populus tomentosa* to reduced nitrogen availability. *Front Plant Sci.* 2021; 12: 769748.

141 Ji H, Ossipov V, Du B, Wen J, Liu C. Differences in the relationship between metabolomic and ionomic traits of *Quercus variabilis* growing at contrasting geologic-phosphorus sites in subtropics. *Plant Soil.* 2019; 439(1, 2): 339–355.

142 Forde BG, Lea PJ. Glutamate in plants: metabolism, regulation, and signalling. *J Exp Bot.* 2007; 58(9): 2339–2358.

143 Polacco JC, Mazzafera P, Tezotto T. Opinion – Nickel and urease in plants: still many knowledge gaps. *Plant Sci.* 2013; 199, 200: 79–90.

144 Uhde-Stone C, Zinn KE, Ramirez-Yáñez M, Li A, Vance CP, Allan DL. Nylon filter arrays reveal differential gene expression in proteoid roots of white lupin in response to phosphorus deficiency. *Plant Physiol.* 2003; 131(3): 1064–1079.

145 Luo X, Li Z, Xiao S, Ye Z, Nie X, Zhang X, et al. Phosphate deficiency enhances cotton resistance to *Verticillium dahliae* through activating jasmonic acid biosynthesis and phenylpropanoid pathway. *Plant Sci.* 2021; 302: 110724.

146 Ployet R, Veneziano Labate MT, Regiani Cataldi T, Christina M, Morel M, San Clemente H, et al. A systems biology view of wood formation in *Eucalyptus grandis* trees submitted to different potassium and water regimes. *New Phytol.* 2019; 223(2): 766–782.

147 Szuba A, Marczak Ł, Ratajczak I. Metabolome adjustments in ectomycorrhizal *Populus × canescens* associated with strong promotion of plant growth by *Paxillus involutus* despite a very low root colonization rate. *Tree Physiol.* 2020; 40(12): 1726–1243.

148 Sivaprakasam Padmanaban PB, Rosenkranz M, Zhu P, Kaling M, Schmidt A, Schmitt-Kopplin P, et al. Mycorrhiza-tree-herbivore interactions: alterations in poplar metabolome and volatilome. *Metabolites.* 2022; 12(2): 93.

149 Baslam M, Mitsui T, Sueyoshi K, Ohyama T. Recent advances in carbon and nitrogen metabolism in C3 plants. *Int J Mol Sci.* 2020; 22(1): 318.

150 Foyer CH, Ferrario-Méry S, Noctor G. Interactions between carbon and nitrogen metabolism. In: *Plant Nitrogen.* Berlin, Heidelberg: Springer; 2001. p. 237–254.

151 Déjardin A, Laurans F, Arnaud D, Breton C, Pilate G, Leplé JC. Wood formation in Angiosperms. *CR Biol.* 2010; 333(4): 325–334.

152 Langer K, Ache P, Geiger D, Stinzing A, Arend M, Wind C, et al. Poplar potassium transporters capable of controlling K^+ homeostasis and K^+ dependent xylogenesis. *Plant J.* 2002; 32(6): 997–1009.

153 Wind C, Arend M, Fromm J. Potassium-dependent cambial growth in poplar. *Plant Biol.* 2004; 6(1): 30–37.

154 Uggla C, Moritz T, Sandberg G, Sundberg B. Auxin as a positional signal in pattern formation in plants. *Proc Natl Acad Sci USA.* 1996; 93(17): 9282–9286.

155 Uggla C, Mellerowicz EJ, Sundberg B. Indole-3-acetic acid controls cambial growth in Scots pine by positional signaling. *Plant Physiol.* 1998; 117(1): 113–121.

156 de Souza MN, de Oliveira Ferreira EV, Junior JCA, Domec JC, Jordan-Meille L, de Moraes Gonçalves JL, et al. The ideal percentage of K substitution by Na in *Eucalyptus* seedlings: evidences from leaf carbon isotopic composition, leaf gas exchanges and plant growth. *Plant Physiol Biochem.* 2019; 137: 102–112.

157 Battie-Laclau P, Laclau JP, Domec JC, Christina M, Bouillet JP, de Cassia PM, et al. Effects of potassium and sodium supply on drought-adaptive mechanisms in *Eucalyptus grandis* plantations. *New Phytol.* 2014; 203(2): 401–413.

158 Almeida JCR, Laclau JP, de Moraes Gonçalves JL, Ranger J, Saint-André L. A positive growth response to NaCl applications in *Eucalyptus* plantations established on K-deficient soils. *For Ecol Manage.* 2010; 259(9): 1786–1795.

10

The Impact of Drought on Plant Metabolism in *Quercus* Species – From Initial Response to Recovery

Juan Sobrino-Plata[1,2], *Francisco Javier Cano*[3,4], *Ismael Aranda*[3], *María Brígida Fernández de Simón*[3], *and Jesús Rodríguez-Calcerrada*[1]

[1] *Research Group Functioning of Forest Systems in a Changing Environment, Department of Natural Systems and Resources, Universidad Politécnica de Madrid, Madrid, Spain*
[2] *Department of Genetics, Physiology and Microbiology, Universidad Complutense de Madrid, Madrid, Spain*
[3] *Instituto de Ciencias Forestales (ICIFOR-INIA), CSIC, Madrid, Spain*
[4] *ARC Centre of Excellence for Translational Photosynthesis, Hawkesbury Institute for the Environment, Western Sydney University, Penrith, NSW, Australia*

10.1 Introduction

There are about 400 oak species (*Quercus* spp.) in Eurasia, the Americas and northern Africa (1, 2). Oaks comprise a group of shrub and mostly tree species with a huge variety of functional strategies inhabiting sites of semiarid, Mediterranean, temperate, subtropical, or subalpine climates, where they form forests of high cultural, ecological, and commercial value (3). However, despite the dominance of oak in many pure and mixed forests, oak regeneration, especially sexual regeneration, is often problematic and limits recruitment into the overstory (4, 5). Land use and climate changes affecting wildfire regimes, coppicing frequency, stand density, herbivory, interspecific competition, and understory irradiance will impact oak recruitment, expanding or contracting their distribution areas (6–8). Although oaks are relatively resistant to drought (9, 10), in some regions, oaks are suffering higher mortality rates than co-occurring tree species that exhibit lower leaf photosynthetic rates and growth, but also lower xylem vulnerability to cavitation and higher hydraulic safety margins (11). The increasing number of reports of oak forest decline worldwide (12) demands action from scientists and managers to ensure oak survival in a drier and hotter world.

Ecometabolomics is emerging as a powerful tool to understand biological responses to stress and assess plant and ecosystem mortality risks. From pioneering biochemical studies on specific metabolites to recent metabolome

Monitoring Forest Damage with Mass Spectrometry-Based Metabolomics Methods, First Edition.
Edited by Carla António.

studies, the analysis of plant metabolism has proved useful in comparing stress resistance on different genetic backgrounds (13–15), especially when combined with measurements scaling from genes to the whole plant (16). However, there is little research in oak metabolomics compared to other tree genera, such as *Pinus*, *Populus*, or *Eucalyptus* (17–20) or to herbaceous genera used in agronomy (e.g. *Oryza* or *Solanum* (21)). The analysis of *Quercus* spp. metabolome within the same study is relatively recent (22–25). Few studies have assessed in detail the expression of compounds other than those related to carbohydrates (26–28) or nitrogen (N) compounds grouped in metabolic classes (29). Moreover, only a few oak species have been amply studied, mostly *Q. suber*, *Q. ilex*, and *Q. pubescens* in the Mediterranean, and *Q. robur* and *Q. petraea* in temperate regions. All these species are European, being notable the little research in American and Asian oaks so far (Table 10.1).

Differences in drought tolerance among oak species are related to their ability to maintain water transport at low water potentials and the metabolic regulation of osmotic adjustment, energy consumption, antioxidant capacity, and/or protein and DNA repair (9, 33, 87, 94, 135, 136). For example, shifts in primary metabolite levels modulate cell turgor and leaf gas exchange rates at a given water potential across species, shifts in pigment levels and antioxidants modulate leaf damage at a given turgor loss, while shifts in amino acid biosynthetic pathways modulate the synthesis and repair of specific proteins during drought stress. At a comparable physiological stress level, quantified by stomatal conductance, intercellular CO_2 level, or more generally leaf water potential across studies (and often categorized as light, moderate, and severe drought stress), the changes in broad metabolite groups are remarkably similar across species.

Drought jeopardizes plant survival at different time scales. In the longer run, undetectable drought legacy effects can increase plant susceptibility to further drought events or other stresses. Increased levels of N and secondary metabolites induced by moderate drought can make oak leaves more attractive but also less nutritious to insects (22, 137), which may need to eat more to satisfy their energy needs. The situation can worsen if prolonged, severe photosynthetic limitations reduce the levels of costly defensive metabolites. The increased susceptibility to disease in drought-weakened trees (50, 130) can contribute to kill adult trees after drought has ended. Drought effects on metabolite pools can have longer, less-predictable consequences on oak forest by modulating litter flammability (82), litter nutritional value for soil heterotrophic microorganisms (119), root exudate composition (57), or acorn viability. Drought stress in adults affects acorn size and viability (138), but can also affect acorn shedding time and chemical composition (139), which might in turn alter acorn predation and germination rates, or even seedling size (140). The study of drought at different time scales is important to clarify these questions. However, research on long-term drought acclimation is limited to a few species, across years or sites of different edaphic conditions, and mostly

Table 10.1 Studies of metabolite contents in oak species in response to natural or experimental drought using classic biochemical methodologies or metabolomics, with indication of studied species and main topic(s) addressed.

Reference	Species	Meta-bolomics	Intensity	Organs	Woody organs	NSC	Taxa	2^{ary} metab.	Transport	Recovery	Long acclimation
(24)	*Q. suber*	**Yes**	No	No	No	No	No	**Yes**	No	**Yes**	No
(30)	*Q. ilex*	**Yes**	No	**Yes**	**Yes (root, wood)**	No	No	**Yes**	No	No	**Yes (sites)**
(31)	*Q. petraea*	**Yes**	No	No	No	No	No	Yes	No	No	**Yes (provenances)**
(25)	*Q. pyrenaica*	**Yes**	No	No	No	**Yes**	No	**Yes**	No	**Yes**	No
(32)	*Q. ilex* *Q. faginea* *Q. pyrenaica* *Q. petraea*	**Yes**	No	No	No	No	**Yes**	No	No	No	No
(33)	*Q. coccifera* *Q. ilex*	No	**Yes**	No	No	No	**Yes**	**Yes**	No	No	No
(34)	*Q. ilex*	No	**Yes**	No	No	No	No	**Yes**	No	**Yes**	**Yes**
(35)	*Q. robur*	No	**Yes**	No	No	**Yes**	No	No	No	No	**Yes (sites)**
(36)	*Q. ilex*	**Yes**	**Yes**	No	No	No	No	**Yes**	No	No	No
(37)	*Q. robur*	**Yes**	No	No	No	No	No	**Yes**	No	No	No
(38)	*Q. ilex*	No	No	No	No	No	No	**Yes**	No	No	No
(39)	*Q. pubescens* *Q. robur*	No	**Yes**	No	No	No	**Yes**	**Yes**	No	**Yes**	No
(40)	*Q. robur*	No	No	**Yes**	**Yes (root)**	No	No	No	No	No	No
(41)	*Q. ilex*	No	No	No	**Yes (stem)**	**Yes (only starch)**	No	**Yes**	No	**Yes**	**Yes (populations)**

(Continued)

Table 10.1 (Continued)

Reference	Species	Metabolomics	Intensity	Organs	Woody organs	NSC	Taxa	2^{ary} metab.	Transport	Recovery	Long acclimation
(26)	*Q. ilex*	No	No	No	No	**Yes**	No	No	No	No	No
(42)	*Q. pubescens*	No	No	No	No	No	No	**Yes**	No	No	No
(43)	*Q. cerris*	No	No	No	No	No	No	**Yes**	No	No	No
(44)	*Q. ilex*	No	No	No	No	No	No	**Yes**	No	No	No
(45)	*Q. suber*	No	No	No	No	**Yes**	**No**	**Yes**	No	No	No
(46)	*Q. suber*	No	**Yes**	No	No	**Yes**	**Yes (populations)**	**Yes**	No	No	No
(47)	*Q. cerris* *Q. robur*	No	**Yes**	No	No	**Yes (only SS)**	**Yes**	No	No	**Yes**	No
(48)	*Q. suber*	No	No	No	**Yes (root)**	**Yes**	No	**Yes**	No	No	No
(49)	*Quercus* spp.	No	**Yes**	**Yes**	**Yes (root)**	**Yes**	**Yes**	No	**Yes**	**Yes**	**Yes**
(50)	*Quercus* spp.	No	No	No	**Yes (phloem)**	**Yes**	**Yes**	**Yes**	**Yes**	No	No
(27)	*Q. petraea*	No	No	No	No	**Yes**	No	No	No	No	No
(51)	*Quercus* spp.	**Yes**	**Yes**	**Yes**	**Yes (root)**	No	**Yes**	**Yes**	No	**Yes**	**Yes**
(52)	*Q. robur*	No	**Yes**	**Yes**	**Yes (root)**	No	No	**Yes (only ABA)**	**Yes**	No	No
(53)	*Q. ilex*	No	No	No	**Yes (lignotuber)**	**Yes**	No	No	No	**Yes**	**Yes**
(54)	*Q. pubescens*	No	**Yes**	No	No	**Yes (only SS)**	No	**Yes**	No	**Yes**	No
(55)	*Q. ilex*	No	No	No	No	**Yes (only SS)**	No	**Yes**	No	**Yes**	No

(56)	*Q. suber*	No	**Yes**	No	No	No	No	**Yes**	No	**Yes**	No
(57)	*Q. ilex*	**Yes**	**Yes**	**Yes**	No	No	No	**Yes**	No	**Yes**	No
(28)	*Q. prinus*	**Yes**	**Yes**	No	No	**Yes**	No	No	No	No	**Yes**
(58)	*Q. pubescens*	No	**Yes**	No	**Yes (branch)**	No	No	**Yes**	No	**Yes**	**Yes**
(59)	*Q. infectoria* *Q. libani*	No	No	No	No	No	**Yes**	**Yes**	No	No	No
(60)	*Q. suber*	**Yes**	**Yes**	No	No	No	No	**Yes**	No	No	No
(61)	*Q. pubescens*	No	No	No	No	**Yes**	No	No	No	No	No
(62)	*Q. robur* *Q. petraea* *Q. pubescens*	No	No	No	No	No	**Yes**	**Yes**	No	**Yes**	**Yes**
(63)	*Q. robur* *Q. petraea* *Q. pubescens*	No	No	No	No	No	**Yes**	No	No	**Yes**	No
(64)	*Q. brantii*	No	**Yes**	No	No	**Yes (only SS)**	**Yes (populations)**	**Yes**	No	No	No
(65)	*Q. ilex*	No	No	No	No	No	No	No	No	No	No
(66)	*Q. alba*	No	No	**Yes**	**Yes (stem, root)**	**Yes**	No	No	No	No	No
(67)	*Q. robur* *Q. cerris*	No	No	No	No	No	**Yes**	**Yes**	No	No	No
(68)	*Q. petraea*	No	No	No	No	No	No	**Yes**	No	No	No
(69)	*Q. suber*	No	**Yes**	No	No	No	No	**Yes**	No	No	No
(70)	*Q. ilex* *Q. faginea*	No	**Yes**	No	No	No	**Yes**	No	No	**Yes**	No

(Continued)

Table 10.1 (Continued)

Reference	Species	Metabolomics	Intensity	Organs	Woody organs	NSC	Taxa	2^{ary} metab.	Transport	Recovery	Long acclimation
(71)	*Q. ilex*	No	**Yes**	No	No	No	No	**Yes**	No	No	**Yes**
(72)	*Q. petraea* *Q. robur*	No	No	No	No	**Yes**	**Yes**	No	No	No	**Yes (provenances)**
(73)	*Q. velutina* *Q. alba*	No	No	No	No	No	**Yes**	**Yes (only ABA)**	**Yes**	**Yes**	No
(74)	*Q. rubra* *Q. velutina*	No	No	No	No	**Yes**	No	No	No	No	No
(75)	*Q. coccifera*	No	**Yes**	No	No	**Yes**	No	**Yes**	No	**Yes**	No
(76)	*Q. garryana*	No	**Yes**	No	No	No	No	No	No	No	**Yes (provenances)**
(77)	*Q. pubescens*	**Yes**	No	No	No	No	No	No	No	No	**Yes**
(78)	*Q. ilex*	No	**Yes**	No	No	No	No	**Yes**	No	**Yes**	**Yes**
(79)	*Q. ilex*	No	No	No	No	No	No	**Yes**	No	No	No
(80)	*Q. ilex*	No	No	No	No	No	No	**Yes**	No	No	**Yes**
(81)	*Q. coccifera*	No	**Yes**	No	No	No	No	**Yes**	No	No	No
(82)	*Q. coccifera*	No	No	No	No	No	No	**Yes**	No	No	**Yes**
(83)	*Q. robur*	No	**Yes**	No	No	No	No	No	No	No	No
(84)	*Q. velutina*	No	No	No	**Yes (root)**	**Yes**	No	**Yes**	No	No	No
(85)	*Q. suber*	**Yes**	**Yes**	No	No	No	No	No	No	**Yes**	No
(86)	*Q. virginiana*	No	**Yes**	No	No	No	No	**Yes**	No	**Yes**	No
(87)	*Q. coccifera* *Q. ilex* *Q. suber*	No	**Yes**	No	No	No	**Yes**	**Yes**	No	No	No

(88)	*Q. ilex*	No	**Yes**	No	No	No	No	**Yes**	No	**Yes**	No
(89)	*Q. ilex*	No	No	No	No	**Yes**	No	**Yes**	No	No	No
(90)	*Q. robur*	**Yes**	No	No	No	No	No	**Yes**	No	No	No
(91)	*Q. robur*	No	**Yes**	No	No	**Yes**	No	No	No	**Yes**	No
(92)	*Q. ilex*	No	**Yes**	No	No	No	No	No	No	**Yes**	No
(93)	*Q. ilex*	No	No	No	No	No	No	No	No	No	No
(94)	*Q. virginiana* *Q. geminata* *Q. fusiformis* *Q. oleoides*	No	No	No	No	No	**Yes**	**Yes**	No	No	**Yes**
(95)	*Q. coccifera*	No	**Yes**	No	No	No	No	**Yes**	No	**Yes**	No
(22)	*Q. ilex*	**Yes**	No	No	No	No	No	**Yes**	No	No	**Yes**
(96)	*Q. ilex*	**Yes**	No	No	No	No	No	**Yes**	No	No	**Yes**
(97)	*Q. petraea*	No	**Yes**	No	No	No	No	No	No	No	No
(98)	*Q. ilex*	No	No	No	No	No	No	No	No	No	**Yes**
(99)	*Q. ilex*	No	**Yes**	No	No	**Yes**	**Yes**	No	No	**Yes**	No
(100)	*Q. pubescens*	No	**Yes**	No	No	**Yes**	No	**Yes**	No	No	No
(101)	*Q. ilex*	No	**Yes**	No	**Yes (stem)**	**Yes**	No	No	No	No	**Yes**
(102)	*Q. ilex*	**Yes**	**Yes**	No	No	No	No	No	No	No	No
(103)	*Q. ilex*	**Yes**	**Yes**	No	**Yes (stem)**	No	No	No	No	No	No
(104)	*Quercus spp.*	No	**Yes**	**Yes**	No	No	**Yes**	**Yes**	No	No	No
(105)	*Q. ilex*	No	No	No	No	**Yes (only SS)**	**Yes (populations)**	**Yes**	No	No	No

(Continued)

Table 10.1 (Continued)

Reference	Species	Metabolomics	Intensity	Organs	Woody organs	NSC	Taxa	2^{ary} metab.	Transport	Recovery	Long acclimation
(106)	*Q. pubescens*	No	No	No	**Yes (branch)**	No	No	**Yes**	No	No	**Yes**
(107)	*Q. pubescens*	**Yes**	**Yes**	No	No	No	No	**Yes**	No	No	**Yes (inter-annual)**
(108)	*Q. pubescens*	No	No	No	**Yes (stem)**	Yes	No	No	No	**Yes**	No
(109)	*Q. robur*	No	No	**Yes**	**Yes (stem, root)**	No	No	**Yes**	**Yes**	No	No
(110)	*Q. robur*	No	**Yes**	No	No	No	No	**Yes**	No	**Yes**	No
(111)	*Q. alba* *Q. prinus*	No	No	No	No	No	**Yes**	**Yes**	No	No	**Yes (inter-annual)**
(112)	*Q. ilex*	No	**Yes**	No	**Yes (root)**	No	No	**Yes**	No	**Yes**	No
(113)	*Q. robur*	No	**Yes**	No	No	**Yes**	No	No	No	**Yes**	**Yes**
(114)	*Q. coccifera* *Q. ilex* *Q. robur*	No	**Yes**	No	No	No	**Yes**	**Yes**	No	No	No
(115)	*Q. ilex*	No	**Yes**	No	No	No	No	**Yes**	No	**Yes**	No
(116)	*Q. suber*	No	**Yes**	No	No	No	**Yes (populations)**	**Yes**	No	**Yes**	No
(117)	*Q. coccifera*	**Yes**	**Yes**	No	No	No	No	**Yes**	No	No	No
(118)	*Q. robur*	No	No	No	No	No	No	**Yes**	No	No	No
(119)	*Q. rubra*	**Yes**	No	No	No	No	No	**Yes**	No	No	**Yes**
(120)	*Q. alba* *Q. rubra*	**Yes**	No	No	**Yes (root)**	No	**Yes**	**Yes**	No	No	No

(121)	*Q. ilex*	**Yes**	No	No	No	No	No	**Yes**	No	No	No
(122)	*Q. virginiana*	No	No	No	No	No	No	**Yes**	No	No	No
(123)	*Q. rubra*	No	No	No	No	No	No	**Yes**	No	No	**Yes**
(124)	*Q. petraea* *Q. robur*	No	No	No	No	No	No	**Yes (only ABA)**	**Yes**	No	**Yes**
(125)	*Q. ilex*	No	No	No	No	**Yes**	No	No	No	**Yes**	No
(126)	*Q. ilex* *Q. sobur*	No	No	No	No	**Yes**	**Yes**	No	No	**Yes**	No
(127)	*Q. coccifera*	No	No	No	**Yes (root)**	**Yes (only starch)**	No	**Yes**	No	No	No
(128)	*Q. coccifera*	No	**Yes**	No	No	No	No	**Yes**	No	No	No
(129)	*Q. ilex*	No	No	**Yes**	**Yes (root)**	**Yes**	No	No	No	No	No
(130)	*Quercus spp.*	No	**Yes**	**Yes**	**Yes (root)**	**Yes**	**Yes**	**Yes**	No	No	No
(131)	*Q. cerris* *Q. frainetto*	No	No	No	No	No	**Yes**	**Yes**	No	**Yes**	No
(132)	*Q. variabilis*	No	**Yes**	No	No	**Yes (only SS)**	No	**Yes**	No	No	No
(133)	*Q. serrata* *Q. fabri* *Q. variabilis* *Q. acutissima*	No	**Yes**	No	No	**Yes (only SS)**	**Yes**	No	No	**Yes**	No
(134)	*Q. rubra*	No	No	No	No	**Yes (only SS)**	No	No	No	No	**Yes**

in Mediterranean *Q. ilex* and *Q. pubescens* stands where rainfall exclusion systems have been set up (Table 10.1).

In this chapter, we address drought effects on plant metabolism and on levels of primary and secondary metabolites in vegetative tissues of oak species. We focus on soil drought resulting from natural or experimental restrictions in soil water available to plants, acknowledging that atmospheric drought (i.e. induced by high air evaporative demand) can interact with or cause different metabolic alterations than soil drought. We focus on short-term droughts, given little literature at long-time scales. Finally, we ignore the interaction of drought with other potential stress factors, such as heat (42, 119), air CO_2 concentration (25, 110), low mineral nutrient availability (36, 42, 62), ozone fumigation (43), or pathogen inoculation (59).

10.2 Primary Metabolic Pathways and Metabolite Levels

10.2.1 Changes in Leaf Metabolism with Drought Stress Intensity

Different metabolic alterations take place as drought persists and plant stress increases. Photosynthetic rates decrease because of lower CO_2 concentration in the chloroplasts and impaired biochemical and photochemical capacity at severe stress levels (49, 55). It is less clear how glycolysis, the tricarboxylic acid (TCA) cycle, and the pentose phosphate pathway (PPP) change with drought stress intensity. Echevarría-Zomeño et al. (141) discussed that the expression of glycolytic enzymes might increase at moderate stress and decrease as stress intensified. The decrease in leaf CO_2 efflux rates observed with increasing drought stress suggests that pyruvate decarboxylation and TCA activity decrease with drought. The decline in respiration can precede that in photosynthesis (66) or lag behind it (99, 102, 142). In any case, non-negligible respiration rates (in leaves and non-photosynthetic organs) when net CO_2 assimilation nears zero can eventually reduce the carbon (C) substrates feeding primary and secondary metabolic routes. We know little about the response of PPP to drought intensity in oak (24). The NADPH formed by glucose-6-phosphate oxidation would add to the pool of NADPH coming from the chloroplast that is not used in CO_2 assimilation, and thus to a growing redox imbalance; however, NADPH and sugar phosphates from PPP can also help in the scavenging of reactive oxygen species (ROS) by promoting the synthesis of non-enzymatic antioxidants (143, 144).

Associated with these metabolic changes, leaf primary metabolite pools (e.g. sugars, organic acids, and amino acids) change (Figure 10.1). Sugars and amino acids often increase linearly or exponentially with drought stress, whereas there is hardly a generalizable pattern for organic acids. Within each broad

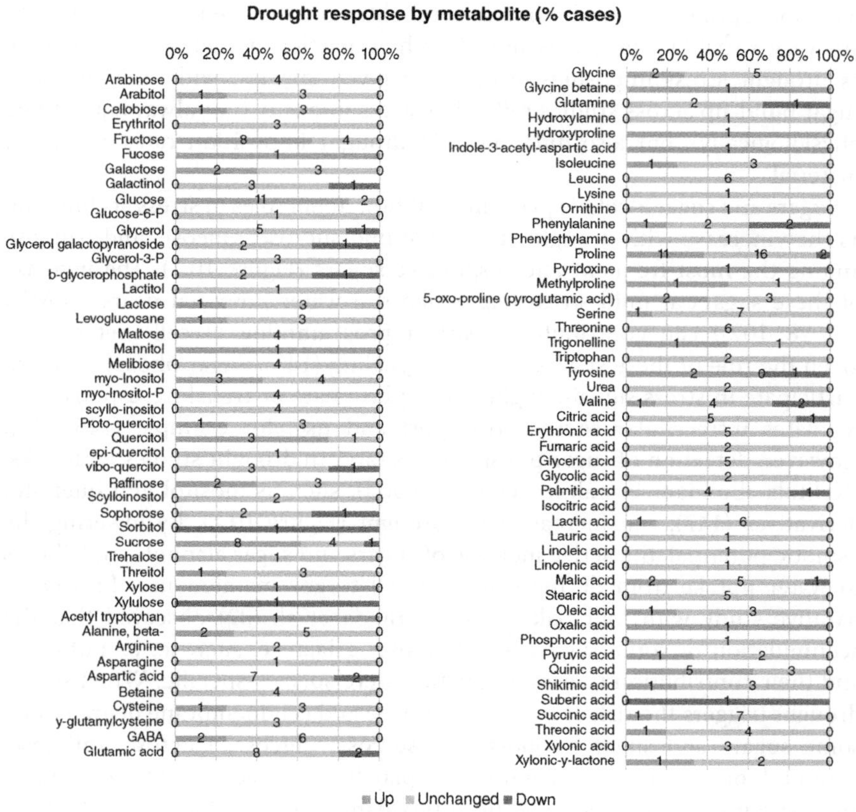

Figure 10.1 Qualitative changes in leaf primary metabolites belonging to sugars, organic acids, and amino acids and derivatives in oak species in response to short-term moderate drought (i.e. approximately −1 to −3 MPa leaf predawn water potential, 80–70% leaf-relative water content, 0.15–0.1 mol m^{-2} s^{-1} stomatal conductance, 50–40% soil water content relative to maximum value, and/or 50% of optimal watering). Results are computed as the percentage of cases with statistically significant "increases" (orange bars) or "decreases" (blue bars) of any specific metabolite, or no significant change ("no change"; white bars), relative to all cases reporting the metabolite. Cases are species studied in one or more studies. Only drought treatments were compared when the study included treatments other than drought. Further studies that evaluate changes in metabolite levels at more severe drought stress can be found in: Almeida et al. (24), Aranda et al. (25, 32), Bessonova and Chonhova (35), Cotrozzi et al. (43), Daoudi et al. (46), Deligöz and Bayar (47), Épron and Dreyer (27), Gebre and Tschaplinski (28), Ghanbary et al. (59), Hu et al. (62, 63), Jafarnia et al. (64), Kwak et al. (69), Lansac et al. (70), Mevy et al. (77), Oufir et al. (83), Picon et al. (91), Rhizopoulou et al. (95), Robakowski et al. (97), Rivas-Ubach et al. (22, 96), Rodríguez-Calcerrada et al. (99, 102), Spieß et al. (113), Stojnićet al. (118), Vilagrosa et al. (128), and Wu et al. (132).

metabolite group, the predominant metabolites and their sensitivity to drought are very variable. Thus, it is not clear how particular metabolites increase as drought stress intensifies, even for a given species, with differences in sugar build-up across studies reflecting stress intensity (not always quantified physiologically), but also differences in drought imposition rates and plant material.

Sugars and sugar alcohols play different functional roles in stress acclimation as they increase with drought stress. We find sucrose, glucose, and fructose among the most frequent and responsive sugars (Figure 10.1). The increase of these sugars at mild stress (e.g. when soil water potential nears −1 MPa (25, 28, 102)) can contribute to early reprogramming of cell metabolism, given the role of these sugars in stress signaling (145). Sugar phosphates also participate in stress signaling; glucose-6-phosphate or trehalose-6-phosphate regulates intracellular sucrose transport and use, for example (145, 146). However, detection of sugar phosphates is rarer in drought studies with oaks. Cyclitols abundant in oak leaves (and wood), such as inositol and quercitol derivatives (147), can increase with drought (Figure 10.1). By lowering the osmotic potential, the active increase of sugars and sugar alcohols contributes to water uptake from drying soils and cell turgor maintenance. In a comparative study with four oak species, Aranda et al. (32) observed that the accumulation of some sugars (e.g. maltose, galactose, sucrose, or raffinose) and their contribution to osmoregulation were species-specific. When severe drought triggers the risk of oxidative damage and cell membrane degradation, some sugars and sugar alcohols (e.g. sucrose, glucose, fructose, raffinose, mannitol, or myo-inositol; Figure 10.1) play helpful roles in ROS scavenging and the synthesis of other powerful antioxidants (e.g. ascorbic acid) (148, 149). At this stage, leaf senescence and nutrient remobilization probably underlie some of the metabolic changes attributed to severe drought itself, including the increase in soluble sugars (61), some of which (e.g. glucose, raffinose, or stachyose) can participate in programmed cell death (150).

Organic acids play various roles in drought stress acclimation. They can accumulate as metabolically cheap solutes, detoxify ROS, initiates jasmonic acid signaling (151), or be exuded into the soil to favor symbiotic associations (see below). There is no consistent drought response of the organic acids reported in oak (Figure 10.1). For example, malic acid accumulates in leaves of *Q. ilex* seedlings subjected to moderate drought (102), but not in other studies on seedlings and adults of the same species (22, 32). Quinic acid shows the most consistent response to drought, increasing in a majority of studies (Figure 10.1), all in Mediterranean oaks, probably for leaf photoprotection during dry and hot summers (85). Organic acids produced by glycolysis and TCA cycle are precursors of amino acids, hormones, or phenolic compounds, among many other metabolite groups. Therefore, the accumulation of TCA organic acids in some studies affects (and reflects) the activity of important metabolic pathways involved in stress acclimation (119). Few fatty acids reported in oak leaves (Figure 10.1) do not show a pattern in their response to drought, being lipids mostly present in acorns (152).

Drought-stress alterations in N metabolism often result in an increase in total free amino acid contents (Figure 10.1). On one hand, the biosynthetic routes of some amino acids (e.g. phenylalanine, tyrosine, and tryptophan) can remain very active during moderate drought stress (24), or even be upregulated (119), probably from C skeletons, energy and reducing power no longer used in growth, a process sensitive to turgor loss. The increase is accompanied by unchanged (127) or increased soluble protein levels (63, 133); in other cases, however, drought can inhibit the synthesis and stimulate the degradation of proteins, especially those not involved in the antioxidant defense, which contributes to rising total free amino acid contents (40). Individually, the amino acids tend to be unresponsive to moderate drought, with only tyrosine increasing in a majority of studies (Figure 10.1). The variable drought response of individual amino acids reflects a variable sensitivity of their biosynthetic pathways to ROS and to the availability of precursors, energy and reducing power. Amino acids are potentially compatible solutes; however, the contribution of amino acids to osmoregulation was much lower than that of sugars and organic acids in a comparative study with four oak species (32). The dramatic increase of some amino acids at severe drought stress conditions (e.g. proline, γ-aminobutyric acid [GABA] or glutamine, reaching up to 60–300 higher concentrations compared to well-watered conditions) (32, 63, 102) suggests they play a main role in ROS detoxification and stabilization of cellular membranes when leaf photochemistry is impaired and the probability of photodamage is high (153). This is particularly evident for proline, which in some Mediterranean oaks does not increase until leaf predawn water potential reaches <-8 MPa (102, 128). Finally, shifts in amino acids, amides, and ureides readily transported in the plant (e.g. arginine, aspartate, glutamate, asparagine, glutamine, or allantoin) (154) likely reflect species-specific drought stress effects on sap pH and internal transport that are related with post-drought recovery of growth (see below). The increase with drought of some amino acids with comparatively high N per amino acid mole (e.g. arginine, glutamine, lysine, histidine, tryptophan, or asparagine) (24, 63, 119)) suggests they have a role in N and C storage that could favor growth upon rehydration (24, 141).

10.2.2 Variation Among Organs – Mobilization of Carbohydrates as a Mechanism of Drought Acclimation

Comprehensive studies comparing organs as drought stress increases and after it ends are rare (Table 10.1). Studies on metabolic reactions to different stressors have very often focused on the response of one single organ, mostly leaves (85, 107, 123, 155, 156). However, aerial and underground organs can have different metabolomes. In some forest tree species, there is a differential partitioning of carbohydrate reserves above- and below-ground (14, 157). Moreover, leaves mainly store C metabolites, including carbohydrates, organic and fatty acids, and polyphenols, while N metabolites, such as amino acids, accumulate differentially in roots (Fernández de Simón et al., unpublished results). Many changes in primary and secondary metabolism encompass modifications that are organ-specific (14, 112, 157, 158). Even within the same organ, there are differences between tissues and cell types (50, 159), as observed for the accumulation of phenolic compounds among cortex, endodermis, and stele cells of *Quercus coccifera* roots in response to osmotic stress (117). Moreover, seasonality in environmental factors and phenological states condition a different partitioning of non-structural N and carbohydrate compounds among roots, stems, twigs, and leaves, as observed in Bazot et al. (160) for *Q. petraea*. Another aspect of the differential accumulation of compounds has to do with the differential sensitivity of organs to drought stress, as is the case for *Quercus pyrenaica* (Figure 10.2). The increase in the levels of amino acids in fine roots and of soluble sugars and organic acids in leaves stand out, although there are also metabolites with a similar behavior in both organs. This is the case of protoquercitol, which experiences important increases in *Q. pyrenaica*, both in fine roots and leaves, due to the effect of drought (Fernández de Simón et al., unpublished results).

Large NSC stored in woody organs of some oaks can help them to cope with drought-induced stomatal limitations of photosynthesis (160, 161). The building up of NSC reserves in woody organs requires maintaining an efficient capacity to export NSC from leaves (26). Thus, NSC increased more in coarse roots and stems than in leaves during periods of water scarcity in *Q. alba* (66). The role of NSC in drought survival has been amply discussed (12, 162). Mobilization of carbohydrates to woody organs can enhance drought tolerance by improving physiological processes such as osmoregulation (163), serving as substrates to restore nutrient uptake and growth after periods of acute stress (112, 164), or facilitating xylem refilling when xylem tension ends (161). In line with this, a decline in root starch levels as drought increases can occur as a result of phloem transport arrestment (see below)

Figure 10.2 (a) The use of chromatographic techniques allows the determination of the differential concentration of metabolites in different organs and environmental conditions. In this example, levels of malic acid were evaluated in leaves (L) and roots (R) of plants submitted to drought (D) or well-watered (C). The integration of the different peaks from the chromatogram translated into different concentrations showing a higher drought reactivity of roots. (b) Some metabolites show a differential concentration among organs depending on water availability (left panel); however, drought sensitivity of some metabolites varies among organs (right panels).

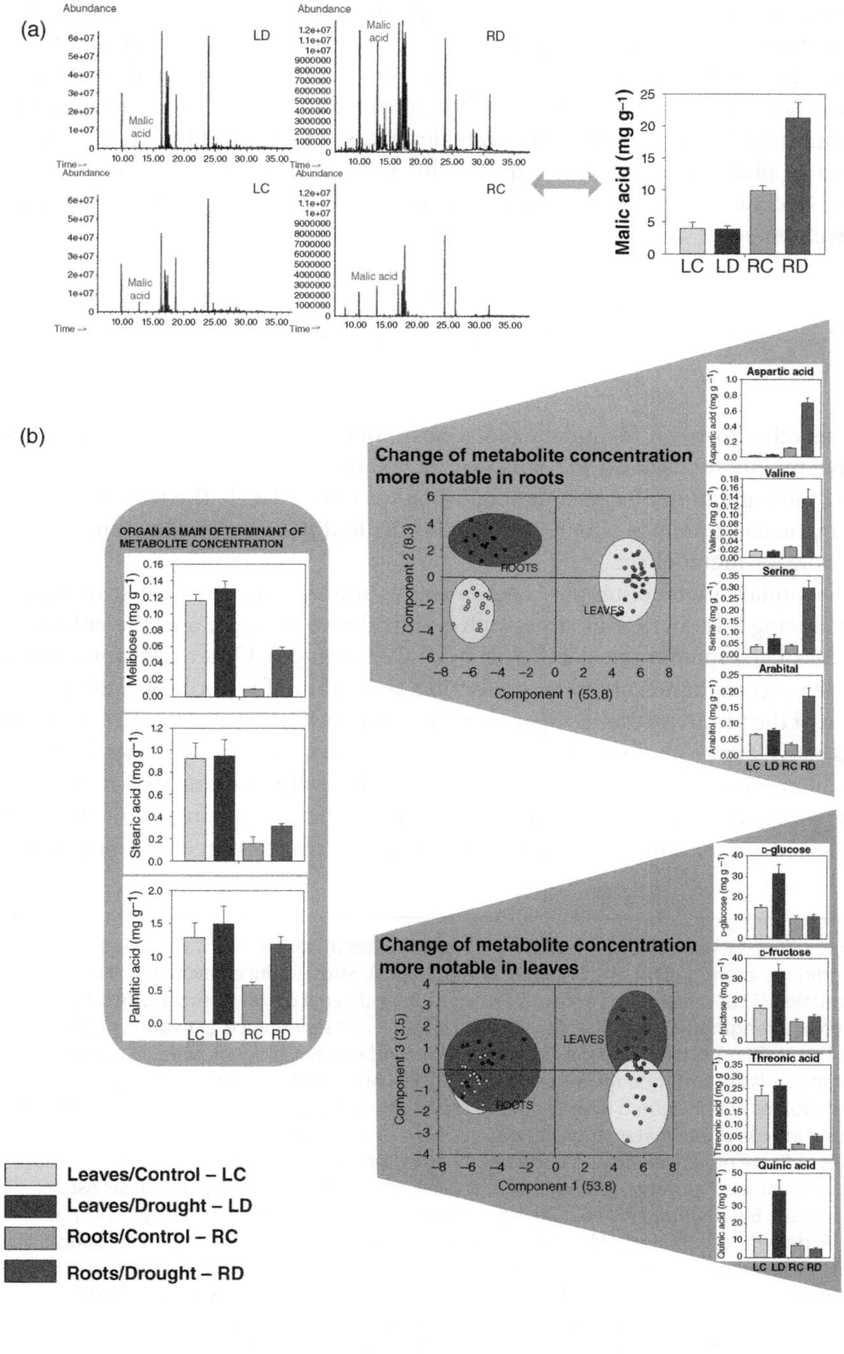
(a)
LD
RD
LC
RC
Malic acid
Malic acid (mg g−1)
LC LD RC RD
(b)
ORGAN AS MAIN DETERMINANT OF METABOLITE CONCENTRATION
Melibiose (mg g−1)
Stearic acid (mg g−1)
Palmitic acid (mg g−1)
Change of metabolite concentration more notable in roots
ROOTS
LEAVES
Component 1 (53.8)
Component 2 (8.3)
Aspartic acid
Valine
Serine
Arabital
Change of metabolite concentration more notable in leaves
Component 3 (3.5)
D-glucose
D-fructose
Threonic acid
Quinic acid
Leaves/Control – LC
Leaves/Drought – LD
Roots/Control – RC
Roots/Drought – RD

and C use in fine-root production, respiratory metabolism, and production of soluble compounds that contribute to osmotic adjustment in the roots (165, 166). NSC partitioning can be related to other functional strategies, such as the regulation of leaf water loss reflected in the isohydric-anysohydric continuum (167). The different strategies across oak species in C partitioning among plant organs could explain the unsuccessful attempts to connect tree decline and mortality to the pool of carbohydrates at mature phases of development.

10.3 Secondary Metabolic Pathways and Metabolite Levels

Research efforts on secondary metabolism in oaks affected by drought stress have mainly focused on the emission of volatile organic compounds (VOCs) and the regulation of antioxidant compounds (Table 10.1). The wide variety of climatic conditions in which oaks live results in different drought responses of secondary metabolism.

Secondary metabolites play a major role as antioxidants and ROS scavengers, producing a reduction of stress-induced oxidative damage, and the enhancement of cell membrane stability and cell wall strength. Other relevant actions of these compounds are osmoprotection, reduction of water loss, and protection of the photosynthetic apparatus (51). Sancho-Knapik et al. (168) described distinct leaf morphological parameters (e.g. density, area, and thickness) related to different adaptive strategies aimed at reducing water loss and oxidative damage in *Quercus* species. Similarly, changes in pigment content are involved in different protection mechanisms depending on whether trees are deciduous

Figure 10.3 General changes in secondary metabolites in oaks under water stress conditions altering plant physiology and metabolism. Stress-induced increase (↑) or reduction (↓) of metabolite concentration in roots and aerial parts are shown. Processes such as strengthening of cell walls, stomatal closure, leaf senescence, activation of defense and antioxidant system, and production of root exudates and biogenic volatile organic compounds are common responses in oaks under water stress conditions. Some secondary metabolites are presented in the figure as representative examples of common compounds found in oaks under water stress conditions. Blue arrows indicate positive effects and red lines indicate negative effects. Dashed black arrows indicate signaling/transport. Abbreviations: Ψ_w, water potential; ABA, abscisic acid; APX, ascorbate peroxidase; Asc, ascorbate; BVOCs, biogenic volatile organic compounds; CAT, catalase; Cyt, cytochrome; DHA, dehydroascorbate; DHAR, dehydroascorbate reductase; GR, glutathione reductase; GSH, reduced glutathione; GSSG, oxidized glutathione; H_2O_2, hydrogen peroxide; JA, jasmonic acid; LHC, light harvesting complex; MDHA, monodehydroascorbate; MeSA, methyl salicylate; $O_2^{\cdot-}$, superoxide anion radical; PCs, phenolic compounds; PS, photosystem; ROS, reactive oxygen species; SOD, superoxide dismutase.

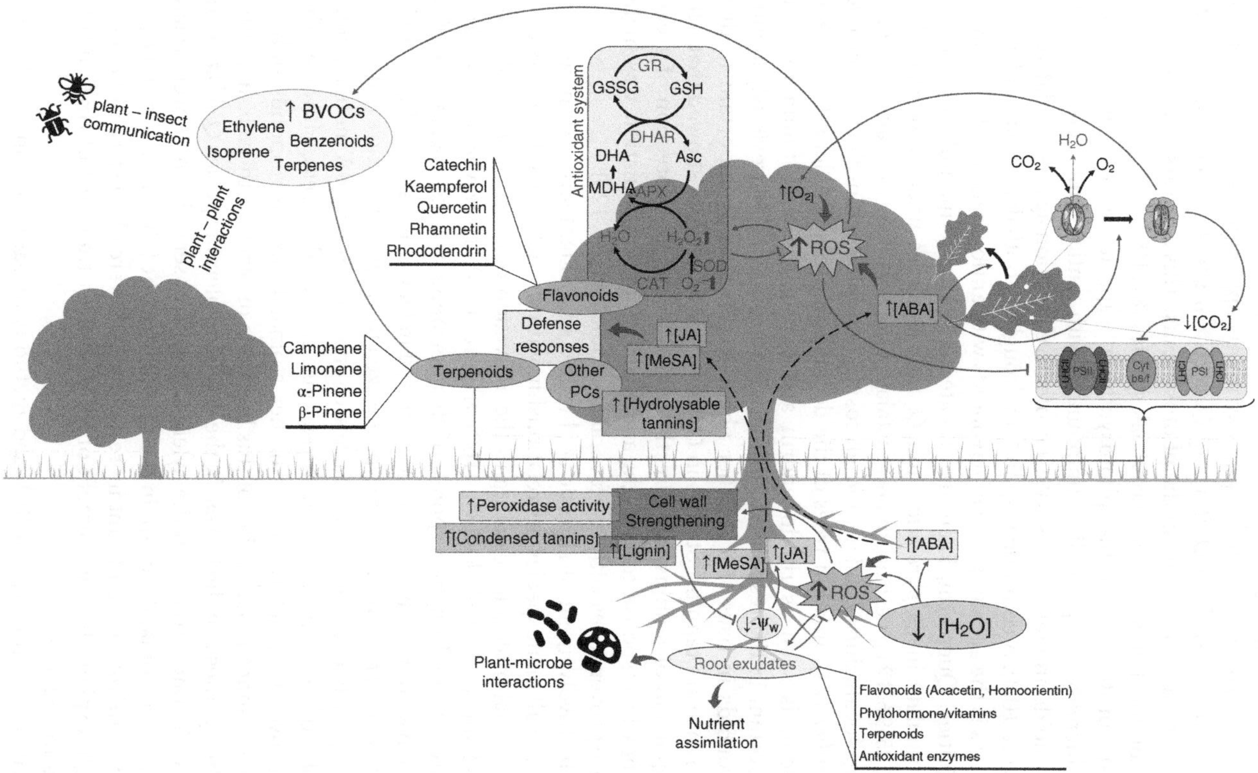

plant – insect communication
↑ BVOCs
Ethylene
Benzenoids
Isoprene
Terpenes
plant – plant interactions
Catechin
Kaempferol
Quercetin
Rhamnetin
Rhododendrin
Camphene
Limonene
α-Pinene
β-Pinene
Antioxidant system
GR
GSSG
GSH
DHAR
DHA
Asc
MDHA
APX
H2O
H2O2
SOD
CAT
O2
Flavonoids
Defense responses
Terpenoids
Other PCs
↑[JA]
↑[MeSA]
↑[Hydrolysable tannins]
↑[O2]
ROS
↑[ABA]
H2O
CO2
O2
↓[CO2]
LHCII
PSII
Cyt b6/f
LHCI
PSI
↑Peroxidase activity
Cell wall Strengthening
↑[Condensed tannins]
↑[Lignin]
↑[MeSA]
↑[JA]
↑[ABA]
ROS
↓-ψw
↓ [H2O]
Plant-microbe interactions
Root exudates
Nutrient assimilation
Flavonoids (Acacetin, Homoorientin)
Phytohormone/vitamins
Terpenoids
Antioxidant enzymes

or evergreen. Thus, in response to drought stress, deciduous species such as *Q. pubescens* and *Q. brantii* reduced chlorophyll and carotenoid contents as a mechanism to protect the photosynthetic apparatus (42, 64), whereas evergreen species such as *Q. ilex* and *Q. coccifera* showed no change in pigment contents but no damage in the photosystems. This suggests that different protection mechanisms exist across oak species (87, 105).

The protection of the photosynthetic capacity is also associated with an increase in phenolic compounds with antioxidant activity, generally flavonoids and terpenoids with a protective function against oxidative damage (Figure 10.3). A common response to drought in both deciduous and evergreen species is the accumulation of flavonoids with antioxidant and ROS scavenging activities. Quercetin, kaempferol, catechin, rhamnetin, and their derivatives accumulate in *Q. pyrenaica* and *Q. ilex* seedlings under moderate drought (22, 25, 96, 121); vescalagin and castalagin in *Q. petraea* (31); and isoscoparine, epigallocatechin 3-gallate and glycoside (−)-epicatechin-3′-*O*-glucuronide in *Q. suber* (24). In adult trees of the deciduous, drought-tolerant *Q. pubescens*, flavonoids such as rhododendrin and derivatives were found to be more important in the response to drought than some of the previously mentioned flavonoids in oaks (107). Other phenolic compounds important in drought tolerance are tannins. Studies in *Q. rubra* and *Q. petraea* showed that a higher proportion of tannins was directly related to greater protection of green leaf tissues in response to drought and that tannin accumulation was higher in those provenances with less favorable conditions (31, 123).

The involvement of abscisic acid (ABA) in drought stress response is common to all plant species. Its relationship with stomatal closure and its function as a signal molecule resulting in the activation of different drought tolerance mechanisms is well known (169) (Figure 10.3). However, few works have analyzed the drought response of this hormone in oaks. Increases in ABA in response to drought have been described in leaves and roots of *Q. ilex* (57, 88) and in *Q. cerris*, where it was hypothesized to induce premature leaf senescence in response to drought (43). There are some studies on ABA mobilization from roots to leaves in xylem sap in response to drought stress. However, there is controversy in determining whether stomatal closure responds only to the root-to-shoot mobilization of this hormone or whether there is an increase in its biosynthesis in the leaves. In the drought-tolerant species *Q. alba*, increases in ABA concentration in the xylem sap were observed, which were related to increased stomatal closure in the face of drought (73). However, in *Q. petraea* and *Q. robur*, no significant increases in ABA were observed in xylem sap that might explain the involvement of root-to-shoot transport of ABA in stomatal closure (52, 124).

Mass spectrometry (MS)-based methods reveal variations in the concentration of other plant hormones. For example, Tienda-Parrilla et al. (121) found that several secondary metabolites were induced by drought stress in an untargeted metabolomics study in the leaves of *Q. ilex* seedlings. The authors proposed a variety of molecular markers of drought stress, among which were auxin, indole-3-acrylic acid, and cytokinin dihydrozeatin-*O*-glucoside. According to recent studies, certain growth hormones could be involved in the regulation of drought stress responses through signaling and cross-talk with other hormonal pathways (170, 171).

Drought sensitivity of secondary metabolites also varies within oak species. For example, the level of glutathione and ascorbic acid, their redox ratios, namely, reduced glutathione (GSH)/oxidized glutathione (GSSG) and ascorbate (Asc)/dehydroascorbate (DHA), and the capacity to eliminate ROS by antioxidant enzymes vary in response to drought depending on soil properties of the seed origin or the site of growth (31, 42, 62) (Figure 10.3). Seedlings of different oak species growing in unfavorable soils are subjected to a higher oxidative pressure, which makes them less able to react to oxidative stress induced by a period of drought. The inefficient antioxidant activity in these trees results in higher ROS production and thus increased lipid peroxidation and damage to the photosynthetic apparatus when subjected to water stress as compared to those trees growing in more favorable soil conditions.

Changes in root secondary metabolism in response to drought are mostly related to root exudates (see below) and cell wall composition (Figure 10.3). A study with *Q. ilex* seedlings subjected to drought found that most of the root enzymes whose presence increased in response to drought were related to secondary metabolism (112). These enzymes included naringenin-chalcone synthase, flavanone 3-hydroxylase and dihydroflavonol 4-reductase (involved in different flavonoid production pathways), shikimate dehydrogenase (involved in the production of phenolic compounds), (+)-neomenthol dehydrogenase (involved in terpene synthesis), and caffeoyl CoA 3-*O*-methyltransferase (involved in cell wall hardening). Changes in root cell wall composition have also been found in other species, such as *Q. rubra* and *Q. alba* (120). In both species, the application of moderate water stress induced an increase in the diameter of fine roots and an increase in the phenolic bonds present in the cell walls. However, drought induced different changes in cell wall structure depending on the tolerance level of each species. Thus, a greater increase in the proportion of condensed tannins was found in the roots of the more drought-tolerant species *Q. alba*, while in *Q. rubra* drought induced a greater production of ellagitannins and guaiacyl in the lignin polymers.

10.4 The Transport of Metabolites within the Plant – Transport Rates and Sap Composition

Few researchers have explored the composition of xylem and phloem sap during drought in oaks. Organic compounds, ions, and phytohormones dissolved in the xylem sap travel to the sites of evapotranspiration in the leaves, according to the theory of cohesion-tension for the ascent of sap in trees (172). Thus, transport rates of these compounds are largely controlled by stomata opening, which are highest at midday (although nocturnal sap flow may account for roughly 20% of whole-day flow in some oaks) (173), and decrease and occur earlier in the day as drought increases. Several organic compounds have been identified in the xylem sap of woody plant, including sugars (mostly sucrose), amino acids and amides and, to a lesser extent, organic acids, phytohormones, and enzymes (174–176). The higher concentration of sugars in the xylem sap occurs during bud break in spring, as a consequence of the conversion of starch stored in roots, branches, and stem to free sugars readily translocated to meristematic tissues; the minimum concentration occurs in late spring to early summer, before the onset of drought stress (176). Few studies have analyzed xylem sap composition in oaks. In *Q. robur* saplings grown under controlled environmental conditions, the total soluble non-protein N compounds (TSNN) were dominated by glutamate, aspartate, and their amides, followed by GABA and arginine (177). The concentrations of soluble sugars (mainly sucrose, followed by much lower concentrations of glucose and fructose) in the xylem sap of *Q. robur* saplings were significantly higher than TSNN concentrations, particularly by the end of July (177). Given that xylem sap was collected several months after leaf flushing, the authors suggested that xylem-transported C derived from phloem-to-xylem exchange of carbohydrates along the stem (i.e. involving a cycling pool of C within the vascular canopy). Previous research also suggested that amino acids produced in the roots were transported to the leaves through the xylem in *Q. rubra* (178). Other organic compounds can move acropetally within oak xylem sap, such as phenolic compounds originated from epiphytic lichens in *Q. pyrenaica* branches (179). Because NSC can decrease in oak roots due to drought (12, 74, 165), it is possible that drought reduces the level of both carbohydrates and amino acids in the sap ascending from roots. Root-derived signals are transported through the xylem to leaves in response to reduced soil moisture (Figure 10.3). For example, increased concentrations of ABA were observed in xylem sap of *Q. robur* (52) and *Q. alba* (73) exposed to drought; nonetheless, total ABA flux may be reduced as a consequence of reduced transpiration rates. Similarly, drought can reduce substantially inorganic N uptake as a consequence of reduced transpiration (180).

Leaf carbohydrates and amino acids need to be translocated in the phloem to sustain metabolism, growth, exudation, and storage of non-photosynthetic organs. The primary uptake of organic compounds from photosynthetic cells to the conducting sieve elements within the leaves is known as phloem loading (181), which can occur via symplast, apoplast, or a combination of both routes depending on the species, the ontogenetic state, and cell type within the phloem (182). Traditionally, symplastic passive load was assumed to occur in tree species with a high number of plasmodesmata connecting bundle sheath or mesophyll cells with phloem sieve elements, whereas sugar transport proteins would be required for those species with few plasmodesmata connections, primarily monocots and herbaceous eudicots (183). Apoplastic phloem loaders transport nearly exclusively sucrose in the phloem, or sucrose together with sugar alcohols (184). Symplastic loading may involve the active biosynthesis of large amounts of the tri- and tetra-saccharides raffinose and stachyose in addition to sucrose (185). Initial studies suggested that oaks may not require an active loading because they have numerous plasmodesmata between companion cells and mesophyll cells and because they often maintain high levels of leaf sugars and sugar-alcohols (186). For example, *Q. robur* was traditionally considered to load carbohydrates passively via symplastic diffusion through the numerous plasmodesmata in the minor vein phloem (187). For the same reasons, mature *Q. coccinea* trees were also considered as passive loaders (186). However, recent works with *Q. robur* plants challenged this view, as phloem sap osmolality was higher than that of whole-leaf sap, indicating that there is a steep uphill sucrose gradient from cytoplasm of mesophyll cells to phloem cells that requires an active apoplastic loading (184, 188). Similar results have been found in the closely related species *Fagus sylvatica*, where the concentration of sucrose was fivefold higher in phloem sap than in the cytosol of mesophyll cells (189). Other studies with *Q. rubra* also supported the use of an active or mixed loading strategy, instead of a purely passive loading strategy (190). Further work is needed to understand the role of plasmodesmata frequency and dynamics in the connectivity among sieve elements and surrounding cells in oaks (191–193), the contribution of active and passive loading, and hence the type of metabolites that travel from mesophyll cells to the phloem. For example, it is unclear whether large raffinose family oligosaccharides (RFOs) (i.e. raffinose, stachyose, and verbascose) and cyclitols (e.g. *myo*-inositol and quercitol) common in leaves and branches of some oaks (159) can be transported by the phloem, as observed in other tree species (194). Also, phloem-mobile information signals and secondary compounds, including phytohormones (195), small RNAs, mRNAs and more than 1000 proteins, have been described in the phloem of other plant species (196).

It is currently accepted that phloem transports metabolites by bulk flow, driven by osmotic gradients between source and sink tissues coupled hydraulically to the xylem (197). The osmotic potential within sieve elements transporting phloem sap is primarily (60–75%) determined by one of several sugar species (sucrose, polyols, or RFOs), with amino acids and inorganic anions and cations (K^+) accounting for most of the remaining osmotic potential (183). Phloem sap contains about 20% sugar by weight (198). In the phloem sap of *Q. robur*, the major sugar was sucrose (1015 mM), with only traces of glucose and fructose and no sugar alcohols such as mannitol or sorbitol or RFOs (184). In *Q. rubra*, phloem sucrose concentration can be as high as 1580 mM and total phloem osmolality as high as 1800 mM, enough to generate a phloem pressure of 3.3 MPa (190). Given that trees have large sugar concentrations in their leaves to offset the low xylem pressure of transpiring leaves, they can maintain small concentration gradients between source and sink tissues even in tall canopies (e.g. 0.2 MPa is sufficient to drive Münch pressure flow in *Q. rubra*) (199). In this oak species, most of photoassimilates are transported out of the leaf during the day, with sucrose being the only soluble carbohydrate exhibiting a clear diurnal trend (200). At midday, especially under mild drought, oak leaves need to maintain mesophyll turgor by increasing mesophyll sucrose concentration (193). This results from reduced rates of C export from leaves in comparison with net photosynthetic rates (200, 201). During more severe drought, phloem cells, including sieve tubes, parenchyma, and fibers may act as additional water sources for the xylem, which may increase phloem sap viscosity, which, together with photosynthetic limitations reducing sucrose pressure gradients, can slow down phloem transport (202). However, most plants can sustain phloem function even during severe drought as long as soil water potential remains above the permanent wilting point (203, 204). The activities of several antioxidant enzymes (superoxide dismutase, dehydroascorbate reductase, or peroxidases) is also expected to increase in the phloem sap of oaks in response to drought stress, as observed in other species (205). Mobile N compounds such as glutamine can be reduced under drought in oak (180). Drought also inhibited basipetal phloem transport of reduced sulfur (e.g. glutathione, GSH; γ-glutamyl-cysteinyl-glycine) in *Q. robur* (109). During recovery from drought stress, transiently stored amino acids in leaves may be translocated to new sinks via phloem (206). Collectively, these results emphasize the importance of the vascular system in drought metabolic adaptation.

10.5 The Release of Metabolites Outside the Plant

A large proportion of photoassimilates exit the plant via leaching, exudation, secretion, translocation to mycorrhizae, losses to parasites, and release of lysates and VOCs (176). Here we focus on root exudation and emission of VOCs. Nonetheless, the other processes can be as important for plant and ecosystem functioning. For example, leaching by the action of rain, dew, or fog affects several compounds, primarily sugars, and can amount up to 86 kg ha^{-1} per month of soluble organic matter leached from *Q. petraea* crowns (207).

10.5.1 Root Exudates

Multitude of compounds are released by roots into the rhizosphere, known collectively as rhizodeposits. They include the secretion of inorganic ions, polysaccharide mucilages, VOCs, soluble lysates of sloughed-off root cap cells, and root exudates that are released from damaged and intact cells (208). However, most of these rhizodeposits are organic compounds (including sugars, amino acids, organic acids, fatty acids, sterols, growth factors and vitamins, enzymes, flavonoids, and other N-rich compounds), which can modify the rhizosphere biota, inhibit the growth of competing plants and alter the physicochemical soil properties (e.g. changing the availability of nutrients for plants) (208–211). Studies using isotopic tracers revealed that 17% of the net C fixed by photosynthesis was emitted by roots and lost as rhizosphere respiration (12%) or remaining in the soil (5%) in young trees and herbaceous species (212). Regarding oak species, C loss via root exudation in mature *Q. serrata* and *Q. glauca* trees was $0.073\,mg\,C\,g^{-1}\,h^{-1}$, about 10% of the C loss via root respiration, which was equivalent to 3% of the annual net primary production (213). Rhizodeposits are highly dependent on plant physiological activity and plant metabolome. Thus, drought affects the quantity and quality of root exudates. There is an increase in rhizodeposition under moderate drought due to the uncoupling of photosynthesis and plant growth rates, but also to damage to cell root membranes increasing lysates (214, 215). Some plants exude more secondary compounds during drought (e.g. phenolic compounds, antioxidants, and especially flavonoids) (216) (Figure 10.3), or phytohormones, soluble sugars, and metabolites that act as osmolytes and buffer the cellular redox status (209). These changes correspond to shifts in plant metabolism under drought and optimize plant relationship with heterotrophic microorganisms once drought has ended, for example improving nutrient cycling (217–219).

The few studies with oaks suggest they are not among the tree species with higher root exudation rates, in comparison with companion tree species. For example, *Q. rubra* had comparatively lower diversity and quantity of phenolic compounds in root exudates, including catechin, the most abundant phenol, than other five tree species grown in a glasshouse, albeit the number of compounds in root tissues was similar for all six species (220). In a comparative study, *Q. ilex* had a significantly higher root total organic C (TOC) exudation rate but exhibited a higher reduction in TOC exudation rates compared to *Pinus sylvestris,* which kept TOC exudation rates constant for more than 14 weeks of mild water deficit (93). In a previous experiment with *Q. ilex* submitted to a fast water deprivation experiment for 21 days, the exudation of TOC by roots increased only initially (92). The apparent contradiction among these studies was explained by fast stress imposition rapidly stopping growth in the later study, so that a higher proportion of photosynthetic C could be allocated to exudates rather than biomass production (92). Eight weeks after the prolonged mild drought ended, *Q. ilex* showed the same TOC exudation rate with respect to the drought period and lower to control plants, suggesting carry-over drought effects (93), in comparison with the return to non-drought levels of TOC exudation rates after the most severe, shorter drought event in Preece et al. (92).

Of the 601 metabolites that were detected in root exudates of *Q. ilex* plants studied by Preece et al. (92), 63 were identified. Amino acids were the most abundant compounds, followed by carbohydrates and organic acids. On the contrary, secondary metabolites comprised the majority of root exudates under drought (71% of total metabolites) in Gargallo-Garriga et al. (57), with primary metabolites dominating the chemical profile of root exudates when recovering from drought stress. The severity of drought stress and the degree of stress recovery had strong effects on the composition of root exudate metabolites in this study. Under initial mild stress, concentrations of lactic acid, pyruvate, malic and citric acids, ABA, and glucose increased, suggesting stimulation of glycolysis and Krebs metabolism derived from an excess of carbohydrates supplied to the roots (57). Under more severe drought, there was a significant decrease of adenosine, lysine, and jasmonic acid, a significant decrease of organic acids from the Krebs cycle (e.g. succinic, malic, and citric acids), and associated biochemical pathways (e.g. pyruvate) that were previously induced, and higher levels of compounds that can be converted into acetyl-CoA and feed the synthesis of alkaloids and terpenoids, including ABA (57)). Interestingly, the amount and composition of root metabolites exuded during drought and subsequent recovery were independent of soil biota (57, 92). The addition of phosphorous (P) to the substrate increased root exudation in *Q. ilex* during and after drought, but not the addition of N or the addition of P together with N (93). However, nutrient deficiency under well-watered conditions usually enhances exudation of phenolic and organic acid compounds to enhance P absorption by roots (221). In summary, root exudates derive from the most abundant root metabolites (220) and change in amount and composition depending on drought stress effects on plant metabolism and species drought stress tolerance (57, 217).

10.5.2 Volatile Organic Compounds

Methane (CH_4) is an important, short-lived greenhouse gas that accounts for almost half of the total net global warming (0.5 °C out of 1.1 °C increase, relative to 1850–1900) (222). The emission of methane by trees may be induced by UV radiation acting on pectin and other plant compounds. However, it can also have a microbial origin. Microbial methanogenesis can occur on the surface of the trunks and the heartwood of aging trees. Moreover, a significant proportion of methane is produced by soil microbes in anoxic soils (e.g. wetlands), transported in the xylem transpiration stream, and then emitted through stem lenticels (223, 224). The emission of methane by some temperate oak stems is small, with a peak of emission of 0.031 nmol CH_4 m^{-2} s^{-1} in spring, but still lower than that from other upland tree species such as *Pinus* and *Fagus* (225).

Non-methane biogenic volatile organic compounds (BVOCs) comprise a large variety of molecules (>30,000) emitted from different plant tissues (226, 227) that differ in size and physicochemical properties as well as in their metabolic origin (228). A number of low-molecular-weight ($C < 5$) BVOCs are emitted by plants, such as methanol and formaldehyde (derived from glucose), ethylene (from Krebs cycle and the amino acid methionine), ethanol and acetaldehyde (from pyruvate), and acetone and hydrogen cyanide (from the amino acid valine) (229). BVOCs are involved in defense and reproduction processes (e.g. phenylpropanoids and benzenoids, and ethylene), plant–plant interactions (e.g. with priming effects of mono-, sesqui-, and homoterpene compounds inducing defense), plant–pathogen/insect communication (e.g. to repel an herbivore or attract herbivore predators through the release of alcohols, esters, or terpenes), and abiotic stress tolerance (e.g. with isoprene protecting leaf photochemistry from heat and oxidative damage) (226, 229, 230) (Figure 10.3).

BVOCs synthesis and rates are largely controlled by gene expression, but post-translational regulatory mechanisms and substrate availability (e.g. of dimethylallyl diphosphate DMAPP) also play key roles in adjusting BVOCs emission rates to shifting environmental conditions, including light, temperature, and drought (39, 226, 231–234). Oaks can release large amounts of isoprene (e.g., 50–100 $nmol\,m^{-2}\,s^{-1}$, representing up to 2–5% of the photosynthetic net C uptake), but they cannot store it (235, 236). On the contrary, many plants store terpenoids, including some oaks. For example, *Q. suber* and *Q. ilex* are strong terpenoid emitters; however, they do not possess specialized storage organs in their leaves (60, 78). In contrast, *Q. robur*, besides being a strong isoprene emitter, increased the accumulation of different monoterpenes and sesquiterpenes in response to drought in specific storage structures (237). The emission of these BVOCs is directly related to current photosynthetic activity (238, 239). Terpenoid synthesis rate follows the classical peaked enzymatic curve in response to temperature, with lower optimum temperatures for monoterpene (40.6 °C) than for isoprene (44 °C), in *Q. ilex* and *Q. rubra*, respectively (240, 241). This reflects the role of isoprenoids in enhancing thermotolerance and protecting leaves against oxidative damage (231, 235, 242, 243). Although oaks are among the biggest isoprene emitters species, there are several modern oak species distributed mostly around the Mediterranean (and mostly evergreen) that do not produce isoprene, but alternatively emit large amounts of light-dependent monoterpene (e.g. *Ilex* subgenus) (71, 115, 244). There are several chemotypes described in *Q. ilex* that differ in the main terpenoid emitted (244–246), which include linalool, α-thujene, camphene, and γ-terpinene, all emitted at low rates (238, 245). Terpene composition changes little across seasons, canopy positions, daytime, or leaf age in *Q. ilex* (71, 115, 244, 247). Other oaks belonging to the *Cerris* subgenus do not emit either type of these compounds (236, 248).

A recent metanalysis of BVOCs emission rates under 50% (moderate) water deficit shows that isoprene and photosynthesis respond negatively to drought (−15% and −42% relative to controls well-watered), whereas monoterpenes are not significantly affected (249). In oaks, the level of isoprene and its derivatives can increase with mild short-term water stress (37,106). However, leaf isoprene emission rates usually decline with decreasing water potential, comparatively less than net photosynthesis, which results in an increasing percentage of fixed C lost as isoprene as plants become more stressed (86, 100, 104, 249). Prolonged water deprivation reduced to nearly zero leaf gas exchange rates at predawn water potential of −1.5 MPa in *Q. robur* and −2.5 MPa in *Q. pubescens*, but isoprene emission rates did not decline significantly until 1 MPa lower predawn water potential in both species (250). Drought effects on monoterpene emission are more variable. Moderate water stress significantly increased total terpene concentrations in *Q. ilex* leaves (36). Similar results were reached for the monoterpene emitter and drought-resistant *Q. coccinea*, which presented constant monoterpene and sesquiterpene emission rates from −2 to −5 MPa midday leaf water potential (81). By contrast, drought reduced the basal emission of monoterpenes in *Q. ilex* in other studies (34, 115, 180, 251). In a natural *Q. ilex* stand, monoterpene emission increased by 82% during moderate drought, but rapidly declined when predawn water potential dropped below −2 MPa and photosynthesis resulted severely inhibited (71). Severe water stress may limit terpenoid synthesis as a consequence of limiting C substrate allocated to the chloroplastic methylerythritol phosphate (MEP) pathway, increased resistance to diffusion out of the leaf due to stomatal closure, and/or low energy and reducing power (ATP and NADPH) supply derived from limited photosynthetic electron transport rate; however, none of these factors occur usually under mild-to-moderate drought stress (230, 252, 253). Further, severe water stress may limit the positive scaling between temperature and monoterpene or isoprene emissions (34, 254). The emission of other BVOCs may increase under drought. For example, there was a correlation between the emission of membrane lipoxygenation products ((*Z*)-3-hexenal, (*Z*)-3-hexenol, and (*E*)-2-hexenol) and ion leakage in *Q. ilex* leaves (255). Despite drought, the opening of stomata in early mornings may release large pools of oxygenated BVOCs, acetone, and methanol accumulated within mesophyll cells in *Q. ilex* (256). Terpene leaf composition was rather constant under drought conditions in *Q. ilex*, with leaves showing only significant increments in α-Pinene, β-Pinene, Δ^3-Carene, and β-Ocimene (36). Compounds α-pinene, β-pinene, camphene or limonene have been described to be necessary for the protection of the photosynthetic apparatus in the leaves of *Q. ilex* or *Q. suber* (36, 60, 78).

Emission of BVOCs exhibits a variable response to rewatering. Monoterpene and isoprene emission can recover rapidly together with leaf gas exchange, but sometimes not to the same level as before stress (34, 39, 116). Monoterpene emission was stimulated in *Q. ilex*, *Q. coccinea*, and *Q. suber* after recovering from moderate drought (116, 246). Biosynthesis and emission of terpenoids after recovering from drought depend on sugar accumulation during drought and the availability of electron transport chain products upon recovery.

10.6 Conclusions

Information on plant metabolism under drought is fragmentary for oak species. Most studies measure particular metabolites, such as proline, ABA, or the total concentration of sugars, amino acids, or phenolic compounds. However, further studies on targeted and untargeted metabolomics, combined with ecophysiological and other omics approaches, will help depict oak responses to drought more comprehensively. Re-adjustments of primary and secondary metabolic pathways increase the osmoregulatory, antioxidant, and cell-repairing capacity, and can help to maintain long-distance transport of water and metabolites in vascular tissues. These responses play an important role in acclimating to and recovering from short-term (i.e. weeks to months) drought. However, we largely ignore how exposure to recurrent, chronic drought affects plant metabolism and its capacity to govern long-term drought acclimation by reducing drought stress and/or increasing plant resilience. Until now, there is insufficient information to compare short- and long-term drought effects on the metabolome of oaks. In addition, more research that compares plant organs, species, provenances, ontogenetic stages, and regeneration origins (e.g. seedlings versus resprouts) is needed.

References

1 Denk T, Grimm GW, Manos PS, Deng M, Hipp AL. An updated infrageneric classification of the oaks: review of previous taxonomic schemes and synthesis of evolutionary patterns. In: Gil-Peregrin E, Peguero-Pina JJ, & Sancho-Knapik D, editors. *Oaks Physiological Ecology. Exploring the Functional Diversity of Genus* Quercus *L.* Cham (CH): Springer; 2017. p. 13–38.
2 Kremer A, Hipp AL. Oaks: an evolutionary success story. *New Phytol.* 2020; 226(4): 987–1011.
3 Gil-Pelegrín E, Peguero-Pina JJ, Sancho-Knapik D. Oaks and people: a long journey together. In: Gil-Peregrin E, Peguero-Pina JJ, & Sancho-Knapik D, editors. *Oaks Physiological Ecology. Exploring the Functional Diversity of Genus* Quercus *L.* Cham (CH): Springer; 2017. p. 1–11.
4 Dey DC. Sustaining oak forests in Eastern North America: regeneration and recruitment, the pillars of sustainability. *For. Sci.* 2014; 60(5): 926–942.

5 Pemán J, Chirino E, Espelta JM, Jacobs DF, Martín-Gómez P, Navarro-Cerrillo R, et al. Physiological keys for natural and artificial regeneration of oaks. In: Gil-Peregrin E, Peguero-Pina JJ, & Sancho-Knapik D, editors. *Oaks Physiological Ecology. Exploring the Functional Diversity of Genus* Quercus *L.* Cham (CH): Springer; 2017. p. 453–511.
6 Zhu K, Woodall CW, Ghosh S, Gelfand AE, Clark JS. Dual impacts of climate change: forest migration and turnover through life history. *Global Change Biol.* 2014; 20(1): 251–264.
7 Rigling A, Bigler C, Eilmann B, Feldmeyer-Christe E, Gimmi U, Ginzler C, et al. Driving factors of a vegetation shift from Scots pine to pubescent oak in dry Alpine forests. *Global Change Biol.* 2013; 19(1): 229–240.
8 Sun S, Zhang Y, Huang D, Wang H, Cao Q, Fan P, et al. The effect of climate change on the richness distribution pattern of oaks (*Quercus* L.) in China. *Sci. Total Environ.* 2020; 744: 140786.
9 Abrams MD. Adaptations and responses to drought in *Quercus* species of North America. *Tree Physiol.* 1990; 7(1–4): 227–238.
10 Niinemets Ü, Valladares F. Tolerance to shade, drought, and waterlogging of temperate northern hemisphere trees and shrubs. *Ecol. Monogr.* 2006; 76(4): 521–547.
11 Novick K, Jo I, D'Orangeville L, Benson M, Au TF, Barnes M, et al. The drought response of eastern US oaks in the context of their declining abundance. *Bioscience.* 2022; 72(4): 333–346.
12 Rodríguez-Calcerrada J, Sancho-Knapik D, Martin-StPaul NK, Limousin JM, McDowell NG, Gil-Pelegrín E. Drought-induced oak decline—factors involved, physiological dysfunctions, and potential attenuation by forestry practices. In: Gil-Peregrin E, Peguero-Pina JJ, & Sancho-Knapik D, editors. *Oaks Physiological Ecology. Exploring the Functional Diversity of Genus* Quercus *L.* Cham (CH): Springer; 2017. p. 419–451.
13 Aranda I, Sánchez-Gómez D, Cadahía E, de Simón BF. Ecophysiological and metabolic response patterns to drought under controlled condition in open-pollinated maternal families from a *Fagus sylvatica* L. population. *Environ. Exp. Bot.* 2018; 150: 209–221.
14 de Miguel M, Guevara MÁ, Sánchez-Gómez D, de María N, Díaz LM, Mancha JA, et al. Organ-specific metabolic responses to drought in *Pinus pinaster* Ait. *Plant Physiol. Biochem.* 2016; 102: 17–26.
15 Warren CR, Aranda I, Cano FJ. Metabolomics demonstrates divergent responses of two *Eucalyptus* species to water stress. *Metabolomics.* 2012; 8(2): 186–200.
16 Sardans J, Peñuelas J, Rivas-Ubach A. Ecological metabolomics: overview of current developments and future challenges. *Chemoecology.* 2011; 21(4): 191–225.
17 de Simón BF, Sanz M, Cervera MT, Pinto E, Aranda I, Cadahía E. Leaf metabolic response to water deficit in *Pinus pinaster* Ait. relies upon ontogeny and genotype. *Environ. Exp. Bot.* 2017; 140: 41–55.

18 Merchant A, Tausz M, Arndt SK, Adams MA. Cyclitols and carbohydrates in leaves and roots of 13 *Eucalyptus* species suggest contrasting physiological responses to water deficit. *Plant Cell Environ.* 2006; 29(11): 2017–2029.

19 Tschaplinski TJ, Abraham PE, Jawdy SS, Gunter LE, Martin MZ, Engle NL, et al. The nature of the progression of drought stress drives differential metabolomic responses in *Populus deltoides*. *Ann. Bot.* 2019; 124(4): 617–626.

20 Warren CR, Aranda I, Cano FJ. Responses to water stress of gas exchange and metabolites in *Eucalyptus* and *Acacia* spp. *Plant Cell Environ.* 2011; 34(10): 1609–1629.

21 Rodrigues AM, Miguel C, Chaves I, António C. Mass spectrometry-based forest tree metabolomics. *Mass Spectrom. Rev.* 2021; 40(2): 126–157.

22 Rivas-Ubach A, Gargallo-Garriga A, Sardans J, Oravec M, Mateu-Castell L, Pérez-Trujillo M, et al. Drought enhances folivory by shifting foliar metabolomes in *Quercus ilex* trees. *New Phytol.* 2014; 202(3): 874–885.

23 López-Hidalgo C, Guerrero-Sánchez VM, Gómez-Gálvez I, Sánchez-Lucas R, Castillejo-Sánchez MA, Maldonado-Alconada AM, et al. A multi-omics analysis pipeline for the metabolic pathway reconstruction in the orphan species *Quercus ilex*. *Front. Plant Sci.* 2018; 9: 935.

24 Almeida T, Pinto G, Correia B, Gonçalves S, Meijón M, Escandón M. In-depth analysis of the *Quercus suber* metabolome under drought stress and recovery reveals potential key metabolic players. *Plant Sci.* 2020; 299: 110606.

25 Aranda I, Cadahía E, Fernández de Simón B. Leaf ecophysiological and metabolic response in *Quercus pyrenaica* Willd seedlings to moderate drought under enriched CO_2 atmosphere. *J. Plant Physiol.* 2020; 244: 153083.

26 Chaves MM, Pereira JS, Cerasoli S, Clifton-Brown J, Miglietta F, Raschi A. Leaf metabolism during summer drought in *Quercus ilex* trees with lifetime exposure to elevated CO_2. *J. Biogeogr.* 1995; 22(2/3): 255.

27 Épron D, Dreyer E. Starch and soluble carbohydrates in leaves of water-stressed oak saplings. *Ann. For. Sci.* 1996; 53(2–3): 263–268.

28 Gebre GM, Tschaplinski TJ. Solute accumulation of chestnut oak and dogwood leaves in response to throughfall manipulation of an upland oak forest. *Tree Physiol.* 2002; 22(4): 251–260.

29 Gilson A, Barthes L, Delpierre N, Dufrene E, Fresneau C, Bazot S. Seasonal changes in carbon and nitrogen compound concentrations in a *Quercus petraea* chronosequence. *Tree Physiol.* 2014; 34(7): 716–729.

30 Andreetta A, Dignac MF, Carnicelli S. Biological and physico-chemical processes influence cutin and suberin biomarker distribution in two Mediterranean forest soil profiles. *Biogeochemistry.* 2013; 112(1–3): 41–58.

31 Arab L, Seegmueller S, Dannenmann M, Eiblmeier M, Albasher G, Alfarraj S, et al. Foliar traits of sessile oak (*Quercus petraea* Liebl)

seedlings are largely determined by site properties rather than seed origin. *Tree Physiol.* 2020; 40(12): 1648–1667.
32 Aranda I, Cadahía E, Fernández de Simón B. Specific leaf metabolic changes that underlie adjustment of osmotic potential in response to drought by four *Quercus* species. *Tree Physiol.* 2021; 41(5): 728–743.
33 Baquedano FJ, Castillo FJ. Comparative ecophysiological effects of drought on seedlings of the Mediterranean water-saver Pinus halepensis and water-spenders *Quercus coccifera* and *Quercus ilex*. *Trees*. 2006; 20(6): 689–700.
34 Bertin N, Staudt M. Effect of water stress on monoterpene emissions from young potted holm oak (*Quercus ilex* L.) trees. *Oecologia*. 1996; 107(4): 456–462.
35 Bessonova VP, Chonhova AS. Influence of soil moisture level on metabolism of non-structural carbohydrates in *Quercus robur* leaves. *Regul. Mech. Biosyst.* 2022; 12(4): 628–634.
36 Blanch JS, Peñuelas J, Sardans J, Llusià J. Drought, warming and soil fertilization effects on leaf volatile terpene concentrations in *Pinus halepensis* and *Quercus ilex*. *Acta Physiol. Plant*. 2009; 31(1): 207–218.
37 Bourtsoukidis E, Kawaletz H, Radacki D, Schütz S, Hakola H, Hellén H, et al. Impact of flooding and drought conditions on the emission of volatile organic compounds of *Quercus robur* and *Prunus serotina*. *Trees*. 2014; 28(1): 193–204.
38 Brossa R, Casals I, Pintó-Marijuan M, Fleck I. Leaf flavonoid content in *Quercus ilex* L. resprouts and its seasonal variation. *Trees*. 2009; 23(2): 401–408.
39 Brüggemann N, Schnitzler JP. Diurnal variation of dimethylallyl diphosphate concentrations in oak (*Quercus robur*) leaves. *Physiol. Plant*. 2002; 115(2): 190–196.
40 Bueno A, Pritsch K, Simon J. Responses of native and invasive woody seedlings to combined competition and drought are species-specific. *Tree Physiol.* 2021; 41(3): 343–357.
41 Bussotti F, Bettini D, Grossoni P, Mansuino S, Nibbi R, Soda C, et al. Structural and functional traits of *Quercus ilex* in response to water availability. *Environ. Exp. Bot.* 2002; 47(1): 11–23.
42 Contran N, Günthardt-Goerg MS, Kuster TM, Cerana R, Crosti P, Paoletti E. Physiological and biochemical responses of *Quercus pubescens* to air warming and drought on acidic and calcareous soils. *Plant Biol.* 2013; 15(Suppl 1): 157–168.
43 Cotrozzi L, Remorini D, Pellegrini E, Guidi L, Lorenzini G, Massai R, et al. Cross-talk between physiological and metabolic adjustments adopted by *Quercus cerris* to mitigate the effects of severe drought and realistic future ozone concentrations. *Forests*. 2017; 8(5): 148.

44 Cotrozzi L, Pellegrini E, Guidi L, Landi M, Lorenzini G, Massai R, et al. Losing the warning signal: drought compromises the cross-talk of signaling molecules in *Quercus ilex* exposed to ozone. *Front. Plant Sci.* 2017; 8: 1020.

45 Daoudi H, Derridj A, Hannachi L, Mevy JP. Growth, ectomycorrhization and biochemical parameters of *Quercus suber* L. seedlings under drought conditions. *Afr. J. Biotechnol.* 2016; 15(38): 2082–2090.

46 Daoudi H, Derridj A, Hannachi L, Mévy JP. Comparative drought responses of *Quercus suber* seedlings of three algerian provenances under greenhouse conditions. *Rev. Ecol. (Terre Vie).* 2018; 73(1): 57–70.

47 Deligöz A, Bayar E. Drought stress responses of seedlings of two oak species (*Quercus cerris* and *Quercus robur*). *Turk. J. Agric. For.* 2018; 42(2): 114–123.

48 Díaz-Guerra L, Llorens L, Bell TL, Font J, González JA, Verdaguer D. Physiological, growth and root biochemical responses of *Arbutus unedo* and *Quercus suber* seedlings to UV radiation and water availability before and after aboveground biomass removal. *Environ. Exp. Bot.* 2019; 168: 103861.

49 Dickson R, Tomlinson P. Oak growth, development and carbon metabolism in response to water stress. *Ann. For. Sci.* 1996; 53(2–3): 181–196.

50 Dunn JP, Potter DA, Kimmerer TW. Carbohydrate reserves, radial growth, and mechanisms of resistance of oak trees to phloem-boring insects. *Oecologia.* 1990; 83(4): 458–468.

51 Escandón M, Castillejo MÁ, Jorrín-Novo JV, Rey MD. Molecular research on stress responses in *Quercus* spp.: from classical biochemistry to systems biology through omics analysis. *Forests.* 2021; 12(3): 364.

52 Fort C, Fauveau ML, Muller F, Label P, Granier A, Dreyer E. Stomatal conductance, growth and root signaling in young oak seedlings subjected to partial soil drying. *Tree Physiol.* 1997; 17(5): 281–289.

53 Galiano L, Martínez-Vilalta J, Sabaté S, Lloret F. Determinants of drought effects on crown condition and their relationship with depletion of carbon reserves in a Mediterranean holm oak forest. *Tree Physiol.* 2012; 32(4): 478–489.

54 Gallé A, Haldimann P, Feller U. Photosynthetic performance and water relations in young pubescent oak (*Quercus pubescens*) trees during drought stress and recovery. *New Phytol.* 2007; 174(4): 799–810.

55 Gallé A, Florez-Sarasa I, el Aououad H, Flexas J. The Mediterranean evergreen *Quercus ilex* and the semi-deciduous *Cistus albidus* differ in their leaf gas exchange regulation and acclimation to repeated drought and re-watering cycles. *J. Exp. Bot.* 2011; 62(14): 5207–5216.

56 García-Plazaola JI, Faria T, Abadía J, Abadía A, Chaves MM, Pereira JS. Seasonal changes in xanthophyll composition and photosynthesis of cork

oak (*Quercus suber* L.) leaves under mediterranean climate. *J. Exp. Bot.* 1997; 48(9): 1667–1674.

57 Gargallo-Garriga A, Preece C, Sardans J, Oravec M, Urban O, Peñuelas J. Root exudate metabolomes change under drought and show limited capacity for recovery. *Sci. Rep.* 2018; 8(1): 12696.

58 Genard-Zielinski AC, Boissard C, Ormeño E, Lathière J, Reiter IM, Wortham H, et al. Seasonal variations of *Quercus pubescens* isoprene emissions from an in natura forest under drought stress and sensitivity to future climate change in the Mediterranean area. *Biogeosciences.* 2018; 15(15): 4711–4730.

59 Ghanbary E, Tabari Kouchaksaraei M, Zarafshar M, Bader KM, Mirabolfathy M, Ziaei M. Differential physiological and biochemical responses of *Quercus infectoria* and *Q. libani* to drought and charcoal disease. *Physiol. Plant.* 2020; 168(4): 876–892.

60 Haberstroh S, Kreuzwieser J, Lobo-Do-Vale R, Caldeira MC, Dubbert M, Werner C. Terpenoid emissions of two mediterranean woody species in response to drought stress. *Front. Plant Sci.* 2018; 9: 1071.

61 Holland V, Koller S, Lukas S, Brüggemann W. Drought- and frost-induced accumulation of soluble carbohydrates during accelerated senescence in *Quercus pubescens*. *Trees.* 2016; 30(1): 215–226.

62 Hu B, Simon J, Rennenberg H. Drought and air warming affect the species-specific levels of stress-related foliar metabolites of three oak species on acidic and calcareous soil. *Tree Physiol.* 2013; 33(5): 489–504.

63 Hu B, Simon J, Günthardt-Goerg MS, Arend M, Kuster TM, Rennenberg H. Changes in the dynamics of foliar N metabolites in oak saplings by drought and air warming depend on species and soil type. *PLoS One.* 2015; 10(5): e0126701.

64 Jafarnia S, Akbarinia M, Hosseinpour B, Modarres Sanavi S, Salami S. Effect of drought stress on some growth, morphological, physiological, and biochemical parameters of two different populations of *Quercus brantii*. *iForest.* 2018; 11(2): 212–220.

65 Jorge I, Navarro RM, Lenz C, Ariza D, Jorrín J. Variation in the holm oak leaf proteome at different plant developmental stages, between provenances and in response to drought stress. *Proteomics.* 2006; 6(S1): S207–S214.

66 Kannenberg SA, Novick KA, Phillips RP. Coarse roots prevent declines in whole-tree non-structural carbohydrate pools during drought in an isohydric and an anisohydric species. *Tree Physiol.* 2018; 38(4): 582–590.

67 Kebert M, Vuksanović V, Stefels J, Bojović M, Horák R, Kostić S, et al. Species-level differences in osmoprotectants and antioxidants contribute to stress tolerance of *Quercus robur* L., and *Q. cerris* L. seedlings under water deficit and high temperatures. *Plants.* 2022; 11(13): 1744.

68 Kurz C, Schmieden U, Strobel P, Wild A. The combined effect of CO_2, ozone, and drought on the radical scavenging system of young oak trees (*Quercus petraea*) – a phytothron study. *Chemosphere.* 1998; 36(4–5): 783–788.

69 Kwak MJ, Lee SH, Woo SY. Physiological and biochemical traits of different water and light intensities on cork oak (*Quercus suber* L.) seedlings. *Afr. J. Biotechnol.* 2012; 10(68): 15305–15319.

70 Lansac AR, Zaballos JP, Martin A. Seasonal water potential changes and proline accumulation in mediterranean shrubland species. *Vegetatio.* 1994; 113(2): 141–154.

71 Lavoir AV, Staudt M, Schnitzler JP, Landais D, Massol F, Rocheteau A, et al. Drought reduced monoterpene emissions from the evergreen Mediterranean oak *Quercus ilex*: results from a throughfall displacement experiment. *Biogeosciences.* 2009; 6(7): 1167–1180.

72 Li MH, Cherubini P, Dobbertin M, Arend M, Xiao WF, Rigling A. Responses of leaf nitrogen and mobile carbohydrates in different *Quercus* species/provenances to moderate climate changes. *Plant Biol.* 2013; 15(Suppl 1): 177–184.

73 Loewenstein NJ, Pallardy SG. Influence of a drying cycle on post-drought xylem sap abscisic acid and stomatal responses in young temperate deciduous angiosperms. *New Phytol.* 2002; 156(3): 351–361.

74 Maguire AJ, Kobe RK. Drought and shade deplete nonstructural carbohydrate reserves in seedlings of five temperate tree species. *Ecol. Evol.* 2015; 5(23): 5711–5721.

75 Meletiou-Christou MS, Rhizopoulou S, Diamantoglou S. Seasonal changes of carbohydrates, lipids and nitrogen content in sun and shade leaves from four mediterranean evergreen sclerophylls. *Environ. Exp. Bot.* 1994; 34(2): 129–140.

76 Merz MA, Donahue RA, Poulson ME. Physiological response of Garry Oak (*Quercus garryana*) seedlings to drought. *Northwest Sci.* 2017; 91(2): 140–159.

77 Mevy JP, Loriod B, Liu X, Corre E, Torres M, Büttner M, et al. Response of Downy Oak (*Quercus pubescens* Willd.) to climate change: transcriptome assembly, differential gene analysis and targeted metabolomics. *Plants.* 2020; 9(9): 1149.

78 Mu Z, Llusià J, Liu D, Ogaya R, Asensio D, Zhang C, et al. Seasonal and diurnal variations of plant isoprenoid emissions from two dominant species in Mediterranean shrubland and forest submitted to experimental drought. *Atmos. Environ.* 2018; 191: 105–115.

79 Munné-Bosch S, Peñuelas J, Asensio D, Llusià J. Airborne ethylene may alter antioxidant protection and reduce tolerance of Holm Oak to heat and drought dtress. *Plant Physiol.* 2004; 136(2): 2937–2947.

80 Nogués I, Llusià J, Ogaya R, Munné-Bosch S, Sardans J, Peñuelas J, et al. Physiological and antioxidant responses of *Quercus ilex* to drought in two different seasons. *Plant Biosyst.* 2013; 148(2): 268–278.
81 Ormeño E, Mévy JP, Vila B, Bousquet-Mélou A, Greff S, Bonin G, et al. Water deficit stress induces different monoterpene and sesquiterpene emission changes in Mediterranean species. Relationship between terpene emissions and plant water potential. *Chemosphere.* 2007; 67(2): 276–284.
82 Ormeño E, Ruffault J, Gutigny C, Madrigal J, Guijarro M, Hernando C, et al. Increasing cuticular wax concentrations in a drier climate promote litter flammability. *For. Ecol. Manage.* 2020; 473: 118242.
83 Oufir M, Schulz N, Sha Vallikhan PS, Wilhelm E, Burg K, Hausman JF, et al. Simultaneous measurement of proline and related compounds in oak leaves by high-performance ligand-exchange chromatography and electrospray ionization mass spectrometry for environmental stress studies. *J. Chromatogr. A.* 2009; 1216(7): 1094–1099.
84 Parker J, Patton RL. Effects of drought and defoliation on some metabolites in roots of Black Oak seedlings. *Can. J. For. Res.* 2011; 5(3): 457–463.
85 Passarinho JAP, Lamosa P, Baeta JP, Santos H, Ricardo CPP. Annual changes in the concentration of minerals and organic compounds of *Quercus suber* leaves. *Physiol. Plant.* 2006; 127(1): 100–110.
86 Pegoraro E, Rey A, Greenberg J, Harley P, Grace J, Malhi Y, et al. Effect of drought on isoprene emission rates from leaves of *Quercus virginiana* Mill. *Atmos. Environ.* 2004; 38(36): 6149–6156.
87 Peguero-Pina JJ, Sancho-Knapik D, Morales F, Flexas J, Gil-Pelegrín E. Differential photosynthetic performance and photoprotection mechanisms of three Mediterranean evergreen oaks under severe drought stress. *Funct. Plant Biol.* 2009; 36(5): 453.
88 Peguero-Pina JJ, Mendoza-Herrer Ó, Gil-Pelegrín E, Sancho-Knapik D. Cavitation Limits the Recovery of Gas Exchange after Severe Drought Stress in Holm Oak (*Quercus ilex* L.). *Forests.* 2018; 9(8): 443.
89 Peña-Rojas K, Aranda X, Joffre R, Fleck I. Leaf morphology, photochemistry and water status changes in resprouting *Quercus ilex* during drought. *Funct. Plant Biol.* 2005; 32(2): 117–130.
90 Peron A, Kaser L, Charlott Fitzky A, Graus M, Halbwirth H, Greiner J, et al. Combined effects of ozone and drought stress on the emission of biogenic volatile organic compounds from *Quercus robur* L. *Biogeosciences.* 2021; 18(2): 535–556.
91 Picon C, Ferhi A, Guehl JM. Concentration and $\sigma^{13}C$ of leaf carbohydrates in relation to gas exchange in *Quercus robur* under elevated CO_2 and drought. *J. Exp. Bot.* 1997; 48(8): 1547–1556.
92 Preece C, Farré-Armengol G, Llusià J, Peñuelas J. Thirsty tree roots exude more carbon. *Tree Physiol.* 2018; 38(5): 690–695.

93 Preece C, Farré-Armengol G, Verbruggen E, Peñuelas J. Interactive effects of soil water content and nutrients on root exudation in two Mediterranean tree species. *Soil Biol. Biochem.* 2021; 163: 108453.

94 Ramirez-Valiente JA, Koehler K, Cavender-Bares J. Climatic origins predict variation in photoprotective leaf pigments in response to drought and low temperatures in live oaks (*Quercus* series *Virentes*). *Tree Physiol.* 2015; 35(5): 521–534.

95 Rhizopoulou S, Meletiou-Christou MS, Diamantoglou S. Water relations for sun and shade leaves of four Mediterranean evergreen sclerophylls. *J. Exp. Bot.* 1991; 42(5): 627–635.

96 Rivas-Ubach A, Barbeta A, Sardans J, Guenther A, Ogaya R, Oravec M, et al. Topsoil depth substantially influences the responses to drought of the foliar metabolomes of Mediterranean forests. *Perspect. Plant Ecol. Evol. Syst.* 2016; 21: 41–54.

97 Robakowski P, Wyka TP, Kowalkowski W, Barzdajn W, Pers-Kamczyc E, Jankowski A, et al. Practical implications of different phenotypic and molecular responses of evergreen conifer and broadleaf deciduous forest tree species to regulated water deficit in a container nursery. *Forests.* 2020; 11(9): 1011.

98 Rodríguez-Calcerrada J, Jaeger C, Limousin JM, Ourcival JM, Joffre R, Rambal S. Leaf CO_2 efflux is attenuated by acclimation of respiration to heat and drought in a Mediterranean tree. *Funct. Ecol.* 2011; 25(5): 983–995.

99 Rodríguez-Calcerrada J, Shahin O, del Carmen del Rey M, Rambal S. Opposite changes in leaf dark respiration and soluble sugars with drought in two Mediterranean oaks. *Funct. Plant Biol.* 2011; 38(12): 1004.

100 Rodríguez-Calcerrada J, Buatois B, Chiche E, Shahin O, Staudt M. Leaf isoprene emission declines in *Quercus pubescens* seedlings experiencing drought – any implication of soluble sugars and mitochondrial respiration? *Environ. Exp. Bot.* 2013; 85: 36–42.

101 Rodríguez-Calcerrada J, Martin-StPaul NK, Lempereur M, Ourcival JM, del Rey MDC, Joffre R, et al. Stem CO_2 efflux and its contribution to ecosystem CO_2 efflux decrease with drought in a Mediterranean forest stand. *Agric. For. Meteorol.* 2014; 195–196: 61–72.

102 Rodríguez-Calcerrada J, Rodrigues AM, Perdiguero P, António C, Atkin OK, Li M, et al. A molecular approach to drought-induced reduction in leaf CO_2 exchange in drought-resistant *Quercus ilex*. *Physiol. Plant.* 2018; 162(4): 394–408.

103 Rodríguez-Calcerrada J, Rodrigues AM, António C, Perdiguero P, Pita P, Collada C, et al. Stem metabolism under drought stress – a paradox of increasing respiratory substrates and decreasing respiratory rates. *Physiol. Plant.* 2021; 172(2): 391–404.

104 Salomón RL, Rodríguez-Calcerrada J, Staudt M. Carbon losses from respiration and emission of volatile organic compounds—he overlooked side of tree carbon budgets. In: Gil-Peregrin E, Peguero-Pina JJ, & Sancho-Knapik D, editors. *Oaks Physiological Ecology. Exploring the Functional Diversity of Genus* Quercus *L.* Cham (CH): Springer; 2017. p. 327–359.
105 San-Eufrasio B, Sánchez-Lucas R, López-Hidalgo C, Guerrero-Sánchez VM, Castillejo MÁ, Maldonado-Alconada AM, et al. Responses and differences in tolerance to water shortage under climatic dryness conditions in seedlings from *Quercus* spp. and Andalusian *Q. ilex* populations. *Forests.* 2020; 11(6): 707.
106 Saunier A, Ormeño E, Wortham H, Temime-Roussel B, Lecareux C, Boissard C, et al. Chronic drought decreases anabolic and catabolic BVOC emissions of *Quercus pubescens* in a Mediterranean forest. *Front. Plant Sci.* 2017; 8: 71.
107 Saunier A, Greff S, Blande JD, Lecareux C, Baldy V, Fernandez C, et al. Amplified drought and seasonal cycle modulate *Quercus pubescens* leaf metabolome. *Metabolites.* 2022; 12(4): 307.
108 Savi T, Casolo V, Luglio J, Bertuzzi S, Trifilo' P, Lo Gullo MA, et al. Species-specific reversal of stem xylem embolism after a prolonged drought correlates to endpoint concentration of soluble sugars. *Plant Physiol. Biochem.* 2016; 106: 198–207.
109 Schulte M, Herschbach C, Rennenberg H. Interactive effects of elevated atmospheric CO_2, mycorrhization and drought on long-distance transport of reduced sulphur in young pedunculate oak trees (*Quercus robur* L.). *Plant Cell Environ.* 1998; 21(8): 917–926.
110 Schwanz P, Polle A. Differential stress responses of antioxidative systems to drought in pendunculate oak (*Quercus robur*) and maritime pine (*Pinus pinaster*) grown under high CO_2 concentrations. *J. Exp. Bot.* 2001; 52(354): 133–143.
111 Shure DJ, Mooreside PD, Ogle SM. Rainfall effects on plant–herbivore processes in an upland oak forest. *Ecology.* 1998; 79(2): 604–617.
112 Simova-Stoilova LP, Romero-Rodríguez MC, Sánchez-Lucas R, Navarro-Cerrillo RM, Medina-Aunon JA, Jorrín-Novo JV. 2-DE proteomics analysis of drought treated seedlings of *Quercus ilex* supports a root active strategy for metabolic adaptation in response to water shortage. *Front. Plant Sci.* 2015; 6: 627.
113 Spieß N, Oufir M, Matušíková I, Stierschneider M, Kopecky D, Homolka A, et al. Ecophysiological and transcriptomic responses of oak (*Quercus robur*) to long-term drought exposure and rewatering. *Environ. Exp. Bot.* 2012; 77: 117–126.
114 Spyropoulos CG, Mavrommatis M. Effect of water stress on pigment formation in *Quercus* species. *J. Exp. Bot.* 1978; 29(2): 473–477.

115 Staudt M, Rambal S, Joffre R, Kesselmeier J. Impact of drought on seasonal monoterpene emissions from *Quercus ilex* in southern France. *J. Geophys. Res.* 2002; 107(D21): 15/1–15/9.

116 Staudt M, Ennajah A, Mouillot F, Joffre R. Do volatile organic compound emissions of Tunisian cork oak populations originating from contrasting climatic conditions differ in their responses to summer drought? *Can. J. For. Res.* 2008; 38(12): 2965–2975.

117 Stefi AL, Nikou T, Papadopoulou S, Kalaboka M, Vassilacopoulou D, Halabalaki M, et al. The response of the laboratory cultivated *Quercus coccifera* plants to an artificial water stress. *Plant Stress.* 2022; 4: 100077.

118 Stojnić S, Pekeč S, Kebert M, Pilipović A, Stojanović D, Stojanović M, et al. Drought effects on physiology and biochemistry of Pedunculate Oak (*Quercus robur* L.) and Hornbeam (*Carpinus betulus* L.) saplings grown in urban area of Novi Sad, Serbia. *SEEFOR.* 2016; 7(1): 57–63.

119 Suseela V, Tharayil N, Xing B, Dukes JS. Warming and drought differentially influence the production and resorption of elemental and metabolic nitrogen pools in *Quercus rubra*. *Global Change Biol.* 2015; 21(11): 4177–4195.

120 Suseela V, Tharayil N, Orr G, Hu D. Chemical plasticity in the fine root construct of *Quercus* spp. varies with root order and drought. *New Phytol.* 2020; 228(6): 1835–1851.

121 Tienda-Parrilla M, López-Hidalgo C, Guerrero-Sanchez VM, Infantes-González Á, Valderrama-Fernández R, Castillejo MÁ, et al. Untargeted MS-based metabolomics analysis of the responses to drought stress in *Quercus ilex* L. leaf seedlings and the identification of putative compounds related to tolerance. *Forests.* 2022; 13(4): 551.

122 Tingey DT, Evans R, Gumpertz M. Effects of environmental conditions on isoprene emission from live oak. *Planta.* 1981; 152(6): 565–570.

123 Top SM, Preston CM, Dukes JS, Tharayil N. Climate influences the content and chemical composition of foliar tannins in green and senesced tissues of *Quercus rubra*. *Front. Plant Sci.* 2017; 8: 423.

124 Triboulot M, Fauveau M, Bréda N, Label P, Dreyer E. Stomatal conductance and xylem-sap abscisic acid (ABA) in adult oak trees during a gradually imposed drought. *Ann. For. Sci.* 1996; 53(2–3): 207–220.

125 Turnbull MH, Ogaya R, Barbeta A, Peñuelas J, Zaragoza-Castells J, Atkin OK, et al. Light inhibition of foliar respiration in response to soil water availability and seasonal changes in temperature in Mediterranean holm oak (*Quercus ilex*) forest. *Funct. Plant Biol.* 2017; 44(12): 1178–1193.

126 Vaz M, Pereira JS, Gazarini LC, David TS, David JS, Rodrigues A, et al. Drought-induced photosynthetic inhibition and autumn recovery in two Mediterranean oak species (*Quercus ilex* and *Quercus suber*). *Tree Physiol.* 2010; 30(8): 946–956.

127 Verdaguer D, Vilagran J, Lloansi S, Fleck I. Morphological and physiological acclimation of *Quercus coccifera* L. seedlings to water availability and growing medium. *New For.* 2011; 42(3): 363–381.
128 Vilagrosa A, Morales F, Abadía A, Bellot J, Cochard H, Gil-Pelegrin E. Are symplast tolerance to intense drought conditions and xylem vulnerability to cavitation coordinated? An integrated analysis of photosynthetic, hydraulic and leaf level processes in two Mediterranean drought-resistant species. *Environ. Exp. Bot.* 2010; 69(3): 233–242.
129 Villar-Salvador P, Planelles R, Oliet J, Peñuelas-Rubira JL, Jacobs DF, González M. Drought tolerance and transplanting performance of holm oak (*Quercus ilex*) seedlings after drought hardening in the nursery. *Tree Physiol.* 2004; 24(10): 1147–1155.
130 Wargo P. Consequences of environmental stress on oak: predisposition to pathogens. *Ann. Sci. For.* 1996; 53(2–3): 359–368.
131 Wolkerstorfer SV, Wonisch A, Stankova T, Tsvetkova N, Tausz M. Seasonal variations of gas exchange, photosynthetic pigments, and antioxidants in Turkey oak (*Quercus cerris* L.) and Hungarian oak (*Quercus frainetto* Ten.) of different age. *Trees.* 2011; 25(6): 1043–1052.
132 Wu M, Zhang WH, Ma C, Zhou JY. Changes in morphological, physiological, and biochemical responses to different levels of drought stress in chinese cork oak (*Quercus variabilis* Bl.) seedlings. *Russ. J. Plant Physiol.* 2013 Sep; 60(5): 681–692.
133 Xiong S, Wang Y, Chen Y, Gao M, Zhao Y, Wu L. Effects of drought stress and rehydration on physiological and biochemical properties of four oak species in China. *Plants.* 2022; 11(5): 679.
134 Xu CY, Griffin KL. Seasonal variation in the temperature response of leaf respiration in *Quercus rubra*: foliage respiration and leaf properties. *Funct. Ecol.* 2006; 20(5): 778–789.
135 Bartlett MK, Zhang Y, Kreidler N, Sun S, Ardy R, Cao K, et al. Global analysis of plasticity in turgor loss point, a key drought tolerance trait. *Ecol. Lett.* 2014; 17(12): 1580–1590.
136 Ramírez-Valiente JA, Cavender-Bares J. Evolutionary trade-offs between drought resistance mechanisms across a precipitation gradient in a seasonally dry tropical oak (*Quercus oleoides*). *Tree Physiol.* 2017; 37(7): 889–901.
137 Rouault G, Candau JN, Lieutier F, Nageleisen LM, Martin JC, Warzée N. Effects of drought and heat on forest insect populations in relation to the 2003 drought in Western Europe. *Ann. For. Sci.* 2006; 63(6): 613–624.
138 Rodríguez-Calcerrada J, Pérez-Ramos IM, Ourcival JM, Limousin JM, Joffre R, Rambal S. Is selective thinning an adequate practice for adapting *Quercus ilex* coppices to climate change? *Ann. For. Sci.* 2011; 68(3): 575–585.

139 Ferraz de Oliveira MI, Machado M, Cancela d'Abreu M. Acorn chemical composition depending on shedding date and *Quercus* species. In: De Pedro EJ, & Cabezas AB, editors. *Options Méditerranéennes: Série A. Séminaires Méditerranéens.* Seventh Edition. International Symposium on the Mediterranean Pig. Zaragoza: CIHEAM; 2012. p. 229–234; n. 101.

140 Zolfaghari R, Dalvand F, Fayyaz P, Solla A. Maternal drought stress on Persian oak (*Quercus brantii* Lindl.) affects susceptibility to single and combined drought and biotic stress in offspring. *Environ. Exp. Bot.* 2022; 194: 104716.

141 Echevarría-Zomeño S, Ariza D, Jorge I, Lenz C, del Campo A, Jorrín JV, et al. Changes in the protein profile of *Quercus ilex* leaves in response to drought stress and recovery. *J. Plant Physiol.* 2009; 166(3): 233–245.

142 Varone L, Gratani L. Leaf respiration responsiveness to induced water stress in Mediterranean species. *Environ. Exp. Bot.* 2015; 109: 141–150.

143 Pinheiro C, Chaves MM. Photosynthesis and drought: can we make metabolic connections from available data? *J. Exp. Bot.* 2011; 62(3): 869–882.

144 Hasanuzzaman M, Bhuyan MHM, Zulfiqar F, Raza A, Mohsin S, Mahmud J, et al. Reactive oxygen species and antioxidant defense in plants under abiotic stress: revisiting the crucial role of a universal defense regulator. *Antioxidants.* 2020; 9(8): 681.

145 Cho YH, Yoo SD. Signaling role of fructose mediated by FINS1/FBP in *Arabidopsis thaliana. PLos Genet.* 2011; 7(1): e1001263.

146 Rébeillé F, Bligny R, Martin JB, Douce R. Effect of sucrose starvation on sycamore (*Acer pseudoplatanus*) cell carbohydrate and Pi status. *Biochem. J.* 1985; 226(3): 679–684.

147 Rodríguez-Sánchez S, Ruiz-Matute AI, Alañón ME, Pérez-Coello MS, de Julio-Torres LF, Morales R, et al. Analysis of cyclitols in different *Quercus* species by gas chromatography-mass spectrometry. *J. Sci. Food Agric.* 2010; 90(10): 1735–1738.

148 Nishizawa-Yokoi A, Yabuta Y, Shigeoka S. The contribution of carbohydrates including raffinose family oligosaccharides and sugar alcohols to protection of plant cells from oxidative damage. *Plant Signaling Behav.* 2008; 3(11): 1016–1018.

149 Tan J, Wang C, Xiang B, Han R, Guo Z. Hydrogen peroxide and nitric oxide mediated cold- and dehydration-induced *myo*-inositol phosphate synthase that confers multiple resistances to abiotic stresses. *Plant Cell Environ.* 2013; 36(2): 288–299.

150 Okemo P, Long H, Cheng Y, Mundree S, Williams B. Stachyose triggers apoptotic like cell death in drought sensitive but not resilient plants. *Sci. Rep.* 2021; 11(1): 7099.

151 Panchal P, Miller AJ, Giri J. Organic acids: versatile stress-response roles in plants. *J. Exp. Bot.* 2021; 72(11): 4038–4052.

152 Shimada T, Saitoh T. Re-evaluation of the relationship between rodent populations and acorn masting: a review from the aspect of nutrients and defensive chemicals in acorns. *Popul. Ecol.* 2006; 48(4): 341–352.

153 Ain-Lhout F, Zunzunegui M, Diaz Barradas MC, Tirado R, Clavijo A, Garcia NF. Comparison of proline accumulation in two mediterranean shrubs subjected to natural and experimental water deficit. *Plant Soil.* 2001; 230(2): 175–183.

154 Dickson RE. Carbon and nitrogen allocation in trees. *Ann. Sci. For.* 1989; 46(Suppl): 631s–647s.

155 Salminen JP, Roslin T, Karonen M, Sinkkonen J, Pihlaja K, Pulkkinen P. Seasonal variation in the content of hydrolyzable tannins, flavonoid glycosides, and proanthocyanidins in oak leaves. *J. Chem. Ecol.* 2004; 30(9): 1693–1711.

156 Taib M, Rezzak Y, Bouyazza L, Lyoussi B. Medicinal uses, phytochemistry, and pharmacological activities of *Quercus* species. *Evid. Based Complement Alternat Med.* 2020; 2020: 1920683.

157 Fernández de Simón B, Aranda I, López-Hinojosa M, Miguel L, Cervera MT. Scion-rootstock interaction and drought systemic effect modulate the organ-specific terpene profiles in grafted *Pinus pinaster* Ait. *Environ. Exp. Bot.* 2021; 186: 104437.

158 Molina-García L, Martínez-Expósito R, Fernández-de Córdova ML, Llorent-Martínez EJ. Determination of the phenolic profile and antioxidant activity of leaves and fruits of Spanish *Quercus coccifera*. *J. Chem.* 2018; 2018: 1–9.

159 Hoch G, Richter A, Körner C. Non-structural carbon compounds in temperate forest trees. *Plant Cell Environ.* 2003; 26(7): 1067–1081.

160 Bazot S, Barthes L, Blanot D, Fresneau C. Distribution of non-structural nitrogen and carbohydrate compounds in mature oak trees in a temperate forest at four key phenological stages. *Trees.* 2013; 27(4): 1023–1034.

161 Robert EMR, Mencuccini M, Martínez-Vilalta J. The anatomy and functioning of the xylem in oaks. In: Gil-Peregrin E, Peguero-Pina JJ, & Sancho-Knapik D, editors. *Oaks Physiological Ecology. Exploring the Functional Diversity of Genus* Quercus *L.* Cham (CH): Springer; 2017. p. 261–302.

162 Canham CD, Kobe RK, Latty EF, Chazdon RL. Interspecific and intraspecific variation in tree seedling survival: Effects of allocation to roots versus carbohydrate reserves. *Oecologia.* 1999; 121(1): 1–11.

163 Osonubi O, Davies WJ. Solute accumulation in leaves and roots of woody plants subjected to water stress. *Oecologia.* 1978; 32(3): 323–332.

164 Kabeya D, Sakai S. The relative importance of carbohydrate and nitrogen for the resprouting ability of *Quercus crispula* seedlings. *Ann. Bot.* 2005; 96(3): 479–488.

165 He W, Liu H, Qi Y, Liu F, Zhu X. Patterns in nonstructural carbohydrate contents at the tree organ level in response to drought duration. *Global Change Biol.* 2020; 26(6): 3627–3638.

166 Brunner I, Herzog C, Dawes MA, Arend M, Sperisen C. How tree roots respond to drought. *Front. Plant Sci.* 2015; 6: 547.

167 Woodruff DR, Meinzer FC, Marias DE, Sevanto S, Jenkins MW, McDowell NG. Linking nonstructural carbohydrate dynamics to gas exchange and leaf hydraulic behavior in *Pinus edulis* and *Juniperus monosperma*. *New Phytol.* 2015; 206(1): 411–421.

168 Sancho-Knapik D, Escudero A, Mediavilla S, Scoffoni C, Zailaa J, Cavender-Bares J, et al. Deciduous and evergreen oaks show contrasting adaptive responses in leaf mass per area across environments. *New Phytol.* 2021; 230(2): 521–534.

169 Daszkowska-Golec A. The role of abscisic acid in drought stress: How ABA helps plants to cope with drought stress. In: Hossain MA, Wani SH, Bhattacharjee S, Burritt DJ, & Tran LSP, editors. *Drought Stress Tolerance in Plants Volume 2. Molecular and Genetic Perspectives.* Cham (CH): Springer; 2016. p. 123–151.

170 Salvi P, Manna M, Kaur H, Thakur T, Gandass N, Bhatt D, et al. Phytohormone signaling and crosstalk in regulating drought stress response in plants. *Plant Cell Rep.* 2021; 40(8): 1305–1329.

171 Castander-Olarieta A, Moncaleán P, Pereira C, Pěnčík A, Petřík I, Pavlović I, et al. Cytokinins are involved in drought tolerance of *Pinus radiata* plants originating from embryonal masses induced at high temperatures. *Tree Physiol.* 2021; 41(6): 912–926.

172 Brown HR. The theory of the rise of sap in trees: some historical and conceptual remarks. *Phys. Perspect.* 2014; 15(3): 320–358.

173 Barbeta A, Ogaya R, Peñuelas J. Comparative study of diurnal and nocturnal sap flow of *Quercus ilex* and *Phillyrea latifolia* in a Mediterranean holm oak forest in Prades (Catalonia, NE Spain). *Trees.* 2012; 26(5): 1651–1659.

174 Glad C, Regnard JL, Querou Y, Brun O, Morot-Gaudry JF. Flux and chemical composition of xylem exudates from Chardonnay grapevines: temporal evolution and effect of recut. *Am. J. Enol. Vitic.* 1992; 43(3): 275–282.

175 Patonnier MP, Peltier JP, Marigo G. Drought-induced increase in xylem malate and mannitol concentrations and closure of *Fraxinus excelsior* L. stomata. *J. Exp. Bot.* 1999; 50(336): 1223–1229.

176 Kozlowski TT. Carbohydrate sources and sinks in woody plants. *Bot. Rev.* 1992; 58(2): 107–222.

177 Heizmann U, Kreuzwieser J, Schnitzler JP, Brüggemann N, Rennenberg H. Assimilate transport in the xylem sap of pedunculate oak (*Quercus robur*) saplings. *Plant Biol.* 2001; 3(2): 132–138.

178 Dickson RE, Isebrands JS, Tomlinson PT. Distribution and metabolism of current photosynthate by single-flush northern red oak seedlings. *Tree Physiol.* 1990; 7(1–4): 65–77.

179 Avalos A, Legaz ME, Vicente C. The occurrence of lichen phenolics in the xylem sap of *Quercus pyrenaica*, their translocation to leaves and biological significance. *Biochem. Syst. Ecol.* 1986; 14(4): 381–384.

180 Rennenberg H, Loreto F, Polle A, Brilli F, Fares S, Beniwal RS, et al. Physiological responses of forest trees to heat and drought. *Plant Biol.* 2006; 8(5): 556–571.

181 Zhang C, Turgeon R. Mechanisms of phloem loading. *Curr. Opin. Plant Biol.* 2018; 43: 71–75.

182 Slewinski TL, Zhang C, Turgeon R. Structural and functional heterogeneity in phloem loading and transport. *Front. Plant Sci.* 2013; 4: 244.

183 Turgeon R, Wolf S. Phloem transport: cellular pathways and molecular trafficking. *Annu. Rev. Plant Biol.* 2009; 60(1): 207–221.

184 Öner-Sieben S, Lohaus G. Apoplastic and symplastic phloem loading in *Quercus robur* and *Fraxinus excelsior*. *J. Exp. Bot.* 2014; 65(7): 1905–1916.

185 Rennie EA, Turgeon R. A comprehensive picture of phloem loading strategies. *Proc. Natl. Acad. Sci.* 2009; 106(33): 14162–14167.

186 Fu Q, Cheng L, Guo Y, Turgeon R. Phloem loading strategies and water relations in trees and herbaceous plants. *Plant Physiol.* 2011; 157(3): 1518–1527.

187 Gamalei Y. Structure and function of leaf minor veins in trees and herbs. *Trees.* 1989; 3(2): 96–110.

188 Miehe W, Czempik L, Klebl F, Lohaus G. Sugar concentrations and expression of SUTs suggest active phloem loading in tall trees of *Fagus sylvatica* and *Quercus robur*. *Tree Physiol.* 2022; 43: tpac152.

189 Fink D, Dobbelstein E, Barbian A, Lohaus G. Ratio of sugar concentrations in the phloem sap and the cytosol of mesophyll cells in different tree species as an indicator of the phloem loading mechanism. *Planta.* 2018; 248(3): 661–673.

190 Gersony JT, McClelland A, Holbrook NM. Raman spectroscopy reveals high phloem sugar content in leaves of canopy red oak trees. *New Phytol.* 2021; 232(1): 418–424.

191 Oparka KJ, Prior DAM. Direct evidence for pressure-generated closure of plasmodesmata. *Plant J.* 1992; 2(5): 741–750.

192 Park K, Knoblauch J, Oparka K, Jensen KH. Controlling intercellular flow through mechanosensitive plasmodesmata nanopores. *Nat. Commun.* 2019; 10(1): 3564.

193 Liesche J. Sucrose transporters and plasmodesmal regulation in passive phloem loading. *J. Integr. Plant Biol.* 2017; 59(5): 311–321.

194 Arndt SK, Livesley SJ, Merchant A, Bleby TM, Grierson PF. Quercitol and osmotic adaptation of field-grown Eucalyptus under seasonal drought stress. *Plant Cell Environ.* 2008; 31(7): 915–924.

195 Hoad GV. Transport of hormones in the phloem of higher plants. *Plant Growth Regul.* 1995; 16(2): 173–182.

196 Lucas WJ, Groover A, Lichtenberger R, Furuta K, Yadav SR, Helariutta Y, et al. The plant vascular system: evolution, development and functions. *J. Integr. Plant Biol.* 2013; 55(4): 294–388.

197 Münch E. *Stoffbewegungen in der Pflanze.* G. Fischer; 1930: 234pp.

198 Jensen KH, Savage JA, Holbrook NM. Optimal concentration for sugar transport in plants. *J. R. Soc. Interface.* 2013; 10(83): 20130055.

199 Savage JA, Beecher SD, Clerx L, Gersony JT, Knoblauch J, Losada JM, et al. Maintenance of carbohydrate transport in tall trees. *Nat. Plants.* 2017; 3(12): 965–972.

200 Gersony JT, Hochberg U, Rockwell FE, Park M, Gauthier PPG, Holbrook NM. Leaf carbon export and nonstructural carbohydrates in relation to diurnal water dynamics in mature oak trees. *Plant Physiol.* 2020; 183(4): 1612–1621.

201 Faria T, Wilkins D, Besford RT, Vaz M, Pereira JS, Chaves MM. Growth at elevated CO_2 leads to down-regulation of photosynthesis and altered response to high temperature in *Quercus suber* L. seedlings. *J. Exp. Bot.* 1996; 47(11): 1755–1761.

202 Savage JA, Clearwater MJ, Haines DF, Klein T, Mencuccini M, Sevanto S, et al. Allocation, stress tolerance and carbon transport in plants: how does phloem physiology affect plant ecology? *Plant Cell Environ.* 2016; 39(4): 709–725.

203 Sevanto S. Phloem transport and drought. *J. Exp. Bot.* 2014; 65(7): 1751–1759.

204 Konrad W, Katul G, Roth-Nebelsick A, Jensen KH. Xylem functioning, dysfunction and repair: a physical perspective and implications for phloem transport. *Tree Physiol.* 2019; 39(2): 243–261.

205 Walz C, Juenger M, Schad M, Kehr J. Evidence for the presence and activity of a complete antioxidant defence system in mature sieve tubes. *Plant J.* 2002; 31(2): 189–197.

206 Tegeder M. Transporters involved in source to sink partitioning of amino acids and ureides: opportunities for crop improvement. *J. Exp. Bot.* 2014; 65(7): 1865–1878.

207 Carlisle A. Carbohydrates in the precipitation beneath a sessile oak *Quercus petraea* (Mattushka) Liebl. Canopy. *Plant Soil.* 1965; 22(3): 399–400.

208 Dennis PG, Miller AJ, Hirsch PR. Are root exudates more important than other sources of rhizodeposits in structuring rhizosphere bacterial communities? *FEMS Microbiol. Ecol.* 2010; 72(3): 313–327.

209 van Dam NM, Bouwmeester HJ. Metabolomics in the rhizosphere: tapping into belowground chemical communication. *Trends Plant Sci.* 2016; 21(3): 256–265.

210 Bais HP, Weir TL, Perry LG, Gilroy S, Vivanco JM. The role of root exudates in rhizosphere interactions with plants and other organisms. *Annu. Rev. Plant Biol.* 2006; 57(1): 233–266.

211 Pugnaire FI, Morillo JA, Peñuelas J, Reich PB, Bardgett RD, Gaxiola A, et al. Climate change effects on plant-soil feedbacks and consequences for biodiversity and functioning of terrestrial ecosystems. *Sci. Adv.* 2019; 5(11): eaaz1834.

212 Nguyen C. Rhizodeposition of organic C by plant: mechanisms and controls. In: Lichtfouse E, Navarrete M, Debaeke P, Véronique S, & Alberola C, editors. *Sustainable Agriculture.* Dordrecht (NL): Springer; 2009. p. 97–123.

213 Sun L, Ataka M, Kominami Y, Yoshimura K. Relationship between fine-root exudation and respiration of two *Quercus* species in a Japanese temperate forest. *Tree Physiol.* 2017; 37(8): 1011–1020.

214 Preece C, Peñuelas J. Rhizodeposition under drought and consequences for soil communities and ecosystem resilience. *Plant Soil.* 2016; 409(1–2): 1–17.

215 Jones DL, Hodge A, Kuzyakov Y. Plant and mycorrhizal regulation of rhizodeposition. *New Phytol.* 2004; 163(3): 459–480.

216 Cesco S, Mimmo T, Tonon G, Tomasi N, Pinton R, Terzano R, et al. Plant-borne flavonoids released into the rhizosphere: impact on soil bio-activities related to plant nutrition. A review. *Biol. Fertil. Soils.* 2012; 48(2): 123–149.

217 Williams A, Vries FT. Plant root exudation under drought: implications for ecosystem functioning. *New Phytol.* 2020; 225(5): 1899–1905.

218 Prescott CE, Grayston SJ, Helmisaari HS, Kaštovská E, Körner C, Lambers H, et al. Surplus carbon drives allocation and plant–soil interactions. *Trends Ecol. Evol.* 2020; 35(12): 1110–1118.

219 Naylor D, Coleman-Derr D. Drought stress and root-associated bacterial communities. *Front. Plant Sci.* 2018; 8: 2223.

220 Zwetsloot MJ, Kessler A, Bauerle TL. Phenolic root exudate and tissue compounds vary widely among temperate forest tree species and have contrasting effects on soil microbial respiration. *New Phytol.* 2018; 218(2): 530–541.

221 Hinsinger P. Bioavailability of soil inorganic P in the rhizosphere as affected by root-induced chemical changes: a review. *Plant Soil.* 2001; 237(2): 173–195.

222 IPCC (2023). Climate change 2021: the physical science basis. Contribution of Working Group I to the Sixth Assessment Report of the Intergovernmental Panel on Climate Change [Internet]. Available from (https://www.ipcc.ch/report/ar6/wg1/), accessed 3 March 2023.

223 Barba J, Bradford MA, Brewer PE, Bruhn D, Covey K, Haren J, et al. Methane emissions from tree stems: a new frontier in the global carbon cycle. *New Phytol.* 2019; 222(1): 18–28.

224 Covey KR, Megonigal JP. Methane production and emissions in trees and forests. *New Phytol.* 2019; 222(1): 35–51.
225 Plain C, Ndiaye F, Bonnaud P, Ranger J, Epron D. Impact of vegetation on the methane budget of a temperate forest. *New Phytol.* 2019; 221(3): 1447–1456.
226 Dudareva N, Negre F, Nagegowda DA, Orlova I. Plant volatiles: recent advances and future perspectives. *Crit. Rev. Plant Sci.* 2006; 25(5): 417–440.
227 Vrkočová P, Valterová I, Vrkoč J, Koutek B. Volatiles released from oak, a host tree for the bark beetle *Scolytus intricatus*. *Biochem. Syst. Ecol.* 2000; 28(10): 933–947.
228 Peñuelas J, Llusià J. Plant VOC emissions: making use of the unavoidable. *Trends Ecol. Evol.* 2004; 19(8): 402–404.
229 Laothawornkitkul J, Taylor JE, Paul ND, Hewitt CN. Biogenic volatile organic compounds in the Earth system. *New Phytol.* 2009; 183(1): 27–51.
230 Peñuelas J, Staudt M. BVOCs and global change. *Trends Plant Sci.* 2010; 15(3): 133–144.
231 Monson RK, Jaeger CH, Adams WW, Driggers EM, Silver GM, Fall R. Relationships among isoprene emission rate, photosynthesis, and isoprene synthase activity as influenced by temperature. *Plant Physiol.* 1992; 98(3): 1175–1180.
232 Staudt M, Bertin N. Light and temperature dependence of the emission of cyclic and acyclic monoterpenes from holm oak (*Quercus ilex* L.) leaves. *Plant Cell Environ.* 1998; 21(4): 385–395.
233 Grote R, Niinemets Ü. Modeling volatile isoprenoid emissions – a story with split ends. *Plant Biol.* 2008; 10(1): 8–28.
234 Niinemets Ü, Arneth A, Kuhn U, Monson RK, Peñuelas J, Staudt M. The emission factor of volatile isoprenoids: stress, acclimation, and developmental responses. *Trends Plant Sci.* 2010; 7(7): 2203–2223.
235 Loreto F, Schnitzler JP. Abiotic stresses and induced BVOCs. *Trends Plant Sci.* 2010; 15(3): 154–166.
236 Monson RK, Jones RT, Rosenstiel TN, Schnitzler JP. Why only some plants emit isoprene. *Plant Cell Environ.* 2013; 36(3): 503–516.
237 Karl M, Guenther A, Köble R, Leip A, Seufert G. A new European plant-specific emission inventory of biogenic volatile organic compounds for use in atmospheric transport models. *Biogeosciences.* 2009; 6(6): 1059–1087.
238 Loreto F, Ciccioli P, Brancaleoni E, Cecinato A, Frattoni M, Sharkey TD. Different sources of reduced carbon contribute to form three classes of terpenoid emitted by *Quercus ilex* L. leaves. *Proc. Natl. Acad. Sci. U.S.A.* 1996; 93(18): 9966–9969.
239 Guenther A, Hewitt CN, Erickson D, Fall R, Geron C, Graedel T, et al. A global model of natural volatile organic compound emissions. *J. Geophys. Res.* 1995; 100(D5): 8873.

240 Fischbach RJ, Zimmer I, Steinbrecher R, Pfichner A, Schnitzler JP. Monoterpene synthase activities in leaves of *Picea abies* (L.) Karst. and *Quercus ilex* L. *Phytochemistry*. 2000; 54(3): 257–265.

241 Singsaas EL, Sharkey TD. The effects of high temperature on isoprene synthesis in oak leaves. *Plant Cell Environ.* 2000; 23(7): 751–757.

242 Copolovici LO, Filella I, Llusià J, Niinemets U, Peñuelas J. The capacity for thermal protection of photosynthetic electron transport varies for different monoterpenes in *Quercus ilex*. *Plant Physiol.* 2005; 139(1): 485–496.

243 Velikova V, Várkonyi Z, Szabó M, Maslenkova L, Nogues I, Kovács L, et al. Increased thermostability of thylakoid membranes in isoprene-emitting leaves probed with three biophysical techniques. *Plant Physiol.* 2011; 157(2): 905–916.

244 Bsaibes S, Piel F, Gros V, Truong F, Lafouge F, Ciuraru R, et al. Monoterpene chemical speciation with high time resolution using a FastGC/PTR-MS: results from the COV3ER experiment on *Quercus ilex*. *Atmosphere (Basel)*. 2020; 11(7): 690.

245 Staudt M, Mandl N, Joffre R, Rambal S. Intraspecific variability of monoterpene composition emitted by *Quercus ilex* leaves. *Can. J. For. Res.* 2011; 31(1): 174–180.

246 Niinemets U, Seufert G, Steinbrecher R, Tenhunen JD. A model coupling foliar monoterpene emissions to leaf photosynthetic characteristics in Mediterranean evergreen *Quercus* species. *New Phytol.* 2002; 153(2): 257–275.

247 Fischbach RJ, Staudt M, Zimmer I, Rambal S, Schnitzler JP. Seasonal pattern of monoterpene synthase activities in leaves of the evergreen tree *Quercus ilex*. *Physiol. Plant.* 2002; 114(3): 354–360.

248 Loreto F. Distribution of isoprenoid emitters in the *Quercus* genus around the world: chemo-taxonomical implications and evolutionary considerations based on the ecological function of the trait. *Perspect. Plant Ecol. Evol. Syst.* 2002; 5(3): 185–192.

249 Feng Z, Yuan X, Fares S, Loreto F, Li P, Hoshika Y, et al. Isoprene is more affected by climate drivers than monoterpenes: a meta-analytic review on plant isoprenoid emissions. *Plant Cell Environ.* 2019; 42(6): 1939–1949.

250 Brüggemann N, Schnitzler JP. Comparison of isoprene emission, intercellular isoprene concentration and photosynthetic performance in water-limited oak (*Quercus pubescens* Willd. and *Quercus robur* L.) saplings. *Plant Biol.* 2002; 4(4): 456–463.

251 Loreto F, Fischbach RJ, Schnitzler JP, Ciccioli P, Brancaleoni E, Calfapietra C, et al. Monoterpene emission and monoterpene synthase activities in the Mediterranean evergreen oak *Quercus ilex* L. grown at elevated CO_2 concentrations. *Global Change Biol.* 2001; 7(6): 709–717.

252 Sharkey TD, Yeh S. Isoprene emission from plants. *Annu. Rev. Plant Physiol. Plant Mol. Biol.* 2001; 52(1): 407–436.

253 Niinemets Ü. Mild versus severe stress and BVOCs: thresholds, priming and consequences. *Trends Plant Sci.* 2010; 15(3): 145–153.
254 Fortunati A, Barta C, Brilli F, Centritto M, Zimmer I, Schnitzler JP, et al. Isoprene emission is not temperature-dependent during and after severe drought-stress: a physiological and biochemical analysis. *Plant J.* 2008; 55(4): 687–697.
255 Capitani D, Brilli F, Mannina L, Proietti N, Loreto F. In situ investigation of leaf water status by portable unilateral nuclear magnetic resonance. *Plant Physiol.* 2009; 149(4): 1638–1647.
256 Fares S, Mereu S, Scarascia Mugnozza G, Vitale M, Manes F, Frattoni M, et al. The ACCENT-VOCBAS field campaign on biosphere-atmosphere interactions in a Mediterranean ecosystem of Castelporziano (Rome): site characteristics, climatic and meteorological conditions, and eco-physiology of vegetation. *Biogeosciences.* 2009; 6(6): 1043–1058.

Further Reading/Resources

Aranda I, Gil-Pelegrín E, Gascó A, Guevara MA, Cano JF, Miguel M, et al. Drought response in forest trees: from the species to the gene. In: Aroca R, editor. *Plant Responses to Drought Stress.* Berlin Heidelberg: Springer; 2012. p. 293–333.
Blumstein M, Gersony J, Martínez-Vilalta J, Sala A. Global variation in nonstructural carbohydrate stores in response to climate. *Global Change Biol.* 2023; 29(7): 1854–1869.
Hartmann H, Bastos A, Das AJ, Esquivel-Muelbert A, Hammond WM, Martínez-Vilalta J, et al. Climate change risks to global forest health: emergence of unexpected events of elevated tree mortality worldwide. *Annu. Rev. Plant Biol.* 2022; 73(5): 673–702.
Jogawat A, Yadav B, Lakra N, Singh AK, Narayan OP. Crosstalk between phytohormones and secondary metabolites in the drought stress tolerance of crop plants: a review. *Physiol. Plant.* 2021; 172(2): 1106–1132.
Laxa M, Liebthal M, Telman W, Chibani K, Dietz K-J. The role of the plant antioxidant system in drought tolerance. *Antioxidants.* 2019; 8(4): 94.

11

Metabolomics of Forest Tree Responses to Fluctuations of Temperature and Elevated Atmospheric CO_2

Fernanda Rezende Castro-Moretti[1], *Daniela Feltrim*[2], *Sara Adrián Lopez de Andrade*[2], *and Paulo Mazzafera*[2]

[1] *Department of Plant Pathology, School of Agriculture Luiz de Queiroz, University of São Paulo, Piracicaba, SP, Brazil*
[2] *Department of Plant Biology, Institute of Biology, University of Campinas, Campinas, SP, Brazil*

11.1 Introduction

Metabolome is the group of components, or metabolites, resulting from gene expression which defines the plant biochemical phenotype (1). Metabolomics is a state-of-the-art technique to study the metabolome, usually performed by gas or liquid chromatography (GC and LC, respectively) coupled to mass spectrometry- (MS or MS/MS if in tandem) or also by using nuclear magnetic resonance (NMR)-based approaches. Because metabolite concentration in plant tissues can be found in the micromole scale (μmol, 10^{-6}) and even lower, the sensitivity and precision staged by metabolomics analyses allow the detection of minimal perturbations in the metabolome (2).

Metabolomics, however, generates large amounts of data that need processing and curation before statistical analysis. The use of spectral libraries and standards accredit metabolite identification. However, forest tree metabolomics research has advanced to a lesser extent regarding other crops because of the difficulties involved in tree studies, such as long life cycles and less availability of genomic information (3). Additionally, forest tree species have challenging cultivation in greenhouse/laboratory conditions and difficult access for sample collection. *Eucalyptus, Pinus, Populus,* and *Quercus* are the most studied genera because of their economic importance (3–6). The usual pipelines using metabolomics for tree studies are illustrated in Figure 11.1.

Forests have been exposed to significantly higher atmospheric CO_2 levels and more drastic temperature alterations imposed by climate change, reducing their chance to overcome such restrictions to healthy growth and reproduction (7, 8). Even though trees can adapt to stress, it is still unknown if they

Monitoring Forest Damage with Mass Spectrometry-Based Metabolomics Methods, First Edition.
Edited by Carla António.

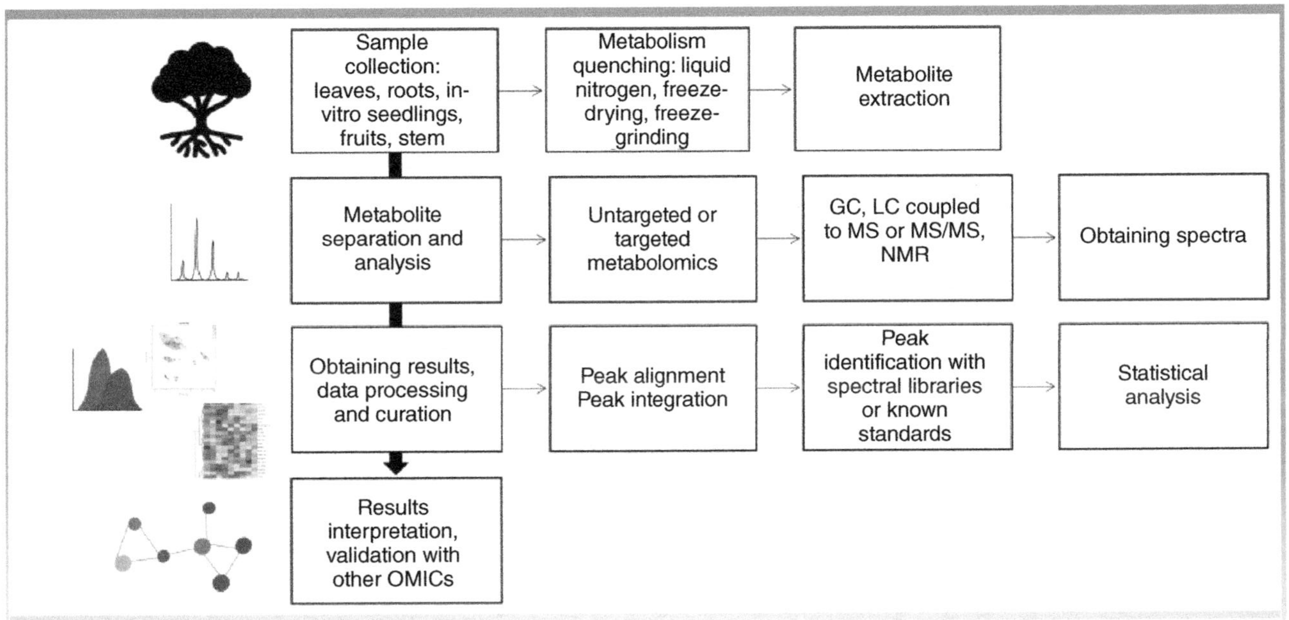

Figure 11.1 Pipeline for metabolomics studies in trees.

will be able to endure the consequences of climate change. Some studies affirm that trees cannot overcome those effects, despite their phenological plasticity (9, 10). A study performed on poplar compared tree recovery after heat stress, and they concluded that the trees could recover when the stress was applied from three to six hours, but there was irreparable damage in the photosynthetic machinery when the stress was applied for longer periods (11). More intense and frequent wildfires, below- and above-average temperature levels, and higher concentrations of atmospheric CO_2 levels are some of the consequences of climate change that could drastically affect the tree metabolome.

Climate change is defined as significant shifts in the weather pattern for an extended period, which can affect local temperature and rainfall (12). Consequently, changes in the atmosphere are expected, such as variations in atmospheric CO_2 levels. Additionally, above-average surface and ocean temperatures have been reported to increase yearly (13). Because forests significantly impact the carbon cycle and local climate, it is imperative to understand how tree metabolism will cope with temperature and atmospheric CO_2 alterations.

Extreme temperatures and CO_2 level increments are the two most important effects of global warming and climate change. They exert influence on plant metabolism and plant physiology. It is challenging to exclude one factor from another as high CO_2 is usually associated with increasing air temperature (14, 15). Temperature stress in plants, for instance, is not easy to define, as temperature variation is a natural event that fluctuates during the daytime and through seasons during the year (16). However, fluctuations and increasing temperatures have been intensified by climate change (17, 18). A study on three *Eucalyptus* species grown under different temperatures revealed that chlorophyll content is directly correlated with temperature (19). The same work also showed that sucrose is highly concentrated at higher temperatures and has a low abundance when the trees are grown under low temperatures. High temperature is one of the main factors for tree mortality in the natural environment (20). Additionally, stress due to above-average temperature negatively impacts key functions in plant cells, such as photosynthesis, respiration, cell division, growth, and membrane permeability (21). High temperature also affects secondary tree metabolism. A study on *Eucalyptus camaldulensis* showed a higher emission of isoprenes when trees are exposed to heat stress (22).

The atmospheric CO_2 levels, compared to temperature, are more straightforward to measure and have risen since the industrial revolution (23). Ghannoum et al. (24) found that in plants of *Eucalyptus sideroxylon* and *Eucalyptus saligna*, light-saturated photosynthesis (Asat) doubled under elevated CO_2 (eCO_2) (650 ppm), and maximal electron transport was stimulated by 10% (25). These results were further supported by the faster growth of the same species under 600 ppm CO_2 at 30 °C (24). Conversely, Crous et al. (26) showed that photosynthesis was negatively affected in *Eucalyptus globulus* in response to high temperatures during the summer, restored by the seasonal colder temperatures in the winter (26). Quentin et al. (27) found that photosynthesis and biomass accumulation was stimulated under 630 ppm CO_2 but not by warming (27). Aspinwall et al. (28) studied the effects of eCO_2 and temperature in several *E. camaldulensis* genotypes with higher total dry mass production under eCO_2 also increased root mass fraction and photosynthetic nitrogen use efficiency (28). On the other hand, phosphorous deficiency decreased the mass gain of eucalypt plants exposed to eCO_2 for three years (29), while drought suppressed the effects of eCO_2 and high temperatures (30).

11.2 Metabolic Response of Trees to Temperature Changes

Most plants are mesophile organisms that thrive at 10 and 30 °C (16). The negative effects of thermal stress on plants have been extensively studied,

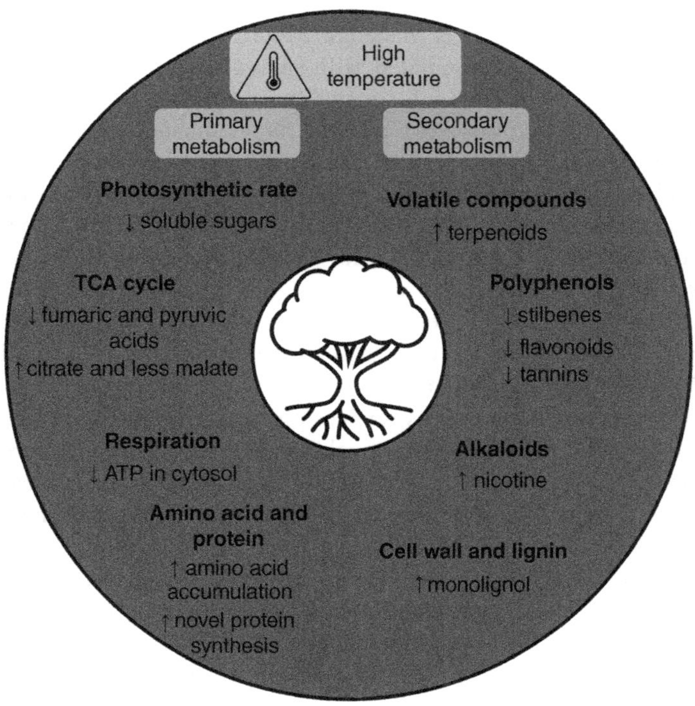

Figure 11.2 Effects of high temperature on tree primary and secondary metabolism. ↑, higher abundance; ↓, less abundance in comparison to non-stressed trees.

as non-optimal temperatures reduce plant fitness and yield. Nonetheless, tree metabolic response to temperature stress has more limited literature. Temperate forests are more exposed to chilling and sub-zero temperatures, which can cause permanent damage to fruit, flower, and leaf tissues. Avoidance of ice formation and tolerance to freezing are the main mechanisms used by trees in cold regions to cope with below- or near-zero temperatures (31). Heat damage is equally detrimental to tree fitness. Protein denaturation results from high-temperature exposure and can lead to cell death (16). Additionally, heat stress disturbs carbon assimilation, causing damage to leaf tissue and irreparable disruption of the photosynthetic machinery if the exposure is too long (32). The most relevant effects of heat stress on tree primary and secondary metabolism are illustrated in Figure 11.2.

11.3 Temperature Effect on Primary Metabolism

Primary metabolism comprises the essential metabolites for plant growth, development, and reproduction. Carbohydrates, phosphorylated sugars, amino acids, and organic acids are traditionally related to primary metabolism and directly linked to cellular processes, such as photosynthesis, respiration, protein synthesis, glycolysis, and other essential processes for cell functioning. Measuring these compounds and their changes when plants are challenged with temperature stress gives us clues to understand how plant cell biosynthetic pathways are affected and, eventually, plant growth and reproduction. For instance, a study on *Halimium halimifolium* L., a shrub from the Mediterranean, showed that extreme heat reduced net CO_2 assimilation and increased day respiration (33). Extreme high temperatures during plant reproduction negatively affect pollination and might result in poor fruit development (34). Regarding secondary growth, a study in *Eucalyptus grandis* revealed that increasing temperatures reduce the proportion of mass to the volume within fiber cells, resulting in higher wood density (35).

11.3.1 Carbohydrates and Photosynthesis

Carbohydrates are essential constituents of plant structure, signaling, and function. Photosynthesis is the main mechanism of carbon acquisition to support growth and reproduction, and it is affected by temperature in trees (36). A study on pistachio (*Pistacia vera*) trees showed that carbon metabolism is affected by temperature gradients (37). In this study, soluble sugar concentration in leaves and roots shifted due to differences in temperature in soil and the canopy. Furthermore, temperature directly correlates with enzyme activity, photorespiration, and cell membrane permeability (32, 38, 39). Therefore, drastic temperature increases could negatively affect the entire tree's metabolism. High temperatures adversely affect the carbon balance in *Thuja occidentalis*, and the effect is even more substantial under drought (40). Similarly, a study with Norway spruce (*Picea abies*) showed that photosynthetic rates and total soluble sugar concentration were reduced when seedlings were exposed to high temperatures (41). A metabolomic and transcriptomics analysis on poplar (*Populus tormentosa*) under heat stress revealed that soluble sugars decreased by more than 30% compared with plants under optimal temperature, which was consistent with the downregulation of genes related to the photosystems PSI and PSII (42).

11.3.2 Respiratory Metabolism

Respiratory metabolism includes glycolysis, the oxidative pentose phosphate pathway, the tricarboxylic acid (TCA) cycle, and the mitochondrial electron transport chain (43). The TCA is central in the catabolism of sugars and fatty acids. Under heat stress, there is an inhibition of the TCA cycle (44). In *P. tormentosa*, an untargeted GC–MS analysis combined with transcriptomics corroborated that TCA was inhibited when the trees grew under prolonged heat periods (42). The TCA citrate lyase gene was downregulated while, consistently, fumaric and pyruvic acid levels were less abundant in trees under heat stress. In another GC–MS study with *E. globulus*, citrate abundance was higher under heat stress, while malate content decreased (45). Interestingly, when combined with drought, the TCA cycle products had a different response: citrate was still more abundant but with a reduction in succinate and no alterations in malate. In Arabidopsis, genes related to the TCA cycle were also downregulated under heat stress (46). Additionally, metabolomics performed on a GC–TOF–MS confirmed the cycle suppression, as its products were less abundant than the control.

Respiration provides energy for plant growth. Under heat stress, plants need higher maintenance of respiration rates, which may result in a reduction of non-structural carbohydrates and less ATP transported to the cytosol (47). However, long-term heat and drought events may decrease photosynthesis may lead to carbon depletion diminishing respiratory demands and decreasing respiration at the ecosystem level (43, 48).

11.3.3 Amino Acids and Proteins

Amino acids have essential roles in plant primary metabolism as components of proteins. They are also important signaling molecules responding to abiotic and biotic stress and are involved in carbon assimilation and energy metabolism, as alternative respiratory substrates, in addition to many other essential functions (49–51). Plants under temperature stress have a higher accumulation of amino acids, probably due to hydrolysis or inhibition of protein synthesis (52). For instance, Arabidopsis plants grown under low temperatures accumulated proline, ornithine, citrulline, and putrescine, which are metabolites related to cold tolerance (53). Similarly, low-temperature stress up-regulated genes related to amino acid pathways and induced amino acid-derivatives accumulation in apple trees, such as γ-aminobutyric acid (GABA) (54). High temperatures also induced proline accumulation in the roots and leaves of palm seedlings (55). Interestingly, grain-filling and fruit quality parallels can be established regarding high temperatures and amino acid content. In rice, temperature stress impaired grain filling, reduced starch accumulation, and induced the accumulation of amino acid derivatives (i.e. GABA, putrescine, and glutathione) (56). In mandarins, heat stress reduced fruit quality and increased the pulp's amino acid. Glycine, alanine, serine, proline, valine, threonine, and leucine accumulated in trees exposed to high temperatures (57). *E. grandis* plants accumulated proline in the shoot of both cold and heat stress exposure (58). Shoot SDS/PAGE analysis revealed differences between acclimated and heat-stroke treatments, suggesting that novel proteins were synthesized under stress. In poplar leaves, a proteomics study identified 11 proteins that were differentially affected in concentration and synthesis under heat stress compared to the control, and most of them were related to photosynthesis (59).

11.4 Temperature Effect on Secondary Metabolism

Secondary metabolites are small molecules produced by plants and derived from the primary metabolism. Secondary metabolism, also known as specialized metabolism, includes three major categories: phenolics, terpenes, and alkaloids. Secondary metabolites are essential for plant growth, defense against abiotic and biotic stress, and reproduction (60–64). Specialized metabolites can be volatile or non-volatile and are comprised of three major groups: phenolics, terpenes, and alkaloids (65–67). In trees, apart from the functions known in other plants, secondary metabolites are involved in the production of heartwood (68). Also, trees produce specialized metabolites with medicinal and nutritional properties. Caffeine obtained from *Coffea* spp., flavonoids glycosides from *Ginkgo biloba,* antibacterial components and spice obtained from *Cinnamomum* spp. are some examples of substances from trees with great importance for the pharmaceutical and food industries (69–71). Secondary metabolite production is highly regulated by temperature. Under cold stress, plants produce more phenolic compounds and lignin derivatives to protect them from freeze damage (72). High temperature, in contrast, has been shown to reduce phenolics and flavonoid abundance (73). A study in *Salix* spp. revealed that high-temperature stress reduced apigenin, cyanidin, quercetin, and luteolin production (74). Combined with CO_2 increase, however, the concentration of those compounds was not affected.

11.4.1 Oxidative Stress

The production of reactive oxygen species (ROS) in plants is induced by heat stress. A metabolomics study on poplar (*P. tormentosa*) showed an increase in H_2O_2 production when the trees were subjected to high temperatures (43). Also, the activity of the enzymes directly related to ROS, superoxide dismutase, and catalase, increased. Consequently, proline, which acts in ROS detoxification, was accumulated in *P. tormentosa* under high-temperature stress. Similarly, *Pinus radiata* had proline accumulation as a response to heat stress, so did serine and phenylalanine (75). Phenylalanine was considered a possible marker for thermotolerance, as it was strongly correlated to high-temperature treatments and *P. radiata* adaptation. Also, a short-term response to high temperatures was indicated by the accumulation of ascorbic acid and glutathione, both antioxidant molecules in the ROS scavenging mechanisms plants have.

11.4.2 Phenylpropanoid Pathway

The phenylpropanoid pathway is a class of phenolic compounds and is one of the most studied metabolic routes in secondary plant metabolism. It produces a large array of secondary metabolites involved in plant response to environmental and biotic stresses (76). Flavonoids, lignin, and coumarins are synthesized in this route (77). The enzyme phenylalanine ammonia-lyase (PAL) plays a central role in the phenylpropanoid pathway. It converts phenylalanine to trans-cinnamic acid, the precursor of all phenylpropanoid compounds (78, 79). Plants of *P. radiata* exposed to heat for six hours for five days had a decrease in the phenylalanine levels, while PAL activity increased, indicating an acclimation to long-term temperature stress (75). Also, products from the phenylpropanoid pathway inhibit ROS formation, which explains the accumulation of phenylalanine in *P. radiata*. Similarly, *E. globulus* exposed to high temperatures exhibited an accumulation of phenylalanine and shikimic acid, a precursor of phenylalanine (45). Interestingly, cinnamic acid only accumulated when the trees were exposed to a combination of heat and water stress. On the contrary, cold stress also led to the accumulation of phenolic compounds and anthocyanins in eucalypts (80), indicating a general response to temperature stress.

11.4.3 Lignin and Cell Wall

Lignin is a complex plant polymer constituted by phenolic compounds and is essential to the cell wall structure (81). The plant cell wall determines cell shape and size, thus controlling tissue and organ morphology (82). In trees, lignin composition also plays an important role in wood quality (83).

The control of lignin accumulation is highly complex and tree responses to stress occur in different ways. For example, among distinct eucalyptus species exposed to drought, *E. urograndis* had less lignin accumulation in the apical stem, while *E. globulus* presented higher abundance in the same region (84). 4-vinylsyringol, coniferyl alcohol, coniferaldehyde, 4-vinylguaiacol, and 4-methylguaiacol are the main constituents of the lignin of teak (*Tectona grandis*) (82). In coffee leaves, plants exposed to heat changed lignin formation, presenting varied monomer compositions compared to plants under normal conditions (85). The contents of guaiacyl and syringyl monolignols increased in the leaves of plants under heat stress. Hemicellulose composition was also affected in the same plants, as xylose decreased and galactose increased.

11.4.4 Disease Response to Abiotic (Heat) Stress

Studies have shown that biotic stress associated with plant pathogens often increases disease appearance and severity (86–88). For instance, a coffee leaf rust outbreak in Colombia was associated with an increase in the local temperature (89). Maize plants infected with leaf blight (*Cochliobolus heterostrophus*) and under heat stress showed larger lesioned areas than diseased plants under normal temperature, with marked changes in the metabolome (90). One of the consequences of increased-disease infection during heat stress might be the loss of resistance (91). In wheat, heat stress provoked genetic alterations related to damage in the cell membrane and accumulated metabolites negatively correlated to desired traits (92). Metabolomics studies on forest damage are scarce, and they are even more rare when combinations with abiotic factors are included. The nearest case of study to this topic was research on citrus affected by greening (Huanglongbing disease), which showed how trees under shade (consequently, lower temperatures) were more resilient to the disease (93). Less starch accumulation in leaves and lower levels of sucrose and hormone regulation were promoted by moderate shade conditions, which also reduced leaf flushing and downregulated photosynthesis. More studies of this nature will help elucidate the metabolic responses of trees (and plants) to the effects of abiotic factors and pathogens.

11.5 Effects of Elevated CO_2 on Tree Metabolism

Global climate change has generally been described as long-term temperature and weather patterns shifts. Burning fossil fuels, cutting down forests, and farming livestock have significantly impacted the climate and the Earth's temperatures. These practices add enormous amounts of greenhouse gases to those naturally occurring in the atmosphere, increasing greenhouse effects and global warming (94). Temperature and CO_2 are the two most important effects of global warming and climate change. They exert an influence on plant metabolism and plant physiology (36, 95). It is usually challenging to exclude one factor from another since high CO_2 is generally associated with high temperature.

11.5.1 CO_2 Fixation in Plants and Climate Predictions about Atmospheric CO_2

Photosynthesis is the primary process in plant metabolism to produce carbohydrates and energy for cells besides releasing O_2 (96, 97). The biochemical fixation of the CO_2-carbon is catalyzed by the enzyme Rubisco (ribulose-1, 5-bisphosphate carboxylase/oxygenase) (98). This enzyme also has oxygenase activity in a process known as photorespiration, which competes with CO_2 fixation (99). Some plants evolved an adaptive mechanism named the C4 pathway in which photorespiration is almost absent (100). Another additional pathway is called Crassulacean acid metabolism (CAM) (101). In CAM plants, stomata open during the night, and CO_2 is first trapped in malic acid, which is stored in the vacuole. During the day, a malic enzyme releases CO_2 for Rubisco activity (101, 102).

According to climate models, it is expected that the CO_2 concentration will rise in the atmosphere to 550 ppm by the year 2050 (69) and may double by 2100, provoking global warming (70). The increase in CO_2 levels has notable impacts on global carbon fluxes and the processes governing carbon cycles on Earth. Because CO_2 is a well-mixed gas, only minor spatial and temporal heterogeneity in CO_2 concentration across the Earth's surface is expected at any time (around 10 ppm). These differences are mainly due to regional weather patterns and plant and soil activity (103). The atmosphere will likely become drier and directly influence plant evapotranspiration (104). Plant performance will be affected by increasing CO_2 levels and changes in air temperature and water availability, which are associated with drought stress (36). In terrestrial ecosystems, plant-related processes, such as CO_2 assimilation and plant respiration, have profound effects on: (i) net primary productivity; (ii) carbon allocation and sequestration; and (iii) belowground processes, such as those related to organic matter decomposition and nutrient cycling (105).

11.5.2 Elevated CO_2 and Plant Metabolism

Despite the physiological data obtained throughout the years about the physiological responses of trees to elevated CO_2 (eCO_2), metabolomics data are relatively scarce and mainly found in model plants or a few crop species like rice, maize, and sorghum (76). For instance, the effect of eCO_2 in six crops significantly decreased in great of wheat, rice, and soybean (106). In maize, high CO_2 increases carbohydrate metabolism (i.e. increased levels of glucose, mannose, and galactose) (107). Spring wheat (*Triticum aestivum* L. cv. Triso) exposed to 573 ppm CO_2 had a decrease in total grain protein, whereas raffinose, fructose, and fructan content increased (108). A study using seedlings of bread wheat (*T. aestivum*, cv. Yocoro rojo) grown in three different CO_2 concentrations (400 ppm, ambient; 1500 ppm, high; and 10,000 ppm, extremely high) revealed that the metabolome profile under high CO_2 was similar to that of adult plants under ambient CO_2. At high CO_2, seedlings accumulated high levels of hexose phosphates. Total starch content increased transiently in plants grown at extremely high CO_2 than in high CO_2 treatment (109). Under high CO_2, the amount of carbohydrates in perennial ryegrass leaves increased at low N, indicating an increase in C/N ratios (110). Tomato plants grown at eCO_2 showed higher leaf dark respiration than ambient CO_2, which was attributed to the abundance of proteins, carbohydrates, and transcripts involved in carbohydrate metabolism, glycolysis, TCA, and mitochondrial electron transport energy metabolism (111).

Because of the current climate change and its consequences worldwide, integrating data from different OMICs has become imperative (77). Most metabolomics studies involving high CO_2 are based on LC–MS untargeted metabolite analysis. In this technique, materials are separated based on their different chemical properties using distinct columns, and there is no need for sample derivatization. Therefore, LC–MS allows an accurate analysis using a higher amount and chemically diverse metabolites (78). However, compound identification requires aid from spectral/homemade libraries (112).

11.5.3 Elevated CO_2 and Tree Metabolism

Plant metabolic responses are expected to change after exposition to eCO_2. For instance, plants are expected to improve water use efficiency, primarily because of stomatal closure and/or soluble sugar accumulation in the mesophyll cells (113, 114) and increase carbon assimilation due to the higher CO_2 availability for Rubisco carboxylase activity (113). Subsequently, the rise of internal CO_2 partial pressure (Ci) and the accumulation of specific metabolites can result in a series of negative feedback regulations, such as a decrease in stomatal conductance (*gs*) and stomatal density and the inhibition of photosynthesis-related proteins. The C/N ratio related to the inhibition of photorespiration in plants exposed to eCO_2 increases non-structural carbohydrates and decreases phenolic compounds (115). Most studies on plant metabolite responses use conventional methodologies (43, 71, 116–118); however, metabolite responses to eCO_2 are fairly poorly cataloged and understood (75).

Ghannoum et al. (24) found that in plants of *E. sideroxylon* and *E. saligna*, light-saturated photosynthesis (*Asat*) doubled under eCO_2 (650 ppm), and maximal electron transport was stimulated by 10%. These results were further supported by the faster growth of the same species under 600 ppm CO_2 at 30 °C (25). Conversely, Crous et al. (26) showed that photosynthesis was negatively affected in *E. globulus* in response to high temperatures during the summer, restored by the seasonal colder temperatures in the winter. Quentin et al. (27) found that photosynthesis and biomass accumulation was stimulated under 630 ppm CO_2 but not by warming. Aspinwall et al. (28) studied the effects of eCO_2 and temperature in several genotypes of *E. camaldulensis*. Genotypes with higher total dry mass under eCO_2 also increased root mass fraction and photosynthetic nitrogen use efficiency (28). On the other hand, phosphorus deficiency decreased the mass gain of *Eucalyptus* plants exposed to eCO_2 for three years (29), and drought suppressed the effects of eCO_2 and high temperatures (30). In a work using *Pinus pinaster* Aiton (79), the effect of eCO_2 on leaf metabolism was evaluated. In conifers, leaves are the major sink of assimilated carbon, allowing proper growth and increase of leaf biomass. Exposing these trees to eCO_2 led to changes in primary and secondary metabolism. There was a decrease in cyclitols (myo-inositol e pinitol); however, no variation was observed in the most abundant soluble carbohydrates (glucose, fructose, galactose, sucrose, maltose, and isomaltose) (79). Nonetheless, there was an accumulation of malic acid and a decline in oxalic acid. Malate can be readily metabolized, and it is important in stomatic aperture (119). The stimulation of the photosynthetic rate did not enhance carbohydrate or starch synthesis. There was, however, an increase in secondary metabolites, which could act as chemical defense compounds. An illustration of eCO_2 effects on plant metabolism is depicted in Figure 11.3.

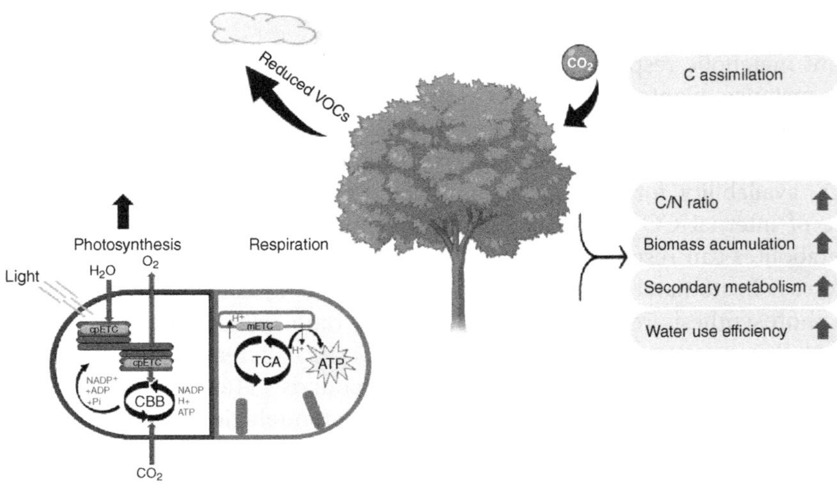

Figure 11.3 Schematic view of elevated CO_2 effects in plant metabolism. The black upright arrow indicates upregulation. The plus signal indicates an increase in the activity of metabolic processes.

11.6 CO_2 Effects on Isoprene Emissions

An important class of compounds, the volatile organic compounds (VOCs, mainly isoprenoids), is emitted by global vegetation landscapes as protective compounds against biotic and abiotic stresses (115). This isoprene affects atmospheric chemistry and climate (80). On the other hand, changes in atmospheric CO_2 affect hydrocarbon load and change ozone and aerosol formation (81). Plants respond differently in isoprene emissions rates under eCO_2, but in general, they reduce the emission of these volatile under eCO_2 (82). In Mediterranean oak, high CO_2 did not significantly affect VOCs emission after a two-year exposition under 1000 ppm. Other factors like an increment in foliar temperature or other stresses compensate for the CO_2 effect in Mediterranean ecosystems. In contrast, *Populus tremuloides Michx* grown under 640 ppm decreased isoprene emission (60–70% relative to ambient CO_2 at 395 ppm) (83). *Populus deltoides* Bart. ex-Marsh also diminished isoprene emission by 41% when CO_2 was increased to 1200 ppm. Since increased CO_2 levels reduced the isoprene precursor dimethylallyl diphosphate (DMAPP) in leaves and leaf protoplast, it seems that intracellular competition for phosphoenolpyruvate was the central point in the suppression of isoprene emission under high CO_2 conditions (84). Other reports described an increase in the emission of isoprene molecules under high CO_2 in *Populus canescens* when CO_2 was increased up to 590 ppm (80).

Young leaves of white poplar (*Populus alba*) emitted less isoprene and had high dark respiration in ambient and elevated CO_2, indicating that isoprene emission and dark respiration are inversely correlated (85).

11.7 CO_2 and Plant Productivity

Plant productivity is also affected by eCO_2 (86,87). In deciduous trees (*P. tremuloides, Quercus rubra, Acer saccharum*) grown under 650 ppm, an increase in starch content was observed. Also, there was an increase in condensed tannins and ellagitannins compared to trees grown under ambient CO_2 (88). The triterpenoid content grew in European white birch (*Betula pendula Roth*) stem under elevated (700 ppm) or high CO_2 (1050 ppm) compared to ambient CO_2 (350 ppm). However, the proanthocyanidin content and flavonoids were reduced in coffee leaves under eCO_2 (89). The differences in triterpenoids and phenylpropanoid contents may be due to their different biosynthetic origin. The former derives from C-based pyruvate metabolism, while the latter derives from N-based phenylalanine metabolism. Plantlets of Norway spruce (*Picea abies* (L.) Karst.) grown under high CO_2 increased the terpenoid content, while the phenolic content did not change significantly (90).

A meta-analysis of ionomics data from 130 plant species, including crops and trees, showed that eCO_2 decreased plant mineral content by 8%, while there was an increase in the ratio of carbohydrate/protein (91, 120). The reduction of photorespiration under eCO_2 has also been related to lower cytoplasmic NADH levels, lower nitrate reductase activities, and nitrogen contents in Eucalyptus trees (121).

11.8 Acclimation After a Long Period of CO_2 Exposure

Long exposure to CO_2 can lead to a process known as CO_2 acclimation (92, 93). This response is probably a downregulation o some photosynthesizing enzymes like Rubisco (94). One hypothesis to explain this phenomenon is the accumulation of photosynthates on source leaves because of low utilization by sink tissues, leading to negative feedback and decreasing the activity of photosynthetic enzymes (36, 95). The second hypothesis is related to the dilution/acquisition of nutrients, mainly nitrogen and phosphorous, resulting from rapid growth or allocation to sink tissues (96, 97). Feng et al. (98) reported a negative correlation between nitrogen concentration and ecosystems, including croplands, grasslands, and forests. The increase in biomass belowground and soil respiration might be why trees exposed to eCO_2 maintain high CO_2 assimilation, stabilizing a connection between photosynthesis and ecosystem respiration (99, 100). However, the eCO_2 effect in the respiration process at the transcriptional level is still poorly investigated in trees, especially in roots. Certainly, trees will remodel their metabolism under eCO_2, diminishing the expression of key photosynthetic proteins, adjusting and distributing the nutrients between the tissues, and increasing CO_2 assimilatory capacity (122).

The metabolomics studies in plants are in infancy. About 200,000 compounds are estimated in the plant kingdom and exhibit diverse chemical properties, increasing analysis complexity. Quantitative advances in metabolomics analysis could help comprehend the cellular biochemical responses in response to different CO_2. The data collected could help predict how trees will respond to future climate change scenarios and provide the tools to preserve biological communities and mitigate the hazardous impacts of higher temperatures and CO_2 concentration.

11.9 The Interactive Effect of Elevated CO_2 and High Temperature in Trees

Environmental studies using metabolomic techniques involving trees are on the rise. Previous studies mainly focused on physiology and biochemistry, but they suggest that eCO_2 and temperatures would promote growth if water, nutrient supplies, and root growth space were sufficient (101). Heat stress impairs plant growth and productivity by affecting water relations, membrane fluidity, and stability. Consequently, heat stress also has a negative impact on the chloroplast and mitochondria reactions and whole metabolism (102, 103, 123).

Many studies reported that eCO_2 mitigates the negative effect of high temperatures, increasing photosynthesis, water use efficiency, and even biomass (24, 103, 104, 106). Idso and Kimball et al. (124) showed that eCO_2 increased net carbon assimilation by 200% in orange trees at 42 °C, relative to ambient CO_2. Leaf cooling is also reported in aspen clones and birch trees under elevated CO_2 compared to ambient CO_2 (108). Darbah et al. (125) also observed a large increase in the maximum velocity of carboxylation, electron transport rate, total phosphate use, and stomatal conductance under elevated CO_2 at 40–41 °C in aspen.

A study with seedlings of the eucalyptus *E. saligna* and *E. sideroxylon* exposed to high CO_2 (650 ppm) and high temperature (ambient +4 °C) stimulated photosynthesis and growth, but with no interaction between eCO_2 and high temperature (24, 25). Despite the lack of interaction, this study shows that in future scenarios, plants like eucalypts could enhance their growth and increase their carbon sequestration if water and nutrients were sufficient.

The associated effects of temperature and eCO_2 on physiology and molecular aspects of wood formation were studied in recent work with *E. grandis* and *E. globulus* (126). Plants were grown for 35 days under three temperature ranges (10–12 °C low; 20–22 °C control; and 33–35 °C high) combined with two CO_2 concentrations (390 and 700 ppm). *E. grandis* has a subtropical origin in the Australian mainland and *E. globulus* has a temperate origin in Tasmania. Maltose and sucrose increased in *E. globulus* under high-temperature whereas in *E. grandis*, maltose increased at low temperatures, and sucrose increased at high temperatures combined with eCO_2 (126). Mannitol, a well-known osmoprotectant (127, 128), increased in *E. globulus* under low temperature and high temperature combined with eCO_2, suggesting in both cases a stressing situation. Many photosynthetic genes increased under high and low temperatures combined with eCO_2 in both species (126). Lignin biosynthesis genes were up-regulated under low and high temperatures combined with eCO_2 in both species. Cellulose and hemicellulose biosynthesis genes were up-regulated under low and high temperatures in *E. globulus* (126). Photosynthesis, transpiration, and stomatal conductance were also determined in this study. eCO_2 had a more positive impact than the temperature in photosynthetic activity in both species, even in low temperatures. The greatest effect under low temperatures was in *E. globulus*, as this species is originally from a cold-tempered region. Transpiration increased in high temperature and high temperature combined with high CO_2 in both species, without a clear distinction of whether temperature or eCO_2 had the strongest influence. Stomatal conductance did not change according to photosynthesis or transpiration (126).

Several studies on trees showed a beneficial compensatory effect of eCO_2 in elevated temperatures, preventing water loss and hydraulic failure (36, 113, 129, 130). These results highlight the plasticity of *Eucalyptus* to future scenarios with increased CO_2 concentration (131). *Eucalyptus globulus* and *E. grandis* were studied regarding their response to high transpiration demand and water stress using transcriptomics and proteomics data (130,132). Ten aquaporin genes with water transport properties (6 PIPs and 4 TIPs) were identified; among them, seven were up-regulated, and three were down-regulated in *E. globulus* exposed to 30 °C and high vapor-pressure deficit (132). There were no aquaporins differentially expressed in *E. grandis.* Several proteins possibly related to surfactant production or action in both species were also identified (132). These results may reflect the origin of each species, as *E. grandis* is better adapted to warmer climates. Thus, only a basal expression of aquaporins might be sufficient to keep plants of this species in an adequate physiological performance even under high transpiration demands. Regarding surfactant-related gene expression, *E. globulus* is well adapted to cold places, and it could be a mechanism to keep water transport safe under high transpiration demand in hot climates. It was proposed a function to these proteins in the regulation of water flow and embolism avoidance in these species of *Eucalyptus* and suggested future studies to help explain the role of aquaporins and surfactants in the radial hydraulic conductance and lowering surface tension, respectively and their contribution to water transport (132).

eCO_2 ameliorates the negative impact of high temperatures on the coffee species *Coffea arabica* and *Coffea canephora* (106, 109, 110). The combination of these stresses improved photochemical efficiency, energy use, and biochemical functioning, especially in warmer conditions compared to plants grown at ambient CO_2. The better performance of these plants was due to the induction of protective molecules against ROS (99). Different classes of genes were differentially expressed, mostly related to specialized metabolism, like lignin. Genes involved in the photosynthetic process were up-regulated at 37 °C and down-regulated at 42 °C (109).

Synthesizing woody plants might take advantage of eCO_2 under high temperatures to enhance: (i) photosynthetic machinery; (ii) cell wall composition; and (iii) specialized metabolites involved in stress signaling and defense. Figure 11.4 resumes the interaction of temperature stress and high CO_2 levels and their combined impact on tree metabolism.

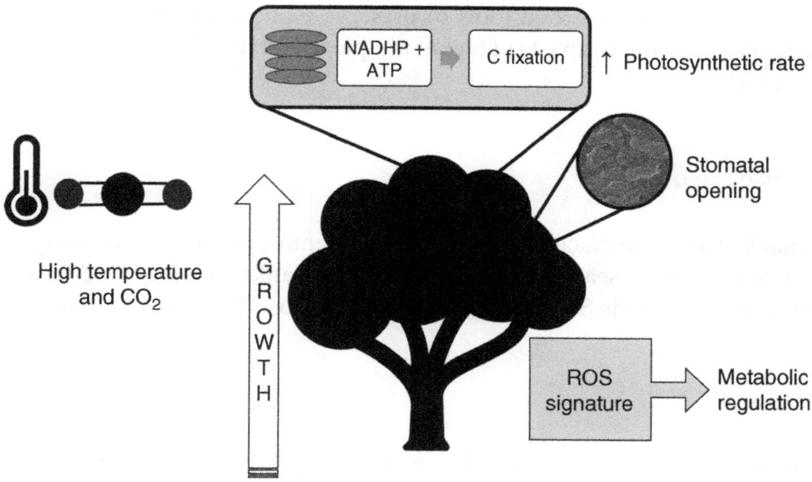

Figure 11.4 Combined impact of temperature stress and high CO_2 levels on tree metabolism. Upward arrows indicate higher activity.

11.10 Conclusions and Future Perspectives

Temperature and CO_2 are the two most important effects of global warming and climate change. They exert influence on plant tree metabolism and plant physiology. More intense and frequent wildfires, below and above-average temperature levels, and high concentrations of atmospheric CO_2 levels are some of the consequences of climate change that could drastically affect the tree metabolome. Stresses due to above-average temperatures in plant cells negatively impact key functions, such as photosynthesis, cell division, growth, and membrane permeability. High temperature also impacts secondary metabolism. CO_2 levels, compared to temperature, are more straightforward to measure and have risen in our atmosphere since the industrial revolution. Under high CO_2, forest trees are expected to improve water use efficiency, increase carbon assimilation, C/N ratio, increase non-structural carbohydrates and decrease allocation of phenolic compounds. When temperature stress and higher CO_2 are combined, eCO_2 mitigates the negative impact of high temperatures, increasing photosynthesis, water use efficiency, and even biomass. Quantitative advances in metabolomics analysis could help comprehend the impact of global climate changes on plant cell metabolism and the responses to eCO_2 and extreme temperatures. Moreover, the data collected could help us predict how forest tree species respond to future climate change scenarios and provide the much-needed tools to preserve biological communities and mitigate the hazardous impacts of higher temperatures and CO_2 concentration in our forests.

Acknowledgments

P.M. thanks Conselho Nacional de Desenvolvimento Científico e Tecnológico (CNPq-Brazil) for a research fellowship. F.R.C.M. thanks Fundação de Amparo à Pesquisa do Estado de São Paulo (FAPESP) for a post-doctoral fellowship.

References

1 Sumner LW, Mendes P, Dixon RA. Plant metabolomics: large-scale phytochemistry in the functional genomics era. *Phytochemistry*. 2003; 62(6): 817–836.
2 Hall RD. Plant metabolomics: from holistic hope, to hype, to hot topic. *New Phytol*. 2006; 169(3): 453–468.
3 Rodrigues AM, Miguel C, Chaves I, António C. Mass spectrometry-based forest tree metabolomics. *Mass Spectrom Rev*. 2021; 40(2): 126–157.
4 Almeida T, Pinto G, Correia B, Gonçalves S, Meijón M, Escandón M. In-depth analysis of the *Quercus suber* metabolome under drought stress and recovery reveals potential key metabolic players. *Plant Sci*. 2020; 299: 110606.
5 Meijõn M, Feito I, Oravec M, Delatorre C, Weckwerth W, Majada J, et al. Exploring natural variation of *Pinus pinaster* Aiton using metabolomics: is it possible to identify the region of origin of a pine from its metabolites? *Mol Ecol*. 2016; 25(4): 959–976.
6 Morreel K, Goeminne G, Storme V, Sterck L, Ralph J, Coppieters W, et al. Genetical metabolomics of flavonoid biosynthesis in Populus: a case study. *Plant J*. 2006; 47(2): 224–237.
7 Jactel H, Koricheva J, Castagneyrol B. Responses of forest insect pests to climate change: not so simple. *Curr Opin Insect Sci*. 2019; 35: 103–108.
8 Holopainen JK, Virjamo V, Ghimire RP, Blande JD, Julkunen-Tiitto R, Kivimäenpää M. Climate change effects on secondary compounds of forest trees in the northern hemisphere. *Front Plant Sci*. 2018; 9: 1445.
9 Duputié A, Rutschmann A, Ronce O, Chuine I. Phenological plasticity will not help all species adapt to climate change. *Global Change Biol*. 2015; 21(8): 3062–3073.
10 Franks SJ, Weber JJ, Aitken SN. Evolutionary and plastic responses to climate change in terrestrial plant populations. *Evol Appl*. 2014; 7(1): 123–139.

11 Song Y, Chen Q, Ci D, Shao X, Zhang D. Effects of high temperature on photosynthesis and related gene expression in poplar. *BMC Plant Biol.* 2014; 14(111): 1–20.

12 Hansen J, Ruedy R, Sato M, Lo K. Global surface temperature change. *Rev Geophys.* 2010; 48(4): RG4004.

13 Zhou DK, Larar AM, Liu X. Global surface skin temperature analysis from recent decadal IASI observations. In: *Multispectral, Hyperspectral, and Ultraspectral Remote Sensing Technology, Techniques and Applications VII.* Volume 10780. SPIE; 2018. p. 8–16.

14 Zhou R, Yu X, Wen J, Jensen NB, Dos Santos TM, Wu Z, et al. Interactive effects of elevated CO_2 concentration and combined heat and drought stress on tomato photosynthesis. *BMC Plant Biol.* 2020; 20(1): 1–12.

15 Wang X, Liu F. Effects of elevated CO_2 and heat on wheat grain quality. *Plants.* 2021; 10(5): 1027.

16 Źróbek-Sokolnik A. Temperature stress and responses of plants. In: Ahmad P & Prasad MNV, editors. *Environmental Adaptations and Stress Tolerance of Plants in the Era of Climate Change.* New York: Springer; 2012: p. 113–134.

17 Heal G, Park J. Temperature stress and the direct impact of climate change: a review of an emerging literature. *Review of Environmental Economics and Policy.* 2016; 10(2): 347–362.

18 Anderegg WRL, Kane JM, Anderegg LDL. Consequences of widespread tree mortality triggered by drought and temperature stress. *Nat Clim Chang.* 2012; 3(1): 30–36.

19 Mokochinski JB, Mazzafera P, Sawaya ACHF, Mumm R, de Vos RCH, Hall RD. Metabolic responses of eucalyptus species to different temperature regimes. *J Integr Plant Biol.* 2018; 60(5): 397–411.

20 McDowell N, Allen CD, Anderson-Teixeira K, Brando P, Brienen R, Chambers J, et al. Drivers and mechanisms of tree mortality in moist tropical forests. *New Phytol.* 2018; 219(3): 851–869.

21 Estravis-Barcala M, Mattera MG, Soliani C, Bellora N, Opgenoorth L, Heer K, et al. Molecular bases of responses to abiotic stress in trees. *J Exp Bot.* 2020; 71(13): 3765–3779.

22 Guidolotti G, Pallozzi E, Gavrichkova O, Scartazza A, Mattioni M, Loreto F, et al. Emission of constitutive isoprene, induced monoterpenes, and other volatiles under high temperatures in *Eucalyptus camaldulensis*: a ^{13}C labelling study. *Plant Cell Environ.* 2019; 42(6): 1929–1938.

23 Medina S, Vicente R, Amador A, Araus JL. Interactive effects of elevated CO_2 and water stress on physiological traits and gene expression during vegetative growth in four durum wheat genotypes. *Front Plant Sci.* 2016; 7: 1738.

24 Ghannoum O, Phillips NG, Sears MA, Logan BA, Lewis JD, Conroy JP, et al. Photosynthetic responses of two eucalypts to industrial-age changes

in atmospheric CO_2 and temperature. *Plant Cell Environ.* 2010; 33(10): 1671–1681.
25 Ghannoum O, Phillips NG, Conroy JP, Smith RA, Attard RD, Woodfield R, et al. Exposure to preindustrial, current and future atmospheric CO_2 and temperature differentially affects growth and photosynthesis in eucalyptus. *Global Change Biol.* 2010; 16(1): 303–319.
26 Crous KY, Zaragoza-Castells J, Löw M, Ellsworth DS, Tissue DT, Tjoelker MG, et al. Seasonal acclimation of leaf respiration in *Eucalyptus saligna* trees: impacts of elevated atmospheric CO_2 and summer drought. *Global Change Biol.* 2011; 17(4): 1560–1576.
27 Quentin AG, Crous KY, Barton CVM, Ellsworth DS. Photosynthetic enhancement by elevated CO_2 depends on seasonal temperatures for warmed and non-warmed *Eucalyptus globulus* trees. *Tree Physiol.* 2015; 35(11): 1249–1263.
28 Aspinwall MJ, Drake JE, Campany C, Vårhammar A, Ghannoum O, Tissue DT, et al. Convergent acclimation of leaf photosynthesis and respiration to prevailing ambient temperatures under current and warmer climates in *Eucalyptus tereticornis*. *New Phytol.* 2016; 212(2): 354–367.
29 Ellsworth DS, Anderson IC, Crous KY, Cooke J, Drake JE, Gherlenda AN, et al. Elevated CO_2 does not increase eucalypt forest productivity on a low-phosphorus soil. *Nat Clim Chang.* 2017; 7(4): 279–282.
30 Duan H, Amthor JS, Duursma RA, O'Grady AP, Choat B, Tissue DT. Carbon dynamics of eucalypt seedlings exposed to progressive drought in elevated CO_2 and elevated temperature. *Tree Physiol.* 2013; 33(8): 779–792.
31 Körner C. The cold range limit of trees. *Trends Ecol Evol.* 2021; 36(11): 979–989.
32 Tarvainen L, Wittemann M, Mujawamariya M, Manishimwe A, Zibera E, Ntirugulirwa B, et al. Handling the heat – photosynthetic thermal stress in tropical trees. *New Phytol.* 2022; 233(1): 236–250.
33 Werner C, Fasbender L, Romek KM, Yáñez-Serrano AM, Kreuzwieser J. Heat waves change plant carbon allocation among primary and secondary metabolism altering CO_2 assimilation, respiration, and VOC emissions. *Front Plant Sci.* 2020; 11: 1242.
34 Hatfield JL, Prueger JH. Temperature extremes: effect on plant growth and development. *Weather Clim Extrem.* 2015; 10: 4–10.
35 Thomas DS, Montagu KD, Conroy JP. Temperature effects on wood anatomy, wood density, photosynthesis and biomass partitioning of Eucalyptus grandis seedlings. *Tree Physiol.* 2007; 27(2): 251–260.
36 Dusenge ME, Duarte AG, Way DA. Plant carbon metabolism and climate change: elevated CO_2 and temperature impacts on photosynthesis, photorespiration and respiration. *New Phytol.* 2019; 221(1): 32–49.

37 Sperling O, Silva LCR, Tixier A, Théroux-Rancourt G, Zwieniecki MA. Temperature gradients assist carbohydrate allocation within trees. *Sci Rep.* 2017; 7: 3265.

38 Drake JE, Tjoelker MG, Vårhammar A, Medlyn BE, Reich PB, Leigh A, et al. Trees tolerate an extreme heatwave via sustained transpirational cooling and increased leaf thermal tolerance. *Global Change Biol.* 2018; 24(6): 2390–2402.

39 Crous KY, Quentin AG, Lin YS, Medlyn BE, Williams DG, Barton CVM, et al. Photosynthesis of temperate *Eucalyptus globulus* trees outside their native range has limited adjustment to elevated CO_2 and climate warming. *Global Change Biol.* 2013; 19(12): 3790–3807.

40 Zhao J, Hartmann H, Trumbore S, Ziegler W, Zhang Y. High temperature causes negative whole-plant carbon balance under mild drought. *New Phytol.* 2013; 200(2): 330–339.

41 Riikonen J, Kontunen-Soppela S, Ossipov V, Tervahauta A, Tuomainen M, Oksanen E, et al. Needle metabolome, freezing tolerance and gas exchange in Norway spruce seedlings exposed to elevated temperature and ozone concentration. *Tree Physiol.* 2012; 32(9): 1102–1112.

42 Ren S, Ma K, Lu Z, Chen G, Cui J, Tong P, et al. Transcriptomic and metabolomic analysis of the heat-stress response of *Populus tomentosa* Carr. *Forests.* 2019; 10(5): 383.

43 O'Leary BM, Asao S, Millar AH, Atkin OK. Core principles which explain variation in respiration across biological scales. *New Phytol.* 2019; 222(2): 670–686.

44 Xu Y, Fu X, Tran P, Patel MK, Xu Y, Fu X. Reprogramming of plant central metabolism in response to abiotic stresses: a metabolomics view. *Int J Mol Sci.* 2022; 23(10): 5716.

45 Correia B, Hancock RD, Amaral J, Gomez-Cadenas A, Valledor L, Pinto G. Combined drought and heat activates protective responses in *Eucalyptus globulus* that are not activated when subjected to drought or heat stress alone. *Front Plant Sci.* 2018; 9: 819.

46 Wang L, Ma KB, Lu ZG, Ren SX, Jiang HR, Cui JW, et al. Differential physiological, transcriptomic and metabolomic responses of Arabidopsis leaves under prolonged warming and heat shock. *BMC Plant Biol.* 2020; 20(1).

47 Sharma N, Thakur M, Suryakumar P, Mukherjee P, Raza A, Prakash CS, et al. 'Breathing out' under heat stress – respiratory control of crop yield under high temperature. *Agron.* 2022; 12(4): 806.

48 Ciais P, Reichstein M, Viovy N, et al. Europe-wide reduction in primary productivity caused by the heat and drought in 2003. *Nature.* 2005; 437: 529–533.

49 Batista-Silva W, Heinemann B, Rugen N, Nunes-Nesi A, Araújo WL, Braun HP, et al. The role of amino acid metabolism during abiotic stress release. *Plant Cell Environ.* 2019; 42(5): 1630–1644.

50 Galili G, Amir R, Fernie AR. The regulation of essential amino acid synthesis and accumulation in plants. *Annu Rev Plant Biol.* 2016; 67: 153–178.

51 Yang Q, Zhao D, Liu Q. Connections between amino acid metabolisms in plants: lysine as an example. *Front Plant Sci.* 2020; 11: 928.

52 Botta A. Enhancing plant tolerance to temperature stress with amino acids: an approach to their mode of action. *Acta Hortic.* 2012; 1009: 29–36.

53 Guy C, Kaplan F, Kopka J, Selbig J, Hincha DK. Metabolomics of temperature stress. *Physiol Plant.* 2008; 132(2): 220–235.

54 Xu G, Li L, Zhou J, Lv D, Zhao D, Qin S. Comparison of transcriptome and metabolome analysis revealed differences in cold resistant metabolic pathways in different apple cultivars under low temperature stress. *Hortic Plant J.* 2023; 9(2): 183–198.

55 Yaish M. Proline accumulation is a general response to abiotic stress in the date palm tree (*Phoenix dactylifera* L.). *Mol Res.* 2015; 14(3): 9943–9950.

56 Yamakawa H, Hakata M. Atlas of rice grain filling-related metabolism under high temperature: joint analysis of metabolome and transcriptome demonstrated inhibition of starch accumulation and induction of amino acid accumulation. *Plant Cell Physiol.* 2010; 51(5): 795–809.

57 Kim M, Kang SB, Yun SK, Kim SS, Joa J, Park Y. Influence of excessively high temperatures on the fruit growth and physicochemical properties of Shiranuhi Mandarin in plastic-film greenhouse cultivation. *Plants.* 2021; 10(8).

58 Souza GM, Cardoso VJM, Gonçalves AN. Proline content and protein patterns in *Eucalyptus grandis* shoots submitted to high and low temperature shocks. *Brazilian Arch Biol Technol.* 2004; 47(3): 355–362.

59 He C, Zhang J, Duan A, Zheng S, Sun H, Fu L. Proteins responding to drought and high-temperature stress in *Populus x euramericana* cv. '74/76'. *Trees - Struct Funct.* 2008; 22(6): 803–813.

60 Noctor G, Lelarge-Trouverie C, Mhamdi A. The metabolomics of oxidative stress. *Phytochemistry.* 2015; 112: 33–53.

61 Luckner M. *Secondary Metabolism in Microorganisms, Plants, and Animals.* Springer-Verlag; 1984 576 p.

62 Bixenmann RJ, Coley PD, Weinhold A, Kursar TA. High herbivore pressure favors constitutive over induced defense. *Ecol Evol.* 2016; 6(17): 6037–6049.

63 Nakabayashi R, Saito K. Integrated metabolomics for abiotic stress responses in plants. *Curr Opin Plant Biol.* 2015; 24: 10–16.

64 Dixon RA, Achnine L, Kota P, Liu C-J, Reddy MSS, Wang L. The phenylpropanoid pathway and plant defence-a genomics perspective. *Mol Plant Pathol.* 2002; 3(5): 371–390.

65 Dixon RA. Natural products and plant disease resistance. *Nature.* 2001; 411(6839): 843–847.

66 Zwenger S, Basu C. Plant terpenoids : applications and future potentials. *Biotechnol Mol Biol Rev.* 2008; 3: 1–7.

67 Li Z-H, Wang Q, Ruan X, Pan C-D, Jiang D-A. Phenolics and plant allelopathy. *Molecules.* 2010; 15(12): 8933–8952.

68 Celedon JM, Bohlmann J. An extended model of heartwood secondary metabolism informed by functional genomics. *Tree Physiol.* 2018; 38(3): 311–319.

69 Vasconcelos NG, Croda J, Simionatto S. Antibacterial mechanisms of cinnamon and its constituents: a review. *Microb Pathog.* 2018; 120: 198–203.

70 Bedir E, Tatli II, Khan RA, Zhao J, Takamatsu S, Walker LA, et al. Biologically active secondary metabolites from *Ginkgo biloba. J Agric Food Chem.* 2002; 50(11): 3150–3155.

71 de Melo PGV, de Carvalho Neto DP, Magalhães Júnior AI, do Prado FG, MGB P, Karp SG, et al. Chemical composition and health properties of coffee and coffee by-products. *Adv Food Nutr Res.* 2020; 91: 65–96.

72 Ramakrishna A, Ravishankar GA. Influence of abiotic stress signals on secondary metabolites in plants. *Plant Signaling Behav.* 2011; 6(11): 1720–1731.

73 Yang L, Wen KS, Ruan X, Zhao YX, Wei F, Wang Q. Response of plant secondary metabolites to environmental factors. *Molecules.* 2018; 23(4): 1–26.

74 Austen N, Walker HJ, Lake JA, Phoenix GK, Cameron DD. The regulation of plant secondary metabolism in response to abiotic stress: interactions between heat shock and elevated CO_2. *Front Plant Sci.* 2019; 10: 1463.

75 Escandón M, Meijón M, Valledor L, Pascual J, Pinto G, Cañal MJ. Metabolome integrated analysis of high-temperature response in *Pinus radiata. Front Plant Sci.* 2018; 9: 485.

76 Biała W, Jasiński M. The phenylpropanoid case – it is transport that matters. *Front Plant Sci.* 2018; 9: 1610.

77 Douglas CJ. Phenylpropanoid metabolism and lignin biosynthesis: from weeds to trees. *Trends Plant Sci.* 1996; 1(6): 171–178.

78 MacDonald MJ, D'Cunha GB. A modern view of phenylalanine ammonia lyase. *Biochem Cell Biol.* 2007; 85(3): 273–282.

79 Jun SY, Sattler SA, Cortez GS, Vermerris W, Sattler SE, Kang CH. Biochemical and structural analysis of substrate specificity of a phenylalanine ammonia-lyase. *Plant Physiol.* 2018; 176(2): 1452–1468.

80 Oberschelp GPJ, Guarnaschelli AB, Teson N, Harrand L, Podestá FE, Margarit E. Cold acclimation and freezing tolerance in three *Eucalyptus* species: a metabolomic and proteomic approach. *Plant Physiol Biochem.* 2020; 154: 316–327.
81 Dixon RA, Barros J. Lignin biosynthesis: old roads revisited and new roads explored. *Open Biol.* 2019; 9(12).
82 Le Gall H, Philippe F, Domon JM, Gillet F, Pelloux J, Rayon C. Cell wall metabolism in response to abiotic stress. *Plants.* 2015; 4(1): 112–166.
83 Lourenço A, Neiva DM, Gominho J, Marques AV, Pereira H. Characterization of lignin in heartwood, sapwood and bark from *Tectona grandis* using Py–GC–MS/FID. *Wood Sci Technol.* 2015; 49(1): 159–175.
84 Moura-Sobczak J, Souza U, Mazzafera P. Drought stress and changes in the lignin content and composition in Eucalyptus. *BMC Proc.* 2011; 5(S7): 13083.
85 Lima RB, Dos Santos TB, Vieira LGE, Ferrarese MDLL, Ferrarese-Filho O, Donatti L, et al. Heat stress causes alterations in the cell-wall polymers and anatomy of coffee leaves (Coffea arabica L.). *Carbohydr Polym.* 2013; 93(1): 135–143.
86 Sun Q, Miao C, Hanel M, Borthwick AGL, Duan Q, Ji D, et al. Global heat stress on health, wildfires, and agricultural crops under different levels of climate warming. *Environ Int.* 2019; 128: 125–136.
87 Janni M, Gullì M, Maestri E, Marmiroli M, Valliyodan B, Nguyen HT, et al. Molecular and genetic bases of heat stress responses in crop plants and breeding for increased resilience and productivity. *J Exp Bot.* 2020; 71(13): 3780–3802.
88 Thapa R, Zhang K, Snavely N, Belongie S, Khan A. The plant pathology challenge 2020 data set to classify foliar disease of apples. *Appl Plant Sci.* 2020; 8(9): e11390.
89 Marraccini P, Villain L, Breitler J, Etienne H. Healthy tropical plants to mitigate the impact of climate change—as exemplified in coffee. In: Torquebiau E, editor. *Climate Change and Agriculture Worldwide.* Dordrecht (NL): Springer; 2016. p. 83–95.
90 Christensen SA, Santana EA, Alborn HT, Block AK, Chamberlain CA. Metabolomics by UHPLC–HRMS reveals the impact of heat stress on pathogen-elicited immunity in maize. *Metabolomics.* 2021; 17: 6.
91 Cohen SP, Leach JE. High temperature-induced plant disease susceptibility: more than the sum of its parts. *Curr Opin Plant Biol.* 2020; 56: 235–241.
92 Thomason K, Babar MA, Erickson JE, Mulvaney M, Beecher C, MacDonald G. Comparative physiological and metabolomics analysis of wheat (*Triticum aestivum* L.) following post-anthesis heat stress. *PLoS One.* 2018; 13(6): e0197919.

93 Suh JH, Guha A, Wang Z, Li SY, Killiny N, Vincent C, et al. Metabolomic analysis elucidates how shade conditions ameliorate the deleterious effects of greening (Huanglongbing) disease in citrus. *Plant J.* 2021; 108(6): 1798–1814.

94 IPCC. Sixth assessment report of the intergovernmental panel on climate change. Climate Change 2021: The Physical Science Basis. 2021. pp. 1–1927.

95 Mndela M, Tjelele JT, Madakadze IC, Mangwane M, Samuels IM, Muller F, et al. A global meta-analysis of woody plant responses to elevated CO_2: implications on biomass, growth, leaf N content, photosynthesis and water relations. *Ecol Process.* 2022; 11(1): 52.

96 Taiz L, Møller IM, Murphy A. *Plant Physiology and Development.* Seventh Edition. Oxford (UK): Oxford University Press; 2022. p. 864.

97 Kruger NJ, von Schaewen A. The oxidative pentose phosphate pathway: structure and organisation. *Curr Opin Plant Biol.* 2003; 6(3): 236–246.

98 Taiz L, Zeiger E, Møller IM, Murphy A. *Plant Physiology and Development.* Sixth Edition. Sunderland CT: Sinauer Associates; 2015. p. 761.

99 Salvucci ME, Ogren WL. The mechanism of Rubisco activase: insights from studies of the properties and structure of the enzyme. *Photosynth Res.* 1996; 47(1): 1–11.

100 Jensen RG. Photosynthesis: C3, C4. Mechanisms, and cellular and environmental regulation, of photosynthesis. *Science.* 1983; 222(4627): 1009.

101 Cushman JC. Crassulacean acid metabolism. A plastic photosynthetic adaptation to arid environments. *Plant Physiol.* 2001; 127(4): 1439–1448.

102 Magel E, Einig W, Hampp R. Carbohydrates in trees. *Dev Crop Sci.* 2000; 26(C): 317–336.

103 Chahine MT, Chen L, Dimotakis P, Jiang X, Li Q, Olsen ET, et al. Satellite remote sounding of mid-tropospheric CO_2. *Geophys Res Lett.* 2008; 35(17): L17807.

104 Ficklin DL, Novick KA. Historic and projected changes in vapor pressure deficit suggest a continental-scale drying of the United States atmosphere. *J Geophys Res.* 2017; 122(4): 2061–2079.

105 Walker AP, De Kauwe MG, Bastos A, Belmecheri S, Georgiou K, Keeling RF, et al. Integrating the evidence for a terrestrial carbon sink caused by increasing atmospheric CO_2. *New Phytol.* 2021; 229(5): 2413–2445.

106 Myers SS, Zanobetti A, Kloog I, Huybers P, Leakey ADB, Bloom AJ, et al. Increasing CO_2 threatens human nutrition. *Nature.* 2014; 510(7503): 139–142.

107 Prins A, Mukubi JM, Pellny TK, Verrier PJ, Beyene G, Lopes MS, et al. Acclimation to high CO_2 in maize is related to water status and dependent on leaf rank. *Plant, Cell Environ.* 2011; 34(2): 314–331.

108 Högy P, Brunnbauer M, Koehler P, Schwadorf K, Breuer J, Franzaring J, et al. Grain quality characteristics of spring wheat (*Triticum aestivum*) as affected by free-air CO_2 enrichment. *Environ Exp Bot.* 2013; 88: 11–18.

109 Levine LH, Kasahara H, Kopka J, Erban A, Fehrl I, Kaplan F, et al. Physiologic and metabolic responses of wheat seedlings to elevated and super-elevated carbon dioxide. *Adv Sp Res.* 2008; 42(12): 1917–1928.

110 Fischer BU, Frehner M, Hebeisen T, Zanetti S, Stadelmann F, Lüscher A, et al. Source-sink relations in *Lolium perenne* L. as reflected by carbohydrate concentrations in leaves and pseudo-stems during regrowth in a free air carbon dioxide enrichment (FACE) experiment. *Plant, Cell Environ.* 1997; 20(7): 945–952.

111 Li X, Zhang G, Sun B, Zhang S, Zhang Y, Liao Y, et al. Stimulated leaf dark respiration in tomato in an elevated carbon dioxide atmosphere. *Sci Rep.* 2013; 3: 3433.

112 Perez de Souza L, Alseekh S, Naake T, Fernie A. Mass spectrometry-based untargeted plant metabolomics. *Curr Protoc Plant Biol.* 2019; 4(4): e20100.

113 Ainsworth EA, Rogers A. The response of photosynthesis and stomatal conductance to rising CO_2: mechanisms and environmental interactions. *Plant Cell Environ.* 2007; 30(3): 258–270.

114 Lobo AKM, de Oliveira MM, Lima Neto MC, Machado EC, Ribeiro RV, Silveira JAG. Exogenous sucrose supply changes sugar metabolism and reduces photosynthesis of sugarcane through the down-regulation of Rubisco abundance and activity. *J Plant Physiol.* 2015; 179: 113–121.

115 Misra BB, Chen S. Advances in understanding CO_2 responsive plant metabolomes in the era of climate change. *Metabolomics.* 2015; 11(6): 1478–1491.

116 Rapparini F, Baraldi R, Miglietta F, Loreto F. Isoprenoid emission in trees of *Quercus pubescens* and *Quercus ilex* with lifetime exposure to naturally high CO_2 environment. *Plant Cell Environ.* 2004; 27(4): 381–391.

117 Ainsworth EA. The effects of elevated CO_2 concentration on soybean gene expression. An analysis of growing and mature leaves. *Plant Physiol.* 2006; 142(1): 135–147.

118 Wang SY, Bunce JA, Maas JL. Elevated carbon dioxide increases contents of antioxidant compounds in field-grown strawberries. *J Agric Food Chem.* 2003; 51(15): 4315–4320.

119 Maurino VG, Engqvist MKM. 2-Hydroxy acids in plant metabolism. *Arab B.* 2015; 13(13): e0182.

120 Gojon A, Cassan O, Bach L, Lejay L, Martin A. The decline of plant mineral nutrition under rising CO_2: physiological and molecular aspects of a bad deal. *Trends Plant Sci.* 2023; 28(2): 185–198.

121 Wujeska-Klause A, Crous KY, Oula G, Ellsworth DS. Lower photorespiration in elevated CO_2 reduces leaf N concentrations in mature Eucalyptus trees in the field. *Global Change Biol.* 2019; 25: 1282–1295.

122 Lobo AKM, Catarino ICA, Silva EA, Centeno DC, Domingues DS. Physiological and molecular responses of woody plants exposed to future atmospheric CO_2 levels under abiotic stresses. *Plants.* 2022; 11(14): 1880.

123 Feng Z, Rütting T, Pleijel H, Wallin G, Reich PB, Kammann CI, et al. Constraints to nitrogen acquisition of terrestrial plants under elevated CO_2. *Global Change Biol.* 2015; 21(8): 3152–3168.

124 Idso SB, Kimball BA. Effects of atmospheric CO_2 enrichment on photosynthesis, respiration, and growth of sour Orange trees. *Plant Physiol.* 1992; 99(1): 341–343.

125 Darbah JNT, Sharkey TD, Calfapietra C, Karnosky DF. Differential response of aspen and birch trees to heat stress under elevated carbon dioxide. *Environ Pollut.* 2010; 158(4): 1008–1014.

126 Feltrim D, Gupta B, Gundimeda S, Kiyota E, Júnior APD, Cintra LC, et al. Exposure of Eucalyptus to varied temperature and CO_2 has a profound effect on the physiology and expression of genes related to cell wall formation and remodeling. *Tree Genet Genomes.* 2022; 18(1): 1–23.

127 Fotopoulos V. Plant invertases: structure, function and regulation. *Gene.* 2005; 127–137.

128 Theocharis A, Clément C, Barka EA. Physiological and molecular changes in plants grown at low temperatures. *Planta.* 2012; 235(6): 1091–1105.

129 Leakey ADB, Ainsworth EA, Bernacchi CJ, Rogers A, Long SP, Ort DR. Elevated CO_2 effects on plant carbon, nitrogen, and water relations: six important lessons from FACE. *J Exp Bot.* 2009; 60(10): 2859–2876.

130 Booth TH, Broadhurst LM, Pinkard E, Prober SM, Dillon SK, Bush D, et al. Native forests and climate change: lessons from eucalypts. *For Ecol Manage.* 2015; 347: 18–29.

131 Feltrim D, Pereira L, de Santana Costa MG, Balbuena TS, Mazzafera P. Stem aquaporins and surfactant-related genes are differentially expressed in two *Eucalyptus* species in response to water stress. *Plant Stress.* 2021; 1: 100003.

132 de Santana Costa MG, Feltrim D, Mazzafera P, Balbuena TS. Revisiting the stem proteome of *Eucalyptus grandis* and *Eucalyptus globulus*: Identification of temperature-induced changes. *Biochim Biophys Acta, Proteins Proteomics.* 2020: 140530.

12

Integration of Primary Metabolism with Physiological and Anatomical Data to Assess Dutch Elm Disease Susceptibility in Three Elm Species – A Case Study

Jesús Rodríguez-Calcerrada[1], Ana Margarida Rodrigues[2], Carla António[2], Rosana López[1], Jorge Domínguez[1], Juan Sobrino-Plata[1,3], Luis Gil[1], and Juan Antonio Martín[1]

[1] *Department of Natural Systems and Resources, Research Group Functioning of Forest Systems in a Changing Environment, Universidad Politécnica de Madrid, Madrid, Spain*
[2] *Plant Metabolomics Lab Portugal, Centro de Estudos Florestais (CEF), Instituto Superior de Agronomia (ISA), Universidade de Lisboa (ULisboa), Lisboa, Portugal*
[3] *Facultad de Ciencias Biológicas, Department of Genetics, Physiology and Microbiology, Universidad Complutense de Madrid, Madrid, Spain*

12.1 Impacts of Dutch Elm Disease on Plant Metabolism and Its Modulation by Climate

There are three elms indigenous to Europe: *Ulmus minor* Miller, *Ulmus laevis* Pallas, and *Ulmus glabra* Hudson. The three species are sensitive to the fungal pathogens *Ophiostoma ulmi* and *Ophiostoma novo-ulmi* causing Dutch elm disease (DED), responsible for widespread, massive mortality during the last decades. Several European programs have been launched for the conservation of natural elm populations and the search of resistant genotypes. However, large-scale restoration of elm natural habitats with resistant elm genotypes remains challenging. Genotypes resistant to DED are few and threatened by pathogen evolution (1, 2). Moreover, global climate change is favoring insect pests and tree decline worldwide, which can enhance the susceptibility of damaged trees to disease and infestation (3, 4). European elms and other riparian species that rely on relatively high water tables to grow and develop can suffer a considerable decrease in vitality and productivity, considering the increased incidence of extreme heat and drought events (5, 6). Conservation programs will benefit from a better knowledge of plant physiology, anatomy, and metabolism of *U. minor*, *U. laevis*, and *U. glabra* upon pathogen infection. These features play a role in DED resistance and their modulation by abiotic stresses should be the focus of research too.

Monitoring Forest Damage with Mass Spectrometry-Based Metabolomics Methods, First Edition.
Edited by Carla António.

Recent research suggests that primary metabolism is a key factor in DED resistance. In a transcriptomics study (7), the authors found tradeoffs between growth and defense in the response of Atinian elm (*Ulmus procera*) to *O. novo-ulmi*. In another transcriptomics study of resistant and susceptible *Ulmus americana* trees, the authors found that, before inoculation, the levels of primary metabolites increased in the susceptible cultivar, suggesting again tradeoffs between growth and defense in DED resistance (8). In fact, under certain field experimental conditions, it has been demonstrated that tree growth correlates positively with susceptibility to DED (9).

During their spread throughout the vascular system, DED pathogens secrete a variety of cell wall-degrading enzymes (10). These enzymes allow penetration of fungal hyphae through the plant cell walls, mainly through pits (11, 12). The pathogen then can easily move to contiguous vessels or other xylem cells. It can also contribute to sap flow cessation through the release of products (e.g. simple sugars or degrading substances) metabolized by the fungus or the host cells. The DED pathogens are probably hemibiotrophs (12, 13), with an initial phase of plant colonization as biotrophs and a subsequent phase as necrotrophs. Early after penetration into xylem tissues, the fungus relies on sugars (fructose, glucose and sucrose) and nitrogenous compounds contained in the xylem sap for their growth (14). In more advanced stages of plant colonization, easily assimilable substrates, including soluble sugars, lipids, peptides, and starch, found almost exclusively in parenchyma cells, possibly serve as supplementary carbon sources for pathogen growth (15). Since parenchyma cells are living cells, the pathogen should kill them before colonizing them during the necrotrophic phase. Although microscopic research on *U. americana* revealed that parenchyma cells are rarely invaded by *O. ulmi* (16), subsequent research on *U. minor* revealed direct penetration of fungal hyphae into ray and paratracheal parenchyma cells by the more aggressive *O. novo-ulmi* (15).

One of the first cytological reactions occurring in elms after DED infection is indeed the alteration of parenchymatic paratracheal and ray cells, as well as cells at the cambial area (17). These alterations include rapid increases in density and opacity of cytoplasmic contents, and modifications of cell wall layers, including the deposition of suberin-like compounds. These early reactions can lead to the death of parenchyma cells, particularly in susceptible hosts (17, and refs. therein). Parenchyma cells are also involved in the production of tyloses, gels, alveolar reticula, and other coating materials into invaded vessels, substances mainly composed of phenolic, pectinic, and lipidic compounds (17), some of which may originate from the pathogen (18). The exact role of these reactions in DED resistance is not well understood as they can be observed in both resistant and susceptible hosts, but in general, they represent an attempt of the tree to restrict vertical and horizontal pathogen spread, and to limit harmful pathogen effects in the living cells (19). The success of these reactions most likely relies on their temporal and spatial coordination with other defense mechanisms.

Concerning primary metabolites in the xylem sap, early research on DED suggested that the concentration of sugars may have an influence on the resistance to DED. For instance, keeping the plants in the dark decreased the sugar content and the DED resistance, while the application of sugar solutions to the tree increased the resistance (20, 21). However, no differences in constitutive sugar concentration were subsequently found between resistant and susceptible hosts (22). Regarding nitrogenous compounds, it has been evidenced that pathogen inoculation increases the levels of nitrogenous compounds in elms. Thus, the content of γ-aminobutyric acid (GABA), proline, and alanine increased in *U. americana* after *O. ulmi* inoculation (14). It is remarkable that proline, which was present in trace amounts in healthy sap, accounted for 14–38% of the total nitrogenous compounds in infected trees. Furthermore, it was reported that the xylem sap of *Ulmaceae* spp. resistant to DED contained considerable amounts of constitutive proline, while susceptible species contained only trace amounts (23). It should be noted that proline usually plays important protective roles in plants under stress, acting as an antioxidative and signaling molecule (24). Proline regulates concentrations of reactive oxygen species (ROS) within normal ranges, thus preventing oxidative bursts in plants. In *U. minor*, a high level of oxidative stress was detected in a resistant genotype during the first hours after pathogen inoculation, but values returned to normal ranges in the following days (25); on the contrary, a susceptible genotype showed a high level of oxidative damage in the long run. This suggests an earlier and more efficient control of oxidative stress in resistant trees.

The role of hormones in DED pathogenesis is not well understood. An increase in auxin concentration was reported in response to DED infection (26), as it has also been reported in other wilt diseases. Indole acetic acid (IAA) might increase the plasticity of pit membranes, favoring the formation of tyloses (27). Furthermore, since auxins are involved in the division and differentiation of cambial cells, higher auxin levels could promote the formation of barrier zones. These barriers are made of layers of specialized (often suberized) cells formed by the cambium to compartmentalize the pathogen and favor the formation of new healthy xylem (17, 28, 29). Nevertheless, little is known about the role of other hormones in disease resistance, although molecular works on *U. minor* suggested a key role of salicylic acid (SA) in triggering defense pathways against the pathogen, possibly during the initial biotrophic phase (7, 30). Other works on *U. americana* also evidenced the involvement of jasmonic acid (JA) in the induction of defense mechanisms, suggesting that the coordinated action of SA and JA might be crucial in DED resistance (8, 13).

The SA-mediated defense is related to the activation of phenylalanine ammonia-lyase (PAL), a key enzyme in the phenylpropanoid pathway that produces a broad range of plant phenolic compounds, including SA itself, coumarins, anthocyanins, phytoalexins and flavonoids, and structural components like lignin, lignans, suberin or callose (31). In elms, DED promotes the activation of this pathway, and thus, the accumulation of several phenolic compounds (32, 33). Increases in catechins, condensed tannins, and the flavonol quercetin, are part of the common response in *Ulmus* spp. to different stressors (34–36). Indeed, the accumulation of defensive compounds (e.g. coumarin phenolics, scopoletin, and elm mansonones) was well correlated to DED resistance in different *Ulmus* spp. On the one hand, a higher production of scopoletin was found in cell cultures of the DED-resistant species *Ulmus pumila* in the presence of *O. ulmi* spores, compared to cells of the susceptible species *Ulmus campestris* (a.k.a. *U.* minor) (37). On the other hand, the accumulation of mansonones conferred *Ulmus hollandica, U. americana, U. pumila,* and *U. glabra* resistance against *O. ulmi* (38–42). Finally, a recent study presented the emission of volatile organic compounds (VOCs) as a resistance factor (43). These compounds are a group of primary and secondary metabolites that serve in chemical communication within the plant, between plants, or between them and other organisms like insects and microbes (44). *O. novo-ulmi* inoculation in *U. laevis* stems prompted stronger VOC emission from leaves, and therefore, stronger elicitation of plant defense than in *U. glabra,* suggesting a higher susceptibility to DED in this last species (43).

Although the embolism of xylem vessels caused by the pathogen is one of the main causes of tree physiological failure upon *O. novo-ulmi* infection, carbon starvation appears as another factor contributing to tree death (45). Tree infection with the DED pathogen drastically reduces the starch reserves stored in xylem tissues (46). Reduction of starch reserves is most likely related to the activation of defense mechanisms, mainly the formation of barrier zones by the cambium. Cells that constitute the barrier zones are rich in phenolic compounds and suberin, and their formation is energetically costly. Furthermore, the barrier zones decrease the space for the storage of starch by walling off large volumes of wood (46). The critical role of carbon reserves in tree survival is exacerbated by the fact that peak susceptibility to DED infection occurs after fructification and leaf expansion, processes that reduce the levels of starch to a minimum. Under these circumstances, a restriction in photosynthesis caused by pathogen infection (45) can further deplete starch levels and kill the tree (47). In a study with *U. Americana,* authors found that trees survived better a DED infection when they had higher levels of starch in the year previous to the infection (46). Interestingly, the levels of extractable sugars were similar between inoculated and control elms, whereas starch reserves were not replenished in infected trees, which suggests starch reserves were hydrolyzed to maintain metabolic activity and did not build up during infection. Similar results were found by FT–IR analyses of *U. minor* wood after DED infection (29). Starch proportion in the xylem appeared as the main chemical trait discriminating control and inoculated elms.

Furthermore, inoculated-resistant genotypes showed higher starch reserves than inoculated-susceptible genotypes 15 days after infection, suggesting an earlier depletion of starch reserves in the susceptible genotypes.

Drought can affect DED susceptibility in different ways, for example via alterations of plant metabolism and xylem anatomy. The glycolytic, pentose-phosphate, and tricarboxylic acid cycle pathways have a primary role in providing substrates, reducing power and energy to secondary pathways involved in defense. Severe drought-induced limitations to photosynthesis via stomata closure, photoinhibition, or leaf shedding may eventually reduce the concentration and pool of carbohydrate reserves (3). A reduction in the pool of non-structural carbohydrates (NSC), in turn, can reduce carbon sources for fungal nutrition, but also the availability of the plant to maintain basic physiological functions via respiration, which tends to increase during pathogen infection to sustain defense costs (47, 48). For example, in some tree species, chronic carbon limitations due to drought can reduce xylem cell wall thickening and resin canals, favoring pathogen spread and tree death (49). On the other hand, cell growth and cell-wall thickening are strong carbon sinks very sensitive to drought stress (50), which can attenuate the depletion of carbon reserves during drought. According to the growth–differentiation balance hypothesis (51), high growth sensitivity to drought stress may increase carbon available for defense metabolism, which has otherwise a low priority under stress-free conditions. Anyway, shifts in plant vulnerability to pathogen infection caused by drought stress are not merely mediated by abundance of NSC. Drought affects deeply the operation of primary metabolic pathways (52) to provide cells with needed metabolites. Such re-organization can deliver primary metabolites to defensive pathways involved in the production of ROS and associated defensive genes, enzymes, hormones, and many other secondary metabolites; for example, alkaloids, phytoalexins, glucosinolates, SA, flavonoids, and other phenolic compounds from aromatic amino acids (31, 53–55). Indeed, drought results in increased concentrations of secondary metabolites or expression of genes involved in their biosynthetic pathways: suberin, alkaloids, salicylic acid, lignans, flavonoids, and other phenolic compounds (56–58), reported to participate directly or indirectly in defense against DED (see above). On the contrary, drought can reduce the mobilization of carbon compounds to and the relative water content in sites of pathogen attack (47, 59), which can compromise the local defense response. Few studies have examined drought effects on elm primary metabolism (46, 60–64). The different sensitivity and re-organization of the primary (and secondary) metabolism in response to drought stress among elm species will likely determine their ability to combat *O. novo-ulmi*.

Studies show that drought stress enhances (65, 66), decreases (67, 68) or does not clearly affect (69) susceptibility to vascular diseases, and mostly discuss these results in relation to drought stress effects on xylem hydraulic conductance, water status, and carbon metabolism. In relation to DED, existing evidences point to a beneficial effect of drought stress in resisting the disease. The timing of drought occurrence relative to tree infection is critical to the appearance of symptoms. Higher symptoms of DED were found in potted *U. minor* plants that had been well watered before inoculation than in those that had been subjected to drought (70). Because stressed plants had formed smaller vessels, these authors discussed that reduced drought-induced DED susceptibility was related to lower velocity of pathogen propagation through xylem, in support of observations that higher vessel size, hydraulic conductance, and transpiration favors the appearance of DED symptoms (71–73). Xylem vessel size has been shown to be an important factor determining the proliferation of other pathogens causing vascular wilt diseases in other plant species (74).

Here we evaluated the effect of a long-term drought on the susceptibility of elm species to the pathogen *O. novo-ulmi*. To this end, five-year-old trees of *U. minor*, *U. laevis*, and *U. glabra* subjected to either moderate drought or well-watered conditions over two growing seasons were inoculated with either *O. novo-ulmi* or water in the spring of the next year. Thereafter, samples at the base of the stem, where the inoculation was performed (i.e. local response), and in mid-canopy branches (i.e. systemic response) were taken for the analysis of primary metabolites with gas chromatography time-of-flight mass spectrometry (GC-TOF-MS). Stem anatomical features were also analyzed to compare how drought-induced changes in metabolic and anatomic traits affect tree susceptibility to DED. The general hypotheses of this work are that: (i) *O. novo-ulmi* impacts stem and less so branch primary metabolite levels and (ii) such impacts vary across species according to their resistance to DED.

12.2 Material and Methods

12.2.1 Plant Material and Experimental Design

Seedlings of *U. glabra*, *U. minor*, and *U. laevis* were planted in 2010 in a field plot at "Puerta de Hierro" Forest Breeding Centre (Madrid, Spain; 40°27′24″ N; 3°45′0″ W; 600 m above the sea level, inland Mediterranean climate). Forty-six *U. minor* seedlings were propagated by conventional crossing techniques (75, 76) from parents considered moderately resistant to DED based on previous susceptibility tests performed within the Spanish Elm Breeding Program. Thirty-six *U. laevis* seedlings came from nine mother trees in three Spanish locations, and 34 *U. glabra* seedlings from eight mother trees in two Spanish locations. The plants were spaced 0.55 × 2.0 m and distributed in four blocks, in an incomplete randomized block design. There was an additional line of trees surrounding the plot, and four rows of trees separating the blocks to minimize border effects; these trees were not selected for any of the measurements described below. The soil was covered with an anti-weed mesh.

Each plant was irrigated with 8 L of water during summer months, four days a week since 2010–2014. In the summers of 2014 and 2015, irrigation was stopped in two of the blocks during one month each year (Drought; D), and was kept as in previous years in the other two blocks (Watered; W). Periods without irrigation were separated by at least one week of irrigation. All plants were again equally irrigated in 2016 and 2017.

12.2.2 *O. novo-ulmi* Inoculation

Since maximum susceptibility of European elms to DED starts 40–50 days after bud burst (77), bud burst, and leaf phenology were monitored since March 2016 until complete leaf development to decide the day for inoculation. Ninety-two trees were inoculated on 18 May 2016 with *O. novo-ulmi* (Inoculated; I), leaving 24 trees (four trees per species and watering treatment) as negative controls inoculated with distilled water (Control; C). Inoculation was performed with the *O. novo-ulmi* ssp. *americana* M-SO1, isolated in 2014 from a DED-infected 168-year-old *U. minor* tree in El Pardo forest (Madrid, Spain). The identity of the pathogen was determined morphologically and later confirmed by molecular analysis (2). The pathogen grew on a 2% malt extract agar medium (MEA), in 9 cm Petri dishes kept at 22 °C in dark conditions, and sub-cultured every week for two months before inoculation. The inoculum was produced from a blastospore suspension grown in Tchernoff's liquid medium (78), which was centrifuged and suspended in distilled sterilized water. Concentration of spores was adjusted to 10^6 spores mL^{-1} using a Thoma hemocytometer. Inoculations were performed introducing 0.1 mL of the blastospore suspension into the outer xylem at the base of the main stem using a sharp blade, taking advantage of the transpiration-pulling force (79).

12.2.3 Measurements

12.2.3.1 Stomatal Conductance

Leaf stomatal conductance (g_s) was measured during the summers of 2014 and 2015. This variable was intended as a non-destructive proxy of drought stress suffered by trees before the inoculation. We used a leaf porometer (model SC-1, Decagon Devices, Pullman, WA, USA) to measure one leaf per tree in all trees (excepting border trees separating D from C blocks and those forming the border of the plot). The leaf chosen for measurement was in the mid part of the crown (1–2 m height) and oriented to the southeast.

12.2.3.2 Crown Wilting

DED symptoms were evaluated by measuring the percentage of tree crown exhibiting foliar wilting. Three independent observers estimated this value at 14, 30, 60, and 120 days post inoculation (dpi), to finally calculate an average value per tree in each date. All trees were evaluated (except the border trees).

12.2.3.3 Primary Metabolites

Fourteen days post inoculation, four trees per species and treatment were selected for primary metabolite profiling with gas chromatography time-of-flight mass spectrometry (GC–TOF–MS). To target systemic and local responses to *O. novo-ulmi*, respectively, branches and the main stem close to the inoculation incision were sampled. Sampling at 14 dpi aimed at assessing early tree responses to *O. novo-ulmi* when there are no DED symptoms (80), rather than the drought stress associated with this vascular disease. The selection of trees was based on the mean g_s values exhibited by trees in the year before inoculation, trying to maximize differences in g_s between W and D trees and minimize differences in g_s between (future) C and I trees (Table 12.1). Sampling was conducted between 10.00 and 12.00 hours local time to minimize diurnal chemical fluctuations. Stems were sampled with an increment borer, approximately 2 cm above the inoculation incision, and branches were cut from the southeast mid-crown. At 14 dpi, a few *U. minor* trees already exhibited symptoms of crown wilting. However, to keep sampling representative of overall tree health status, healthy branches with no symptoms of foliar wilting were sampled, as no sampled tree had an average crown wilting higher than 30%. Samples were immediately placed in a polystyrene box (e.g. Styrofoam or similar) with liquid nitrogen and stored at −80 °C until further GC–TOF–MS primary metabolite profiling analysis.

Primary metabolites were extracted from 100 mg fresh weight (FW) of freeze-ground powder in 1400 μL ice-cold methanol containing ribitol as internal standard (0.2 mg mL^{-1} ribitol in water) (81). Samples were vortex-mixed and incubated in a shaker (ThermoMixer, Eppendorf, Hamburg, Germany) for 15 minutes at 70 °C and 950 rpm, and subsequently centrifuged at 12,000 *g* for 10 minutes. The supernatant was collected, mixed with 750 μL chloroform and 1500 μL water, and vortex-mixed. After centrifugation at 2200 *g* for 15 minutes, an aliquot of 150 μL of the polar aqueous/methanol phase was evaporated to dryness for 3 hours at 30 °C using

Table 12.1 Mean (± SE) stomatal conductance averaged over 2015 in elm trees sampled for stem and branch GC–TOF–MS metabolite profiling, 14 dpi with *O. novo-ulmi* (I) or with distilled water (C) in May 2016. Trees were exposed to regular summer watering (W) or drought cycles with no watering (D) for two years before inoculation.

	U. glabra		*U. laevis*		*U. minor*	
	C	I	C	I	C	I
D	89 ± 20	93 ± 15	115 ± 8	101 ± 11	80 ± 21	104 ± 15
W	195 ± 25	197 ± 26	155 ± 18	160 ± 23	214 ± 30	267 ± 69

a centrifugal concentrator (Vacufuge Plus, Eppendorf, Hamburg, Germany), and stored at −80 °C prior to GC–TOF–MS analysis. Dried polar extracts were derivatized (81) with 40 μL of 20 mg mL^{-1} methoxyamine hydrochloride in pyridine, vortex-mixed, and incubated in a shaker (ThermoMixer, Eppendorf, Hamburg, Germany) for 2 hours at 37 °C and 950 rpm. Subsequently, 70 μL of *N*-methyl-*N*-trimethylsilyltrifluoroacetamide (TMS derivatization) and 20 μL mL^{-1} of a mixture of fatty acid methyl esters (FAMES) were added to sample mixtures, vortex-mixed, and incubated in a shaker (ThermoMixer, Eppendorf, Hamburg, Germany) for 30 minutes at 37 °C and 950 rpm. Primary metabolite profiling analysis of the derivatized samples (1 μL injection) was performed on an Agilent 6890N gas chromatograph (Agilent Technologies, Böblingen, Germany) coupled to a LECO Pegasus III TOF-MS running in electron ionization (EI) mode (LECO Instrumente, Mönchengladbach, Germany). The chromatographic separation was performed on a VF-5MS column (Varian Inc., 30 m length, 0.25 mm inner diameter, and 0.25 mm film thickness). The temperature program was set as follows: isothermal for 2 minutes at 85 °C, followed by a 15 °C min^{-1} ramp to 360 °C, and hold at this temperature for six minutes. The injector and transfer line temperatures were set to 230 and 250 °C, respectively, and the helium carrier gas flow set to 2 mL min^{-1}. After a solvent delay of 180 seconds, mass spectra were scanned from *m/z* 70 to 600 with an acquisition rate of 20 spectra s^{-1} and a detector voltage between 1700 and 1850 V. GC–TOF–MS data were evaluated using AMDIS (Automated Mass Spectral Deconvolution and Identification System) software (v. 2.71) (www.amdis.net) that allows the separation and identification of overlapping peaks. Primary metabolites were annotated using TagFinder 4.0 software (82) and a reference library of ambient mass spectra and retention indices from the Golm Metabolome Database (83, 84). Metadata information of this experiment following minimum reporting standard guidelines of the Metabolomics Standard Initiative (81, 85, 86) can be found in Table 12.2.

12.2.3.4 Xylem Anatomy

After felling the trees in 2017, one year after the inoculation, a wood slice was obtained from the base of the tree. Stem radial growth and wood anatomical traits were measured for 2014 and 2015 to assess the effect of drought, and after the inoculation with *O. novo-ulmi* in 2016, to assess the effect of drought on susceptibility to DED. Four trees per species, drought treatment, and inoculation treatment were selected for this purpose, occasionally three, if a tree died between 2016 and 2017.

Wood samples were covered with formaldehyde:acetic acid:70% ethanol (5 : 5 : 90, v:v:v) fixative for one week and then kept in 70% ethanol until cross sections were prepared for light microscopy analysis following standard

Table 12.2 Metabolomics Standards Initiative (MSI) Compliant Metadata: metadata information of the GC-TOF-MS primary metabolite profiling analysis.

BioSource		
	Species	*Ulmus glabra, U. laevis, U. minor*
	Genotype	n.a.
	Organ	Branches and stems
	Organ specification	n.a.
	Amount	100 mg fresh weight (FW) per sample
Growth		
	Support	Puerta de Hierro Forest Breeding Center
	Location	Madrid
	Plot design	Incomplete randomized block design
	Light period	Full sunlight
	Humidity day	Variable (20–100%)
	Humidity night	Variable (20–100%)
	Temperature (day)	Variable (−2 to 40 °C)
	Temperature (night)	Variable (−10 to 25 °C)
	Watering	Rainfall plus half trees supplemented with 8 L water during summer
	Nutritional	Sandy soil
	Date of plant establishment	2010
Treatment		
	Abiotic treatment	Drought stress
	Biotic treatment	Inoculation with *Ophiostoma novo-ulmi*
	Dose	Two-year summer drought followed by single inoculation with *O. novo-ulmi*
	Duration	3 y
Harvest		
	Date	1st June 2016
	Time	Mid-morning
	Growth stage	Saplings
	Metabolism quenching	Liquid nitrogen
	Harvest method	Prunning
	Storage	−80 °C

(Continued)

Table 12.2 (Continued)

Sample processing, extraction, and protocols		
	Tissue processing	Fresh tissues
	Replicate sampling and analyses	Three biological replicates per treatment
	Extraction of Samples: GC-MS	Methanol:water:chloroform solution containing 0.2 mg mL^{-1} ribitol (IS) was used for extraction of polar metabolites for 15 min at 70 °C, 150 μL of the polar fraction was evaporated to dryness (81)
	Extract derivatization: GC-MS	Dried polar extracts were derivatized with 40 μL of 20 mg mL^{-1} of methoxyamine hydrochloride in pyridine followed by TMS derivatization using 70 μL of *N*-methyl-*N*-trimethylsilyltrifluoroacetamide and 20 μL of a mixture of fatty acid methyl esters (FAMES) (81)
	Extract clean-up	None
	Extract storage	At −80 °C
GC-MS instrument		
Gas Chromatography	Sample preparation	Samples were used as described above
	Auto injector	CTC Combi PAL, CTC Analytics AG, Zwingen, Switzerland
	Chromatography Instrument	6890N Agilent, Böblingen, Germany
	Separation column	VF-5MS column (Varian Inc., 30 m length, 0.25 mm inner diameter, and 0.25 μm film thickness)
	Separation parameters	The injection temperature was 230 °C. The initial temperature of the oven (85 °C) increased at a rate of 15 °C min^{-1} up to a final temperature of 360 °C
Mass spectrometry		
	Instrument	Pegasus III TOF, Leco Instruments, St. Joseph, USA
	Sample Introduction	GC-MS
	Ionization	Electron ionization
	Polarity	Positive ionization
	Mass Analyzer	Time-of-flight (TOF)
	Data acquisition	The transfer line to the mass spectrometer and ion source were set to 250 °C. After a solvent delay of 180 s mass spectra were scanned from *m/z* 70–600 with the acquisition rate of 20 spectra s^{-1}

(Continued)

Table 12.2 (Continued)

Instrument performance and method validation		
	Instrument calibration	Internal mass calibration was performed using Agilent calibrant (PFTBA)
	Mass resolution	Unit mass
	Mass accuracy	Nominal mass
	Quantification	Relative quantification by internal standard ribitol and sample fresh weight
	Variation	Based upon the internal standard
	Sample replications	1 analytical (same analytical sample preparation) and three biological replicates per treatment
	Internal standard	Ribitol
	Quality controls	FAMES internal standard markers and a QC standard solution of 41 pure reference compounds (i.e. the most detected and abundant metabolites) throughout the analysis
	Blanks	Multiple blanks and wash analyses were included in the experimental design
Data preprocessing		
	File format	Data were exported in a net.cdf format using Agilent MSD ChemStation
	Pre-processing details	Data files were deconvoluted using AMDIS (Automated Mass Spectral Deconvolution and Identification System) software (v. 2.71) (www .amdis.net)
	Statistics	Data were normalized by dividing each peak intensity by the corresponding internal standard ribitol peak intensity, followed by dividing by the fresh weight of each sample.
Metabolite identifications		
	Unknown compounds	n.a.
	Identified compounds	Primary metabolites were annotated using TagFinder 4.0 software (82) and a reference library of ambient mass spectra and retention indices from the Golm Metabolome Database (83,84)

Source: Lisec et al. (81), Fiehn et al. (85), and Sumner et al. (86).

protocols (87). Transversal sections 15 µm thick were cut using a sliding microtome (Leica SM 2400) and stained with a solution of 1% safranin. Sections were mounted on slides, and digitalized with a camera attached to a light microscope (Olympus BX50) for image analysis with ImageJ (88). Three radial sectors were randomly selected per sample and growth-ring length for measuring percentage of latewood, individual vessel lumen area in earlywood and latewood, groups of vessels, and parenchyma ray frequency on a per tree-ring basis from 2014 to 2016.

12.2.4 Data Analysis

The combination of the two groups with differential watering (W and D) and the two groups of differential inoculation (I and C) resulted in the following treatments: W+I (plants inoculated with *O. novo-ulmi* that were previously well-watered), W+C (plants mock-inoculated that were previously well-watered), D+I (plants inoculated with *O. novo-ulmi* that were previously exposed to two years of drought) and D+C (plants mock-inoculated that were previously exposed to two years of drought). The comparisons among these treatments were made separately by species.

Means of g_s, and stem radial growth and wood anatomical variables were analyzed using repeated-measures ANOVA analyses, including species and drought treatment for the years 2014 and 2015. In addition, means of g_s at each date were compared with *t* tests, or with Welch tests when the variances of C and W plants were unequal. A general linear model (GLM) was used to compare stem radial growth and wood anatomical variables in 2016, with species, inoculation, and drought treatment (in the previous two years) as fixed factors. Means were compared using Fisher's least significant difference test (LSD, $\alpha = 0.05$). Crown wilting at 14, 30, 60, and 120 dpi was compared among treatments using non-parametric statistics due to non-normal distribution of residuals. We used a Kruskal–Wallis test to determine general differences between treatments, followed by a Wilcoxon rank-sum test to make multiple pairwise comparisons, using the R packages "kruskal.test" and "pairwise.wilcox.test," respectively (89).

Multivariate statistical analysis of primary metabolite data was performed using the R packages "gplots" (90) for heatmap plotting, and "mixOmics" (91) for partial least squares discriminant analysis (PLS–DA). The relative abundance of primary metabolites was normalized to the internal standard (ribitol) and the FW of the samples. One-way ANOVA was used to assess differences between treatments at $\alpha = 0.05$. Subsequently, fold-changes were determined and Log_{10}-transformed for heatmap plotting. Supervised PLS–DA was performed using the leave-one-out cross-validation embedded in the "mixOmics" package.

12.3 Results

12.3.1 Drought Effects on Stomatal Conductance, Xylem Anatomy, and Primary Metabolism

Stomatal conductance (g_s) was lower in D than W trees over 2014 and 2015, with differences depending on species and dates ($P < 0.05$ for drought main effect and drought interaction effects with species and time). The three species showed significant differences in g_s between W and D trees at the end of summer 2014 and at the end of the two experimental drought cycles imposed in 2015, but differences were less marked for *U. laevis* (Figure 12.1). Surprisingly, g_s was higher in D than W trees of *U. laevis* and *U. glabra* upon rewatering in 2015 (Figure 12.1).

Ulmus glabra showed the lowest stem radial growth of the three species during the course of the study, mainly due to lower latewood production (46% of the growth ring), with no effect of irrigation on these variables (Table 12.3). Stopping irrigation had a significant effect on *U. laevis* and *U. minor* in 2014, when D trees of both species grew c. 40% less than W trees. However, growth was not affected by irrigation in 2015 or 2016. Mean area of earlywood vessels was similar in the three species and was not affected by irrigation, whereas latewood vessel area was consistently lower in *U. glabra* than in the other two species. No differences between species or irrigation treatments were found in vessel grouping or frequency of parenchyma rays.

The analysis of the primary metabolome of stems and branches sampled in spring 2016 showed that the previous drought in 2014 and 2015 had barely any effect on primary metabolite concentrations (Figures 12.2–12.4; Table 12.4). Branch and stem samples of W and D trees that were mock inoculated showed an overlapping distribution in the two-dimensional plane generated by PLS-DA, which explained 55% and 35% of variance for branches and stems, respectively (Figures 12.2 and 12.3). On the contrary, the PLS-DA discriminated the three species better in support of their different ecology. The heatmap showed that only three metabolites exhibited a significant difference between mock-inoculated W and D trees: guanidine, sucrose, and aspartate (Figure 12.4).

12.3.2 Impact of *O. novo-ulmi* on Crown Wilting, Xylem Anatomy, and Primary Metabolism

Inoculation with *O. novo-ulmi* had no effect on crown foliar wilting at 14 dpi on any species (Figure 12.5). At 30, 60, and 120 dpi, there were significant differences in foliar wilting between trees inoculated with *O. novo-ulmi*

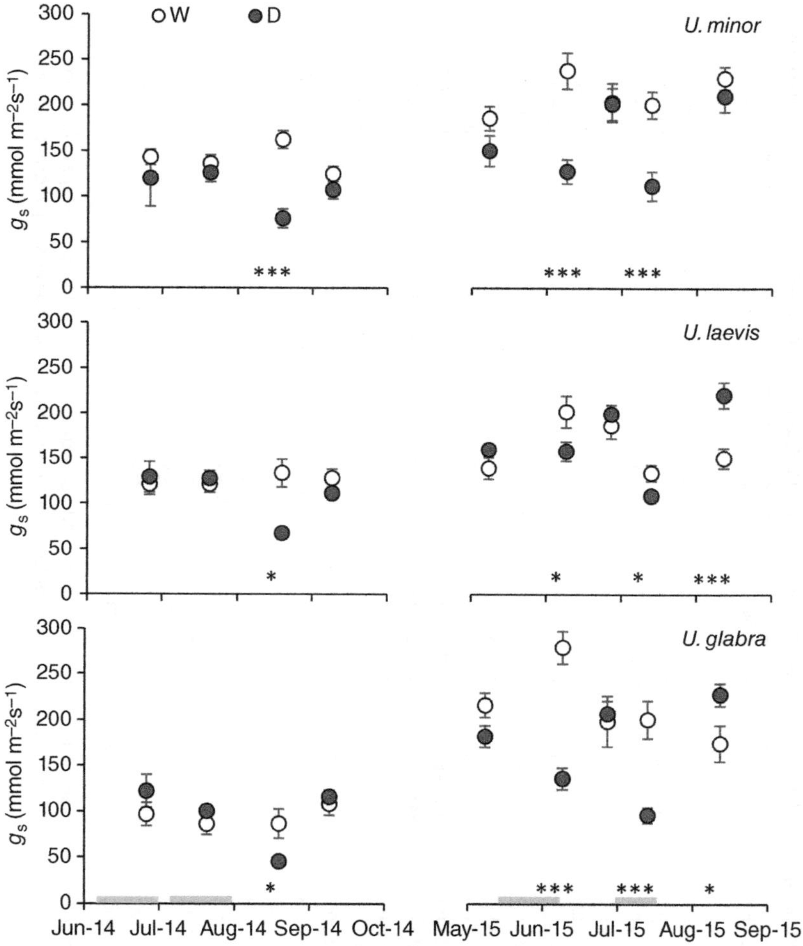

Figure 12.1 Mean (± SE) leaf stomatal conductance (g_s) in elm trees watered (W) or non-watered (D) during summers of 2014 and 2015, before the inoculation with *O. novo-ulmi* in May 2016. Yellow lines in the lower panel correspond to periods of watering restriction (W). Asterisks denote significant differences between means of W and D plants at $P < 0.05$ (*) and $P < 0.001$ (***).

and trees inoculated with water, for all elm species. Symptoms of foliar wilting started earlier in *U. minor* (Figure 12.2). However, symptomatology was higher in *U. glabra* than *U. minor* at 120 dpi ($P = 0.053$), when median wilting exceeded 50% and barely surpassed 15% in inoculated *U. glabra* and

Table 12.3 Mean (± SE) annual radial growth and wood anatomical traits in elm trees watered (W) or non-watered (D) during the summers of 2014 and 2015, and inoculated with *O. novo-ulmi* in May 2016.

Species	Year	Treatment	GRL (mm)	%LW	VA-EW (μm^2)	VA-LW (μm^2)	VG (vessels group^{-1})	RD (rays mm^{-2})
Ulmus glabra	2014	W	0.4 ± 0.1	58.5 ± 6.5	4414 ± 746	274 ± 33	5.9 ± 1.0	7.0 ± 0.7
		D	0.4 ± 0.1	39.0 ± 6.0	3757 ± 691	219 ± 29	5.1 ± 0.9	6.0 ± 0.6
	2015	W	0.3 ± 0.1	53.9 ± 7.0	3896 ± 979	302 ± 41	4.4 ± 1.2	7.0 ± 0.7
		D	0.4 ± 0.1	36.7 ± 6.5	4935 ± 899	208 ± 37	4.5 ± 1.2	6.0 ± 0.7
	2016	WC	0.5 ± 0.1	60.5 ± 9.0	3782 ± 965	267 ± 45	5.5 ± 1.1	7.0 ± 0.7
		WI	0.2 ± 0.1	30.1 ± 8.0	2020 ± 715	167 ± 32	4.7 ± 1.1	7.0 ± 0.7
		DC	0.7 ± 0.2	50.8 ± 8.0	3933 ± 822	336 ± 69	6.5 ± 1.3	5.0 ± 0.6
		DI	0.5 ± 0.1	64.0 ± 9.0	4028 ± 989	259 ± 39	3.2 ± 1.0	7.0 ± 0.7
Ulmus laevis	2014	W	1.8 ± 0.3	79.0 ± 6.5	4233 ± 740	491 ± 53	5.4 ± 1.1	8.0 ± 0.7
		D	1.1 ± 0.3	76.9 ± 6.0	3724 ± 695	438 ± 49	4.3 ± 0.9	9.0 ± 0.7
	2015	W	1.1 ± 0.3	76.8 ± 7.0	4816 ± 1070	504 ± 51	7.2 ± 1.4	8.0 ± 0.7
		D	1.3 ± 0.2	82.3 ± 6.6	4659 ± 999	506 ± 47	4.4 ± 1.2	9.0 ± 0.7
	2016	WC	1.0 ± 0.3	84.9 ± 9.0	3171 ± 1000	445 ± 66	3.8 ± 1.1	8.0 ± 0.7
		WI	1.0 ± 0.2	76.2 ± 9.0	2560 ± 990	470 ± 81	6.4 ± 1.3	8.0 ± 0.7
		DC	1.6 ± 0.3	79.8 ± 11.0	4256 ± 988	561 ± 58	4.6 ± 1.2	8.0 ± 0.7
		DI	0.4 ± 0.1	56.9 ± 11.0	1653 ± 1104	379 ± 66	4.0 ± 1.2	9.0 ± 0.7
Ulmus minor	2014	W	2.0 ± 0.3	74.8 ± 6.5	3395 ± 785	384 ± 48	7.7 ± 1.0	8.0 ± 0.7
		D	1.3 ± 0.3	69.5 ± 6.0	4372 ± 690	459 ± 51	4.7 ± 0.9	8.0 ± 0.7
	2015	W	1.5 ± 0.3	64.1 ± 7.0	3356 ± 1065	382 ± 48	7.8 ± 1.2	7.0 ± 0.7
		D	1.3 ± 0.2	68.8 ± 6.7	5849 ± 1002	490 ± 45	4.9 ± 1.1	8.0 ± 0.7
	2016	WC	0.9 ± 0.2	57.2 ± 8.0	3430 ± 985	405 ± 62	4.9 ± 1.1	7.0 ± 0.7
		WI	0.9 ± 0.2	57.9 ± 8.0	2046 ± 825	385 ± 47	6.8 ± 1.3	8.0 ± 0.7
		DC	1.4 ± 0.3	68.7 ± 10.0	7892 ± 1141	548 ± 68	4.2 ± 1.1	8.0 ± 0.7
		DI	0.7 ± 0.2	63.9 ± 10.0	1908 ± 853	316 ± 46	4.5 ± 1.1	8.0 ± 0.7

Abbreviations: C, control (inoculation with distilled water); I, inoculated with *O. novo-ulmi*; GRL, growth-ring length; %LW, percentage of latewood in the growth ring; VA-EW, mean vessel area of earlywood vessels; VA-LW, mean vessel area of latewood vessels; VG, number of vessels per group of vessels; RD, parenchyma ray density.

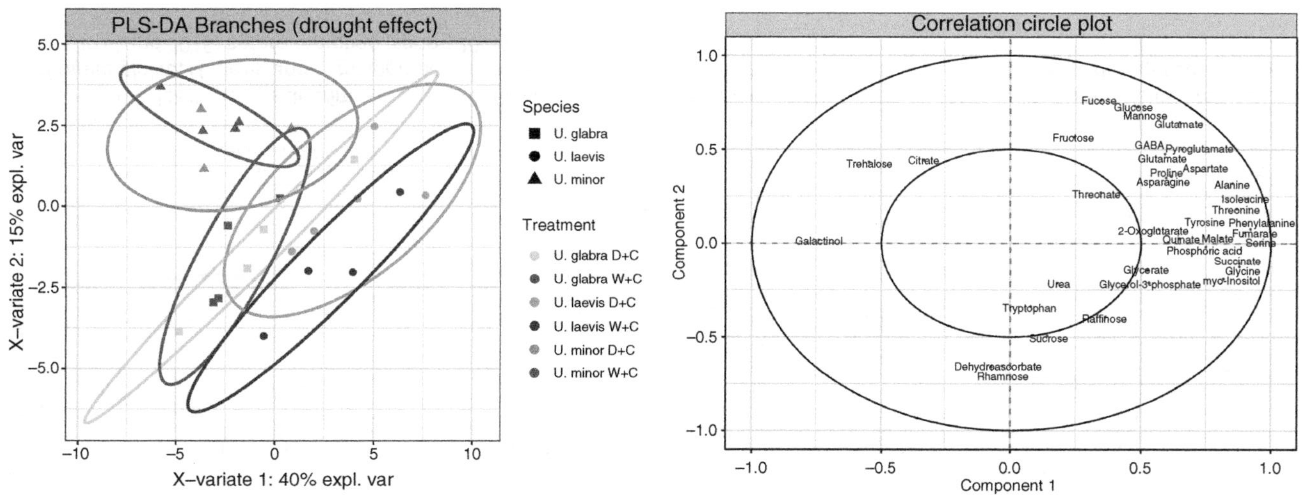

Figure 12.2 Partial least squares discriminant analysis (PLS-DA) score plots (a) and correlation plots (b) of the primary metabolite profile of branches of *Ulmus glabra*, *U. laevis*, and *U. minor* trees that were either well-watered (W+C) or exposed to drought stress (D+C) for two years (2014 and 2015) before sampling (spring 2016).

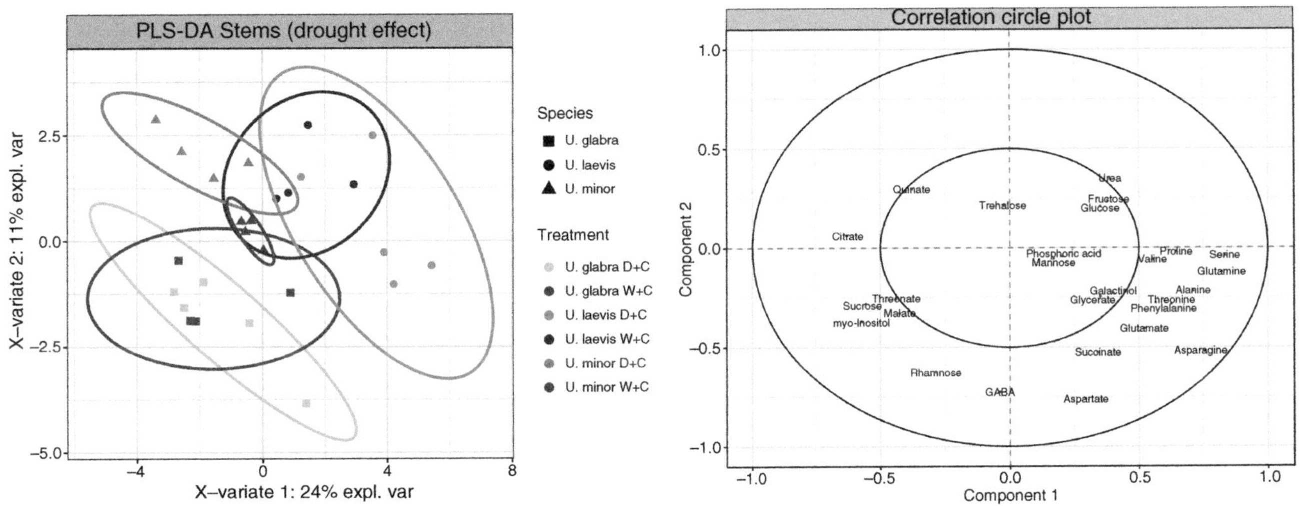

Figure 12.3 Partial least squares discriminant analysis (PLS-DA) score plots (a) and correlation plots (b) of the primary metabolite profile of main stems of *Ulmus glabra*, *U. laevis*, and *U. minor* mock-inoculated trees that were either well-watered (W+C) or exposed to drought stress (D+C) for two years (2014 and 2015) before sampling (spring 2016).

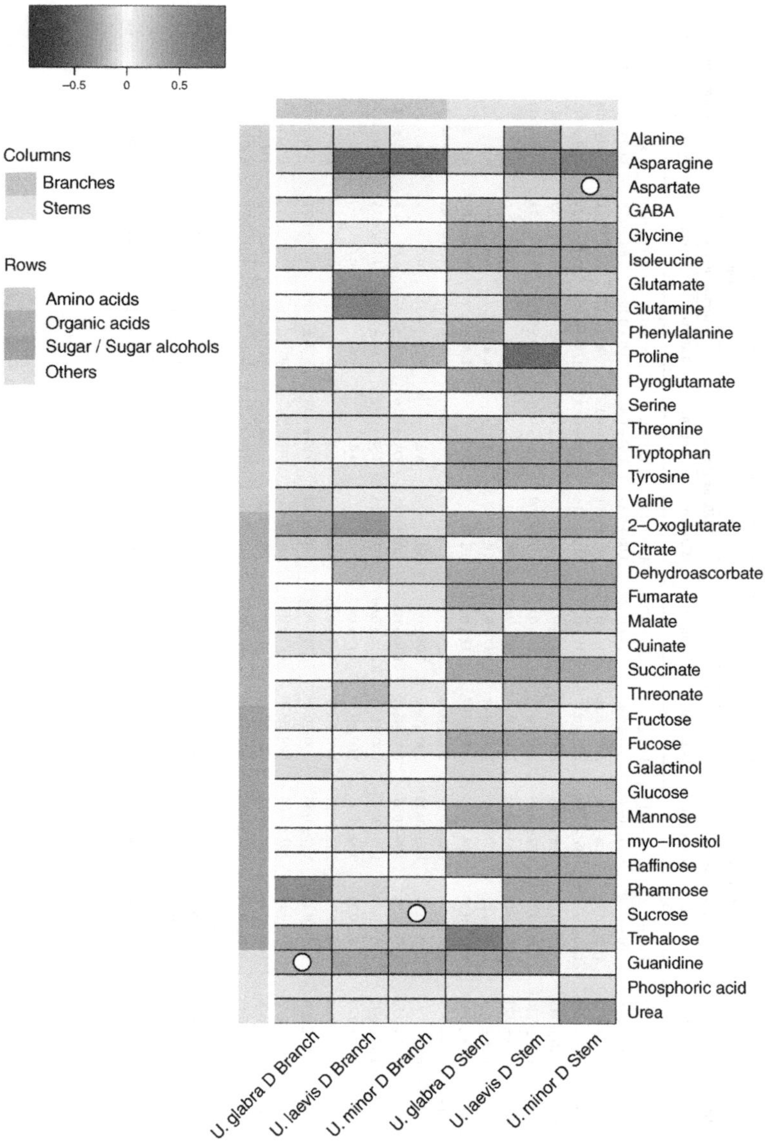

Figure 12.4 Heatmap representing the changes in relative levels of primary metabolites in branches and stems of *Ulmus glabra*, *U. laevis*, and *U. minor* subjected to drought stress. Relative values (as means of three to four independent measurements) were normalized to the internal standard ribitol and fresh weight of the samples; false color imaging was performed on $\log_{10}$-transformed GC–TOF–MS metabolite data. Fold changes were calculated in relation to the watered controls. Significant changes using one-way ANOVA are indicated as ○ $P < 0.05$. Gray-color squares represent undetected metabolites.

Table 12.4 Fold changes in relative levels of primary metabolites in stem and branch tissues of *Ulmus* spp. exposed to drought stress.

		Branch						Stem					
		Ulmus glabra		*Ulmus laevis*		*Ulmus minor*		*Ulmus glabra*		*Ulmus laevis*		*Ulmus minor*	
		Watered	Drought	Watered	Drought	Watered	Drought	Watered	Drought	Watered	Drought	Watered	Drought
Amino acids	Alanine	1.00 ± 0.03	1.43 ± 0.33	1.00 ± 0.21	1.46 ± 0.32	1.00 ± 0.11	1.03 ± 0.23	1.00 ± 0.15	1.09 ± 0.13	1.00 ± 0.04	**2.27 ± 0.25**	1.00 ± 0.09	0.81 ± 0.21
	Asparagine	1.00 ± 0.20	1.45 ± 0.79	1.00 ± 0.77	0.34 ± 0.06	1.00 ± 0.48	1.13 ± 0.14	1.00 ± 0.39	1.54 ± 0.46	1.00 ± 0.40	2.86 ± 0.77	1.00 ± 0.31	0.39 ± 0.11
	Aspartate	1.00 ± 0.10	0.90 ± 0.26	1.00 ± 0.40	2.08 ± 0.46	1.00 ± 0.19	1.18 ± 0.32	1.00 ± 0.11	1.01 ± 0.17	1.00 ± 0.08	1.49 ± 0.21	1.00 ± 0.08	**0.61 ± 0.07**
	GABA	1.00 ± 0.04	1.46 ± 0.34	1.00 ± 0.16	0.96 ± 0.14	1.00 ± 0.17	1.12 ± 0.19	1.00 ± 0.14	1.83 ± 0.63	1.00 ± 0.15	0.93 ± 0.24	1.00 ± 0.15	0.68 ± 0.04
	Glutamate	1.00 ± 0.07	1.10 ± 0.19	1.00 ± 0.27	2.68 ± 0.65	1.00 ± 0.21	1.16 ± 0.31	1.00 ± 0.17	0.82 ± 0.09	1.00 ± 0.17	2.02 ± 0.39	1.00 ± 0.10	0.70 ± 0.12
	Glutamine	1.00 ± 0.07	1.07 ± 0.30	1.00 ± 0.27	3.72 ± 1.18	1.00 ± 0.18	1.19 ± 0.46	1.00 ± 0.09	0.84 ± 0.07	1.00 ± 0.17	2.23 ± 0.73	1.00 ± 0.11	0.62 ± 0.16
	Glycine	1.00 ± 0.08	0.96 ± 0.21	1.00 ± 0.19	0.85 ± 0.10	1.00 ± 0.21	1.10 ± 0.29	n.d.	n.d.	n.d.	n.d.	n.d.	n.d.
	Isoleucine	1.00 ± 0.12	1.46 ± 0.61	1.00 ± 0.22	1.00 ± 0.27	1.00 ± 0.11	1.17 ± 0.36	n.d.	n.d.	n.d.	n.d.	n.d.	n.d.
	Phenylalanine	1.00 ± 0.13	1.32 ± 0.31	1.00 ± 0.27	0.83 ± 0.08	1.00 ± 0.07	1.11 ± 0.21	n.d.	n.d.	1.00 ± 0.17	1.15 ± 0.07	n.d.	n.d.
	Proline	1.00 ± 0.19	0.99 ± 0.14	1.00 ± 0.16	1.48 ± 0.42	1.00 ± 0.08	1.13 ± 0.18	1.00 ± 0.07	0.97 ± 0.10	1.00 ± 0.07	8.49 ± 3.93	1.00 ± 0.35	0.98 ± 0.27
	Pyroglutamate	n.d.	n.d.	1.00 ± 0.64	1.13 ± 0.22	1.00 ± 0.06	1.06 ± 0.43	n.d.	n.d.	n.d.	n.d.	n.d.	n.d.
	Serine	1.00 ± 0.04	1.22 ± 0.29	1.00 ± 0.27	1.35 ± 0.31	1.00 ± 0.21	1.10 ± 0.38	1.00 ± 0.21	1.08 ± 0.16	1.00 ± 0.07	1.38 ± 0.19	1.00 ± 0.12	0.93 ± 0.16
	Threonine	1.00 ± 0.12	1.36 ± 0.42	1.00 ± 0.29	1.32 ± 0.30	1.00 ± 0.11	1.16 ± 0.60	1.00 ± 0.15	1.38 ± 0.32	1.00 ± 0.15	1.13 ± 0.15	1.00 ± 0.23	0.78 ± 0.12
	Tryptophan	1.00 ± 0.54	1.23 ± 0.22	1.00 ± 0.29	0.92 ± 0.12	1.00 ± 0.33	1.19 ± 0.53	n.d.	n.d.	n.d.	n.d.	n.d.	n.d.
	Tyrosine	1.00 ± 0.12	1.19 ± 0.41	1.00 ± 0.12	1.25 ± 0.21	1.00 ± 0.17	1.14 ± 0.28	n.d.	n.d.	n.d.	n.d.	n.d.	n.d.
	Valine	1.00 ± 0.08	1.43 ± 0.49	1.00 ± 0.24	1.17 ± 0.38	1.00 ± 0.18	1.14 ± 0.32	1.00 ± 0.22	0.92 ± 0.14	1.00 ± 0.16	1.10 ± 0.21	1.00 ± 0.16	0.97 ± 0.16
Organic acids	2-Oxoglutarate	n.d.	n.d.	1.00 ± 0.14	2.51 ± 1.64	1.00 ± 0.29	1.24 ± 0.15	n.d.	n.d.	n.d.	n.d.	n.d.	n.d.
	Citrate	1.00 ± 0.23	0.65 ± 0.05	1.00 ± 0.18	1.54 ± 0.31	1.00 ± 0.22	1.10 ± 0.64	1.00 ± 0.20	0.96 ± 0.22	1.00 ± 0.19	1.82 ± 0.42	1.00 ± 0.26	1.68 ± 0.40
	Dehydroascorbate	1.00 ± 0.18	1.09 ± 0.18	1.00 ± 0.24	0.63 ± 0.13	1.00 ± 0.18	1.10 ± 0.23	n.d.	n.d.	n.d.	n.d.	n.d.	n.d.
	Fumarate	1.00 ± 0.19	0.99 ± 0.19	1.00 ± 0.34	1.02 ± 0.13	1.00 ± 0.15	1.17 ± 0.29	n.d.	n.d.	n.d.	n.d.	n.d.	n.d.
	Malate	1.00 ± 0.15	1.24 ± 0.24	1.00 ± 0.18	1.03 ± 0.12	1.00 ± 0.08	1.19 ± 0.22	1.00 ± 0.18	1.51 ± 0.47	1.00 ± 0.20	0.99 ± 0.21	1.00 ± 0.07	1.49 ± 0.31
	Quinate	1.00 ± 0.10	1.24 ± 0.23	1.00 ± 0.12	0.90 ± 0.06	1.00 ± 0.02	1.15 ± 0.29	1.00 ± 0.18	0.99 ± 0.18	1.00 ± 0.19	0.57 ± 0.14	1.00 ± 0.13	1.32 ± 0.72
	Succinate	1.00 ± 0.21	0.93 ± 0.15	1.00 ± 0.17	1.23 ± 0.19	1.00 ± 0.28	1.10 ± 0.33	n.d.	n.d.	n.d.	n.d.	n.d.	n.d.
	Threonate	1.00 ± 0.20	1.23 ± 0.29	1.00 ± 0.26	1.76 ± 0.55	1.00 ± 0.21	1.10 ± 0.04	1.00 ± 0.32	0.99 ± 0.18	1.00 ± 0.16	1.55 ± 0.36	1.00 ± 0.14	1.28 ± 0.27

Sugar & Sugar alcohols	Fructose	1.00 ± 0.27	1.24 ± 0.43	1.00 ± 0.32	1.24 ± 0.26	1.00 ± 0.17	1.13 ± 0.26	1.00 ± 0.21	1.43 ± 0.45	1.00 ± 0.21	1.42 ± 0.37	1.00 ± 0.35	0.92 ± 0.30
	Fucose	1.00 ± 0.24	1.01 ± 0.21	1.00 ± 0.26	0.96 ± 0.09	1.00 ± 0.11	1.18 ± 0.21	n.d.	n.d.	n.d.	n.d.	n.d.	n.d.
	Galactinol	1.00 ± 0.12	0.75 ± 0.07	1.00 ± 0.23	1.22 ± 0.31	1.00 ± 0.12	1.16 ± 0.35	1.00 ± 0.07	0.78 ± 0.15	1.00 ± 0.22	1.31 ± 0.15	1.00 ± 0.18	1.25 ± 0.10
	Glucose	1.00 ± 0.18	1.01 ± 0.22	1.00 ± 0.14	1.26 ± 0.08	1.00 ± 0.08	1.12 ± 0.03	1.00 ± 0.18	1.38 ± 0.29	1.00 ± 0.32	1.15 ± 0.35	1.00 ± 0.21	1.74 ± 0.31
	Mannose	1.00 ± 0.21	1.06 ± 0.27	1.00 ± 0.15	1.27 ± 0.08	1.00 ± 0.08	1.11 ± 0.02	n.d.	n.d.	n.d.	n.d.	n.d.	n.d.
	myo-Inositol	1.00 ± 0.10	0.96 ± 0.14	1.00 ± 0.08	1.26 ± 0.16	1.00 ± 0.12	1.11 ± 0.49	1.00 ± 0.14	1.24 ± 0.15	1.00 ± 0.17	1.19 ± 0.10	1.00 ± 0.07	1.04 ± 0.16
	Raffinose	1.00 ± 0.22	1.07 ± 0.13	1.00 ± 0.22	1.00 ± 0.23	1.00 ± 0.27	1.13 ± 0.29	n.d.	n.d.	n.d.	n.d.	n.d.	n.d.
	Rhamnose	1.00 ± 0.29	0.46 ± 0.13	1.00 ± 0.17	0.73 ± 0.10	1.00 ± 0.12	1.18 ± 0.17	1.00 ± 0.27	1.06 ± 0.16	n.d.	n.d.	n.d.	n.d.
	Sucrose	1.00 ± 0.13	0.94 ± 0.07	1.00 ± 0.04	0.88 ± 0.11	1.00 ± 0.11	**1.15 ± 0.12**	1.00 ± 0.13	1.16 ± 0.25	1.00 ± 0.09	0.79 ± 0.10	1.00 ± 0.06	1.28 ± 0.30
	Trehalose	1.00 ± 0.25	0.54 ± 0.07	1.00 ± 0.15	1.62 ± 0.51	1.00 ± 0.34	1.18 ± 0.26	1.00 ± 0.45	0.41 ± 0.14	n.d.	n.d.	1.00 ± 0.50	0.72 ± 0.13
Others	Guanidine	1.00 ± 0.09	**1.94 ± 0.28**	n.d.	n.d.	n.d.	n.d.	n.d.	n.d.	n.d.	n.d.	1.00 ± 0.09	0.99 ± 0.11
	Phosphoric acid	1.00 ± 0.26	1.36 ± 0.37	1.00 ± 0.19	0.90 ± 0.16	1.00 ± 0.14	1.19 ± 0.21	1.00 ± 0.10	0.81 ± 0.22	1.00 ± 0.07	0.97 ± 0.19	1.00 ± 0.21	1.33 ± 0.16
	Urea	1.00 ± 0.19	0.68 ± 0.15	1.00 ± 0.26	0.84 ± 0.10	1.00 ± 0.24	1.11 ± 0.10	1.00 ± 0.13	1.89 ± 0.28	1.00 ± 0.22	0.98 ± 0.21	1.00 ± 0.19	2.52 ± 1.19

Relative values are normalized to the internal standard (ribitol) and fresh weight (FW) of the samples. Data are presented as means ± SE of at least three independent measurements. Values in bold type indicate significant differences calculated using one-way ANOVA ($P < 0.05$) with respect to the watered control. n.d. not detected.

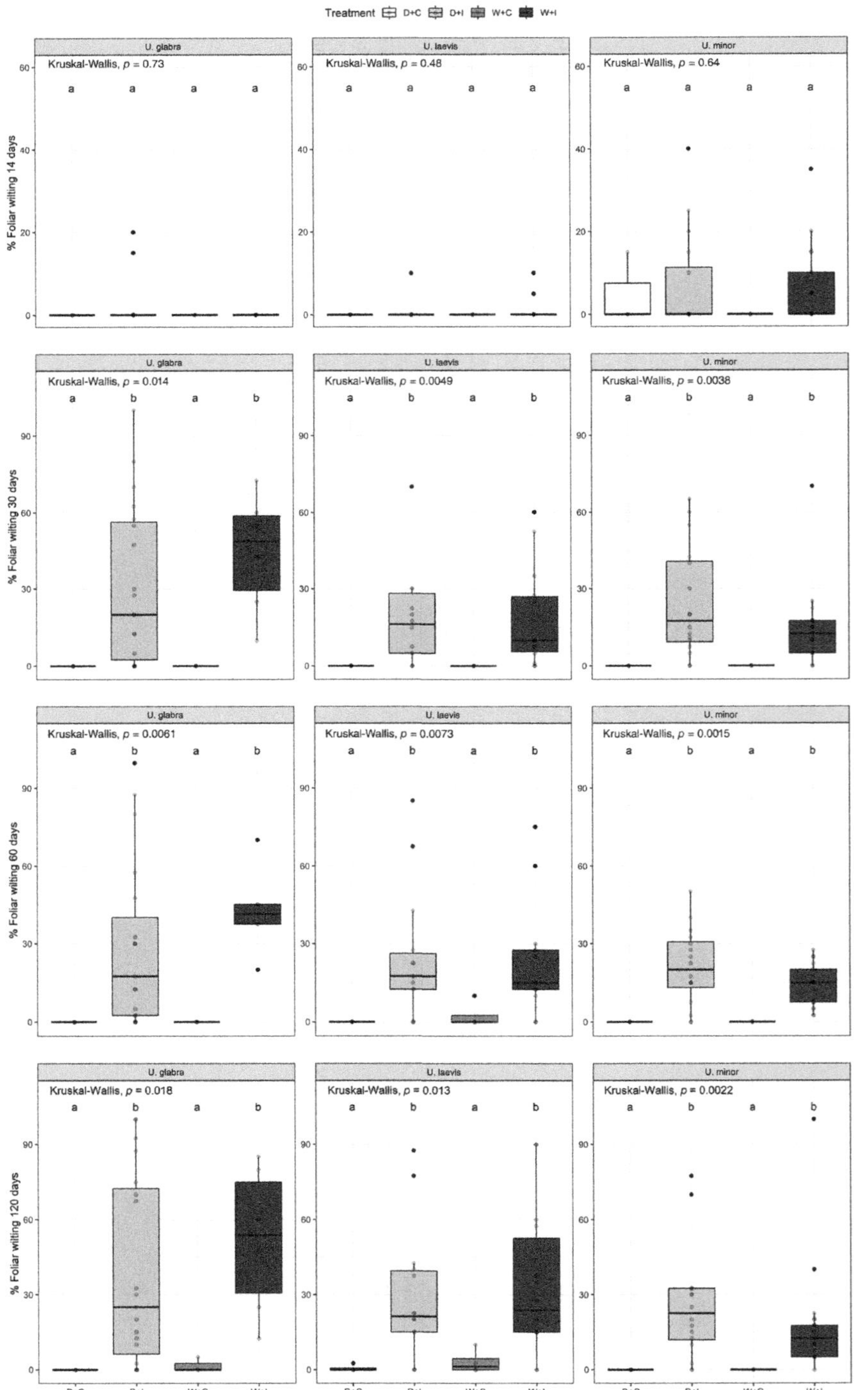
Treatment D+C D+I W+C W+I
U. glabra
U. laevis
U. minor
Kruskal-Wallis, p = 0.73
Kruskal-Wallis, p = 0.48
Kruskal-Wallis, p = 0.64
% Foliar wilting 14 days
Kruskal-Wallis, p = 0.014
Kruskal-Wallis, p = 0.0049
Kruskal-Wallis, p = 0.0038
% Foliar wilting 30 days
Kruskal-Wallis, p = 0.0061
Kruskal-Wallis, p = 0.0073
Kruskal-Wallis, p = 0.0015
% Foliar wilting 60 days
Kruskal-Wallis, p = 0.018
Kruskal-Wallis, p = 0.013
Kruskal-Wallis, p = 0.0022
% Foliar wilting 120 days
D+C D+I W+C W+I

Figure 12.5 Box-plots of percentage foliar wilting at 14, 30, 60, and 120 dpi. Data are separated by species and treatments: D+C (plants mock inoculated that were exposed to two years of drought), D+I (plants inoculated with *O. novo-ulmi* that were exposed to two years of drought), W+C (plants mock inoculated that were previously well watered) and W+I (plants inoculated with *O. novo-ulmi* that were previously well watered). The significance of a Kruskal-Wallis test comparing all treatments is given. Different letters mark significantly different treatments according to pairwise Wilcoxon rank-sum tests.

U. minor trees, respectively. None of these results were affected by previous differences in irrigation regime between trees.

Inoculation with *O. novo-ulmi* decreased radial growth in *U. laevis* by 50% ($P = 0.01$) and marginally in *U. minor* ($P = 0.10$; Table 12.3). The proportion of latewood in these two species was higher than 70% of the growth ring and was not affected by any of the treatments. Pathogen inoculation consistently decreased the diameter of earlywood ($P = 0.007$) and latewood ($P = 0.022$) vessels, particularly of *U. minor*. Trees of this species inoculated with *O. novo-ulmi* formed earlywood vessels 65% smaller and latewood vessels 26% smaller than mock-inoculated trees. Inoculation did not affect any other of the wood anatomical traits measured. None of these results were affected by irrigation.

Inoculation with *O. novo-ulmi* had little effect on primary metabolism at 14 dpi, an effect that varied among species, organs, and drought stress before inoculation. Within species, samples of I and C well-watered trees overlapped in the PLS-DA plane, which explained 47% and 33% of variance for branches and stems, respectively (Figures 12.6 and 12.7). There were significant differences in the levels of some metabolites in the branches of *U. glabra* and *U. minor* relative to mock-inoculated controls. In the case of *U. glabra*, only three metabolites exhibited a significant change, and only in trees that had been previously well watered: phenylalanine and raffinose increased and urea decreased (Figure 12.8; Table 12.5). In the case of *U. minor*, the response to *O. novo-ulmi* was different in trees depending on previous drought stress. Phenylalanine, tryptophan, *myo*-inositol, and raffinose increased in W+I trees, and GABA, glutamate, quinate, glucose, mannose, and sucrose decreased in D+I trees (relative to their respective mock-inoculated controls; Figure 12.8). Primary metabolite levels did not change in response to *O. novo-ulmi* at the stem base.

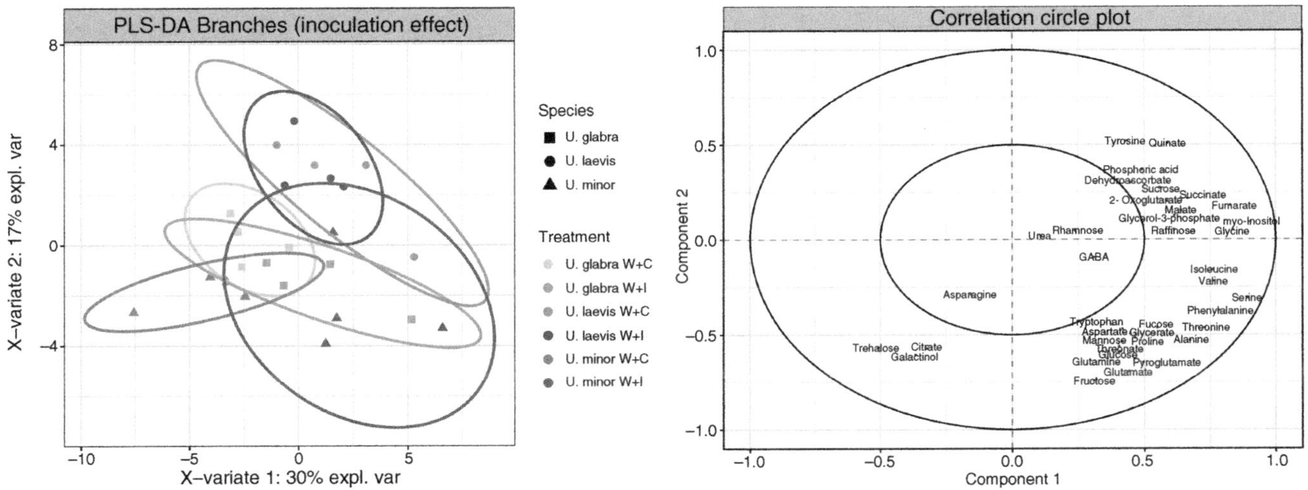

Figure 12.6 Partial least squares discriminant analysis (PLS-DA) score plots (a) and correlation plots (b) of the primary metabolite profile of branches of *Ulmus glabra*, *U. laevis*, and *U. minor* well-watered trees that were either inoculated with *O. novo-ulmi* (W+I) or mock inoculated (W+C) 14 days before sampling (spring 2016).

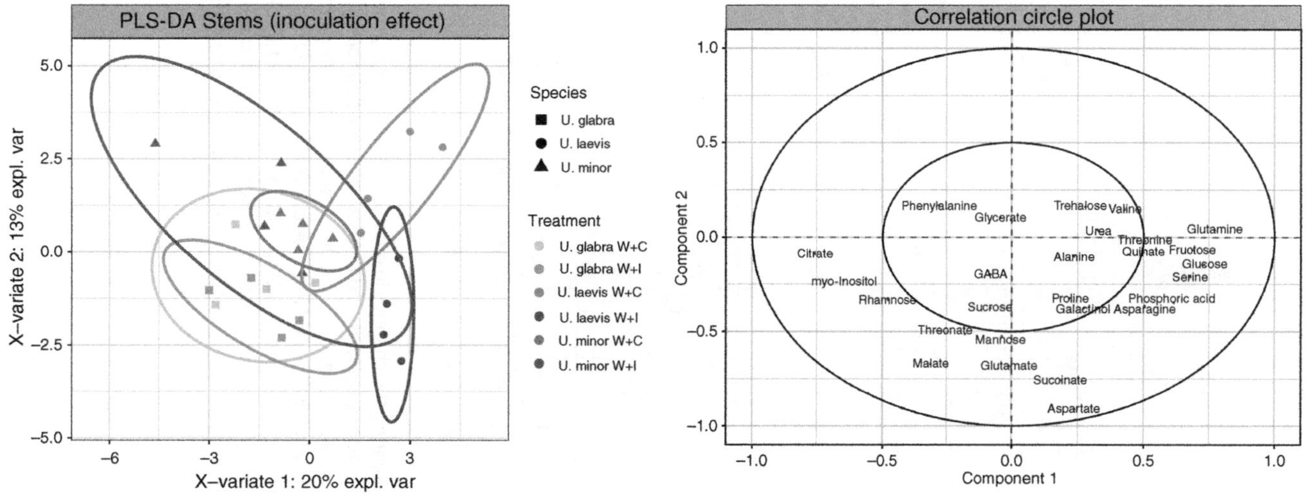

Figure 12.7 Partial least squares discriminant analysis (PLS-DA) score plots (a) and correlation plots (b) of the primary metabolite profile of main stems of *Ulmus glabra*, *U. laevis* and *U. minor* well-watered trees that were either inoculated with *O. novo-ulmi* (W+I) or mock inoculated (W+C) 14 days before sampling (spring 2016).

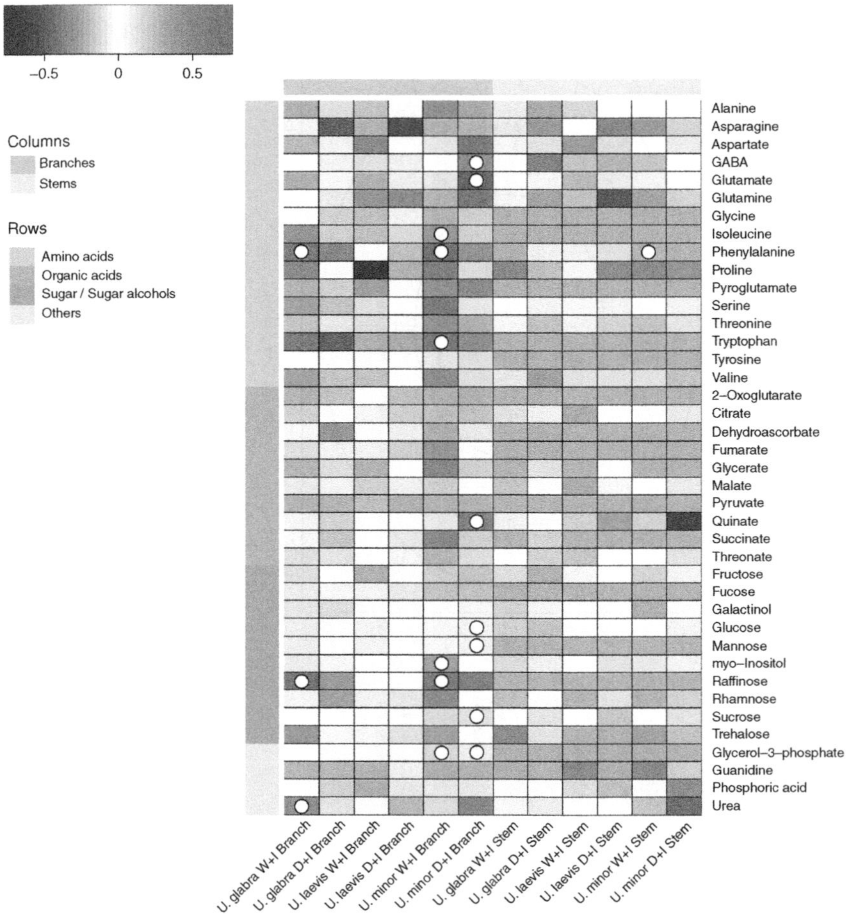

Figure 12.8 Heatmap representing the changes in relative levels of primary metabolites in branches and stems of *Ulmus glabra*, *U. laevis*, and *U. minor* after inoculation with *O. novo-ulmi*. Relative values (as means of three to four independent measurements) were normalized to the internal standard ribitol and fresh weight of the samples; false color imaging was performed on $\log_{10}$-transformed GC–TOF–MS metabolite data. Fold changes were calculated in relation to the mock-inoculated controls. Significant changes using one-way ANOVA are indicated as ○ $P < 0.05$. Gray-color squares represent undetected metabolites.

Table 12.5 Fold changes in relative levels of primary metabolites in branch and stem tissues of *Ulmus* spp. inoculated with *O. novo-ulmi* and exposed to drought stress, or in well-watered plants.

		Branch											
		U. glabra				*U. laevis*				*U. minor*			
		Watered Control	Watered Inoculated	Drought Control	Drought Inoculated	Watered Control	Watered Inoculated	Drought Control	Drought Inoculated	Watered Control	Watered Inoculated	Drought Control	Drought Inoculated
Amino acids	Alanine	1.00 ± 0.03	1.96 ± 0.77	1.00 ± 0.23	1.32 ± 0.52	1.00 ± 0.21	0.70 ± 0.19	1.00 ± 0.22	0.98 ± 0.31	1.00 ± 0.11	2.75 ± 0.94	1.00 ± 0.22	0.54 ± 0.06
	Asparagine	1.00 ± 0.20	0.85 ± 0.53	1.00 ± 0.65	0.34 ± 0.22	n.d.	n.d.	1.00 ± 0.64	0.28 ± 0.25	n.d.	n.d.	n.d.	n.d.
	Aspartate	1.00 ± 0.10	1.74 ± 0.39	1.00 ± 0.42	0.88 ± 0.35	1.00 ± 0.52	0.49 ± 0.13	1.00 ± 0.22	0.97 ± 0.49	1.00 ± 0.19	1.26 ± 0.56	1.00 ± 0.33	0.39 ± 0.11
	GABA	1.00 ± 0.04	1.04 ± 0.09	1.00 ± 0.23	0.90 ± 0.16	1.00 ± 0.16	0.80 ± 0.18	1.00 ± 0.15	0.85 ± 0.13	1.00 ± 0.17	1.01 ± 0.27	1.00 ± 0.19	**0.46 ± 0.02**
	Glutamate	1.00 ± 0.07	1.94 ± 0.54	1.00 ± 0.27	1.19 ± 0.56	1.00 ± 0.45	0.63 ± 0.17	1.00 ± 0.24	0.90 ± 0.43	1.00 ± 0.21	1.27 ± 0.38	1.00 ± 0.29	**0.33 ± 0.04**
	Glutamine	1.00 ± 0.07	1.01 ± 0.03	1.00 ± 0.36	1.22 ± 0.68	1.00 ± 0.51	0.51 ± 0.17	1.00 ± 0.42	0.47 ± 0.23	1.00 ± 0.18	1.78 ± 0.80	1.00 ± 0.39	0.38 ± 0.03
	Glycine	1.00 ± 0.08	1.09 ± 0.39	1.00 ± 0.22	1.48 ± 0.40	1.00 ± 0.19	0.68 ± 0.04	1.00 ± 0.12	1.17 ± 0.19	1.00 ± 0.21	1.80 ± 0.50	1.00 ± 0.27	1.47 ± 0.39
	Isoleucine	1.00 ± 0.12	2.65 ± 1.84	1.00 ± 0.42	1.52 ± 0.89	1.00 ± 0.22	0.67 ± 0.09	1.00 ± 0.33	1.36 ± 0.56	1.00 ± 0.11	**2.23 ± 0.39**	1.00 ± 0.30	0.72 ± 0.12
	Phenylalanine	1.00 ± 0.13	**4.35 ± 0.81**	1.00 ± 0.23	2.96 ± 0.96	1.00 ± 0.27	1.09 ± 0.22	1.00 ± 0.46	1.64 ± 0.30	1.00 ± 0.06	**5.71 ± 0.85**	1.00 ± 0.22	2.46 ± 0.85
	Proline	1.00 ± 0.19	2.94 ± 2.47	1.00 ± 0.14	0.95 ± 0.23	1.00 ± 0.84	0.17 ± 0.04	1.00 ± 0.54	1.76 ± 0.86	1.00 ± 0.08	3.22 ± 2.18	1.00 ± 0.29	0.83 ± 0.04
	Pyroglutamate	n.d.	n.d.	1.00 ± 0.42	0.74 ± 0.20	1.00 ± 0.55	0.55 ± 0.09	1.00 ± 0.19	1.05 ± 0.31	1.00 ± 0.06	2.50 ± 1.44	1.00 ± 0.40	0.46 ± 0.05
	Serine	1.00 ± 0.04	2.17 ± 0.76	1.00 ± 0.24	1.61 ± 0.46	1.00 ± 0.27	0.79 ± 0.13	1.00 ± 0.23	1.06 ± 0.28	1.00 ± 0.21	4.03 ± 1.73	1.00 ± 0.34	1.27 ± 0.14
	Threonine	1.00 ± 0.12	1.69 ± 0.80	1.00 ± 0.31	1.26 ± 0.54	1.00 ± 0.29	0.79 ± 0.12	1.00 ± 0.29	0.94 ± 0.29	1.00 ± 0.11	2.44 ± 0.87	1.00 ± 0.48	0.59 ± 0.10
	Tryptophan	1.00 ± 0.54	4.22 ± 1.53	1.00 ± 0.59	0.33 ± 0.13	1.00 ± 0.36	0.60 ± 0.05	1.00 ± 0.28	1.92 ± 1.02	1.00 ± 0.33	**4.24 ± 1.16**	1.00 ± 0.49	2.80 ± 2.13
	Tyrosine	1.00 ± 0.12	1.05 ± 0.18	1.00 ± 0.34	1.02 ± 0.30	1.00 ± 0.12	0.95 ± 0.10	1.00 ± 0.17	1.09 ± 0.33	1.00 ± 0.17	1.24 ± 0.15	1.00 ± 0.25	0.78 ± 0.09
	Valine	1.00 ± 0.08	2.08 ± 1.26	1.00 ± 0.34	1.61 ± 0.99	1.00 ± 0.24	0.67 ± 0.12	1.00 ± 0.41	1.01 ± 0.38	1.00 ± 0.18	2.80 ± 0.99	1.00 ± 0.28	0.80 ± 0.07
Organic acids	2-Oxoglutarate	n.d.	n.d.	1.00 ± 0.11	1.57 ± 0.38	1.00 ± 0.14	1.01 ± 0.23	1.00 ± 0.51	0.70 ± 0.14	n.d.	n.d.	n.d.	n.d.
	Citrate	1.00 ± 0.38	1.54 ± 0.49	1.00 ± 0.30	0.95 ± 0.06	1.00 ± 0.18	1.18 ± 0.42	1.00 ± 0.20	1.55 ± 0.72	1.00 ± 0.22	0.65 ± 0.17	1.00 ± 0.46	0.67 ± 0.26
	Dehydroascorbate	1.00 ± 0.18	1.18 ± 0.32	1.00 ± 0.16	0.53 ± 0.10	1.00 ± 0.24	0.95 ± 0.19	1.00 ± 0.20	0.84 ± 0.07	n.d.	n.d.	1.00 ± 0.16	1.23 ± 0.25
	Fumarate	1.00 ± 0.19	1.34 ± 0.17	1.00 ± 0.46	0.89 ± 0.26	1.00 ± 0.34	0.89 ± 0.07	1.00 ± 0.13	1.53 ± 0.51	1.00 ± 0.15	2.33 ± 0.58	1.00 ± 0.23	0.95 ± 0.17
	Glycerate	n.d.	n.d.	1.00 ± 0.16	1.38 ± 0.27	n.d.	n.d.	1.00 ± 0.07	1.04 ± 0.23	1.00 ± 0.21	3.11 ± 1.03	1.00 ± 0.06	1.46 ± 0.32
	Malate	1.00 ± 0.15	1.27 ± 0.24	1.00 ± 0.19	0.85 ± 0.19	1.00 ± 0.18	0.75 ± 0.17	1.00 ± 0.11	0.89 ± 0.26	1.00 ± 0.22	1.43 ± 0.35	1.00 ± 0.20	0.92 ± 0.11
	Pyruvate	n.d.	n.d.	n.d.	n.d.	1.00 ± 0.11	0.69 ± 0.10	n.d.	n.d.	n.d.	n.d.	n.d.	n.d.
	Quinate	1.00 ± 0.10	0.91 ± 0.13	1.00 ± 0.18	0.70 ± 0.15	1.00 ± 0.12	1.03 ± 0.11	1.00 ± 0.06	0.92 ± 0.11	1.00 ± 0.22	1.25 ± 0.20	1.00 ± 0.22	**0.45 ± 0.03**
	Succinate	1.00 ± 0.21	1.18 ± 0.10	1.00 ± 0.17	1.66 ± 0.28	1.00 ± 0.17	0.94 ± 0.15	1.00 ± 0.15	1.22 ± 0.48	1.00 ± 0.28	3.11 ± 1.19	1.00 ± 0.30	1.45 ± 0.21
	Threonate	1.00 ± 0.20	1.41 ± 0.20	1.00 ± 0.23	0.82 ± 0.07	1.00 ± 0.26	1.03 ± 0.18	1.00 ± 0.31	1.13 ± 0.45	1.00 ± 0.21	1.72 ± 0.45	1.00 ± 0.05	1.33 ± 0.44

(Continued)

Table 12.5 (Continued)

		Branch											
		U. glabra				U. laevis				U. minor			
		Watered Control	Watered Inoculated	Drought Control	Drought Inoculated	Watered Control	Watered Inoculated	Drought Control	Drought Inoculated	Watered Control	Watered Inoculated	Drought Control	Drought Inoculated
Sugar & Sugar alcohols	Fructose	1.00 ± 0.38	1.33 ± 0.33	1.00 ± 0.34	1.03 ± 0.08	1.00 ± 0.32	0.63 ± 0.07	1.00 ± 0.21	0.99 ± 0.29	1.00 ± 0.17	1.47 ± 0.29	1.00 ± 0.28	1.65 ± 0.08
	Fucose	1.00 ± 0.24	1.30 ± 0.30	1.00 ± 0.21	1.13 ± 0.16	1.00 ± 0.26	0.90 ± 0.09	1.00 ± 0.09	1.28 ± 0.26	1.00 ± 0.11	1.69 ± 0.31	1.00 ± 0.15	0.66 ± 0.05
	Galactinol	1.00 ± 0.12	0.78 ± 0.11	1.00 ± 0.09	1.39 ± 0.16	1.00 ± 0.23	1.07 ± 0.22	1.00 ± 0.20	0.97 ± 0.33	1.00 ± 0.12	0.96 ± 0.12	1.00 ± 0.33	1.11 ± 0.13
	Glucose	1.00 ± 0.18	1.22 ± 0.25	1.00 ± 0.22	0.98 ± 0.15	1.00 ± 0.14	0.84 ± 0.10	1.00 ± 0.07	1.01 ± 0.05	1.00 ± 0.08	1.18 ± 0.06	1.00 ± 0.03	**0.91 ± 0.02**
	Mannose	1.00 ± 0.24	1.25 ± 0.26	1.00 ± 0.26	0.97 ± 0.16	1.00 ± 0.15	0.88 ± 0.10	1.00 ± 0.06	1.01 ± 0.04	1.00 ± 0.08	1.17 ± 0.07	1.00 ± 0.02	**0.91 ± 0.02**
	myo-Inositol	1.00 ± 0.10	1.21 ± 0.22	1.00 ± 0.14	1.14 ± 0.15	1.00 ± 0.08	1.05 ± 0.13	1.00 ± 0.13	1.00 ± 0.18	1.00 ± 0.12	**2.16 ± 0.30**	1.00 ± 0.38	1.04 ± 0.12
	Raffinose	1.00 ± 0.22	**5.43 ± 1.64**	1.00 ± 0.13	2.40 ± 0.60	1.00 ± 0.22	0.96 ± 0.18	1.00 ± 0.21	1.01 ± 0.25	1.00 ± 0.27	**3.90 ± 0.73**	1.00 ± 0.32	3.78 ± 0.90
	Rhamnose	1.00 ± 0.34	0.90 ± 0.24	1.00 ± 0.39	2.10 ± 0.62	1.00 ± 0.17	0.91 ± 0.09	1.00 ± 0.13	1.20 ± 0.20	1.00 ± 0.12	2.74 ± 0.89	1.00 ± 0.19	1.09 ± 0.20
	Sucrose	1.00 ± 0.13	0.92 ± 0.13	1.00 ± 0.07	1.01 ± 0.13	1.00 ± 0.04	1.02 ± 0.11	1.00 ± 0.13	1.03 ± 0.07	1.00 ± 0.11	1.33 ± 0.17	1.00 ± 0.08	**0.78 ± 0.04**
	Trehalose	1.00 ± 0.25	0.52 ± 0.17	1.00 ± 0.14	1.13 ± 0.47	1.00 ± 0.15	1.17 ± 0.36	1.00 ± 0.31	0.80 ± 0.11	1.00 ± 0.34	0.58 ± 0.08	1.00 ± 0.39	0.98 ± 0.20
Others	Glycerol-3-phosphate	1.00 ± 0.06	1.01 ± 0.05	1.00 ± 0.07	1.00 ± 0.11	1.00 ± 0.08	0.95 ± 0.02	1.00 ± 0.06	1.02 ± 0.08	1.00 ± 0.09	**1.35 ± 0.05**	1.00 ± 0.03	**0.86 ± 0.03**
	Guanidine	n.d.	n.d.	1.00 ± 0.14	1.76 ± 0.58	n.d.	n.d.	1.00 ± 0.41	0.86 ± 0.17	n.d.	n.d.	1.00 ± 0.16	1.59 ± 0.83
	Phosphoric acid	1.00 ± 0.30	1.04 ± 0.14	1.00 ± 0.27	0.71 ± 0.19	1.00 ± 0.19	0.59 ± 0.04	1.00 ± 0.18	1.28 ± 0.29	1.00 ± 0.23	1.26 ± 0.38	1.00 ± 0.24	1.22 ± 0.26
	Urea	1.00 ± 0.19	**0.49 ± 0.04**	1.00 ± 0.22	0.83 ± 0.08	1.00 ± 0.26	0.94 ± 0.17	1.00 ± 0.12	1.77 ± 0.38	1.00 ± 0.40	1.42 ± 0.54	1.00 ± 0.09	3.19 ± 1.55

Table 12.5 (Continued)

		Stem											
		U. glabra				*U. laevis*				*U. minor*			
		Watered Control	**Watered Inoculated**	**Drought Control**	**Drought Inoculated**	**Watered Control**	**Watered Inoculated**	**Drought Control**	**Drought Inoculated**	**Watered Control**	**Watered Inoculated**	**Drought Control**	**Drought Inoculated**
Amino acids	Alanine	1.00 ± 0.15	1.28 ± 0.41	1.00 ± 0.33	0.58 ± 0.11	1.00 ± 0.04	1.44 ± 0.24	1.00 ± 0.11	1.06 ± 0.11	1.00 ± 0.09	0.92 ± 0.08	1.00 ± 0.26	0.97 ± 0.27
	Asparagine	1.00 ± 0.39	1.24 ± 0.48	1.00 ± 0.30	0.53 ± 0.24	1.00 ± 0.40	1.01 ± 0.17	1.00 ± 0.24	0.45 ± 0.12	1.00 ± 0.31	0.52 ± 0.10	1.00 ± 0.29	1.43 ± 0.41
	Aspartate	1.00 ± 0.11	1.11 ± 0.11	1.00 ± 0.17	0.83 ± 0.16	1.00 ± 0.08	**2.29 ± 0.35**	1.00 ± 0.14	1.40 ± 0.07	1.00 ± 0.08	1.01 ± 0.16	1.00 ± 0.12	1.20 ± 0.14
	GABA	1.00 ± 0.14	1.07 ± 0.34	1.00 ± 0.34	0.44 ± 0.06	1.00 ± 0.15	0.69 ± 0.08	1.00 ± 0.23	0.59 ± 0.07	1.00 ± 0.15	0.68 ± 0.16	1.00 ± 0.06	1.00 ± 0.35
	Glutamate	1.00 ± 0.17	1.04 ± 0.09	1.00 ± 0.12	1.16 ± 0.36	1.00 ± 0.17	1.85 ± 0.34	1.00 ± 0.19	1.20 ± 0.29	1.00 ± 0.10	0.87 ± 0.06	1.00 ± 0.16	1.07 ± 0.11
	Glutamine	1.00 ± 0.09	1.15 ± 0.40	1.00 ± 0.51	0.55 ± 0.11	1.00 ± 0.62	0.69 ± 0.08	1.00 ± 0.47	0.31 ± 0.06	1.00 ± 0.11	0.58 ± 0.14	1.00 ± 0.26	0.73 ± 0.23
	Glycine	n.d.	n.d.	n.d.	n.d.	n.d.	n.d.	n.d.	n.d.	n.d.	n.d.	n.d.	n.d.
	Isoleucine	n.d.	n.d.	n.d.	n.d.	n.d.	n.d.	n.d.	n.d.	n.d.	n.d.	n.d.	n.d.
	Phenylalanine			1.00 ± 0.35	0.89 ± 0.05	1.00 ± 0.17	0.84 ± 0.06	1.00 ± 0.05	0.83 ± 0.02	1.00 ± 0.14	**1.71 ± 0.23**		
	Proline	1.00 ± 0.07	2.94 ± 1.80	1.00 ± 0.38	0.66 ± 0.15	1.00 ± 0.07	1.19 ± 0.10	1.00 ± 0.46	0.48 ± 0.20	1.00 ± 0.35	0.48 ± 0.11	1.00 ± 0.30	0.53 ± 0.12
	Pyroglutamate	n.d.	n.d.	n.d.	n.d.	n.d.	n.d.	n.d.	n.d.	n.d.	n.d.	n.d.	n.d.
	Serine	1.00 ± 0.21	1.06 ± 0.20	1.00 ± 0.15	0.86 ± 0.15	1.00 ± 0.07	1.16 ± 0.11	1.00 ± 0.14	0.90 ± 0.11	1.00 ± 0.12	0.92 ± 0.13	1.00 ± 0.18	0.85 ± 0.19
	Threonine	1.00 ± 0.15	1.15 ± 0.24	1.00 ± 0.23	0.66 ± 0.10	1.00 ± 0.15	0.87 ± 0.05	1.00 ± 0.14	0.75 ± 0.09	1.00 ± 0.23	0.68 ± 0.05	1.00 ± 0.15	0.79 ± 0.11
	Tryptophan	n.d.	n.d.	n.d.	n.d.	n.d.	n.d.	n.d.	n.d.	n.d.	n.d.	n.d.	n.d.
	Tyrosine	n.d.	n.d.	n.d.	n.d.	n.d.	n.d.	n.d.	n.d.	n.d.	n.d.	n.d.	n.d.
	Valine	1.00 ± 0.22	1.21 ± 0.25	1.00 ± 0.42	0.55 ± 0.07	1.00 ± 0.16	0.77 ± 0.04	1.00 ± 0.19	0.78 ± 0.11	1.00 ± 0.16	0.83 ± 0.24	1.00 ± 0.17	0.65 ± 0.09
Organic acids	2-Oxoglutarate	n.d.	n.d.	n.d.	n.d.	n.d.	n.d.	n.d.	n.d.	n.d.	n.d.	n.d.	n.d.
	Citrate	1.00 ± 0.38	0.79 ± 0.13	1.00 ± 0.38	0.89 ± 0.19	1.00 ± 0.19	1.88 ± 0.32	1.00 ± 0.21	1.07 ± 0.28	1.00 ± 0.26	1.03 ± 0.14	1.00 ± 0.24	1.29 ± 0.13
	Dehydroascorbate	n.d.	n.d.	n.d.	n.d.	n.d.	n.d.	n.d.	n.d.	n.d.	n.d.	n.d.	n.d.
	Fumarate	n.d.	n.d.	n.d.	n.d.	n.d.	n.d.	n.d.	n.d.	n.d.	n.d.	n.d.	n.d.
	Glycerate			1.00 ± 0.15	0.79 ± 0.04	n.d.	n.d.	1.00 ± 0.24	0.99 ± 0.02	n.d.	n.d.	n.d.	n.d.
	Malate	1.00 ± 0.18	1.68 ± 0.34	1.00 ± 0.31	0.94 ± 0.15	1.00 ± 0.20	1.95 ± 0.39	1.00 ± 0.19	1.17 ± 0.29	1.00 ± 0.07	1.03 ± 0.12	1.00 ± 0.21	1.23 ± 0.39
	Pyruvate	n.d.	n.d.	n.d.	n.d.	n.d.	n.d.	n.d.	n.d.	n.d.	n.d.	n.d.	n.d.
	Quinate	1.00 ± 0.25	0.84 ± 0.15	1.00 ± 0.18	0.93 ± 0.29	1.00 ± 0.19	1.46 ± 0.62	1.00 ± 0.25	0.57 ± 0.28	1.00 ± 0.13	0.72 ± 0.26	1.00 ± 0.67	0.22 ± 0.06
	Succinate	n.d.	n.d.	1.00 ± 0.29	0.75 ± 0.09	n.d.	n.d.	n.d.	n.d.	n.d.	n.d.	n.d.	n.d.

(Continued)

Table 12.5 (Continued)

		Stem											
		U. glabra				U. laevis				U. minor			
		Watered Control	Watered Inoculated	Drought Control	Drought Inoculated	Watered Control	Watered Inoculated	Drought Control	Drought Inoculated	Watered Control	Watered Inoculated	Drought Control	Drought Inoculated
Sugar & Sugar alcohols	Threonate	1.00 ± 0.32	1.06 ± 0.15	1.00 ± 0.32	0.73 ± 0.28	1.00 ± 0.50	1.59 ± 0.60	1.00 ± 0.21	1.09 ± 0.13	1.00 ± 0.14	1.09 ± 0.11	1.00 ± 0.21	0.82 ± 0.05
	Fructose	1.00 ± 0.21	1.37 ± 0.32	1.00 ± 0.39	0.60 ± 0.09	1.00 ± 0.19	1.16 ± 0.19	1.00 ± 0.23	0.96 ± 0.22	1.00 ± 0.35	0.72 ± 0.24	1.00 ± 0.33	0.89 ± 0.67
	Fucose	n.d.	n.d.	n.d.	n.d.	n.d.	n.d.	n.d.	n.d.	n.d.	n.d.	n.d.	n.d.
	Galactinol	1.00 ± 0.23	0.72 ± 0.15	1.00 ± 0.19	1.15 ± 0.16	1.00 ± 0.22	1.05 ± 0.10	1.00 ± 0.12	1.02 ± 0.16	1.00 ± 0.18	0.64 ± 0.21	1.00 ± 0.23	0.97 ± 0.23
	Glucose	1.00 ± 0.18	1.45 ± 0.32	1.00 ± 0.34	0.67 ± 0.09	1.00 ± 0.28	0.99 ± 0.13	1.00 ± 0.27	1.03 ± 0.19	1.00 ± 0.21	1.00 ± 0.27	1.00 ± 0.29	0.85 ± 0.50
	Mannose	n.d.	n.d.	n.d.	n.d.	n.d.	n.d.	n.d.	n.d.	n.d.	n.d.	n.d.	n.d.
	myo-Inositol	1.00 ± 0.14	1.39 ± 0.20	1.00 ± 0.12	0.90 ± 0.16	1.00 ± 0.17	1.01 ± 0.11	1.00 ± 0.09	0.83 ± 0.16	1.00 ± 0.07	1.20 ± 0.27	1.00 ± 0.16	0.90 ± 0.08
	Raffinose	n.d.	n.d.	n.d.	n.d.	n.d.	n.d.	n.d.	n.d.	n.d.	n.d.	n.d.	n.d.
	Rhamnose	1.00 ± 0.27	0.70 ± 0.08	1.00 ± 0.15	0.98 ± 0.20	n.d.	n.d.	1.00 ± 0.16	0.82 ± 0.12	n.d.	n.d.	1.00 ± 0.30	0.72 ± 0.06
	Sucrose	1.00 ± 0.13	0.93 ± 0.08	1.00 ± 0.21	0.78 ± 0.09	1.00 ± 0.09	1.08 ± 0.16	1.00 ± 0.12	1.59 ± 0.51	1.00 ± 0.06	0.96 ± 0.21	1.00 ± 0.23	0.77 ± 0.05
	Trehalose	1.00 ± 0.39	0.47 ± 0.12	1.00 ± 0.31	0.82 ± 0.28	n.d.	n.d.	n.d.	n.d.	1.00 ± 0.40	0.56 ± 0.12	1.00 ± 0.19	0.69 ± 0.16
Others	Glycerol-3-phosphate	n.d.	n.d.	n.d.	n.d.	n.d.	n.d.	n.d.	n.d.	n.d.	n.d.	n.d.	n.d.
	Guanidine	n.d.	n.d.	n.d.	n.d.	1.00 ± 0.21	0.49 ± 0.07	n.d.	n.d.	1.00 ± 0.57	0.47 ± 0.06	1.00 ± 0.42	1.51 ± 0.92
	Phosphoric acid	1.00 ± 0.24	0.99 ± 0.15	1.00 ± 0.25	1.03 ± 0.28	1.00 ± 0.07	1.41 ± 0.18	1.00 ± 0.20	0.68 ± 0.05	1.00 ± 0.31	0.98 ± 0.43	1.00 ± 0.23	0.49 ± 0.07
	Urea	1.00 ± 0.62	1.17 ± 0.33	1.00 ± 0.26	1.27 ± 0.24	1.00 ± 0.22	1.05 ± 0.14	1.00 ± 0.33	1.04 ± 0.28	1.00 ± 0.19	1.60 ± 0.52	1.00 ± 0.47	0.35 ± 0.08

Relative values are normalized to the internal standard (ribitol) and fresh weight (FW) of the samples. Data are presented as means ± SE of at least three independent measurements. Values in bold type indicate significant differences calculated using one-way ANOVA ($P < 0.05$) with respect to their respective watered or drought control. n.d. not detected.

12.4 Discussion

12.4.1 No Legacy Effect of Two-Year Moderate Drought on Primary Metabolism

Data on leaf stomatal conductance (g_s) and tree-ring width evidenced that restricting summer watering stressed the trees during 2014 and 2015 before the inoculation with *O. novo-ulmi* in the spring of 2016. The degree of stress was moderate, based on the extent of the reduction of g_s in D trees compared to W trees, which ranged from 44 to 54% across dates for *U. minor* and *U. glabra*, and 19% to 49% for *U. laevis*. Two studies found that photosynthetic and respiratory metabolic activity declined markedly in several plant species below 100–150 mmol m^{-2} s^{-1} g_s (92, 93). Although W trees reached these values in 2014, suggesting that the control watering regime was not enough to prevent natural drought stress in these trees, this g_s threshold was trespassed more often and deeply by D trees, especially of *U. minor* and *U. glabra* in 2015. Unexpectedly, g_s was higher in D than W trees of *U. laevis* and *U. glabra* upon rewatering in 2015. This could be related to previous leaf shedding caused by drought (62, 94) and remaining leaves in branches spared from xylem cavitation benefiting from higher leaf-specific hydraulic conductivity, especially once new xylem has been produced (95).

The following year, trees from all species appeared to recover from drought stress. Similar tree-ring width of D and W trees during 2016 supports the recovery of tree normal functioning after two years of summer drought, which is usually the case unless drought has not been severe and/or sustained for several years (96). Stem and branch primary metabolite levels in spring were also remarkably similar in D and W trees of any species. Only two metabolites changed significantly in D trees of *U. minor* relative to W trees. The increase in sucrose in branches of D trees was consistent with seasonal drought stress usually enhancing osmoregulation at the cost of starch reserves in non-fat storing tree species (97). The need to adjust actively the osmotic content can persist long after seasonal drought ends in trees suffering chronic drought stress, as seen for example in leaves of evergreen oak trees exposed to partial rainfall exclusion for more than nine years (98). However, sucrose did not increase in the stem of *U. minor* seedlings during drought (64), supporting that metabolic shifts can respond to different needs during acute, short-term drought stress and after moderate, long-term drought stress. The fact that changes in some metabolites did not occur in the other elm species, and were not accompanied by a clear re-organization of the primary metabolome, led us to conclude that there was no major legacy of moderate drought on primary metabolism.

12.4.2 Early Re-organization of Primary Metabolism upon Inoculation with *O. novo-ulmi* Is Moderate and Species- and Organ-Specific

The recovery of trees from drought stress and the early sampling for GC–TOF–MS primary metabolite profiling analysis at 14 dpi both explain that inoculation had a small effect on the concentrations of primary metabolites in all trees, irrespective of the species, and the drought stress before the inoculation. Nonetheless, *O. novo-ulmi* triggered a significant change in the concentrations of several metabolites in trees of *U. glabra* and especially of *U. minor*, only in the branches.

Unexpectedly, *O. novo-ulmi* induced a stronger change in metabolite pools in mid-canopy branches than in the main stem close to the pathogen entry point. It is possible that *O. novo-ulmi* reached the upper crown within a few days of inoculation. The spores travel in the transpiration stream upward easily in the wide, long earlywood vessels produced by elms in spring and can germinate within 96 hours inside the xylem (99), causing the degradation of pit membranes and the metabolic reorganization of live parenchyma cells. However, the inoculation of *O. novo-ulmi* in the stem did not affect the pool of primary metabolites in the stem. Shigo et al. (46) also found that the levels of extractable sugars in the main stem were similar in inoculated and control elms, suggesting the conversion of starch to simple sugars to maintain growth processes, and observing the depletion of starch reserves several weeks after the infection. It is possible that higher starch reserves in the main stem in comparison with small branches led to lower detectable changes in soluble sugars in the former. Plant local response to pathogen inoculation can occur within hours and shift rapidly to more distal plant locations (100). However, it is unlikely that elm trees responded fast to *O. novo-ulmi* locally and that this response was transient and left no metabolic imprint 14 days after inoculation. Rather, a slow, weak local defense response can favor the expansion of the pathogen and the spread of the disease (17, 101), which caused significant crown wilting in the three elm species. A third hypothesis to explain these results is methodological. Stem samples were obtained with an increment borer and contained more xylem than phloem (in volume and weight), whereas xylem and phloem were more similar in pruned branches (at least in volume). The phloem contains more proportion of live cells than the xylem, and it could be more reactive to the inoculation. This is supported by fewer metabolites detected in stem than in branch samples (Figure 12.8).

12.4.3 Drought Affects the Primary Metabolic Response of *U. minor* Branches to *O. novo-ulmi*

Changes in branch primary metabolite levels in response to *O. novo-ulmi* inoculation depended on previous watering regime for *U. minor* trees. This did not occur at the base of the stem, close to the pathogen inoculation scar, or in any organ in *U. glabra* or *U. laevis*. Although *U. laevis* and *U. glabra* are more sensitive to drought than *U. minor* in natural conditions (6, 94), in this study, *U. minor* was the only species that showed a reduction in both stem growth and g_s, suggesting that modulation of early response to *O. novo-ulmi* was related to the intensity of previous drought stress. Thus, more severe and/or longer droughts affecting primary metabolism and xylem anatomy (both barely affected by previous drought here) could elicit a stronger (positive or negative) effect on DED sensitivity, as observed in the study of other diseases (102, 103).

It is worth noting that branch metabolites significantly affected by *O. novo-ulmi* were different in W+I and D+I, and that they showed an opposite change: isoleucine, phenylalanine, tryptophan, *myo*-inositol, and raffinose increased in W+I trees (relative to W+C trees), and GABA, glutamate, quinate, glucose, mannose, and sucrose decreased in D+I trees (relative to D+C trees). These primary metabolites can play different roles in defense. Phenylalanine is a precursor of the phenylpropanoid pathway producing phenolic compounds (104), which reduce sporulation and growth of *O. novo-ulmi* (105), and form eventually the phenolic monomers of suberin (106) involved in pathogen compartmentalization. Raffinose and derived oligosaccharides synthesized from *myo*-inositol (and galactose and sucrose) participate in induced systemic resistance (107, 108). Thus, significant increases in phenylalanine and raffinose could delay the spread of *O. novo-ulmi* in the branches of inoculated well-watered plants (W+I) of *U. minor* and *U. glabra*. Moreover, the tryptophan pathway produces several antimicrobial substances (53), while isoleucine (and phenylalanine, among other amino acids) accumulates in *Arabidopsis* plants inoculated with a virulent strain of *Pseudomonas syringae* (109). Decreases in GABA, glutamate, glucose, mannose, and sucrose observed in inoculated water-stressed trees might reflect a different strategy to starve *O. novo-ulmi*, as these metabolites can support growth of pathogens (110–112), including of *O. novo-ulmi* (113).

However, without further molecular studies, we can only speculate about the implication of these metabolic changes for defense against *O. novo-ulmi*. The fact that crown foliar wilting at 30, 60 or 120 dpi was similar in pathogen-inoculated W and D trees indicates that their different re-structuration of the primary metabolism in response to *O. novo-ulmi* was equally unsuccessful. Moreover, apparently, there was no local response close to the inoculation point. Altogether, these results suggest that distinct metabolic behavior of W+I and D+I (relative to their respective mock-inoculated controls) was a symptom of early water stress caused by *O. novo-ulmi* propagation. The increase in amino acids and oligosaccharides (as observed in W+I trees) is a typical response to water stress that contribute to osmoregulation, removal of ROS, and stability of cellular membranes

(58, 114, 115). The decreases in GABA, glutamate, glucose, mannose, and sucrose in D+I trees are less consistent with an early sign of water stress.

Similar studies should be made on genotypes of contrasted DED resistance (e.g. those recently available for *U. minor* thanks to the Spanish Elm Breeding Program; (116)) to understand the role of primary and secondary metabolism in orchestrating DED resistance. This is the objective of current research testing the effect of drought before and after *O. novo-ulmi* inoculation, which can have different effects on DED susceptibility (70).

12.4.4 High DED Susceptibility of *U. laevis* Was Accompanied by No Change in Stem or Branch Primary Metabolite Concentrations Soon After *O. novo-ulmi* Inoculation

Compared with *U. minor* and *U. glabra*, *U. laevis* was the only species suffering a significant (50%) decline in stem growth upon inoculation with *O. novo-ulmi*, and the only species exhibiting no change in stem or branch primary metabolite profile. This reinforces the idea that a slow response of the tree's primary metabolism to *O. novo-ulmi* favors the spread of the pathogen and the blockage of xylem vessels, which causes massive water stress, stops stem growth, and scorches the crown.

The ranking of DED susceptibility across species based on stem growth arrestment and crown foliar wilting upon inoculation with *O. novo-ulmi* was: *U. laevis* ~ *U. glabra* > *U. minor*. Although *U. minor* showed a marginal decrease in stem radial growth after *O. novo-ulmi* inoculation, we rank *U. glabra* as more susceptible than *U. minor* as the former had higher foliar wilting at 120 dpi. This was in spite of *U. glabra* having the smallest earlywood vessels in this study, which theoretically should offer the highest resistance to *O. novo-ulmi* propagation. DED resistance across species and genotypes of *U. minor* is frequently related to xylem vessel size (59, 72, 117, 118) but also to the ability of trees to compartmentalize the infection soon after pathogen entrance and prevent massive hydraulic failure (17, 119). The compartmentalization involves the synthesis of secondary compounds such as suberin, pectin, and lignin (120–122), which is likely to affect the concentration of primary metabolites being at the root of secondary metabolic pathways (e.g. phenylalanine).

Besides tree resistance mechanisms against the pathogen, the probability of DED transmission under field conditions depends on tree attractiveness to elm bark beetles transporting *O. novo-ulmi* spores, which is likely to change with environmental effects on nutritional quality or emission of VOCs (43, 123). In this regard, *U. laevis* and *U. glabra* are less preferred by elm bark beetles than *U. minor* (124), a reason why their populations remain in general less affected by DED. However, in terms of plant susceptibility to *O. novo-ulmi*, literature usually ranks the species as *U. glabra* > *U. laevis* > *U. minor* (43, 117, 125).

This ranking also changes across sites according to environmental effects on bud burst phenology, xylem features, and defense activity (126–128). Low radial growth and xylem vessel area exhibited by *U. glabra* here could reduce its vulnerability to DED compared to other studies.

12.5 Conclusions

The primary metabolite profile of European elms changed little 14 dpi with *O. novo-ulmi*, when trees still had no DED symptoms (i.e. crown foliar wilting). *U. minor* exhibited the largest response to *O. novo-ulmi*, with one-third of metabolites identified in branches exhibiting a significant change relative to mock-inoculated trees. Of these 12 metabolites, only one (glycerol-3-phosphate) changed significantly irrespectively of previous drought stress; the levels of the remaining metabolites increased (isoleucine, phenylalanine, tryptophan, *myo*-inositol, and raffinose) or decreased (GABA, glutamate, quinate, glucose, mannose, and sucrose) depending on whether trees had been exposed to regular watering or restricting watering during the two years prior to inoculation, respectively. The stronger metabolic reaction in *U. minor* at 14 dpi was associated with lower crown wilting at 120 dpi, compared to *U. laevis* and *U. glabra*, suggesting a defensive role of such metabolic changes. However, these changes could also reflect an early, mild water stress caused by incipient xylem occlusion, rather than a re-orchestration of the metabolism to combat the pathogen. In support of this, the metabolic response to *O. novo-ulmi* was weaker in the main-stem base, close to the inoculation site, than in the branches, suggesting that a slow local response might favor pathogen spread to the canopy. These results indicate that early metabolic response to *O. novo-ulmi* varies among species, organs, and environmental conditions. Stronger and longer droughts that might compromise the drought-stress recovery evidenced in this study should be tested in elm genotypes differing in DED susceptibility.

Acknowledgments

We thank the Dirección General de Biodiversidad, Bosques y Desertificación, Ministerio para la Transición Ecológica y el Reto Demográfico (MITECO) for continuous financial and logistic support within the Spanish Elm Breeding Program. We also thank personnel from the Puerta de Hierro Forest Breeding Center for administrative support and care of plants. Our warm gratitude to Meng Li for measuring stomatal conductance. The project OLMOS (AGL2012-35580) contributed to fund this research.

References

1 Brasier CM, Kirk SA. Rapid emergence of hybrids between the two subspecies of *Ophiostoma novo-ulmi* with a high level of pathogenic fitness. *Plant Pathol.* 2010; 59: 186–199.

2 Martín JA, Domínguez J, Solla A, Brasier CM, Webber JF, Santini A, et al. Complexities underlying the breeding and deployment of Dutch elm disease resistant elms. *New For.* 2021 https://doi.org/10.1007/s11056-021-09865-y.

3 Bréda N, Huc R, Granier A, Dreyer E. Temperate forest trees and stands under severe drought: a review of ecophysiological responses, adaptation processes and long-term consequences. *Ann For Sci.* 2006; 63: 625–644.

4 Allen CD, Breshears DD, McDowell NG. On underestimation of global vulnerability to tree mortality and forest die-off from hotter drought in the Anthropocene. *Ecosphere.* 2015; 6: 129.

5 Horner GJ, Baker PJ, MacNally R, Cunningham SC, Thomson JR, Hamilton F. Mortality of developing floodplain forests subjected to a drying climate and water extraction. *Global Change Biol.* 2009; 15: 2176–2186.

6 Venturas M, Fuentes-Utrilla P, López R, et al. *Ulmus laevis* in the Iberian Peninsula: a review of its ecology and conservation. *iForest.* 2015; 8: 135–142.

7 Perdiguero P, Sobrino-Plata J, Venturas M, Martín JA, Gil L, Collada C. Gene expression tradeoffs between defence and growth in English elm induced by *Ophiostoma novo-ulmi. Plant Cell Environ.* 2018; 41: 198–214.

8 Islam MT, Coutin JF, Shukla M, Dhaliwal AK, Nigg M, Bernier L, et al. Deciphering the genome-wide transcriptomic changes during Interactions of resistant and susceptible genotypes of American elm with *Ophiostoma novo-ulmi. J Fungi.* 2022; 8: 120.

9 Domínguez J, Macaya-Sanz D, Gil L, Martín JA. Excelling the progenitors: breeding for resistance to Dutch elm disease from moderately resistant and susceptible native stock. *For Ecol Manage.* 2022; 511: 120113.

10 Svaldi R, Elgersma DM. Further studies on the activity of cell wall degrading enzymes of aggressive and non-aggressive isolates of *Ophiostoma ulmi. Eur J For Pathol.* 1982; 12: 29–36.

11 Przybył K, Dahm H, Ciesielska A, Moliński K. Cellulolytic activity and virulence of *Ophiostoma ulmi* and *O. novo-ulmi* isolates. *For Pathol.* 2006; 36: 58–67.

12 Martín JA, Solla A, García-Vallejo MC, Gil L. Chemical changes in *Ulmus minor* xylem tissue after salicylic acid or carvacrol treatments are associated with enhanced resistance to *Ophiostoma novo-ulmi. Phytochemistry.* 2012; 83: 104–109.

13 Sherif SM, Shukla MR, Murch SJ, Bernier L, Saxena PK. Simultaneous induction of jasmonic acid and disease-responsive genes signifies tolerance of American elm to Dutch elm disease. *Sci Rep.* 2016; 6: 21934.

14 Singh D, Smalley EB. Nitrogenous compounds in the xylem sap of American elms with Dutch elm disease. *Can J Bot.* 1969a; 47: 1061–1065.
15 Martín JA, Solla A, Esteban LG, de Palacios P, Gil L. Bordered pit and ray morphology involvement in elm resistance to *Ophiostoma novo-ulmi*. *Can J For Res.* 2009; 39: 420–429.
16 Ouellette GB. Comparative ultrastructural observations of reactions of xylem parenchyma cells to *Ceratocystis ulmi* infection in *Ulmus* spp. In: Kondo ES, Hiratsuka Y, & WBG D, editors. *Proceedings, Dutch Elm Disease Symposium and Workshop*. Manitoba Dept. of Natural Resources. Canadian Forestry Service; 1982. p. 267–297.
17 Ouellette GB, Rioux D. Anatomical and physiological aspects of resistance to Dutch elm disease. In: Blanchette RA, & Biggs AR, editors. *Defense mechanisms of woody plants against fungi*. Springer series in wood science. Berlin: Springer; 1992. p. 257–307.
18 Scheffer RJ. *Dutch elm disease: aspects of pathogenesis and control*. PhD thesis. University of Amsterdam; 1984.
19 Kashyap A, Planas-Marquès M, Capellades M, Valls M, Coll NS. Blocking intruders: inducible physico-chemical barriers against plant vascular wilt pathogens. *J Exp Bot.* 2021; 72: 184–198.
20 Zentmeyer GA, Horsfall JG, Wallace PP. Dutch elm disease and its chemotherapy. *Conn Agr Exp Stn Bull.* 1946: 498.
21 Feldman AW, Caroselli NE, Howard FL. Chemicals for Dutch elm disease therapy. *Phytopathology*. 1950; 40: 8–9.
22 Elgersma DM. Factors determining resistance of elms to *Ceratocystis ulmi*. *Phytopathology*. 1967; 57: 641–642.
23 Singh D, Smalley EB. Nitrogenous and carbohydrate compounds in the xylem sap of Ulmaceae species varying in resistance to Dutch elm disease. *Can J Bot.* 1969b; 47: 335–339.
24 Hayat S, Hayat Q, Alyemeni MN, Wani AS, Pichtel J, Ahmad A. Role of proline under changing environments. *Plant Signaling Behav.* 2012; 7: 1456–1466.
25 Martín JA, Sobrino-Plata J, Coira B, Medel D, Collada C, Gil L. Growth resilience and oxidative burst control as tolerance factors to *Ophiostoma novo-ulmi* in *Ulmus minor*. *Tree Physiol.* 2019a; 39: 1512–1524.
26 Beckman CH. Production of pectinase, cellulases, and growth-promoting substances by *Cerastomella ulmi*. *Phytopathology*. 1956; 46: 605–609.
27 Beckman CH, Kuntz JE, Riker AJ, Berbee JG. Host responses associated with the development of oak wilt. *Phytopathology*. 1953; 43: 448–454.
28 Martín JA, Solla A, Woodward S, Gil L. Fourier transform-infrared spectroscopy as a new method for evaluating host resistance in the Dutch elm disease complex. *Tree Physiol.* 2005a; 25: 1331–1338.
29 Martín JA, Solla A, Coimbra MA, Gil L. Metabolic distinction of *Ulmus minor* xylem tissues after inoculation with *Ophiostoma novo-ulmi*. *Phytochemistry*. 2005b; 66: 2458–2467.

30 Martínez-Arias C, Sobrino-Plata J, Gil L, Rodríguez-Calcerrada J, Martín JA. Priming of plant defenses against *Ophiostoma novo-ulmi* by elm (*Ulmus minor* Mill.) fungal endophytes. *J Fungi.* 2021; 7: 687.
31 Maeda H, Dudareva N. The shikimate pathway and aromatic amino acid biosynthesis in plants. *Annu Rev Plant Biol.* 2012; 63: 73–105.
32 Nasmith C, Jeng R, Hubbes M. Targeted gene analysis in *Ulmus americana* and *U. pumila* tissues. *For Pathol.* 2008; 38: 90–103.
33 Nasmith C, Jeng R, Hubbes M. A comparison of in vivo targeted gene expression during fungal colonization of DED-susceptible *Ulmus americana. For Pathol.* 2008; 38: 104–112.
34 Santamour FS, Jr,. Flavonoid distribution in *Ulmus. Bull Torrey Bot Club.* 1972; 99: 127–131.
35 Sherman SL, Giannasi DE. Foliar flavonoids of *Ulmus* in eastern North America. *Biochem Syst Ecol.* 1988; 16: 51–56.
36 Aoun M, Rioux D, Simard M, Bernier L. Fungal colonization and host defense reactions in *Ulmus americana* callus cultures inoculated with *Ophiostoma novo-ulmi. Phytopathology.* 2009; 99: 642–650.
37 Valle T, López JL, Hernández JM, Corchete P. Antifungal activity of scopoletin and its differential accumulation in *Ulmus pumila* and *Ulmus campestris* cell suspension cultures infected with *Ophiostoma ulmi* spores. *Plant Sci.* 1997; 125: 97–101.
38 Elgersma DM, Overeem JC. The relation of mansonones to resistance against dutch elm disease and their accumulation, as induced by several agents. *Neth J Plant Pathol.* 1971; 77: 168–174.
39 Duchesne LC, Jeng RS, Hubbes M. Accumulation of phytoalexins in *Ulmus americana* in response to infection by a nonaggressive and an aggressive strain of *Ophiostoma ulmi. Can J Bot.* 1985; 63: 678–680.
40 Duchesne LC, Hubbes M, Jeng RS. Mansonone E and F accumulation in *Ulmus pumila* resistant to Dutch elm disease. *Can J For Res.* 1986; 16: 410–412.
41 Dumas MT, Straunz GM, Hubbes M, Jeng RS. Inhibition of *Ceratocystis ulmi* by mansonones A, C, D, E, F, and G isolated from *Ulmus americana. Eur J Plant Pathol.* 1986; 16: 217–230.
42 Wu WD, Jeng RS, Hubbes M. Toxic effects of elm phytoalexin mansonones on *Ophiostoma ulmi*, the causal agent of dutch elm disease. *Eur J Plant Pathol.* 1989; 19: 343–357.
43 Kännaste A, Jürisoo L, Runno-Paurson E, Kask K, Talts E, Pärlist P, et al. Impacts of Dutch elm disease-causing fungi on foliage photosynthetic characteristics and volatiles in *Ulmus* species with different pathogen resistance. *Tree Physiol.* 2023; 43: 57–74.
44 Peñuelas J, Llusià J. Plant VOC emissions: making use of the unavoidable. *Trends Ecol Evol.* 2004; 19: 402–404.

45 Li M, López R, Venturas M, Martín JA, Domínguez J, Gordaliza GG, et al. Physiological and biochemical differences among *Ulmus minor* genotypes showing a gradient of resistance to Dutch elm disease. *For Pathol.* 2016; 46: 215–228.

46 Shigo AL, Gregory GF, Campana RJ, Dudzik KR, Zimel DM. Patterns of starch reserves in healthy and diseased American elms. *Can J For Res.* 1986; 16: 204–210.

47 McDowell NG. Mechanisms linking drought, hydraulics, carbon metabolism, and vegetation mortality. *Plant Physiol.* 2011; 155: 1051–1059.

48 Bolton MD. Primary metabolism and plant defense – fuel for the fire. *Mol Plant-Microbe Interact.* 2009; 22: 487–497.

49 López R, Cano FJ, Rodríguez-Calcerrada J, Sangüesa-Barreda G, Gazol A, Camarero JJ, et al. Tree-ring density and carbon isotope composition are early-warning signals of drought-induced mortality in the drought tolerant Canary Island pine. *Agr For Met.* 2021; 310: 108634.

50 Hsiao TC, Acevedo E, Fereres E, Henderson DW. Stress metabolism. Water stress, growth, and osmotic adjustment. *Phil Trans R Soc Lond B.* 1976; 273: 479–500.

51 Herms DA, Mattson WJ. The dilemma of plants: to grow or defend. *Q Rev Biol.* 1992; 67: 283–335.

52 Sweetlove LJ, Beard KFM, Nunes-Nesi A, Fernie AR, Ratcliffe RG. Not just a circle: flux modes in the plant TCA cycle. *Trends Plant Sci.* 2010; 15: 462–470.

53 Radwanski ER, Last RL. Tryptophan biosynthesis and metabolism: biochemical and molecular genetics. *Plant Cell.* 1995; 7: 921–934.

54 War AR, Paulraj MG, War MY, Ignacimuthu S. Role of salicylic acid in induction of plant defense system in chickpea (*Cicer arietinum* L.). *Plant Signaling Behav.* 2011; 6: 1787–1792.

55 Lefevere H, Bauters L, Gheysen G. Salicylic acid biosynthesis in plants. *Front Plant Sci.* 2020; 11: 338.

56 Munné-Bosch S, Peñuelas J. Photo- and antioxidative protection, and a role for salicylic acid during drought and recovery in field-grown *Phillyrea angustifolia* plants. *Planta.* 2003; 217: 758–766.

57 Petridis A, Therios I, Samouris G, Koundouras S, Giannakoula A. Effect of water deficit on leaf phenolic composition, gas exchange, oxidative damage and antioxidant capacity of four Greek olive (*Olea europaea* L.) cultivars. *Plant Physiol Biochem.* 2012; 60: 1–11.

58 Sun T-T, Su Z-H, Wang R, Liu R, Yang T, Zuo W-T, et al. Transcriptome and metabolome analysis reveals the molecular mechanisms of *Tamarix taklamakanensis* under progressive drought and rehydration treatments. *Environ Exp Bot.* 2022; 195: 104766.

59 Pita P, Rodríguez-Calcerrada J, Medel D, Gil L. Further insights into the components of resistance to *Ophiostoma novo-ulmi* in *Ulmus minor*:

hydraulic conductance, stomatal sensitivity and bark dehydration. *Tree Physiol.* 2018; 38: 252–262.

60 Roberts BR. Net Photosynthesis, growth, and transpiration in American elm seedlings as influenced by Dutch elm disease and plant-water stress. *Phytopathology.* 1972; 62: 457–459.

61 Walters MB, Reich PB. Response of *Ulmus americana* seedlings to varying nitrogen and water status. 1 Photosynthesis and growth. *Tree Physiol.* 1989; 5: 159–172.

62 Rodríguez-Calcerrada J, Li M, López R, Cano FJ, Oleksyn J, Atkin OK, et al. Drought-induced shoot dieback starts with massive root xylem embolism and variable depletion of non-structural carbohydrates in seedlings of two tree species. *New Phytol.* 2017; 213: 597–610.

63 Martínez-Arias C, Sobrino-Plata J, Macaya-Sanz D, Aguirre NM, Collada C, Gil L, et al. Changes in plant function and root mycobiome caused by flood and drought in a riparian tree. *Tree Physiol.* 2020; 40: 886–903.

64 Rodríguez-Calcerrada J, Rodrigues AM, António C, Perdiguero P, Pita P, Collada C, et al. Stem metabolism under drought stress – a paradox of increasing respiratory substrates and decreasing respiratory rates. *Physiol Plant.* 2021; 172: 391–404.

65 Morehart AL, Melchior GL. Influence of water stress on *Verticillium* wilt of yellow-poplar. *Can J Bot.* 1982; 60: 201–209.

66 Santos-Rufo A, Hidalgo JJ, Hidalgo JC, Vega V, Rodríguez-Jurado D. Morphophysiological response of Young olive trees to verticillium wilt under different Surface drip irrigation regimes. *Plant Pathol.* 2017; 67: 848–859.

67 Pérez-Rodríguez M, Alcántara E, Amaro M, et al. The influence of irrigation frequency on the onset and development of verticillium wilt of olive. *Plant Dis.* 2015; 99: 488–495.

68 Bortolami G, Gambetta GA, Cassan C, Dayer S, Farolfi E, Ferrer N, et al. Grapevines under drought do not express esca leaf symptoms. *Proc Natl Acad Sci USA.* 2021; 118: e2112825118.

69 Lopisso DT, Knüfer J, Koopmann B, von Tiedemann A. The vascular pathogen *Verticillium longisporum* does not affect water relations and plant responses to drought stress of its host, *Brassica napus. Phytopathology.* 2017; 107: 444–454.

70 Solla A, Gil L. Influence of water stress on Dutch elm disease symptoms in *Ulmus minor. Can J Bot.* 2002a; 80: 810–817.

71 Solla A, Burón M, Gil L. Relationship between gas exchange parameters and initial susceptibility to Dutch elm disease in *U. minor* Mill. *Invest Agrar Sist Recur For.* 2001; 10: 151–163.

72 Solla A, Gil L. Xylem vessel diameter as a factor in resistance of *Ulmus minor* to *Ophiostoma novo-ulmi. For Pathol.* 2002b; 32: 123–134.

73 Solla A, Martín JA, Ouellette GB, Gil L. Influence of plant age on symptom development in *Ulmus minor* following inoculation by *Ophiostoma novo-ulmi*. *Plant Dis*. 2005a; 89: 1035–1040.

74 Pouzoulet J, Pivovaroff AL, Santiago LS, Rolshausen PE. Can vessel dimensión explain tolerance toward fungal vascular wilt diseases in Woody plants? Lessons from Dutch elm disease and esca disease in grapevine. *Front Plant Sci*. 2014; 5: 253.

75 Mittempergher L, La Porta N. Hybridization studies in the Eurasian species of elm (*Ulmus* spp.). *Silvae Genet*. 1991; 40: 237–243.

76 Solla A, Martín JA, Corral P, Gil L. Seasonal changes in wood formation of *Ulmus pumila* and *U. minor* and its relation with Dutch elm disease. *New Phytol*. 2005; 166: 1025–1034.

77 Ghelardini L, Santini A. Avoidance by early flushing: a new perspective on Dutch elm disease research. *iForest*. 2009; 2: 143–153.

78 Tchernoff V. Methods for screening and for the rapid selection of elms for resistance to Dutch Elm Disease 2. *Acta Bot Neerl*. 1965; 14: 409–452.

79 Heybroek HM. The Dutch Elm Breeding Program. In: Sticklen M, & Sherald L, editors. *Dutch elm Disease Research. Cellular and Molecular Approaches*. Springer-Verlag; 1993. p. 16–25.

80 Brasier CM. Comparison of pathogenicity and cultural characteristics in the EAN and NAN aggressive subgroups of *Ophiostoma ulmi*. *Trans Br Mycol Soc*. 1986; 87: 1–13.

81 Lisec J, Schauer N, Kopka J, Willmitzer L, Fernie AR. Gas chromatography mass spectrometry–based metabolite profiling in plants. *Nat Protoc*. 2006; 1: 387–396.

82 Luedemann A, Strassburg K, Erban A, Kopka J. Tag finder for the quantitative analysis of gas chromatography-mass spectrometry (GC-MS)-based metabolite profiling experiments. *Bioinformatics*. 2008; 24: 732–737.

83 Kopka J, Schauer N, Krueger S, Birkemeyer C, Usadel B, Bergmüller E, et al. GMD@CSB.DB: the golm metabolome database. *Bioinformatics*. 2005; 21: 1635–1638.

84 Schauer N, Steinhauser D, Strelkov S, Schomburg D, Allison G, Moritz T, et al. GC-MS libraries for the rapid identification of metabolites in complex biological samples. *FEBS Lett*. 2005; 579: 1332–13378.

85 Fiehn O, Sumner LW, Rhee SY, Ward J, Dickerson J, Lange BM, et al. Minimum reporting standards for plant biology context in metabolomics studies. *Metabolomics*. 2007; 3: 195–201.

86 Sumner LW, Amberg A, Barrett D, Beale MH, Beger R, Daykin CA, et al. Proposed minimum reporting standards for chemical analysis. *Metabolomics*. 2007; 3: 211–221.

87 Rodríguez-García A, Martín JA, López R, Sanz A, Gil L. Effect of four tapping methods on anatomical traits and resin yield in Maritime pine (*Pinus pinaster* Ait.). *Ind Crops Prod.* 2016; 86: 143–154.

88 Schenider CA, Rasband WS, Eliceiri KW. NIH image to ImageJ: 25 years of image analysis. *Nat Methods.* 2012; 9: 671–675.

89 RStudio Team. RStudio: Integrated Development for R. RStudio, PBC, Boston (MA); 2020.

90 Warnes GR, Bolker B, Bonebakker L, Gentleman R, Huber W, Liaw A, et al. Gplots: Various R Programming Tools for Plotting Data. *R Package Version.* 2020; 3.0.3.

91 Rohart F, Gautier B, Singh A, Lê Cao K-A. MixOmics: an R package for 'omics feature selection and multiple data integration. *PLoS Comput Biol.* 2017; 13: e1005752.

92 Flexas J, Bota J, Loreto F, Cornic G, Sharkey TD. Diffusive and metabolic limitations to photosynthesis under drought and salinity in C_3 plants. *Plant Biol.* 2004; 6: 269–279.

93 Galmés J, Ribas-Carbó M, Medrano H, Flexas J. Response of leaf respiration to wáter stress in Mediterranean species with different growth forms. *J Arid Environ.* 2007; 68: 206–222.

94 Venturas M, López R, Gascó A, Gil L. Hydraulic properties of European elms: xylem safety-efficiency tradeoff and species distribution in the Iberian Peninsula. *Trees.* 2013; 27: 1691–1701.

95 López R, Cano FJ, Martin-StPaul NK, Cochard H, Choat B. Coordination of stem and leaf traits define different strategies to regulate water loss and tolerance ranges to aridity. *New Phytol.* 2021b; 230: 497–509.

96 Beloiu M, Stahlmann R, Beierkuhnlein C. Drought impacts in forest canopy and deciduous tree saplings in Central European forests. *For Ecol Manage.* 2022; 509: 120075.

97 Gebre GM, Tschaplinski TJ. Solute accumulation of chestnut oak and dogwood leaves in response to throughfall manipulation of an upland oak forest. *Tree Physiol.* 2002; 22: 251–260.

98 Limousin J-M, Roussel A, Rodríguez-Calcerrada J, Torres-Ruiz JM, Moreno M, García de Jalón L, et al. Drought acclimation of *Quercus ilex* leaves improves tolerance to moderate drought but not resistance to severe water stress. *Plant Cell Environ.* 2022; 45: 1967–1984.

99 Nigg M, de Oliveira TC, Sarmiento-Villamil JL, de la Bastide PY, Hintz WE, Sherif SM, et al. Comparative analysis of transcriptomes of *Ophiostoma novo-ulmi* spp. *americana* colonizing resistant or sensitive genotypes of american elm. *J Fungi.* 2022; 8: 637.

100 Kachroo A, Robin GP. Systemic signalling during plant defense. *Curr Opin Plant Biol.* 2013; 16: 527–533.

101 Martín JA, Sobrino-Plata J, Rodríguez-Calcerrada J, Collada C, Gil L. Breeding and scientific advances in the fight against Dutch elm disease – Will they allow the use of elms in forest restoration? *New For.* 2019b; 50: 183–215.

102 Salle A, Ye H, Yart A, Lieutier F. Seasonal water stress and the resistance of *Pinus yunnanensis* to a bark-beetle associated fungus. *Tree Physiol.* 2008; 28: 679–687.

103 Arango-Velez A, El Kayal W, Copeland CCJ, Zaharia LI, Lusebrink I, Cooke JEK. Differences in defence responses of *Pinus contorta* and *Pinus banksiana* to the mountain pine beetle fungal associate *Grosmannia clavigera* are affected by water deficit. *Plant Cell Environ.* 2016; 39: 726–744.

104 Yadav V, Wang Z, Wei C, Amo A, Ahmed B, Yang X, et al. Phenylpropanoid pathway engineering: an emerging approach towards plant defense. *Pathogens.* 2020; 9: 312.

105 Martín JA, Solla A, Gil L, García-Vallejo MC. Phenological and histochemical changes in *Ulmus minor* due to root absorption of phenol: implications for resistance to DED. *Environ Exp Bot.* 2010; 69: 175–182.

106 Woolfson KN, Esfandiari M, Bernards MA. Suberin biosynthesis, assembly, and regulation. *Plants.* 2022; 11: 555.

107 Valluru R, Van den Ende W. Myo-inositol and beyond–emerging networks under stress. *Plant Sci.* 2011; 181: 387–400.

108 Yan S, Liu Q, Li W, Yan J, Fernie AR. Raffinose family oligosaccharides: crucial regulators of plant development and stress responses. *Crit Rev Plant Sci.* 2022; 41: 286–303.

109 Ward JL, Forcat S, Beckmann M, Bennett M, Miller SJ, Baker JM, et al. The metabolic transition during disease following infection of Arabidopsis thaliana by *Pseudomonas syringae* pv. *tomato*. *Plant J.* 2010; 63: 443–457.

110 Doidy J, Grace E, Kühn C, Simon-Plas F, Casieri L, Wipf D. Sugar transporters in plants and in their interactions with fungi. *Trends Plant Sci.* 2012; 17: 413–422.

111 Mur LA, Simpson C, Kumari A, Gupta AK, Gupta KJ. Moving nitrogen to the centre of plant defence against pathogens. *Ann Bot.* 2017; 119: 703–709.

112 Wang M, Gu Z, Wang R, Guo J, Ling N, Firbank LG, et al. Plant primary metabolism regulated by nitrogen contributes to plant–pathogen interactions. *Plant Cell Physiol.* 2019; 60: 329–342.

113 Martínez-Arias C, Sobrino-Plata J, Ormeño-Moncalvillo S, Gil L, Rodríguez-Calcerrada J, Martín JA. Endophyte inoculation enhances *Ulmus minor* resistance to Dutch elm disease. *Fungal Ecol.* 2021b; 50: 101024.

114 Nishizawa-Yokoi A, Yabuta Y, Shigeoka S. The contribution of carbohydrates including raffinose family oligosaccharides and sugar alcohols to protection of plant cells from oxidative damage. *Plant Signaling Behav.* 2008; 3: 1016–1018.

115 Goufo P, Moutinho-Pereira JM, Jorge TF, Morreia CM, Oliveira MR, Rosa EAS, et al. Cowpea (*Vigna unguiculata* L. Walp.) metabolomics: osmoprotection as a physiological strategy for drought stress resistance and improved yield. Front. *Plant Sci.* 2017; 8: 586.

116 Martín JA, Solla A, Venturas M, Collada C, Domínguez J, Miranda E, et al. Seven *Ulmus minor* clones tolerant to *Ophiostoma novo-ulmi* registered as forest reproductive material in Spain. *iForest.* 2015; 8: 172–180.

117 Solla A, Bohnens J, Collin E, Diamantis S, Franke A, Gil L, et al. Screening European elms for resistance to *Ophiostoma novo-ulmi. For Sci.* 2005c; 51: 134–141.

118 Martín J, Solla A, Oszako T, Gil L. Characterizing offspring of Dutch elm disease resistant trees (*Ulmus minor* Mill.). *Forestry.* 2020; 94: 374–385.

119 Beier GL, Held BW, Giblin CP, Cavender-Bares J, Blanchette RA. American elm cultivars: Variation in compartmentalization of infection by *Ophiostoma novo-ulmi* and its effects on hydraulic conductivity. *For Pathol.* 2017; 47: e12369.

120 Duchesne LC. Mechanisms of resistance: can they help save susceptible elms? In: Sticklen MB, & Sherald JL, editors. *Dutch Elm Disease Research.* New York (NY): Springer; 1993. p. 239–254.

121 Martín JA, Solla A, Domingues MR, Coimbra MA, Gil L. Exogenous phenol increase resistance of *Ulmus minor* to Dutch elm disease through formation of suberin-like compounds on xylem tissues. *Environ Exp Bot.* 2008a; 64: 97–104.

122 Martín JA, Solla A, Coimbra MA, Gil L. Metabolic fingerprinting allows discrimination between *Ulmus pumila* and *U. minor*, and between *U. minor* clones of different susceptibility to Dutch elm disease. *For Pathol.* 2008b; 38: 244–256.

123 Santini A, Faccoli M. Dutch elm disease and elm bark beetles: a century of association. *iForest.* 2015; 8: 126–134.

124 Pajares JA. Elm breeding for resistance against bark beetles. *Invest Agrar Sist Recur For.* 2004; 13: 207–215.

125 Townsend AM. Relative resistance of diploid *Ulmus* species to *Ceratocystis ulmi. Plant Dis Rep.* 1971; 55: 980–982.

126 Smalley EB, Kais AG. Seasonal variations in the resistance of various elm species to Dutch elm disease. In: Gerhold HD, Schreiner EJ, McDermott RE, & Winieski JA, editors. *Breeding Pest-Resistant Trees.* Pergamon; 1966. p. 279–287.

127 Santini A, Fagnani A, Ferrini F, Ghelardini L, Mittempergher L. Variation among Italian and French elm clones in their response to *Ophiostoma novo-ulmi* inoculation. *For Pathol.* 2005; 35: 183–193.
128 Ghelardini L, Falusi M, Santini A. Variation in timing of bud-burst of *Ulmus minor* clones from different geographical origins. *Can J For Res.* 2006; 36: 1982–1991.

Further Reading/Resources

Bernier L, Aoun M, Bouvet G, Comeau A, Dufour J, Naruzawa ES, et al. Genomics of the Dutch elm disease pathosystem: are we there yet? *iForest.* 2015; 8: 149–157.
Blanchette RA, Biggs AR, editors. *Defense mechanisms of woody plants against fungi.* Berlin, Heidelberg: Springer series in wood science. Springer; 1992.
Smalley EB, Guries RP. Breeding elms for resistance to Dutch elm disease. *Annu Rev Phytopathol.* 1993; 31: 325–354.

13

Metabolomics of *Pinus* spp. in Response to Pinewood Nematode Infection

Ana Margarida Rodrigues[1], *Marta Nunes da Silva*[2], *Marta Vasconcelos*[2], *and Carla António*[1]

[1] *Plant Metabolomics Lab Portugal, Centro de Estudos Florestais (CEF), Instituto Superior de Agronomia (ISA), Universidade de Lisboa (ULisboa), Lisboa, Portugal*
[2] *CBQF – Centro de Biotecnologia e Química Fina – Laboratório Associado, Escola Superior de Biotecnologia, Universidade Católica Portuguesa, Porto, Portugal*

13.1 Introduction

The pine wilt disease (PWD) is a serious pathology caused by the pinewood nematode (PWN) *Bursaphelenchus xylophilus* that affects the vascular system of conifers, particularly *Pinus* spp., with severe economic and ecological losses (1, 2). The PWN is native to North America, where it co-exists in ecological balance with indigenous species of conifers, not causing significant or pathogen-associated diseases (3). After the spread of the PWN to the Far East, the rapid infection of native pine species and dissemination of the PWD led to safety measures being implemented for the worldwide wood trade (4, 5). Despite the efforts to contain the spread of the PWD, the PWN was detected in Portugal in 1999 associated with *Pinus pinaster* (6) and the insect vector *Monochamus galloprovincialis* (7). The PWN is currently classified by the European and Mediterranean Plant Protection Organization (EPPO) as an A2 quarantine pest, thereby preventing the open trade of international products that derive from forest areas where its occurrence is documented (8). However, since its detection in mainland Portugal, new PWN outbreaks have been detected in Spain (9) and Madeira Island (10). Because the spread of the PWD poses a serious threat to European conifer forests, it is of the utmost importance to develop strategies for PWD management (11–14). Strategies to mitigate this disease commonly include the destruction of symptomatic trees, trunk injection with nematicides, and the use of traps to capture the insect vector. However, there has been an increasing interest in developing sustainable and environmentally friendly alternatives, such as the creation of breeding programs or the use of essential oils as natural nematicidal

Monitoring Forest Damage with Mass Spectrometry-Based Metabolomics Methods, First Edition.
Edited by Carla António.

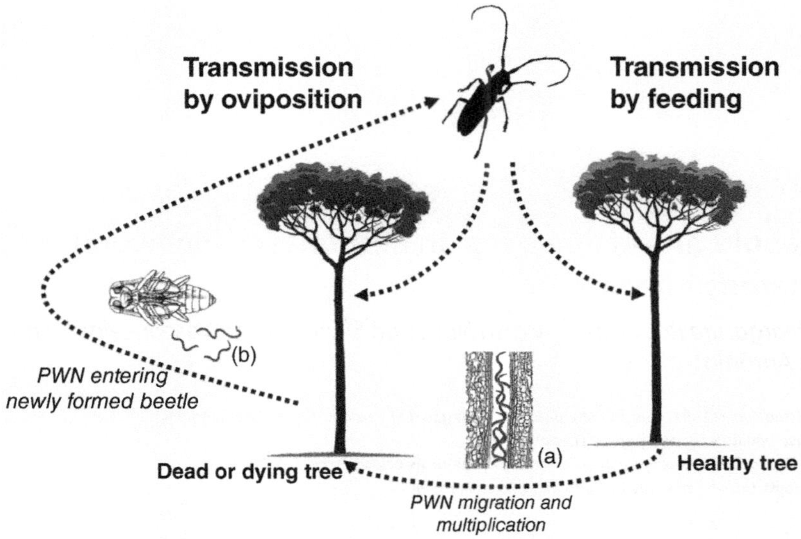

Figure 13.1 Brief description of the relationship between the life cycle of the pinewood nematode (PWN) *Bursaphelenchus xylophilus* and its insect vector from the genus *Monochamus* during host transmission by feeding or oviposition. (a) In transmission by feeding, the nematodes enter the tree and breed in the resin ducts, ultimately leading to the death of the tree. (b) In the transmission by oviposition, the nematodes are released through the slits and as the nematodes and the insect larvae complete their life cycles, the nematodes invade the body of the newly formed beetle that, in turn, will carry with them the nematodes. Source: David Pires / T. Michael Keesey / Public Domain.

compounds (15). Emergent disease management strategies include the use of elicitors of plant defenses, such as phytohormones and chitosan (16) (see Section 13.3).

The short-distance dispersal of the PWN is mediated by the longhorn beetles of the *Monochamus* genus, either by feeding or oviposition (5) (Figure 13.1). Host transmission by feeding occurs when the insect feeds on young shoots of healthy trees and releases the nematodes through the feeding wounds. In a susceptible host, nematodes breed in the resin ducts, ultimately leading to sudden tree death (i.e. in just a few weeks). In the transmission by oviposition, the female insect vector is attracted to dying (or dead) trees to lay eggs, and nematodes are released through the slits. As nematodes and insect larvae complete their life cycles, the lack of food and moisture leads juvenile nematodes to enter a dispersal phase and invade the body of the newly formed beetle. When adult beetles emerge to feed on a healthy tree, they carry with them the nematodes that enter the new tree through the feeding wound sites made by the adult beetle (5, 17).

Once nematodes multiply inside the tree, PWD progression is divided into early and advanced stages. Early-stage symptoms consist mainly of the cavitation and embolism of the tracheids in the inner xylem caused by a flood of terpenoids synthesized in the xylem. Subsequently, as the nematode population rapidly increases, PWD progression enters the advanced stage, which includes: (i) cessation of oleoresin exudation; (ii) needle wilting and chlorosis; (iii) decrease in leaf water potential, photosynthesis, and transpiration; (iv) embolism in the outermost xylem; (v) traumatic resin canal formation; (vi) cambium destruction; (vii) enhanced production of ethylene (ET); and (viii) enhanced respiration. The induced cavitation and embolism ultimately lead to the blockage of the water-conducting system, and once the trees show external symptoms, tree desiccation and death are rapidly unavoidable (18, 19).

Although differences in the resistance to PWN infection between and within *Pinus* spp. have been documented, little is known about the mechanisms underlying differential tree susceptibility. The transcriptome profile of infected *Pinus* spp. has revealed differentially expressed genes associated with PWD resistance, mostly related to oxidative stress response, cell wall lignification, and biosynthesis of terpenoids and phenylpropanoids (20–24). Lower levels of secondary metabolites, such as phenolic compounds, condensed tannins, and lignin, have been positively correlated with a higher susceptibility to PWN infection, thus indicating a role of these metabolites as constitutive chemical defenses (24, 25). On the other hand, primary metabolism could also have an important role in tree resistance due to the accumulation of the tricarboxylic acid (TCA)-cycle intermediate succinate and γ-aminobutyric acid (GABA) (energy metabolism) observed in resistant plants, which suggests the possible activation of the GABA shunt in these plants for enhanced energy production (26). In the following sections, we first discuss the current state-of-the-art of mass spectrometry (MS)-based forest tree metabolomics to study the responses of *Pinus* spp. to PWN infection followed by the strategies currently in use to mitigate PWD.

13.2 Mass Spectrometry-Based Metabolite Responses to Pinewood Nematode Infection

Mass spectrometry (MS)-based forest tree metabolomics approaches, together with the characterization of differentially expressed genes through transcriptomics analysis, have been used to advance our understanding of the molecular mechanisms underlying the responses of *Pinus* spp. to PWN infection. These studies have revealed contrasting responses between resistant and susceptible plants (20–24, 26). The integration of metabolomics studies with the physiological and anatomical characterization of PWN-inoculated maritime

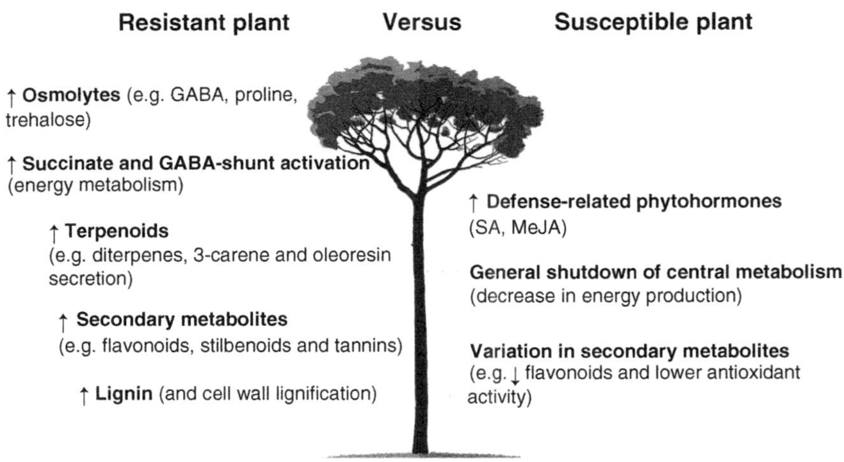

Figure 13.2 Schematic illustration of the main metabolic responses in susceptible and resistant hosts after inoculation with the pinewood nematode (PWN) *Bursaphelenchus xylophilus*. ↑, increase; ↓, decrease; GABA, γ-aminobutyric acid; MeJA, methyl jasmonate; SA, salicylic acid.

pine (*P. pinaster*) plants further allowed shedding some light on the defense mechanisms responsible for tree resistance (26). Changes in primary and secondary metabolism have been associated with enhanced defense responses in resistant plants, from increased osmolyte levels, enhanced energy production, increased levels of secondary (specialized) metabolites, and antioxidant capacity to increased lignin content and cell wall lignification (24, 26) (Figure 13.2). These potential resistance traits provide a valuable resource for the selection of resistant genotypes in breeding programs for PWD mitigation.

13.2.1 Primary Metabolism

The regulation of primary metabolism in biotic stress responses has gained increasing interest mainly due to its key role in fulfilling the energetic demands for plant defense mechanisms as well as a source of signaling molecules (27–30). Recent advances in pine responses to PWN at the transcriptome and metabolome levels have provided important insights into energy-producing primary metabolic pathways and their role in the PWD resistance response. Transcriptomic profiling of *P. pinaster* plants after PWN infection revealed enrichment in amino sugar and nucleotide sugar metabolic pathways in resistant phenotypes (24, 31). Primary metabolite profiling with gas chromatography-time-of-flight-mass spectrometry (GC–TOF–MS) revealed major changes in primary metabolite levels in PWN-inoculated stem tissues of resistant and susceptible phenotypes of two-year-old *P. pinaster* (26). These changes included a decrease in sucrose levels in PWN-infected plants resulting in an accumulation of glucose and fructose at the infection site. The accumulation of these metabolites could be further associated with the induced expression of pathogenesis-related genes (30, 32). However, as the tree enters the advanced stage of the PWD, glucose and fructose levels were shown to decrease in susceptible *P. pinaster* plants (26). Sugar accumulation can be highly affected by changes in photosynthetic activity and metabolite uptake (e.g. hexoses) by the pathogen (e.g. during the phytophagous stage of the PWN) (27, 33). In addition, upon PWN infection, susceptible plants are shown to be more prone to damage in the photosynthetic apparatus than resistant plants (26). In susceptible plants, the disruption of the water conductance caused by xylem embolism leads to a decrease in leaf water potential, inactivation of the water-splitting complex of photosystem II (PSII), and ultimately, to the cessation of photosynthetic activity (18, 34–36). As a result, a general shutdown of central metabolism has been reported in these plants (26). The consequent decrease in energy production, crucial to driving carbon fixation, negatively affects energy-demanding processes related to growth and plant defense, thus contributing to plant susceptibility to PWN (37).

Amino acids were also reported to decrease in response to PWN infection, namely branch chain amino acids (BCAAs; leucine, isoleucine, and valine) and phenylalanine (26). In the absence of photosynthesis, the energy demand under stress conditions can promote the degradation of these metabolites to generate additional energy from protein degradation (38, 39). The decrease in the levels of phenylalanine, at an advanced stage of PWN infection, further suggested a decrease in the production of defense-related secondary metabolites (e.g. lignin) in susceptible plants, and a consequent shortage of enhanced cell wall lignification (26). Cell wall lignification has been described as a crucial physical defense strategy, and several genes related to lignin synthesis were shown to be highly expressed in resistant plants (22, 24, 40). Accordingly, higher lignification in the regions surrounding plant tissues damaged by the PWN has been observed in resistant *Pinus thunbergii* plants and shown to be further associated with limited PWN migration (41).

The importance of cell redox homeostasis has also been considered crucial in plant resistance against PWN infection (24, 26, 31). *Pinus pinaster*-resistant plants were shown to accumulate osmolytes after inoculation (e.g. fucose, GABA, proline, and trehalose), which, in addition to contributing to the maintenance of cell turgor through a decrease in the osmotic potential, are known to be involved in protecting the cells from oxidative damage by restoring cellular redox balance (26). Besides its role as an osmolyte, the accumulation of GABA has been suggested to also play a role in signaling and regulatory mechanisms of stress responses (42–44). The accumulation of GABA, together with the accumulation of alanine and succinate, has been associated with PWN resistance, suggesting the activation of the alanine metabolism and the GABA shunt (26). The GABA shunt is a bypass pathway of the tricarboxylic acid (TCA) cycle from 2-oxoglutarate via succinyl-CoA to succinate (45). During the host-pathogen interaction, the GABA-shunt activation might help to support the costly defense-related metabolic pathways through the up-regulation of the TCA cycle for energy production, feeding succinate and NADH to maintain respiration (28, 42). Accordingly, an increase in the respiration rate in *Pinus massoniana* upon PWN infection has been observed after the first external symptoms, most likely to energetically support defense responses (35). Increased GABA-shunt activity has also been associated with plant resistance due to the regulation of C:N metabolism and hypersensitive responses (HR) to, ultimately, delay infection-induced senescence (42, 44, 46). Immunolocalization of GABA in vascular tissues of pine seedlings (e.g. differentiating xylem, ray parenchyma, and epithelial resin duct cells) further highlights the role of this metabolite not only in vascular processes but also in inducible defense mechanisms, including cell wall lignification (47).

Alanine, GABA, and succinate are also well-known products of hypoxic metabolism in oxygen-deprived plant tissues (48–51). The accumulation of these metabolites could be further associated with oxygen-deprived tissues due to the blockage of the water-conducting system, which in turn hinders oxygen transport via the xylem sap (19, 52, 53). Hypoxia mechanisms in plants have been comprehensively described in response to waterlogging, and during the last decade, these studies have been extended to the involvement of hypoxia in plant development (54) and plant–pathogen interactions (55, 56). In response to pathogen infection, local hypoxia might lead to an activation of specific oxygen-sensitive resistance pathways, or affect the production of elicitor molecules (55).

13.2.2 Secondary Metabolism

Secondary metabolites are well-known regulators of plant-environment interactions and defense responses to biotic stress (57–59). However, the analysis of these metabolites is extremely challenging. Currently, an estimation of 200,000 up to one million metabolites are presumed to occur in the plant kingdom, whereas only a few thousand metabolites (>14,000) can be measured (60). Secondary metabolites are responsible for most of this structural diversity, comprising over 25,000 terpenoids, 12,500 alkaloids, and 8000 polyphenols (mostly flavonoids) (61). In the context of PWD, the recent advances in the transcriptomic profiling of PWN-infected *Pinus* spp. have revealed differentially expressed genes related to secondary metabolism (e.g. jasmonic acid (JA), phenylpropanoid and terpenoid biosynthesis, and reactive oxygen species (ROS)-related genes), which were further associated with PWD resistance (20–24, 31). However, little is known about the plant secondary (specialized) metabolites involved in the plant response to PWN and PWD resistance;

the role of phenolic compounds and volatile organic compounds (VOCs) being the most commonly evaluated.

13.2.2.1 Phenolic Compounds

Polyphenols are the most widely distributed group of plant secondary metabolites in the plant kingdom. These phenolic compounds are generally derived from phenylalanine via shikimate or acetate pathways and can be classified into several sub-groups, including flavonoids, hydroxycinnamates, coumarins, and lignans (62, 63). Flavonoids constitute the largest class of polyphenols and are classified based on their structural properties (e.g. chalcones, flavonols, flavones, anthocyanins, and catechins) and can be further altered through secondary modifications of the backbone, for example by hydroxylation, glycosylation, methylation, prenylation, acylation, dehydration, and polymerization (61). These metabolites play an important role in plant defense against abiotic and biotic stresses and are well known for their strong antioxidant activity (63–66). Regarding plant-nematode interactions, flavonoid synthesis has been shown to be induced upon nematode entry and migration inside the plant. The defense roles of these metabolites include egg-hatching inhibition, movement and fertility impairments, nematode repellent, and nematicidal activity (67).

Other polyphenols, such as stilbenoids and tannins, have also been associated with PWN resistance in *Pinus* spp. by preventing the reproduction and dispersion of nematodes inside the tree (25, 68, 69). Most studies performed in the scope of the PWD have focused on rapid methods to quantify the total phenolic, condensed tannin, and flavonoid contents (16, 25, 70). In susceptible species to PWN infection, the total content of these metabolites has been negatively correlated with PWN growth rate and found to be lower in these plants when compared to resistant species (25, 71). Transcriptomics studies in *P. pinaster* after PWN infection also revealed a high number of differentially expressed genes in the flavonoid and terpenoid backbone biosynthesis pathways (20, 21). Furthermore, the comparison of transcriptional changes between susceptible and resistant *P. pinaster* phenotypes revealed that the genes related to these pathways were more expressed in resistant *P. pinaster* plants (24). When analyzing other *Pinus* spp., the same observations were reported, with a higher expression of flavonoid biosynthesis pathway genes also observed in resistant varieties of *Pinus densiflora* (69) and *P. massoniana* (22). Interestingly, the targeted analysis of salicylic acid (SA) and methyl jasmonate (MeJA) revealed an increase in these phytohormones in susceptible *P. pinaster* plants (24, 26). SA accumulation has a key role in hypersensitive response (HR), mainly characterized by a rapid cell death at the site of pathogen entry (72, 73). In PWN infection, HR includes rapid cell death, production of toxins, and leakage of oleoresin into the tracheids (74). The increase in the levels of SA suggests that after PWN invasion, a series of inefficient HRs are rapidly triggered in susceptible trees, but ultimately lead to tree death (24, 40). In contrast, a moderate HR, combined with an increase in cell wall-related genes to restrict PWN migration, could be a more effective defense against PWN infection (40).

Accumulation of downstream metabolites in the flavonoid pathway has been regarded as a defense response to pathogen infection or herbivory. For example, several flavonoids, including quercetin 3-(6″-acetylglucoside), have been found to accumulate in higher abundance in barley plants resistant to the fusarium head blight (75). The accumulation of catechin and procyanidin B5 (epicatechin-epicatechin dimer) also resulted in increased resistance to rust infection in *Populus nigra* (76, 77), and high concentrations of catechin, procyanidin B1 (catechin-epicatechin dimer), and taxifolin were associated with increased resistance to bark beetle attack in *Picea abies* (78). In addition, flavonol- and dihydroflavonol-type flavonoids (e.g. kaempferol, quercetin, taxifolin, and their derivatives) found in *Pinus sylvestris* and *P. abies* have shown strong anti-bacterial and antioxidant activity (79).

Polyphenols and flavonoids are well known for their antioxidant activity, and the accumulation of these secondary metabolites is strongly promoted upon photo-oxidative stress (80, 81). As the PWNs migrate and feed on epithelial cells lining the ducts, water conductivity is blocked in the stem, successively leading to a decrease in transpiration and photosynthesis in leaves, thereby resulting in generalized oxidative damage (82, 83). Upon oxidative stress, plants activate several antioxidant mechanisms, including the accumulation of antioxidant enzymes, flavonoids, and other polyphenols (16, 70). In addition, resistant phenotypes of *P. pinaster* (24) and *Pinus thunbergii* (40) showed a higher expression of oxidative stress-response genes, and thus, a better protection from oxidative damage was observed in these plants. Furthermore, synthesis and accumulation of polyphenols and terpenes in *Pinus pinea* and *P. pinaster* were shown to be strongly impacted by environmental factors (e.g. lower phytochemical production in field sites with a higher precipitation rate), which in turn may affect host disease susceptibility (71). Metabolomics approaches, for example, based on wide-targeted liquid chromatography coupled to mass spectrometry (LC–MS) (see Chapter 7) would provide additional important information on the plant secondary (specialized) metabolites involved in tree resistance to PWD and defense mechanisms against PWN infection.

13.2.2.2 Volatile Organic Compounds and Essential Oils

The volatile fraction in plants is mainly composed of four compound classes: terpenoids, phenylpropanoids/benzenoids, fatty acid derivatives, and amino acid derivatives (84). This volatile fraction comprises volatile organic compounds (VOCs) and essential oils (EOs), which, given the similarity in their qualitative chemical composition, are often erroneously classified or compared. VOCs are most commonly collected with headspace techniques such as solid-phase microextraction (SPME) (85–87) (see Chapter 6), whereas EOs are mixtures of VOCs obtained exclusively with hydro-, steam-, or dry-distillation, or in the case of Citrus fruits, mechanically without heating (88). Due to the different extraction processes involved, the relative amounts among these volatile fractions can be very different (89–91). EOs are well known for their biological activities and consist mainly of terpene (mono-, sesqui-, and diterpenes) and aromatic compounds, usually with one to three major components with a relative amount ranging, on average, from 20% to 70% (92). Regarding sustainable PWD management strategies, EOs from numerous plant species have been screened successfully as nematicidal compounds, showing nematicidal activities that can surpass those of commonly used synthetic chemical nematicides (93). From the reported EOs, those rich in sulfides or oxygen-containing compounds showed the lowest effective concentrations and are good candidates for *in vivo* testing to assess efficiency against PWD (15, 94).

Pinus spp. produce large amounts of monoterpene hydrocarbons and diterpene resin acids (95). These compounds are major components of the sticky oleoresin secretion that is formed as a constitutive and inducible defense of conifers against invading pests (e.g. bark beetles) and pathogens, in addition to other numerous behavioral and physiological roles in plant-herbivore and plant-pathogen interactions (95, 96). Moreover, these metabolites, particularly monoterpenes, have been shown to increase at the site of attack/entry and to act as a physical barrier against pests and pathogens (97). The role of terpenoid compounds in tree responses to PWN-infection has been assessed through the profiling of these volatile fractions in several *Pinus* spp. (*P. pinaster*, *P. pinea*, *P. halepensis*, and *P. sylvestris*). However, profiling of VOCs emission revealed no differences between control and PWN-infected hosts (98). Similarly, no major differences in EO composition were observed between control and infected *Pinus* spp., apart from the occurrence of EO chemotypes (99, 100). However, the analysis of EOs from the PWN-inoculation area revealed an increase in sesquiterpenes and diterpenic compounds, particularly in resistant species (*P. pinea and P. halepensis*) (100). The analysis of the lipophilic extract of the phloem of *P. pinaster* and *P. pinea* also revealed an accumulation of these compounds after PWN infection, particularly in *P. pinea* (71). Due to their role in the chemical defense of conifers, these studies revealed a possible relationship between the levels of diterpenic acid and tree resistance to herbivores and pathogens. Furthermore, these compounds have also shown a strong nematicidal activity against the PWN, including ingenane-type diterpenoids (i.e. polycyclic diterpenes) extracted from *Euphorbia kansui* roots (101).

Regarding monoterpene composition, a study by Hwang et al. (102) revealed changes in the levels of emitted VOCs as well as in the composition of resin in the needles of five-year-old *P. densiflora* 20 days after PWN inoculation, namely an increase in three-carene levels. Additionally, in this study, the profile of emitted VOCs of *P. densiflora* and *P. koraiensis* was compared, and revealed quantitative and qualitative differences in the emitted monoterpene compounds among *Pinus* species. This observation highlights the need to cautiously explore the volatile composition of the plant response to PWN. Because a high chemical variability between and within species has been reported, it is therefore crucial to carefully annotate all experimental factors, including the method of PWN inoculation, sampling method, VOCs extraction methodology, and environmental conditions, that can allow replication and comparison of the obtained results (100). Moreover, changes in the levels of VOCs are not always only related to plant-PWN interaction. For example, an increase in (−)-α-pinene emission has been reported, not directly related to plant–pathogen interaction but to the insect–plant and insect–pathogen attraction (19, 98, 103, 104).

The chemicals responsible for insect attraction to the host have also been the object of great interest. In a study using susceptible and resistant phenotypes of seven-year-old *P. pinaster* plants, it was observed that, during insect feeding, the susceptible plants released significantly higher amounts of monoterpene β-pinene and sesquiterpene β-caryophyllene, and lower amounts of monoterpene β-myrcene than the resistant phenotypes (105).

13.3 Disease Management

The PWD is characterized by severe disease symptoms and high dissemination potential, making it very difficult to contain when disease outbreaks occur. As such, prevention is undoubtedly the most important approach to controlling *B. xylophilus* and its insect vectors as well as to reducing the incidence and severity of the disease (Figure 13.3). Prevention strategies include monitoring and controlling the insect vector using pheromone traps, a practice extensively applied in Asian countries but limited in European forests due to distinct responses of *Monochamus* spp. to the trapping-pheromones used (106). For *M. galloprovincialis*, one of the most prevalent species in Europe, some degree of success has been obtained with multi-funnel traps baited with pine volatiles (α-pinene and ethanol) and bark beetle pheromones (ipsdienol, ipsenol methyl-butenol, and 2-undecyloxy-1-ethanol) and, in particular, unpainted white traps seem to be the most attractive to this insect (107, 108). Several insecticides based on prothiophos, fenitrothion, fenthion, pyridaphenthion, thiacloprid, or chlorpyrifos-methyl chemicals have also shown high efficacy as preventative sprays against insect vectors (109).

Another chemical approach to controlling the PWN is tree injection with nematicides, such as avermectin (AVM), emamectin benzoate, morantel tartrate, and sulfuryl fluoride (110). This practice is very efficient in protecting the treated tree, but it fails in large-scale applications, such as in extensive forest stands, because it requires periodic applications to maintain its nematicidal

Figure 13.3 Conventional (in black) and emergent (in orange) disease mitigation strategies against the pine wilt disease (PWD). In the last decades, PWD control was largely achieved through the creation of phytosanitary zones, logging of infected trees and treatment of infected material through burning, crushing or burying, trapping of the insect vector, and application of pesticides through spraying or tree injection. Recently, research efforts have focused on more sustainable practices that deliver higher and longer efficacy with lower environmental impacts through less labor-intensive approaches. These include the search for genetic resistance, biological control, and application of plant elicitors, such as phytohormones and chitosan, and nanoagrochemicals.

activity, thus being expensive and labor-intensive (111–114). Moreover, despite their relative effectiveness in mitigating the PWD, the use of traps, insecticides, and nematicides is usually discouraged due to environmental and public health concerns related to impairments in the local biodiversity, lack of specificity to the target species, potential bioaccumulation in the ecosystem, and residual accumulation in humans (108, 109, 115).

When prevention fails and disease outbreaks occur, the first approach for disease management is the eradication of trees showing infection symptoms, as well as all surrounding trees. However, this approach has limited efficacy because disease assessment is usually done by visual observation alone, but the typical disease symptoms only appear when the infection rate is sufficient to compromise the physiology of the tree. This results in the presence of infected asymptomatic trees for long periods, during which no mitigation actions are pursued while the infection keeps spreading. After cutting, infected material can be crushed without further treatment, burnt or buried in the ground, which is labor intensive and expensive, and also requires specific machinery (109).

Given the environmental and practical constraints of these strategies, related to their questionable effectiveness and potential hazard to the environmental and human health, in recent years, research efforts have been focusing on more sustainable strategies, including host resistance and biocontrol agents based on plant EOs and beneficial microorganisms (15, 36, 70, 116–120). However, other emergent strategies, such as the use of phytohormones, chitosan, and agronanochemicals, have been poorly explored, limiting their inclusion in integrated pest management strategies for the effective control of the PWD.

13.3.1 Phytohormones

The activation of plant defenses is mediated by a complex hormonal network in which jasmonates, salicylates, and ET act synergistically or antagonistically with each other, promoting local or systemic defenses in response to pathogen infection or herbivory attacks (121) (see Chapter 8). Jasmonates, including the free-acid JA and MeJA, are important signaling molecules in plants, being generally associated with responses against chewing insects and necrotrophic pathogens through the regulation of secondary metabolites that dissuade feeding or inhibit digestion and induction of volatiles that repel the attacking organism or attract their natural enemies (122). On the other hand, salicylates, particularly SA and its functional analogues methyl salicylic acid (MeSA, a biologically inactive derivative of SA that is converted to SA in distal tissues) and acibenzolar-*s*-methyl (ASM, a synthetic chemical that induces systemic acquired resistance (SAR) through the SA pathway), are widely known to be activated in response to biotrophic pathogens. They mediate the initiation of plant hypersensitive response upon pathogen infection, which is characterized by the production of phytoalexins and pathogen-related proteins that eventually lead to SAR through increased levels of certain defense compounds, including lignin and chitinases (123). On the other hand, the role of ET in the regulation of plant defenses is not as well established because it seems to work in a tight regulation with jasmonates rather than working alone, hampering the recognition of its specific role in the regulation of plant defenses (123, 124).

13.3.1.1 Ethylene

Ethylene controls several developmental and regulatory processes in plants, but pathogens can regulate the ET pathway through microbe-associated molecular patterns, rendering the plant more susceptible to several diseases, including the PWD (125). However, although this phytohormone was the first to be implicated with plant responses against the PWN (126, 127), the role of ET regulation during the PWD has been poorly investigated so far. It has been demonstrated that the PWN impairs plant ET production during infection, resulting in a significant increase of its biosynthesis in two specific moments, which have been correlated to the two stages of the symptom development characteristic of the PWD: the early peak is likely related to the denaturation of xylem parenchyma and was found in both susceptible species and resistant species, while the later peak seems to be related to cambial death, only occurring in susceptible plants (126, 127). Subsequent works have also shown that enhanced ET release only occurs after infection with virulent *B. xylophilus* isolates, resulting in cytological changes in xylem parenchymatous cells, embolism in tracheids, and leaf chlorosis (128). Recently, it was demonstrated that the PWN interferes with the molecular regulation of 1-aminocyclopropane-1-carboxylate oxidase 1 (*ACO1*), involved in ET biosynthesis, through the effector BxSCD1, suppressing cell death and immune responses in *P. thunbergii* (125). Overall, it seems that the PWN hijacks plant defenses through the induction of the ET pathway. Thus, it would be highly relevant to explore the use of plant elicitors that antagonize this hormonal pathway for the development of more sustainable disease management strategies.

13.3.1.2 Jasmonates

It has also been demonstrated that the PWN targets not only the ET pathway to promote infection but also the JA pathway. During infection, it produces a specific effector, Bx-FAR-1, that downregulates the expression of Pt-F-box-1, involved in JA biosynthesis, resulting in impaired plant defenses (129). The importance of the JA pathway in plant responses to the PWN has been acknowledged for some time, with several studies demonstrating that elicitation with MeJA, a plant volatile that accumulates in plant cells due to the methylation of JA, results in lower insect vector feeding rates and impaired PWN colonization (130, 131). Application of MeJA to *P. pinaster* leaves led to lower damage of phloem cells resulting from infestation by the pine weevil *Hylobius abietis*, likely due to the induction of plant secondary metabolites, such as terpenes (132–134). Likewise, exogenous MeJA application also leads to reduced feeding rates by the vector *M. alternatus*, resulting in broad transcriptional responses related to terpenoid and phenolic biosynthesis (131). This study demonstrates that, in *P. massoniana*, *M. alternatus* adults initiate feeding within 12 hours after inoculation, regardless of plant treatment with MeJA, but exogenous MeJA application significantly decreases the debarked area by about 50%. As no differences were observed in the enzymatic activity of the midgut of beetles, the authors hypothesize that this insect may decrease food intake to avoid the toxic compounds resulting from plant elicitation, particularly terpenoids and flavonoids. MeJA treatment also led to an extensive genetic reprogramming related to secondary metabolite biosynthesis (131), which is in agreement with previous studies demonstrating the implication of MeJA in key defensive responses likely involved in plant tolerance to the PWD, such as phenolic compounds, lignin, resin, and formation of traumatic resin ducts (70, 120, 132–135). Foliar treatment of *P. pinaster* with MeJA also resulted in lower foliar damage and PWN colonization rates (130). Without elicitation, 13% of infected plants died and 73% showed extensive disease symptoms, while with MeJA treatment no plant mortality was observed and 73% only showed mild foliar damage. The authors suggest that the higher foliar damage observed in non-elicited plants may be related to mineral losses, particularly of Cu, and lipid peroxidation induced by the high PWN density in stem tissues, while the lower PWN colonization observed in elicited plants likely results from the accumulation of soluble phenols (130). Although these promising results support the adoption of elicitors of the JA pathway in integrated pest management strategies against the PWD, the few studies conducted so far have been performed in nursery plants. Thus, field trials should be dully pursued because several factors inherent to more complex environments, including plant age, presence of other pests and pathogens, and climatic conditions, may influence the actual effectiveness of these compounds.

13.3.1.3 Salicylates

Despite being widely acknowledged that plant responses against necrotrophic pathogens, such as the PWN, are mainly mediated by the JA pathway while biotrophic agents are mainly tackled through the SA pathway, several studies have demonstrated that plant elicitation with salicylates results in the promotion of plant defenses and lower PWN infection rates (130, 136–138). Exogenous SA application to *P. pinaster* plants resulted in only mild leaf chlorosis in 75%, while non-elicited plants had a 13% mortality rate and a 73% incidence of severe disease symptoms, which correlated with a 45% decrease in PWN density inside stem tissues of SA-elicited plants (130). The authors reported that, distinctively to what was observed in JA-elicited plants, SA treatment led to a gradual decrease in phenols accumulation throughout the disease, demonstrating that each hormone modulates plant defenses against the PWN through distinct metabolic pathways. Plant treatment with SA also led to the accumulation of Cu in plant tissues, which was correlated with the lower foliar damage observed. In addition, changes in plant microbiome were also observed following SA elicitation, which increased the relative abundance of *Enterobacter kobei* and *Pantoea alhagi*, implicated in plant growth promotion and tolerance to environmental stresses (130). In addition, treatment of *P. densiflora* calli with SA led to the upregulation of an antimicrobial peptide-encoding gene and another encoding a metallothionein-like protein which has been associated with metal homeostasis and detoxification of heavy metals and reactive oxygen species during plant stress responses (136).

Similar protective effects against the PWN were also observed in *P. thunbergii* and *P. densiflora* plants treated with inducers of the SA pathway, namely MeSA and ASM (137, 138). Foliar treatment with these compounds resulted in significant reductions in PWD severity (up to 73%), particularly after ASM treatment, which was accompanied by important changes in several beneficial microbial taxa that were present at low relative abundance before elicitation (138). Plant treatment with MeSA and ASM decreased the relative abundance of chitinolytic bacteria, and ASM also increased several microbial taxa involved in key developmental and ecological processes, including

Rhodanobacter, Devosia, Bradyrhizobium, Acidibacter, Mesorhizobium, and Hyphomicrobium genera (138). Subsequent transcriptome analysis demonstrated that MeSA and ASM induce the differential regulation of numerous genes, among which 650 showed similar expression patterns between both elicitors. The functional characterization of these co-expressed genes revealed their involvement in plant photosynthesis, carbon fixation, protection against oxidative stress, such as photorespiration, peroxisome metabolism and glycine regulation, and the flavonoids pathway, particularly myricetin and quercetin biosynthesis (139).

More recently, Jeon et al. (137) developed two MeSA formulations for foliar application against the PWN: a dispersible concentrate-type MeSA formulation (MeSA 20 DC, prepared by mixing MeSA with propylene glycol monomethyl ether, propylene glycol, triethanolamine, and polyoxyethylene tristyrylphenyl ether phosphate), and an emulsifiable concentrate-type formulation (MeSA 20 EC, prepared by mixing MeSA with propylene glycol monomethyl ether, ethoxylated castor oil, calcium dodecylbenzenesulfonate, tristyrylphenol ethoxylates, and methylated soybean oil). The authors observed that MeSA 20 EC showed better efficacy against the PWD in *P. densiflora* and *P. thunbergii* (reducing disease incidence up to 58%), while MeSA 20 DC showed a similar protective effect in *P. densiflora* (48% reduction in disease incidence), but not in *P. thunbergii*. Nevertheless, the authors demonstrate that, by including methylated soybean oil in the MeSA 20 EC formulation, the volatility of the formulation decreases, prolonging leaf exposure to MeSA and improving its efficiency in *P. thunbergii*. This evidence highlights the influence of plant genotype and specific chemical formulation of the salicylate-based formulation on their effectiveness against the PWD, supporting the need for research aiming at tailored mitigation strategies to distinct *Pinus* spp. toward the maximization of their protective potential. MeSA led to the upregulation of several pathogenesis-related (PR) proteins commonly used as marker genes for SAR induction, namely *PR-1*, *PR-2*, and *PR-5*, a cell wall biosynthesis-related gene (*ETS*), and *PR-9*, a peroxidase involved in cellular lignification and enhanced resistance against pathogens (137). Therefore, by promoting cell wall integrity and strengthening, salicylates contribute to inhibiting PWN migration and reproduction capacity, supporting their inclusion in efficient and long-lasting mitigation strategies targeting the PWD. However, further research is still needed to support their efficacy in adult forest stands.

13.3.2 Chitosan

Chitosan, a biopolymer obtained by the deacetylation of chitin obtained from the exoskeleton of crustaceans and fungal cell walls, has been receiving attention in sustainable agriculture due to its tremendous biocompatibility and biodegradability (140). It promotes plant growth, productivity, and defenses against various biotic (fungi, bacteria, nematodes, virus, and insects) and abiotic (drought, salinity, cold, and environmental contamination) stresses by modulating key physiological processes, including cell division, nutrient uptake, scavenging of reactive oxygen species, the phenylpropanoids pathway, and plant immunity (140–143).

The nematicidal activity of chitosan against the PWN was first reported by Nunes da Silva et al. (144), which evaluated the effect of high, medium, and low molecular weight (MW) chitosans as soil amendments against the PWD. The authors reported that the low MW chitosan had the highest protective effect, inducing a 38% decrease in nematode density and preventing the loss of total chlorophylls, while the high MW chitosan failed to promote plant protection. This protective effect of low MW chitosan against the PWD was attributed to the modulation of metabolic and transcriptomic mechanisms related to the plant antioxidant system (16). Chitosan-treated *P. pinaster* plants demonstrated higher levels of oxidative stress, which was accompanied by increased antioxidant compounds (catalase, carotenoids, and anthocyanins), secondary metabolites (soluble phenolic compounds and lignin), and upregulation of genes involved in plant antioxidant responses (*TRX*) and plant defenses (*DEF*) (16). Despite these very positive effects, chitosan solutions are generally extremely viscous and need to be applied to each plant individually, which may be difficult to perform in large plantation areas. To overcome this potential limitation, a biofertilizer enriched with diazotrophic bacteria and a chitosan-producing fungus, *Cunninghamella elegans*, was developed (145). In *P. pinaster* plants supplemented with this biofertilizer, PWD symptoms were ameliorated, resulting from lower nematode densities in stem tissues and prevention of water and total chlorophyll loss (145). Seed coating with single and binary mixtures of low and medium MW chitosans was also effective in reducing the incidence of pine pitch canker, caused by *Fusarium circinatum*, in *P. radiata* and *P. sylvestris*, promoting post-emergence survival of the seedlings, higher survival rates, accumulation of phenolic compounds, and plant radical scavenging activity (146). Nevertheless, the efficacy of seed coating with chitosan against the PWN is still to be demonstrated, and the broad application of chitosan in forest management against the PWD is still far from being achieved.

13.3.3 Agronanochemicals

In the last decade, nanotechnology fostered great advances not only in medicine but also agroforestry, through the use of nanoparticles (NPs) that increase the performance of conventional pesticides through improved transport, release, and organism targeting (147–149). The control of the PWN in forest stands continues to largely rely on pesticides, such as avermectin (AVM), due to its high nematicidal activity resulting from suppressed population growth, increased larval lethality, and impaired larval development (150). Therefore, in recent years research efforts have been focusing on the development of AVM-based nanoformulations that deliver higher bioactivity against the PWN and its insect vector *M. alternatus* at lower concentrations (151,152). AVM nanoparticles enriched with bovine serum albumin (AVM@BSA·NPs), a low-cost polymer material with high biocompatibility derived from bovine proteins, showed high rates of cumulative release, loading efficiency, and retention, being continuously released for about 20 days (151). These nanoparticles showed high nematicidal activity and increased AVM ingestion by *M. alternatus* larvae (up to 65%), thus resulting in higher transport to the larval midgut and hindgut and stomach toxicity. AVM@BSA·NPs also proved to strongly adhere to the larval body surface, being subsequently absorbed by their epidermis, delivering higher AVM rates to the target sites (151).

Polymers of copper and trimesic acid were also used to encapsulate AVM (AM@CuBTC·NPs) to control *M. alternatus* (152). This nanoformulation showed a high cumulative release up to 12 days, penetrating the body wall of the larvae and being transported from their body surface to intestinal cells through contact and diffusion mechanisms. AM@CuBTC·NPs showed good dispersibility and efficient AVM delivery, inducing higher cytotoxicity and resulting in mortality rates of up to 90%. The authors propose that, by injecting AM@CuBTC·NPs into the tree parts above the root, AVM is delivered into its interior and transported by layer diffusion through the tree's water transport system, impairing *M. alternatus* activities, including egg hatching, and nematode reproduction. This nanoformulation can, therefore, reduce the incidence of the disease and provide an effective preventive effect against subsequent attacks. It also showed positive effects when injected and sprayed into dead wood, thus being a great solution for the treatment of timber and wood used for industrial and commercial purposes (152). Nonetheless, although the undoubtful usefulness of AVM@BSA·NPs and AM@CuBTC·NPs in reducing the amount of potentially toxic pesticides needed to manage the PWD, the growing concerns on environmental accumulation and pesticide resistance drive the need for more sustainable sylvicultural practices. Thus, recent research efforts have also focused on the inclusion of more sustainable compounds in practical and affordable nanoformulations, based on chitosan, plant extracts, and micronutrients, for the promotion of plant tolerance to the PWD at lower environmental costs (153–156).

The potential of chitosan-based agronanochemicals as sustainable strategies for crop protection against pests and diseases has been attributed to enhanced controlled release, penetration into plant tissues, bioavailability, biocidal activity, and decreased leaching and toxicity to non-target organisms, thus maximizing plant protection while minimizing the negative impacts to the environment and human health commonly observed with conventional agrochemicals (157). Liang et al. (153) demonstrated that by encapsulating AVM within nanoparticles composed of poly-γ-glutamic acid (γ-PGA) and chitosan (AVM-CS/γ-PGA·NPs), it is possible to overcome limitations related to its poor water solubility and high environmental instability. This formulation showed high AVM loading capacity and reduced its losses through photolysis by at least 20%, increasing PWN mortality rates by 29% due to higher AVM accumulation in the intestines and head of the nematodes (153). Therefore, aside from the relevant protective effects of its individual application, chitosan can also be extremely useful in increasing the efficacy of conventional biocides commonly used to control the PWN. Additionally, it can also be used for the encapsulation of more environmentally-friendly formulations, including plant-derived natural compounds, further contributing to reducing the current recurrence of AVM for PWD mitigation.

Chitosan-coated nanoemulsions of two natural compounds derived from *Allium cepa*, dipropyl trisulfide (DPTS) and methyl propyl trisulfide (MPTS), which have strong nematicidal activity against the PWN but low water solubility and stability, demonstrated high long-term storage stability and persistence of nematicidal activity against the PWN, thus supporting the usefulness of chitosan-coated nanoemulsions of sulfides for field application against the PWD (154). As the chitosan-coated DPTS nanoemulsion showed three times stronger nematicidal activity than pure DPTS against the PWN, the authors highlight its potential as a trunk injection agent against this pathogen; nevertheless, further studies are necessary to assess the translocation capacity of this nanoformulation within the tree, as well as potential phytotoxic effects (154). Extracts of *Parkia biglobosa* were also successfully encapsulated in gold nanoparticles (Au·NPs) and tested against key symbiotic bacteria that contribute to the virulence of the PWN (155). These AuNPs inhibited the growth of several bacterial strains isolated from the surface of the PWN, including *Staphylococcus aureus, Pseudomonas syringae, Bacillus anthracis,* and *Escherichia coli*, although their effectiveness was higher against gram-positive bacteria than gram-negative bacteria. The authors hypothesize that the single-layered cell wall of gram-positive bacteria facilitates the entry of the Au·NPs, while the double-layered cell wall of the gram-negative bacteria reduces their bioavailability (155). Still, these AuNPs of *P. biglobosa* extracts possess several secondary metabolites important to plant defenses, such as flavonoids, terpenoids, saponins, glycosides, alkaloids, and tannins, which, together with the antimicrobial activity of the Au·NPs themselves, related to stunted growth and cell damage, support their use as new-generation agrochemicals toward improved PWD mitigation (155).

In recent years, with the rapid development of nanotechnology, selenium nanoparticles have become promising antimicrobial agents, owned to their ability to trigger apoptosis and autophagy through the regulation of reactive oxygen species, low toxicity, degradability, and high bioavailability (158). Despite its high inhibitory activity against distinct bacteria, fungi, and viruses, its nematicidal activity has been poorly explored so far. The combination of selenium and chitosan in nanoformulations (SeNPs@CS) using a simple and environmentally friendly method has demonstrated high nematicidal activity when directly applied to the PWN, also decreasing disease symptoms when used as preventive treatment (156). *In vitro* bioassays showed that SeNPs@CS has high nematicidal activity against the PWN (up to 100%), resulting from structural damage and accumulation of reactive oxygen species throughout their body. This nanoformulation is ingested through the stylets of the PWN within only 30 minutes of treatment, inducing a generalized oxidative stress response in just 12 hours. Additionally, foliar spraying seven days before inoculation resulted in a sharp decrease in leaf wilting and nematode density in stem tissues (156). It has been demonstrated that soils amendments with chitosan nanoparticles and seed-soaking with Se·NPs induces systemic resistance to the root-knot nematode *Meloidogyne incognita* in tomato plants, by stimulating plant growth and development, increasing the phenol content, promoting the activity of peroxidases, polyphenol oxidases and proteinase inhibitors and upregulating the expression of PR genes (159, 160). However, the physiological and molecular aspects underpinning plant responses against the PWN following the application of these nanoparticles have not been described yet, hindering the possibility of exploiting them for an integrated management of the PWD.

13.4 Conclusions and Future Perspectives

The spread of the PWN from Asian countries to Portugal poses a devastating threat to European pine forests. Hence, it is crucial to control the spread of the PWN in territories where its presence has been detected and develop sustainable strategies for disease management to prevent new pest outbreaks. Because metabolomics represents the ultimate response of biological systems to genetic and/or environmental perturbations, MS-based forest tree metabolomics indeed constitutes a unique opportunity to explore the adaptation and tolerance of forest tree species to adverse environmental conditions, particularly in the current climate change scenario. Due to its role in defense responses and as signaling molecules, it would be crucial to further explore the range and chemical diversity of specific plant secondary (specialized) metabolites of PWN-infected plants. For example, a wide-targeted LC–MS-based approach (see Chapter 7) would allow maximizing their biological interpretation by integrating these data into novel metabolic pathways potentially involved in the resistance of *Pinus* spp. against the PWN infection. This information could provide a valuable resource to identify novel nematicidal compounds and contribute to the development of sustainable PWD management strategies.

References

1 Mota M, Vieira PC. Pine wilt disease in Portugal. In: Zhao BG, Futai K, Sutherland JR, & Takeuchi Y, editors. *Pine Wilt Disease*. Tokyo (JP): Springer; 2008. p. 33–38.
2 Soliman T, Mourits MCM, van der Werf W, Hengeveld GM, Robinet C, Lansink AGJMO. Framework for modelling economic impacts of invasive species, applied to pine wood nematode in Europe. *PLoS One*. 2012; 7: e45505.
3 Wingfield MJ, Blanchette RA, Nicholls TH, Robbins K. The pine wood nematode: a comparison of the situation in the United States and Japan. *Can J For Res*. 1982; 12: 71–75.
4 Fielding NJ, Evans HF. The pinewood nematode *Bursaphelenchus xylophilus* (Steiner and Buhrer) Nickle (= *B. lignicolus* Mamiya and Kiyohara): an assessment of the current position. *Forestry*. 1996; 69: 35–46.
5 Evans HF, McNamara DG, Braasch H, Chadoeuf J, Magnusson C. Pest risk analysis (PRA) for the territories of the European Union (as PRA area) on *Bursaphelenchus xylophilus* and its vectors in the genus *Monochamus*. *EPPO Bulletin*. 1996; 26: 199–249.
6 Mota M, Braasch H, Bravo MA, Penas AC, Burgermeister W, Metge K, Sousa E. First report of *Bursaphelenchus xylophilus* in Portugal and in Europe. *Nematology*. 1999; 1: 727–734.
7 Sousa E, Bravo MA, Pires J, Naves P, Penas AC, Bonifácio L, Mota M. *Bursaphelenchus xylophilus* (Nematoda; Aphelenchoididae) associated with *Monochamus galloprovincialis* (Coleoptera; Cerambycidae) in Portugal. *Nematology*. 2001; 3: 89–91.
8 EPPO. PM 9/1 (6) *Bursaphelenchus xylophilus* and its vectors: procedures for official control. *EPPO Bulletin*. 2018; 48: 503–505.
9 Abelleira A, Picoaga A, Mansilla JP, Aguin O. Detection of *Bursaphelenchus xylophilus*, causal agent of pine wilt disease on *Pinus pinaster* in northwestern Spain. *Plant Dis*. 2011; 95: 776.
10 Fonseca L, Cardoso JMS, Lopes A, Pestana M, Abreu F, Nunes N, et al. The pinewood nematode, *Bursaphelenchus xylophilus*, in Madeira Island. *Helminthologia*. 2012; 49: 96–103.
11 de la Fuente B, Beck PSA. Invasive species may disrupt protected area networks: Insights from the pine wood nematode spread in Portugal. *Forests*. 2018; 9: 282.
12 de la Fuente B, Saura S, Beck PSA. Predicting the spread of an invasive tree pest: the pine wood nematode in Southern Europe. *J Appl Ecol*. 2018; 55: 2374–2385.
13 Hirata A, Nakamura K, Nakao K, Kominami Y, Tanaka N, Ohashi H, et al. Potential distribution of pine wilt disease under future climate change scenarios. *PLoS One*. 2017; 12: e0182837.

14 Gruffudd HR, Schröder T, Jenkins TAR, Evans HF. Modelling pine wilt disease (PWD) for current and future climate scenarios as part of a pest risk analysis for pine wood nematode *Bursaphelenchus xylophilus* (Steiner and Buhrer) Nickle in Germany. *J Plant Dis Prot*. 2019; 126: 129–144.

15 Faria JMS, Barbosa P, Vieira P, Vicente CSL, Figueiredo AC, Mota M. Phytochemicals as biopesticides against the pinewood nematode *Bursaphelenchus xylophilus*: a review on essential oils and their volatiles. *Plants*. 2021; 10: 2614.

16 Nunes da Silva M, Santos CS, Cruz A, López-Villamor A, Vasconcelos MW. Chitosan increases *Pinus pinaster* tolerance to the pinewood nematode (*Bursaphelenchus xylophilus*) by promoting plant antioxidative metabolism. *Sci Rep*. 2021; 11: 3781.

17 Evans S, Evans H, Ikegami M. Modeling PWN-Induced wilt expression: a mechanistic approach. In: Mota MM, & Vieira P, editors. *Pine Wilt Disease: A Worldwide Threat to Forest Ecosystems*. Dordrecht (NL): Springer; 2008. p. 259–278.

18 Fukuda K. Physiological process of the symptom development and resistance mechanism in pine wilt disease. *J For Res*. 1997; 2: 171–181.

19 Pine FK, Nematode W. *Bursaphelenchus xylophilus. Annu Rev Phytopathol.* 2013; 51: 61–83.

20 Gaspar D, Trindade C, Usié A, Meireles B, Barbosa P, Fortes AM, et al. Expression profiling in *Pinus pinaster* in response to infection with the pinewood nematode *Bursaphelenchus xylophilus. Forests*. 2017; 8: 279.

21 Gaspar D, Trindade C, Usié A, Meireles B, Fortes AM, Guimarães JG, et al. Comparative transcriptomic response of two Pinus species to infection with the pine wood nematode *Bursaphelenchus xylophilus. Forests*. 2020; 11: 204.

22 Liu Q, Wei Y, Xu L, Hao Y, Chen X, Zhou Z. Transcriptomic profiling reveals differentially expressed genes associated with pine wood nematode resistance in masson pine (*Pinus massoniana* Lamb.). *Sci Rep*. 2017; 7: 4693.

23 Lee IH, Kim J, Woo KS, Jang KH, Kim YH, Shim D. De novo assembly and transcriptome analysis of the *Pinus densiflora* response to pine wilt disease in nature. *Plant Biotechnol Rep*. 2018; 12: 229–236.

24 Modesto I, Sterck L, Arbona V, Gómez-Cadenas A, Carrasquinho I, Van de Peer Y, Miguel CM. Insights into the mechanisms implicated in *Pinus pinaster* resistance to pinewood nematode. *Front Plant Sci*. 2021; 12: 690857.

25 Pimentel CS, Gonçalves EV, Firmino PN, Calvão T, Fonseca L, Abrantes I, et al. Differences in pine species constitutive and inducible defences determining the susceptibility to the pinewood nematode. *Plant Pathol*. 2017; 66: 131–139.

26 Rodrigues AM, Carrasquinho I, António C. Primary metabolite adjustments associated with pinewood nematode resistance in *Pinus pinaster*. *Front Plant Sci*. 2021; 12: 777681.
27 Berger S, Sinha AK, Roitsch T. Plant physiology meets phytopathology: plant primary metabolism and plant–pathogen interactions. *J Exp Bot*. 2007; 58: 4019–4026.
28 Bolton MD. Primary metabolism and plant defense – fuel for the fire. *Mol Plant-Microbe Interact*. 2009; 22: 487–497.
29 Steinbrenner AD, Gómez S, Osorio S, Fernie AR, Orians CM. Herbivore-induced changes in tomato (*Solanum lycopersicum*) primary metabolism: a whole plant perspective. *J Chem Ecol*. 2011; 37: 1294–1303.
30 Rojas CM, Senthil-Kumar M, Tzin V, Mysore KS. Regulation of primary plant metabolism during plant-pathogen interactions and its contribution to plant defense. *Front Plant Sci*. 2014; 5: 17.
31 Modesto I, Inácio V, Van de Peer Y, Miguel CM. MicroRNA-mediated post-transcriptional regulation of *Pinus pinaster* response and resistance to pinewood nematode. *Sci Rep*. 2022; 12: 5160.
32 Proels RK, Hückelhoven R. Cell-wall invertases, key enzymes in the modulation of plant metabolism during defence responses. *Mol Plant Pathol*. 2014; 15: 858–864.
33 Kanwar P, Jha G. Alterations in plant sugar metabolism: signatory of pathogen attack. *Planta*. 2018; 249: 305–318.
34 Melakeberhan H, Toivonen PMA, Vidaver WE, Webster JM, Dube SL. Effect of *Bursaphelenchus xylophilus* on the water potential and water-splitting complex of photosystem II of *Pinus sylvestris* seedlings. *Physiol Mol Plant Pathol*. 1991; 38: 83–91.
35 Gao R, Wang Z, Shi J, Luo Y. Effect of *Bursaphelenchus xylophilus* infection on leaf photosynthetic characteristics and resource-use efficiency of *Pinus massoniana*. *Ecol Evol*. 2017; 7: 3455–3463.
36 Menéndez-Gutiérrez M, Alonso M, Toval G, Diaz R. Variation in pinewood nematode susceptibility among *Pinus pinaster* Ait. provenances from the Iberian Peninsula and France. *Ann For Sci*. 2017; 74: 76.
37 Melakeberhan H, Webster JM. Effect of *Bursapelenchus xylophilus* on the assimilation and translocation of ^{14}C in *Pinus sylvestris*. *J Nematol*. 1990; 22: 506–512.
38 Caldana C, Degenkolbe T, Cuadros-Inostroza A, Klie S, Sulpice R, Leisse A, et al. High-density kinetic analysis of the metabolomic and transcriptomic response of Arabidopsis to eight environmental conditions. *Plant J*. 2011; 67: 869–884.
39 Less H, Angelovici R, Tzin V, Galili G. Coordinated gene networks regulating Arabidopsis plant metabolism in response to various stresses and nutritional cues. *Plant Cell*. 2011; 23: 1264–1271.

40 Hirao T, Fukatsu E, Watanabe A. Characterization of resistance to pine wood nematode infection in *Pinus thunbergii* using suppression subtractive hybridization. *BMC Plant Biol.* 2012; 12: 13.

41 Kusumoto D, Yonemichi T, Inoue H, Hirao T, Watanabe A, Yamada T. Comparison of histological responses and tissue damage expansion between resistant and susceptible *Pinus thunbergii* infected with pine wood nematode *Bursaphelenchus xylophilus*. *J For Res.* 2014; 19: 285–294.

42 Michaeli S, Fromm H. Closing the loop on the GABA shunt in plants: are GABA metabolism and signaling entwined? *Front Plant Sci.* 2015; 6: 419.

43 Carillo P. GABA shunt in durum wheat. *Front Plant Sci.* 2018; 9: 100.

44 Ramos-Ruiz R, Martinez F, Knauf-Beiter G. The effects of GABA in plants. *Cogent Food Agric.* 2019; 5: 1670553.

45 Fait A, Fromm H, Walter D, Galili G, Fernie AR. Highway or byway: the metabolic role of the GABA shunt in plants. *Trends Plant Sci.* 2008; 13: 14–19.

46 Seifi HS, Curvers K, De Vleesschauwer D, Delaere I, Aziz A, Höfte M. Concurrent overactivation of the cytosolic glutamine synthetase and the GABA shunt in the ABA-deficient sitiens mutant of tomato leads to resistance against *Botrytis cinerea*. *New Phytol.* 2013; 199: 490–504.

47 Molina-Rueda JJ, Pascual MB, Pissarra J, Gallardo F. A putative role for γ-aminobutyric acid (GABA) in vascular development in pine seedlings. *Planta.* 2015; 241: 257–267.

48 van Dongen JT, Fröhlich A, Ramírez-Aguilar SJ, Schauer N, Fernie AR, Erban A, et al. Transcript and metabolite profiling of the adaptive response to mild decreases in oxygen concentration in the roots of Arabidopsis plants. *Ann Bot.* 2009; 103: 269–280.

49 Rocha M, Licausi F, Araújo WL, Nunes-Nesi A, Sodek L, Fernie AR, van Dongen JT. Glycolysis and the tricarboxylic acid cycle are linked by alanine aminotransferase during hypoxia induced by waterlogging of *Lotus japonicus*. *Plant Physiol.* 2010; 152: 1501–1513.

50 Zabalza A, Orcaray L, Igal M, Schauer N, Fernie AR, Geingenberger P, et al. Unraveling the role of fermentation in the mode of action of acetolactate synthase inhibitors by metabolic profiling. *J Plant Physiol.* 2011; 168: 1568–1575.

51 António C, Päpke C, Rocha M, Diab H, Limami AM, Obata T, et al. Regulation of primary metabolism in response to low oxygen availability as revealed by carbon and nitrogen isotope redistribution. *Plant Physiol.* 2016; 170: 43–56.

52 Gansert D. Xylem sap flow as a major pathway for oxygen supply to the sapwood of birch (*Betula pubescens* Ehr.). *Plant Cell Environ.* 2003; 26: 1803–1814.

53 Wittmann C, Pfanz H. Bark and woody tissue photosynthesis: a means to avoid hypoxia or anoxia in developing stem tissues. *Funct Plant Biol.* 2014; 41: 940–953.
54 Weits DA, van Dongen JT, Licausi F. Molecular oxygen as a signaling component in plant development. *New Phytol.* 2020; 229: 24–35.
55 Loreti E, Perata P. The many facets of hypoxia in plants. *Plants.* 2020; 9: 745.
56 Valeri MC, Novi G, Weits DA, Mensuali A, Perata P, Loreti E. *Botrytis cinerea* induces local hypoxia in Arabidopsis leaves. *New Phytol.* 2020; 229: 173–185.
57 Piasecka A, Jedrzejczak-Rey N, Bednarek P. Secondary metabolites in plant innate immunity: conserved function of divergent chemicals. *New Phytol.* 2015; 206: 948–964.
58 Zaynab M, Fatima M, Abbas S, Sharif Y, Umair M, Zafar MH, Bahadar K. Role of secondary metabolites in plant defense against pathogens. *Microb Pathog.* 2018; 124: 198–202.
59 Erb M, Kliebenstein DJ. Plant secondary metabolites as defenses, regulators, and primary metabolites: the blurred functional trichotomy. *Plant Physiol.* 2020; 184: 39–52.
60 Alseekh S, Fernie AR. Metabolomics 20 years on: what have we learned and what hurdles remain? *Plant J.* 2018; 94: 933–942.
61 Tohge T, de Souza LP, Fernie AR. Genome-enabled plant metabolomics. *J Chromatogr B.* 2014; 966: 7–20.
62 Tohge T, Watanabe M, Hoefgen R, Fernie AR. Shikimate and phenylalanine biosynthesis in the green lineage. *Front Plant Sci.* 2013; 4: 62.
63 Tohge T, Watanabe M, Hoefgen R, Fernie AR. The evolution of phenylpropanoid metabolism in the green lineage. *Crit Rev Biochem Mol Biol.* 2013; 48: 123–152.
64 Treutter D. Significance of flavonoids in plant resistance and enhancement of their biosynthesis. *Plant Biol.* 2005; 7: 581–591.
65 Ferreyra MLF, Rius SP, Casati P. Flavonoids: biosynthesis, biological functions, and biotechnological applications. *Front Plant Sci.* 2012; 3: 222.
66 Desmedt W, Mangelinckx S, Kyndt T, Vanholme B. A phytochemical perspective on plant defense against nematodes. *Front Plant Sci.* 2020; 11: 602079.
67 Chin S, Behm CA, Mathesius U. Functions of flavonoids in plant–nematode interactions. *Plants.* 2018; 7: 85.
68 Hanawa F, Yamada T, Nakashima T. Phytoalexins from *Pinus strobus* bark infected with pinewood nematode, *Bursaphelenchus xylophilus*. *Phytochemistry.* 2001; 57: 223–228.
69 Kuroda H, Goto S, Kazumi E, Kuroda K. The expressed genes of Japanese red pine (*Pinus densiflora*) involved in the pine wilt disease severity. *BMC Proceedings.* 2011; 5: P92.

70 Nunes da Silva M, Solla A, Sampedro L, Zas R, Vasconcelos MW. Susceptibility to the pinewood nematode (PWN) of four pine species involved in potential range expansion across Europe. *Tree Physiol.* 2015; 00: 1–13.
71 Simões R, Pimentel C, Ferreira-Dias S, Miranda I, Pereira H. Phytochemical characterization of phloem in maritime pine and stone pine in three sites in Portugal. *Heliyon.* 2021; 7: e06718.
72 Radojičić A, Li X, Zhang Y. Salicylic acid: a double-edged sword for programed cell death in plants. *Front Plant Sci.* 2018; 9: 1133.
73 Balint-Kurti P. The plant hypersensitive response: concepts, control and consequences. *Mol Plant Pathol.* 2019; 20: 1163–1178.
74 Myers RF. Pathogenesis in pine wilt caused by pinewood nematode, *Bursaphelenchus xylophilus. J Nematol.* 1988; 20: 236–244.
75 Kumaraswamy KG, Kushalappa AC, Choo TM, Dion Y, Rioux S. Mass spectrometry based metabolomics to identify potential biomarkers for resistance in barley against fusarium head blight (*Fusarium graminearum*). *J Chem Ecol.* 2011; 37: 846–856.
76 Ullah C, Unsicker SB, Fellenberg C, Constabel CP, Schmidt A, Gershenzon J, Hammerbacher A. Flavan-3-ols are an effective chemical defense against rust infection. *Plant Physiol.* 2017; 175: 1560–1578.
77 Ullah C, Unsicker SB, Reichelt M, Gershenzon J, Hammerbacher A. Accumulation of catechin and proanthocyanidins in black poplar stems after infection by *Plectosphaerella populi*: hormonal regulation, biosynthesis and antifungal activity. *Front Plant Sci.* 2019; 10: 1441.
78 Hammerbacher A, Kandasamy D, Ullah C, Schmidt A, Wright LP, Gershenzon J. Flavanone-3-hydroxylase plays an important role in the biosynthesis of spruce phenolic defenses against bark beetles and their fungal associates. *Front Plant Sci.* 2019; 10: 208.
79 Metsämuuronen S, Sirén H. Bioactive phenolic compounds, metabolism and properties: a review on valuable chemical compounds in Scots pine and Norway spruce. *Phytochem Rev.* 2019; 18: 623–664.
80 Agati G, Azzarello E, Pollastri S, Tattini M. Flavonoids as antioxidants in plants: location and functional significance. *Front Plant Sci.* 2012; 196: 67–76.
81 Agati G, Brunetti C, Fini A, Gori A, Guidi L, Landi M, et al. Are flavonoids effective antioxidants in plants? Twenty years of our investigation. *Antioxidants.* 2020; 9: 1098.
82 Kuroda K. Physiological incidences related to symptom development and wilting mechanism. In: Zhao BG, Futai K, Sutherland JR, & Takeuchi Y, editors. *Pine Wilt Disease.* Tokyo(JP): Springer; 2008. p. 204–222.
83 Yamada T. Biochemical responses in pine trees affected by pine wilt disease. In: Zhao BG, Futai K, Sutherland JR, & Takeuchi Y, editors. *Pine Wilt Disease.* Tokyo(JP): Springer; 2008. p. 223–234.

84 Dudareva N, Negre F, Nagegowda D, Orlova I. Plant volatiles: recent advances and future perspectives. *Crit Rev Plant Sci.* 2006; 25: 417–440.
85 Tholl D, Boland W, Hansel A, Loreto F, Röse USR, Schnitzler JP. Practical approaches to plant volatile analysis. *Plant J.* 2006; 45: 540–560.
86 Kusano M, Iizuka Y, Kobayashi M, Fuskushima A, Saito K. Development of a direct headspace collection method from Arabidopsis seedlings using HS-SPME-GC-TOF-MS analysis. *Metabolites.* 2013; 3: 23–242.
87 Vallarino JG, Erban A, Fehrle I, Fernie AR, Kopka J, Osorio S. Acquisition of volatile compounds by gas chromatography–mass spectrometry (GC-MS). In: António C, editor. *Plant Metabolomics: Methods and Protocols. Methods in Molecular Biology.* Volume 1778. New York (NY): Humana Press; 2018. p. 225–239.
88 Council of Europe. *European Pharmacopoeia.* 7th Edition. Strasbourg: France; 2010. p. 241.
89 Rubiolo P, Sgorbini B, Liberto E, Cordero C, Bicchi C. Essential oils and volatiles: sample preparation and analysis. A review. *Flavour Fragance J.* 2010; 25: 282–290.
90 Canbay HS. Effectiveness of liquid-liquid extraction, solid phase extraction, and headspace technique for determination of some volatile water-soluble compounds of rose aromatic water. *Int J Anal Chem.* 2017; 2017: 4870671.
91 Figueiredo AC. Biological properties of essential oils and volatiles: sources of variability. *Nat Volatiles Essent Oils.* 2017; 4: 1–13.
92 Bakkali F, Averbeck S, Averbeck D, Idaomar M. Biological effects of essential oils—a review. *Food Chem Toxicol.* 2008; 46: 446–475.
93 Kong JO, Lee SM, Moon YS, Lee SG, Ahn YJ. Nematicidal activity of plant essential oils against *Bursaphelenchus xylophilus* (Nematoda: Aphelenchoididae). *J Asia-Pac Entomol.* 2006; 9: 173–178.
94 Faria JMS, Barbosa P, Bennett RN, Mota M, Figueiredo AC. Bioactivity against *Bursaphelenchus xylophilus*: Nematotoxics from essential oils, essential oils fractions and decoction waters. *Phytochemistry.* 2013; 94: 220–228.
95 Keeling CI, Bohlmann J. Diterpene resin acids in conifers. *Phytochemistry.* 2006; 67: 2415–2423.
96 Trapp S, Croteau R. Defensive resin biosynthesis in conifers. *Annu Rev Plant Physiol Plant Mol Biol.* 2001; 52: 689–724.
97 Raffa KF, Aukema BH, Erbilgin N, Klepzig KD, Wallin KF. Interactions among conifer terpenoids and bark beetles across multiple levels of scale: an attempt to understand links between population patterns and physiological processes. *Recent Adv Phytochem.* 2005; 39: 79–118.
98 Santos C, Vasconcelos T. Resposta Fisiológica de *Pinus* spp. Nas primeiras horas após infecção com *Bursaphelenchus xylophilus* (Nematoda: Aphelenchoididae). *Silva Lusit.* 2011; 19: 99–110.

99 Faria JMS, Sena I, Vieira da Silva I, Ribeiro B, Barbosa P, Ascensão L, et al. In vitro co-cultures of *Pinus pinaster* with *Bursaphelenchus xylophilus*: a biotechnological approach to study pine wilt disease. *Planta.* 2015; 241: 1325–1326.

100 Rodrigues AM, Mendes MD, Lima AS, Barbosa PM, Ascensão L, Barroso JG, et al. *Pinus halepensis, P. pinaster, P. pinea* and *P. sylvestris* essential oils chemotypes and monoterpene hydrocarbon enantiomers, before and after inoculation with the pinewood nematode *Bursaphelenchus xylophilus. Chem Biodivers.* 2017; 14: e1600153.

101 Shi J, Li Z, Nitoda T, Izumi M, Kanzakia H, Baba N, et al. Antinematodal activities of ingenane diterpenes from *Euphorbia kansui* and their derivatives against the pine wood nematode (*Bursaphelenchus xylophilus*). *Z Naturforsch.* 2008; 63c: 59–65.

102 Hwang H-S, Han J-Y, Choi Y-E. Enhanced emission of monoterpene 3-carene in *Pinus densiflora* infected by pine wood nematode and characterization of 3-carene synthase. *Forests.* 2021; 12: 514.

103 Takeuchi Y, Kanzaki N, Futai K. Volatile compounds in pine stands suffering from pine wilt disease: qualitative and quantitative evaluation. *Nematology.* 2006; 8: 869–879.

104 Zhao LL, Wei W, Kang L, Sun JH. Chemotaxis of the pinewood nematode, *Bursaphelenchus xylophilus,* to volatiles associated with host pine, *Pinus massoniana,* and its vector *Monochamus alternatus. J Chem Ecol.* 2007; 33: 1207–1216.

105 Gonçalves E, Figueiredo AC, Barroso JG, Henriques J, Sousa E, Bonifácio L. Effect of *Monochamus galloprovincialis* feeding on *Pinus pinaster* and *Pinus pinea,* oleoresin and insect volatiles. *Phytochemistry.* 2020; 169: 112159.

106 Schenk M, Loomans A, Kinkar M, Vos S. Pest survey card on non-European *Monochamus* spp. *EFSA Supporting Publ.* 2020; 17(2): 1781E.

107 Naves P, Bonifácio L, Inácio ML, Sousa E. Integrated management of pine wilt disease in Troia. *Rev Ciênc Agrár.* 2018; 41: 4–7.

108 Sukovata L, Dziuk A, Plewa R, Jaworski T. The effect of trap color on catches of *Monochamus galloprovincialis* and three most numerous non-target insect species. *Insects.* 2022; 13: 220.

109 Kim B-N, Kim JH, Ahn J-Y, Kim S, Cho B-K, Kim Y-H, Min J. A short review of the pinewood nematode, *Bursaphelenchus xylophilus. Toxicol Environ Health Sci.* 2020; 12: 297–304.

110 Shin WS, Jung YH, Lee SM, Lee CM, Lee CJ, Kim DS, Lee DW. Development of effective screening method for efficacy test of trunk injection agents against pine wood nematode, *Bursaphelenchus xylophilus* in Japanese Black Pine, *Pinus thunbergii. Korean J Pestic Sci.* 2015; 19: 440–449.

111 Takai K, Suzuki T, Kawazu K. Development and preventative effect against pine wilt disease of a novel liquid formulation of emamectin benzoate. *Pest Manag Sci.* 2003; 59: 365–370.
112 Kwon T-S, Song M-Y, Shin S-C, Park Y-S. Effects of aerial insecticide sprays on ant communities to control pine wilt disease in Korean pine forests. *Appl Entomol Zool.* 2005; 40(4): 563–574.
113 James R, Tisserat N, Todd T. Prevention of pine wilt of scots pine (*Pinus sylvestris*) with systemic abamectin injections. *Arboric Urban For.* 2006; 32: 195.
114 Bonifácio LF, Sousa E, Naves P, Inácio ML, Henriques J, Mota M, et al. Efficacy of sulfuryl fluoride against the pinewood nematode, *Bursaphelenchus xylophilus* (Nematoda: Aphelenchidae), in *Pinus pinaster* boards. *Pest Manag Sci.* 2014; 70: 6–13.
115 Ikenaka Y, Miyabara Y, Ichise T, Nakayama S, Nimako C, Ishizuka M, et al. Exposures of children to neonicotinoids in pine wilt disease control areas. *Environ Toxicol Chem.* 2019; 38: 71–79.
116 Pires D, Vicente CSL, Menéndez E, Faria JMS, Rusinque L, Camacho MJ, Inácio ML. The fight against plant-parasitic nematodes: current status of bacterial and fungal biocontrol agents. *Pathogens.* 2022; 11(10): 1178.
117 Dou G, Yan D-H. Research progress on biocontrol of Pine Wilt Disease by microorganisms. *Forests.* 2022; 13(7): 1047.
118 van Halder I, Sacristan A, Martín-García J, Pajares JA, Jactel H. *Pinus pinea*: a natural barrier for the insect vector of the pine wood nematode? *Annals For Sci.* 2022; 79(1): 43.
119 Carrasquinho I, Lisboa A, Inácio ML, Gonçalves E. Genetic variation in susceptibility to pine wilt disease of maritime pine (*Pinus pinaster* Aiton) half-sib families. *Annals For Sci.* 2018; 75: 85.
120 Zas R, Moreira X, Ramos M, Lima MRM, Nunes da Silva M, Solla A, et al. Intraspecific variation of anatomical and chemical defensive traits in Maritime pine (*Pinus pinaster*) as factors in susceptibility to the pinewood nematode (*Bursaphelenchus xylophilus*). *Trees.* 2015; 29: 663–673.
121 Zhao B, Liu Q, Wang B, Yuan F. Roles of phytohormones and their signaling pathways in leaf development and stress responses. *J Agric Food Chem.* 2021; 69: 3566–35684.
122 Yang J, Duan G, Li C, Liu L, Han G, Zhang Y, Wang C. The crosstalks between jasmonic acid and other plant hormone signaling highlight the involvement of jasmonic acid as a core component in plant response to biotic and abiotic stresses. *Front Plant Sci.* 2019: 10.
123 Koo YM, Heo AY, Choi HW. Salicylic acid as a safe plant protector and growth regulator. *Plant Pathol J.* 2020; 36: 1–10.
124 Li N, Han X, Feng D, Yuan D, Huang L-J. Signaling crosstalk between salicylic acid and ethylene/jasmonate in plant defense: do we understand what they are whispering? *Int J Mol Sci.* 2019; 20(3): 671.

125 Wen TY, Wu XQ, Hu LJ, Qiu YJ, Rui L, Zhang Y, et al. A novel pine wood nematode effector, BxSCD1, suppresses plant immunity and interacts with an ethylene-forming enzyme in pine. *Mol Plant Pathol.* 2021; 22: 1399–1412.

126 Mori T, Inoue T. Pinewood nematode-induced ethylene production in pine stems and cellulase as an inducer. *J Jpn For Soc.* 1986; 68: 43–50.

127 Mori T. Changes of stem respiration during the development of pine wilt disease caused by *Bursaphelechus xylophilus*. In: *Transactions of the 94th Annual Meeting of the Japanese Forestry Society*; 1983. p. 307–308.

128 Fukuda K, Hogetsu T, Suzuki K. Ethylene production during symptom development of pine-wilt disease. *Eur J For Pathol.* 1994; 24: 193–202.

129 Wen TY, Wu XQ, Ye JR, Qiu YJ, Rui L, Zhang Y. A Bursaphelenchus xylophilus pathogenic protein Bx-FAR-1, as potential control target, mediates the jasmonic acid pathway in pines. *Pest Manag Sci.* 2022; 78: 1870–1880.

130 López-Villamor A, Nunes da Silva M, Vasconcelos MW. Evaluation of plant elicitation with methyl-jasmonate, salicylic acid and benzo (1,2,3)-thiadiazole-7-carbothioic acid-S-methyl ester for the sustainable management of the pine wilt disease. *Tree Physiol.* 2022; 42: 2596–2613.

131 Chen R, Huang K, Pan S, Xu T, Tan J, Hao D. Jasmonate induced terpene-based defense in *Pinus massoniana* depresses *Monochamus alternatus* adult feeding. *Pest Manag Sci.* 2021; 77: 731–740.

132 Moreira X, Sampedro L, Zas R. Defensive responses of *Pinus pinaster* seedlings to exogenous application of methyl jasmonate: concentration effect and systemic response. *Environ Exp Bot.* 2009; 67: 94–100.

133 López-Goldar X, Villari C, Bonello P, Borg-Karlson AK, Grivet D, Zas R, Sampedro L. Inducibility of plant secondary metabolites in the stem predicts genetic variation in resistance against a key insect herbivore in maritime pine. *Front Plant Sci.* 2018; 9: 1651.

134 Lundborg L, Sampedro L, Borg-Karlson A-K, Zas R. Effects of methyl jasmonate on the concentration of volatile terpenes in tissues of Maritime pine and Monterey pine and its relation to pine weevil feeding. *Trees.* 2019; 33(1): 53–62.

135 López-Villamor A, Zas R, Pérez A, Cáceres Y, Nunes da Silva M, Vasconcelos M, et al. Traumatic resin ducts induced by methyl jasmonate in *Pinus* spp. *Trees.* 2021; 35: 557–567.

136 Kim N, Jeon HW, Mannaa M, Jeong S-I, Kim J, Kim J, et al. Induction of resistance against pine wilt disease caused by *Bursaphelenchus xylophilus* using selected pine endophytic bacteria. *Plant Pathol.* 2019; 68: 434–444.

137 Jeon HW, Park AR, Sung M, Kim N, Mannaa M, Han G, et al. Systemic acquired resistance-mediated control of pine wilt disease by foliar application with methyl salicylate. *Front Plant Sci.* 2022; 12: 812414.

138 Mannaa M, Han G, Jeon HW, Kim J, Kim N, Park AR, et al. Influence of resistance-inducing chemical elicitors against pine wilt disease on the rhizosphere microbiome. *Microorganisms*. 2020; 8: 884.
139 Park J, Jeon HW, Jung H, Lee H-H, Kim J, Park AR, et al. Comparative transcriptome analysis of pine trees treated with resistance-inducing substances against the nematode *Bursaphelenchus xylophilus*. *Genes*. 2020; 11: 1000.
140 Chakraborty M, Hasanuzzaman M, Rahman M, Khan MAR, Bhowmik P, Mahmud NU, et al. Mechanism of plant growth promotion and disease suppression by chitosan biopolymer. *Agriculture*. 2020; 10: 624.
141 Xing K, Zhu X, Peng X, Qin S. Chitosan antimicrobial and eliciting properties for pest control in agriculture: a review. *Agron Sustainable Dev*. 2015; 35: 569–588.
142 Riseh RS, Hassanisaadi M, Vatankhah M, Babaki SA, Barka EA. Chitosan as a potential natural compound to manage plant diseases. *Int J Biol Macromol*. 2022; 220: 998–1009.
143 Katiyar D, Hemantaranjan A, Singh B. Chitosan as a promising natural compound to enhance potential physiological responses in plant: a review. *Indian J Plant Physiol*. 2015; 20: 1–9.
144 Nunes da Silva M, Cardoso AR, Ferreira S, Brito M, Pintado ME, Vasconcelos MW. Chitosan as a biocontrol agent against the pinewood nematode (*Bursaphelenchus xylophilus*). *For Pathol*. 2014; 44: 420–423.
145 Nunes da Silva M, Pintado ME, Sarmento B, Stamford NP, Vasconcelos MW. A biofertilizer with diazotrophic bacteria and a filamentous fungus increases *Pinus pinaster* tolerance to the pinewood nematode (*Bursaphelenchus xylophilus*). *Biol Control*. 2019; 132: 72–80.
146 Silva-Castro I, Diez JJ, Martín-Ramos P, Pinto G, Alves A, Martín-Gil J, Martín JG. Application of bioactive coatings based on chitosan and propolis for *Pinus* spp. protection against *Fusarium circinatum*. *Forests*. 2018; 9: 685.
147 Singh RP, Handa R, Manchanda G. Nanoparticles in sustainable agriculture: an emerging opportunity. *J Controlled Release*. 2021; 329: 1234–1248.
148 Hazarika A, Yadav M, Yadav DK, Yadav HS. An overview of the role of nanoparticles in sustainable agriculture. *Biocatal Agric Biotechnol*. 2022; 43: 102399.
149 Panpatte DG, Jhala YK, Shelat HN, Vyas RV. Nanoparticles: the next generation technology for sustainable agriculture. In: Singh DP, Singh HB, & Prabha R, editors. *Microbial Inoculants in Sustainable Agricultural Productivity Volume 2: Functional Applications*. New Delhi (India): Springer; 2016. p. 289–300.
150 Bi Z, Gong Y, Huang X, Yu H, Bai L, Hu J. Efficacy of four nematicides against the reproduction and development of pinewood nematode, *Bursaphelenchus xylophilus*. *J Nematol*. 2015; 47: 126–132.

151 Su C, Ji Y, Liu S, Gao S, Cao S, Xu X, et al. Fluorescence-labeled abamectin nanopesticide for comprehensive control of pinewood nematode and *Monochamus alternatus* Hope. *ACS Sustainable Chem Eng*. 2020; 8: 16555–16564.

152 Liu Y, Zhang Y, Xin X, Xu X, Wang G, Gao S, et al. Design and preparation of avermectin nanopesticide for control and prevention of pine wilt disease. *Nanomaterials*. 2022; 12: 1863.

153 Liang W, Yu A, Wang G, Zheng F, Jia J, Xu H. Chitosan-based nanoparticles of avermectin to control pine wood nematodes. *Int J Biol Macromol*. 2018; 112: 258–263.

154 Yang E, Lee J-W, Chang P-S, Park I-K. Development of chitosan-coated nanoemulsions of two sulfides present in onion (*Allium cepa*) essential oil and their nematicidal activities against the pine wood nematode, *Bursaphelenchus xylophilus*. *Environ Sci Pollut Res*. 2021; 28: 69200–69209.

155 Davids JS, Ackah M, Okoampah E, Fometu SS, Guohua W, Jianping Z. Biocontrol of bacteria associated with pine wilt nematode, *Bursaphelenchus xylophilus* by using plant mediated gold nanoparticles. *Int J Agric Biol*. 2021; 26: 517–526.

156 Shang H, Zhang H, Zhao R, Yu M, Ma Y, Sun Z, et al. Selenium nanoparticles are effective in penetrating pine and causing high oxidative damage to *Bursaphelenchus xylophilus* in pine wilt disease control. *Pest Manag Sci*. 2022; 78: 3704–3716.

157 Maluin FN, Hussein MZ. Chitosan-based agronanochemicals as a sustainable alternative in crop protection. *Molecules*. 2020; 25: 1611.

158 Garza-García JJO, Hernández-Díaz JA, Zamudio-Ojeda A, León-Morales JM, Guerrero-Guzmán A, Sánchez-Chiprés DR, et al. The role of selenium nanoparticles in agriculture and food technology. *Biol Trace Elem Res*. 2022; 200: 2528–2548.

159 Udalova ZV, Folmanis GE, Khasanov FK, Zinovieva SV. Selenium nanoparticles – an inducer of tomato resistance to the root-knot nematode *Meloidogyne incognita* (Kofoid et White, 1919) Chitwood 1949. *Dokl Biochem Biophys*. 2018; 482: 264–267.

160 Mouniga R, Anita B, Shanthi A, Lakshmanan A, Karthikeyan G. Phenol and antioxidant enzymatic activity in root knot nematode, *Meloidogyne incognita* infected tomato plants treated with chitosan nanoparticles. *Pharma Innov J*. 2022; 11: 241–245.

Index

Note: Page numbers followed by "*f*" and "*t*" figures and tables, respectively.

Monitoring Forest Damage with Mass Spectrometry-Based Metabolomics Methods, First Edition.
Edited by Carla António.

q

Printed and bound by CPI Group (UK) Ltd, Croydon, CR0 4YY

24/04/2024

14488178-0001